Electric Vehicles for Smart Cities: Trends, Challenges and Opportunities

Electric Vehicles for Smart Cities: Trends, Challenges and Opportunities

Guest Editors

Marianna Jacyna
Mariusz Izdebski
Emilian Szczepański

MDPI

Basel • Beijing • Wuhan • Barcelona • Belgrade • Novi Sad • Cluj • Manchester

Guest Editors

Marianna Jacyna
Faculty of Transport
Warsaw University of Technology
Warsaw
Poland

Mariusz Izdebski
Faculty of Transport
Warsaw University of Technology
Warsaw
Poland

Emilian Szczepański
Faculty of Transport
Warsaw University of Technology
Warsaw
Poland

Editorial Office
MDPI AG
Grosspeteranlage 5
4052 Basel, Switzerland

This is a reprint of the Special Issue, published open access by the journal *Energies* (ISSN 1996-1073), freely accessible at: https://www.mdpi.com/journal/energies/special_issues/6W40E83J75.

For citation purposes, cite each article independently as indicated on the article page online and as indicated below:

Lastname, A.A.; Lastname, B.B. Article Title. *Journal Name* **Year**, *Volume Number*, Page Range.

ISBN 978-3-7258-5003-7 (Hbk)
ISBN 978-3-7258-5004-4 (PDF)
https://doi.org/10.3390/books978-3-7258-5004-4

Contents

energies

MDPI

ditorial

Decision-Making Problems in Urban Transport Decarbonization Strategies: Challenges, Tools, and Methods

milian Szczepański [1], Renata Żochowska [2], Mariusz Izdebski [1] and Marianna Jacyna [1,*]

1 Faculty of Transport, Warsaw University of Technology, 00-662 Warszawa, Poland; emilian.szczepanski@pw.edu.pl (E.S.); mariusz.izdebski@pw.edu.pl (M.I.)
2 Faculty of Transport and Aviation Engineering, Silesian University of Technology, 44-100 Gliwice, Poland; renata.zochowska@polsl.pl
* Correspondence: marianna.jacyna@pw.edu.pl; Tel.: +48-22-234-54-91

1. Introduction

The pursuit of sustainable transport system development and the improvement of residents' quality of life continue to keep these issues relevant and subject to extensive analysis. Researchers are actively seeking methods and tools to assess the implementation of solutions and technologies for transport aimed at reducing the negative environmental impact of mobility. This is particularly relevant to the decarbonization of road transport, given its significant share of overall transport performance [1–4].

An analysis of this topic reveals several key research challenges, including technical and infrastructure-related issues (see, Patil [4], Noussan et al. [5], Yan et al. [6]), difficulties in selecting appropriate optimization methods (see Kozyra et al. [7]), environmental considerations (see Atabaki et al. [8]), social aspects (see Jaździk-Osmólska et al. [9], Nochta et al. [10]), and legal and regulatory constraints, such as Green Bonds and Funds (see Bhandary et al. [11]). Importantly, research indicates that the methods and tools used for transport decarbonization are closely related to those employed in decision-making processes for sustainable transport development [12].

A critical component of decarbonization efforts is the development of charging infrastructure. Numerous studies emphasize that the deployment of such infrastructure is essential for the widespread adoption of electric vehicles [13].

The planning and placement of charging stations must account for a range of factors, including the following:

- Distance, power demand, station availability, and capacity [14];
- Socio-economic and environmental factors such as population density, employment, income levels, and user preferences [13];
- Vehicle and charging station parameters, including battery state of charge (SOC), expected driving range, and station availability [14];
- Geographic and topological conditions [13,14];
- Mobility policies, especially those promoting green mobility [13];
- Forecasting models for electric vehicle demand, incorporating origin–destination (OD) travel patterns and SOC levels [14];
- Types of chargers and public policy strategies for electromobility development, e.g., private (home) chargers, public chargers in residential areas, slow public chargers, and fast public chargers [13].

To enhance the accuracy of charging station location forecasts and improve the efficiency of electric vehicle (EV) infrastructure, various optimization methods are employed. These include Geographic Information Systems (GIS) [13] or transport planning software VISUM [15], multi-objective optimization algorithms [16,17], as well as artificial intelligence techniques or machine learning (ML) [18–20] and the Enhanced Cladistic Neural Network (ECNN) [14].

It should also be noted that the use of zero-emission vehicles, while reducing harmful emissions and noise during travel, also entails negative environmental consequences associated with the vehicle's life cycle. These include the environmental impact of component manufacturing, energy production for charging [21], energy utilization [22], and end-of-life disposal or recycling [23]. Such factors should be considered when analyzing the structure of the transport system and the modal split. In this context, the integration of electric vehicles with the urban power grid is crucial for energy transition and Smart City development [24,25]. These efforts aim to minimize operating costs and optimize the utilization of renewable energy sources. However, challenges related to battery degradation and the need to manage the state of charge (SOC) remain significant [26]. This necessitates the development of methods and tools for electric vehicle route optimization that account for battery capacity constraints [27,28].

It should also be noted that decarbonizing road transport, particularly through the increased adoption of zero-emission vehicles, presents both challenges and opportunities for transport companies [29]. One of the primary challenges is the high upfront cost of purchasing electric vehicles (EVs) and investing in charging infrastructure. However, with adequate financial support and a long-term strategic approach, both transport companies and municipalities can benefit from enhanced operational efficiency, reduced costs, and improved quality of life for residents. Research by the International Energy Agency (IEA) indicates that despite the higher initial purchase price of electric vehicles, their total cost of ownership (TCO) is comparable to that of conventional diesel vehicles. This is primarily due to lower operating expenses, especially reduced fuel and maintenance costs [30,31].

The primary benefit of introducing zero-emission vehicles is the reduction in air pollution and, consequently, an improvement in air quality. According to data from the World Health Organization (WHO) and the European Environment Agency (EEA), in 2020, as much as 96% of the European Union's urban population was exposed to concentrations of particulate matter (PM2.5) exceeding WHO guidelines, contributing to approximately 238,000 premature deaths across the EU-27 [32]. This issue was reaffirmed in the EEA's 2024 report, which highlights its persistence despite ongoing mitigation efforts [33].

In Poland, for example, the widespread adoption of zero-emission vehicles could significantly reduce concentrations of pollutants such as PM2.5 and nitrogen oxides (NOx), directly enhancing urban air quality. Experiences from other European countries demonstrate that, with appropriate investments and supportive public policies, substantial emission reductions are achievable. However, in Poland, the carbon footprint of electricity generation continues to play a major role in determining the environmental performance of electric vehicles. In 2023, it was approximately 660 gCO_2/kWh [34], compared to around 380 gCO_2/kWh in Germany and just 40 gCO_2/kWh in Sweden. Nevertheless, even with higher emissions from electricity generation, life-cycle analyses, such as those based on the methodology of the International Energy Agency, show that electric vehicles can produce up to 30% fewer emissions than internal combustion engine vehicles [35].

In the context of urban energy management and stabilization, it is essential to seek solutions that enhance the energy efficiency of applied technologies or involve reorganizing public transport systems [15]. For instance, in highly urbanized areas, determining optimal traffic control parameters or identifying zones with high parking demand poses a significant challenge. Long-term parking locations, such as shopping centers or workplaces, are particularly well-suited for Vehicle-to-Grid (V2G) hubs, as they offer consistent parking patterns and extended idle times, enabling substantial energy feedback to the grid [36].

There is no doubt that many of these problems are NP-hard due to the vast volume of data and the diversity of user behavior strategies, necessitating the use of advanced heuristic algorithms. Access to detailed and reliable operational data remains a key challenge for many designers and analysts [37]. Support in this area has come from the development of original methods based on heuristic approaches, such as ant colony optimization, genetic algorithms, and hybrid models, which aim to minimize EV energy costs [21]. These approaches also enable the incorporation of uncertainties in charging point operations, such as failure probabilities [38].

Green mobility policies in urban areas require city leaders to adopt a new approach to public transport. Strategic planning for the development of green urban mobility, including the integration of electric vehicles (EVs), is essential for building sustainable and low-emission transport systems. Investments in public transport, such as the modernization of bus fleets with electric vehicles, the expansion of tram and metro networks, and the development of urban rail systems, alongside improvements in service frequency and reliability, are fundamental to reducing traffic congestion. Equally important is the development of safe and inclusive urban infrastructure tailored to the needs of cyclists and pedestrians. In countries such as Denmark and the Netherlands, where public transport is highly developed, residents are significantly more likely to choose sustainable modes of transport over private cars. For instance, in Copenhagen, as of 2020, 49% of all journeys were made by bicycle, an outcome made possible by an extensive network of bike paths and strong pro-cycling policies [39].

Promoting walking and cycling requires continuous investment in infrastructure, including dedicated bicycle lanes, pedestrian crossings, car-free zones, and pedestrian-friendly urban spaces. Cities like Amsterdam and Copenhagen exemplify how the development of comprehensive cycling networks and their integration with other modes of transport can result in cycling becoming the dominant mode for short-distance travel.

Another important step in promoting sustainable urban mobility is the introduction of low-emission zones (LEZs) and restrictions on car traffic in city centers. Such measures have been successfully implemented in cities like London, where the introduction of the Ultra Low Emission Zone (ULEZ) significantly reduced exhaust emissions and car traffic. In fact, the implementation of the ULEZ led to a 10% reduction in the number of vehicles entering central London and a 35% decrease in nitrogen oxide (NOx) emissions within just ten months [40]. These types of regulations can effectively reduce the attractiveness of car use in urban areas while simultaneously promoting alternative modes of transport.

An equally important factor in changing residents' mobility preferences is the integration of different transport modes into a single, coherent network. This involves facilitating seamless transfers between public transport, cycling, and walking, as well as developing digital platforms that enable efficient travel planning using multiple modes of transport [41]. An example is the "WienMobil" system implemented in Vienna [42,43], which allows users to plan trips using public transport, bicycles, rental cars, and taxis within a single app. This integrated approach provides users with the flexibility to choose the most convenient mode of transport and encourages them to reconsider car ownership.

The implementation of flexible transport services like on-demand mobility also contributes to an increased share of public transport usage. Adapting services to passenger needs, as demonstrated in Helsinki, has led to a significant rise in public transport trips, with a 73% share compared to the 48% average across urban passenger transport [44].

Shifting residents' travel preferences toward environmentally friendly modes of transport requires the implementation of a comprehensive set of integrated measures. These include infrastructure development, the introduction of low-emission zones, educational campaigns, financial incentives, and the integration of various transport modes. Improving the availability and attractiveness of sustainable mobility options combined with public education on their health and environmental benefits can effectively influence travel behavior and support sustainable urban development.

In light of the issues outlined above, this article is structured into four main sections. The Section 1 identifies the key problems and challenges associated with transport decarbonization, emphasizing not only research aspects but also the regulatory frameworks shaping this process. The Section 2 presents a critical review of the literature, categorizing recent studies according to the main thematic areas and challenges previously identified. This analysis enables the identification of research gaps and highlights necessary future actions. The Section 3 explores the methods and tools essential for identifying optimal solutions, with particular attention to recent advancements in algorithms and optimization techniques. The role of artificial intelligence and its potential in addressing complex NP-hard problems is also discussed. The Section 4 concludes with a summary of key findings and outlines the most pressing challenges and prospective research directions in the field of transport decarbonization.

2. Literature Analysis

2.1. Strategies for Decarbonizing Urban Transport

Urban transport decarbonization strategies discussed in the scientific literature are multifaceted and encompass a wide range of actions. The most significant of these can be grouped as illustrated in Figure 1.

Key actions aimed at reducing carbon dioxide emissions primarily focus on technological and infrastructural investments. In this context, the electrification of transport, encompassing various subsystems, is of critical importance. The widespread adoption of electric vehicles (EVs) requires the development of appropriately located charging infrastructure tailored to demand. Equally essential is the advancement of battery technologies and the implementation of modern solutions based on hydrogen and fuel cell electric vehicles (FCEVs). Increasing attention is being paid to the optimal management of EV charging profiles, which should be integrated with renewable energy sources as part of broader energy management strategies.

As noted in the Introduction, transport electrification should be complemented by additional measures. These include systemic mobility planning, intelligent traffic and data management, multimodal transport integration (encompassing public transport, cycling, walking, and freight), as well as the introduction of innovative business models and urban policies that promote sustainable transport solutions. A key research area involves the development of intelligent transport systems (ITSs), closely linked to the ongoing process of digitalization. Modern decision–support systems for urban transport management increasingly rely on artificial intelligence and digital twin technologies. Another emerging trend in shaping urban transport infrastructure is the implementation of "green infrastructure" and multimodal approaches. The use of diverse transport systems within urban areas is justified by their ability to adapt to the specific spatial characteristics of the city and the travel behaviors of residents and visitors alike.

Financial and legal instruments play a key role in shaping transport decarbonization strategies and promoting their implementation. These include divestments from the fossil fuel sector, subsidies and tax incentives for low-emission transport, green bonds, and investment funds. The introduction of low-emission zones and restrictions on the use of fossil fuel-powered vehicles in city centers are often integrated into transport policies outlined in Sustainable Urban Mobility Plans (SUMPs). Increasingly, there is also an emphasis on viewing urban mobility as a component of integrated urban management.

Figure 1. Selected strategies for decarbonizing urban transport.

The effective implementation of carbon dioxide reduction strategies in urban areas requires well-structured communication, education initiatives, and public awareness campaigns. Engaging with residents is particularly important when developing transport policies and promoting low-emission mobility options.

Decarbonization strategies should also be addressed at various levels of governance from an organizational perspective. The integration of spatial development planning with transport systems and the broader concept of mobility management calls for the use of diverse decision-making tools to tackle challenges at strategic, tactical, and operational levels. These actions, often designed and implemented through a variety of approaches, are discussed in detail in the following subsections.

2.2. *Infrastructural Strategies*

Many authors emphasize that one of the most practical solutions for decarbonizing transport is the widespread adoption of low-emission vehicles, particularly electric vehicles (EVs). Patil et al. [4] highlight the necessity of expanding charging infrastructure, especially in streets, parking areas, and bus depots. They also stress the importance of consumer perceptions and evaluations of EVs, particularly regarding cost, driving range, and service availability.

Infrastructure development for electric transport is currently focused primarily on road and rail modes. Key priorities include building and expanding networks of charging stations for electric buses and private vehicles, as well as modernizing bus depots and service facilities for electric vehicles [13]. In the rail sector, investments in low-emission and energy-efficient tram and metro systems are of particular importance [29]. However, electrification alone is not sufficient unless it is integrated with renewable energy sources such as photovoltaics installed on depot rooftops, smart grids, and vehicle-to-grid (V2G) technologies to supply clean energy to charging infrastructure [4,41].

The importance of infrastructure supporting micromobility is also receiving increasing attention. Research presented in [41] indicates that the construction of new tram lines in peripheral districts can reduce car dependency, while the modernization of existing infrastructure, such as tracks, stops, and traffic control systems can enhance the efficiency and reliability of the urban transport network. The authors emphasize that strengthening mobility for short distances requires the development of dedicated bicycle lanes and pedestrian pathways, separated from car traffic, as well as the establishment of city bike stations and electric bike service points.

A systemic multimodal approach is essential in contemporary urban mobility planning. Integrated transfer hubs that connect public transport with cycling and pedestrian infrastructure, alongside "Park & Ride" and "Bike & Ride" systems, play a key role in this strategy [45]. These hubs enable seamless transfers and reduce the reliance on private vehicles. Buffer parking lots, typically located on the outskirts of cities, allow commuters to switch modes and continue their journeys via public transport. To encourage the use of these systems, cities are implementing infrastructure enhancements such as shelters, surveillance, and EV charging stations [41].

Efforts to reduce carbon dioxide emissions also involve the deployment of intelligent transport systems (ITSs), which support traffic light prioritization for public transport, real-time passenger information systems, integrated ticketing solutions, and mobile applications that facilitate journey planning and multimodal transfers [46]. Musa et al. [47] propose an infrastructure framework for intelligent traffic management, demonstrating its potential to reduce emissions and improve urban quality of life. The authors also advocate for decarbonization strategies based on the development of so-called "green infrastructure" including transport corridors, green roofs, walls, and streets, which contribute to urban resilience in the face of climate change, economic shifts, and public health challenges. A similar perspective is presented in [48], which highlights the potential of such infrastructure to reduce CO_2 emissions by cooling urban environments, improving air quality, and supporting sustainable mobility. Consequently, the integration of spatial planning with ecological transport solutions is essential [49].

2.3. Technological Strategies

The introduction of low-emission transport is widely regarded as the most important technological strategy for decarbonizing the transport sector. Currently, the market offers various types of electric vehicles. Patil et al. [4] highlight battery electric vehicles (BEVs), hybrid electric vehicles (HEVs), and plug-in hybrid electric vehicles (PHEVs) as key solutions. However, these are not the only available options. Noussan et al. [5] examine the advantages and disadvantages of different propulsion technologies, including electric vehicles, hydrogen, and biofuels, and compare life cycle assessment (LCA) emissions for vehicles powered by various energy sources.

Moura et al. [50] analyze the impact of different EV charging profiles on CO_2 emissions and grid load, depending on the energy mix. Alanazi [51] identifies key technological barriers to the large-scale adoption of electric vehicles, emphasizing limited driving range and an underdeveloped charging infrastructure. The author suggests that advancements in lithium-ion batteries and the development of alternative charging technologies such as solid-state batteries could help address these challenges. Such innovations are expected to improve EV range and reduce charging times, thereby enhancing the overall usability of electric vehicles.

A comprehensive overview of hydrogen technologies, their transport applications, and infrastructure challenges is provided by Soleimani [52]. This review discusses the development of fuel cell technologies (PEMFC and SOFC), the storage and distribution of hydrogen produced from renewable sources (so-called green hydrogen), refueling logistics, and the role of hydrogen in urban transport systems.

The integration of EVs with smart city systems is also of growing importance [50,51]. Such integration can support the EV charging process by aligning it with the availability of renewable energy sources, thereby contributing to optimal energy management and real-time charging solutions (e.g., nighttime charging and dynamic pricing). The necessity of integrating renewable energy sources into the development of EV charging infrastructure is also emphasized in the aforementioned study by Patil et al. [4].

Inderwildi et al. [53] discuss predictive traffic and energy management, as well as route and vehicle energy consumption optimization. The authors analyze how decarbonization of key economic sectors (including urban transport) can be supported through artificial intelligence (AI) and cyber–physical systems (CPSs), such as digital twins. AI technologies can enhance energy efficiency, support traffic and fleet management, and enable advanced modeling and simulation of urban transport systems. They also facilitate broader integration with renewable energy sources. An integrated approach to traffic management in smart cities based on Internet of Things (IoT) technologies and big data-driven intelligent transport systems (ITSs) is also recommended by Musa et al. [47]. Such systems aim to reduce emissions and improve transport efficiency, including traffic flow and travel times. Similarly, Archibald et al. [54] present digital twins as a powerful tool for modeling and optimizing transport systems in the context of decarbonization. They emphasize the importance of integrating data from diverse sources and models to develop more sustainable and low-emission urban transport solutions.

2.4. Financial and Legal Strategies

Positive consumer perception of EVs is crucial for their widespread adoption. As a result, various incentives have been introduced in different countries. Patil et al. [4] present examples of policies implemented in Norway, China, the United States, the European Union, and India, highlighting government subsidies and supportive policies as key drivers accelerating EV uptake.

A significant financial and legal instrument supporting decarbonization is fossil fuel divestment, whereby financial institutions and investors withdraw capital from companies involved in fossil fuel extraction and processing, redirecting investments toward more sustainable sectors, including urban transport. Analyses of divestment impacts show that excluding fossil fuel companies from investment portfolios does not negatively affect financial performance, thus reinforcing the viability of capital redistribution as a decarbonization strategy [55].

Financial and political strategies can play a key role in facilitating EV adoption and mitigating the effects of climate change through mechanisms such as fossil fuel divestment. Green financing of urban transport is emerging as a central element of decarbonization strategies, supported by instruments like green bonds and dedicated investment funds [11].

The issuance of such bonds to finance low-emission projects, such as public transport electrification, bicycle infrastructure development, or modernization of rolling stock, is gaining increasing traction.

Moreover, national and local governments are offering subsidies and tax incentives to public transport operators investing in electric vehicles, hydrogen-powered solutions, and biofuel technologies. Ekins et al. [56] highlight the economic benefits of investing in sustainable development, noting that subsidizing low-emission transport technologies represents a cost-effective approach to climate policy.

Among financial and legal strategies, particular attention should be given to restricting car access in city centers through the establishment of low-emission zones and prioritization of pedestrian and cyclist infrastructure. Regulations that limit the entry of vehicles not meeting specific emission standards like Low Emission Zones (LEZs) encourage users to switch to more environmentally friendly modes of transport. Pfoser [57] analyzes the effectiveness of such legal and regulatory measures, including access restrictions and low-emission zones, in the context of freight transport. These conclusions are also applicable to urban passenger transport, as the underlying legal mechanisms are comparable.

Another important strategy involves the mandatory development of sustainable mobility plans. In many jurisdictions, statutory requirements oblige cities or metropolitan areas to prepare and implement Sustainable Urban Mobility Plans (SUMPs) that incorporate climate and environmental objectives. As noted by Kilic and Demirel [41], governance reform is essential in this context. It requires changes in the legal and institutional frameworks governing urban transport management to enable better integration of climate, transport, and urban planning policies.

The authors analyze decision support systems in public transport planning, emphasizing the role of legal and regulatory instruments in enforcing the adoption of sustainable transport solutions. In their conclusions, they argue that the effective decarbonization of urban transport depends not only on technological advances but also on institutional and legal reforms that facilitate the implementation of integrated urban policies [15].

2.5. *Behavioral Strategies*

Behavioral strategies play a crucial social role and are aimed at shaping sustainable transport behavior. These strategies primarily focus on effective demand management, encouraging changes such as more eco-friendly driving habits and increased vehicle occupancy, often through the promotion of carpooling. Shifting transportation habits can be achieved through targeted education and public awareness campaigns that promote the use of public transport, cycling, walking, and car sharing. Di Ciommo and Shiftan [58] analyzed the impact of social campaigns on the travel behavior of different social groups, noting that one key factor influencing willingness to change is access to private modes of transport. The integration of social and technical dimensions in urban planning, for example, through the use of digital twins, can support more sustainable transport decisions that reflect the needs and preferences of residents.

Nochta et al. [10] emphasize the importance of co-creating transport policies with local communities through participatory planning processes. Involving residents in transport planning enhances public acceptance of measures such as vehicle access restrictions, expansion of cycling infrastructure, and improvements to public transport services, particularly for people with disabilities [59].

Additionally, the growing influence of social media, along with the promotion of environmentally friendly values by social influencers and reward systems for using public transport, should not be underestimated. Gössling et al. [60] examine emission reduction strategies in the tourism sector, such as fleet management, incentive programs for employees and customers, and integration with local policy frameworks, which can be effectively adapted to urban public transport contexts.

2.6. Organizational Strategies

Organizational strategies in transport decarbonization are highly diverse. Some focus on changes in work organization, such as the promotion of teleworking, or the implementation of green human resource management (Green HRM) practices within transport organizations. These practices include recruiting employees with a focus on sustainability, providing training in environmentally friendly practices, and introducing incentive systems that support climate-related goals. Shoaib et al. [61] examine the impact of Green HRM practices on employee organizational commitment in the industrial sector, highlighting their potential applicability to the urban transport sector. The authors underscore the importance of green human capital in achieving environmental objectives.

The implementation of mobility management systems in both public and private organizations also plays a key role. These systems promote low-emission modes of transport among employees and clients, for example, through subsidized employee transport passes, the development of bicycle infrastructure, or carpooling programs [60].

Other organizational measures include route optimization, more efficient use of vehicle space in freight transport, and promoting modal shifts in freight logistics [62,63]. Kilic and Demirel [41] propose the adoption of advanced decision support systems like those based on fuzzy logic in urban transport organizations to enhance route planning, resource allocation, and alignment with climate goals.

Organizational strategies also encompass integrated urban planning based on multi-level governance. This involves coordination across different levels of government (local, regional, and national) as well as cross-sectoral collaboration among transport, environmental, and urban planning authorities to implement coherent decarbonization policies [64]. In some cases, this requires restructuring existing models of transport governance. Argyriou [65] compared two organizational models used in the UK: deregulated and franchised, and analyzed their impact on the deployment of zero-emission buses. His findings indicate that the franchised model (e.g., in London) enables better coordination and stronger investment in decarbonization initiatives.

3. Methods and Tools for Decarbonizing Transport

3.1. Characteristics of Methods and Tools Used in Transport

Given the wide range of issues related to transport decarbonization, such as the electrification of public transport, the literature reveals a broad spectrum of methods and tools applied in this field. Many of these methods are based on artificial intelligence algorithms. For instance, in estimating the energy consumption of electric vehicles, determining optimal routes is essential. Heuristic algorithms are frequently employed to minimize the energy expenditure of electric vehicles [21,66,67].

The literature presents numerous effective methods and algorithms tailored to specific transportation challenges [19,68]. A commonly used approach involves optimization algorithms [38], including genetic algorithms [69] and ant colony optimization algorithms [70]. These techniques are particularly prevalent in vehicle routing problems [71–73]. Their main advantage lies in the ability to generate solutions within relatively short computational times. However, a key limitation is that for highly complex decision-making problems, they often yield near-optimal rather than truly optimal solutions.

Additionally, neural networks and machine learning algorithms are increasingly utilized in the transport sector, particularly for applications such as traffic forecasting [74] and vehicle speed prediction [75].

In general, the algorithms and methods applied in transport decarbonization can be grouped into several thematic areas, including sustainable transport development, intelligent transportation systems, autonomous vehicles, smart cities, and electric vehicles.

3.2. Sustainable Transport Development

The introduction of green mobility [76], including electric vehicles, is one of the key strategies for ensuring sustainable transport development in urban agglomerations. Numerous studies confirm the effectiveness of electric vehicles [77], particularly due to their zero local emissions, which is a significant advantage in densely populated urban areas. Unlike conventional vehicles, electric vehicles require charging at designated stations because of their limited battery capacity [26,78]. To evaluate exhaust emissions and fuel consumption under varying thermal conditions during engine start-up, various tests are conducted under real-world traffic scenarios [79].

Advanced artificial intelligence algorithms, such as neural networks [80] and machine learning methods [81], have demonstrated high effectiveness in supporting sustainable transport development [82].

In the domain of transport electrification, two main decision-making challenges are commonly addressed: determining optimal locations for charging stations [16,83] and minimizing the energy consumption of electric vehicles. In both cases, heuristic algorithms are among the most frequently applied artificial intelligence techniques. For instance, studies [84] and Ref. [85] employed genetic algorithms to determine the optimal placement of charging stations, while study [86] utilized a Multi-Objective Particle Swarm Optimization approach. The problem of minimizing energy expenditure was explored in [87], where a genetic algorithm was combined with a neighborhood search technique. The results confirmed the effectiveness of this hybrid method. A detailed overview of algorithms used for electric vehicle route optimization is provided in [88]. Beyond AI-based solutions, several other optimization methods have been applied in the context of transport electrification. These include linear programming, simulated annealing, tabu search, variable neighborhood search, iterated local search, and adaptive large neighborhood search.

Smart cities also play a crucial role in promoting sustainable transport development [89]. Comprehensive research on smart city systems and their integration with AI technologies is presented in [90]. Seven key research areas were identified where AI algorithms are applicable: safety, quality of life, energy, mobility, health, pollution, and industry. Urban mobility, in particular, has undergone significant innovations thanks to AI applications in technologies such as autonomous vehicles (AVs), electric vehicles (EVs), and unmanned aerial vehicles (UAVs). Study [91] presents the findings of a systematic review of 93 articles, showing that the use of AI in smart cities is an emerging area of both research and practical implementation. AI applications in this context are primarily focused on improving business efficiency, data analytics, education, energy systems, environmental sustainability, healthcare, land use, safety, transportation, and urban management. Further comprehensive analyses and examples of practical implementation in smart cities can be found in [92,93].

3.3. Intelligent Transportation System

Popular algorithms used in Intelligent Transportation Systems (ITSs) are artificial intelligence algorithms [94]. Article [95] summarizes various challenges facing the transportation industry, classified under intelligent transportation systems, such as traffic management, safety management, production management, and logistics. The goal of implementing AI techniques in ITSs is to reduce traffic congestion, shorten travel times, and improve the overall efficiency and cost-effectiveness of the transportation system. AI algorithms assist in identifying road hazards, alleviating traffic congestion [96], reducing greenhouse gas and air pollution emissions, designing and managing transport networks, and analyzing travel demand and pedestrian behavior [97].

AI algorithms in ITSs are applied to traffic management, public transportation systems, and urban safety oversight. The use of AI tools to predict road traffic volumes and congestion levels is described in [98]. This paper presents a range of algorithms, including Extreme Learning Machine (ELM), Recurrent Neural Network, Convolutional Neural Network, Deep Machine Learning, Support Vector Machine, Decision Tree, Regression Model, Shallow Machine Learning, Bayesian Network, Hidden Markov Model, Fuzzy Logic, and Probabilistic Reasoning. Traffic light management using AI algorithms is discussed in [99]. Agent-based programming was applied to determine optimal durations for red and green traffic signals. A comprehensive approach to traffic light control using AI is further detailed in [100,101].

AI algorithms are also used to enhance driver safety. Forecasting driver accidents using machine learning is discussed in [102]. The paper presents intelligent driver health monitoring technologies capable of measuring physiological parameters in real time, transmitting data to the cloud, and analyzing it using machine learning, AI, and big data techniques.

In [103], AI algorithms are applied to the control of autonomous vehicles. A similar study is presented in [104], which examines the use of AI in autonomous braking systems. It was found that vehicles equipped with such systems reduce collision rates by approximately 38 percent. Algorithms such as machine learning (ML), deep learning (DL), deep neural networks (DNNs), and natural language processing (NLP) play a central role in autonomous vehicle technology. Machine learning algorithms are extensively discussed in [105,106].

4. Conclusions, Challenges, and Directions for Further Work

As indicated in the article, the problem of transport decarbonization is both difficult and highly complex. The conclusions drawn from the analysis highlight the need for a holistic approach that integrates technical, operational, environmental, social, legal, and political dimensions to effectively accelerate transport decarbonization and support the development of smart cities [107]. Moreover, the coordinated integration of electromobility with distributed renewable energy sources is of key importance, as it enables a reduction in system costs and increases grid flexibility [108].

Despite the conclusions of numerous studies emphasizing the urgency of action, many regulatory and societal barriers continue to hinder the implementation of environmentally friendly solutions. This is due, among other factors, to the absence of a comprehensive approach to assessing the actual life-cycle impacts of electric vehicles, the substantial investment required to modernize power grids, and the need to develop charging infrastructure for electric vehicles [107,108]. Further research should focus on the development of multi-criteria decision models, the application of advanced artificial intelligence algorithms, and improved coordination of policies and regulations [14,109].

The analysis of transport decarbonization strategies and the potential application of various methods and tools to support sustainable transport development reveals several challenges related to the selection of the following:

- Modes of transport (subsystems) for the implementation of sustainable development strategies, including public land transport, public maritime transport, public air transport, bicycle transport, pedestrian transport, and mixed modes of transport such as park and ride [110].
- Appropriate locations for infrastructure, such as vehicle and bicycle sharing points, energy storage facilities, and charging stations [13,36,111].
- Appropriate strategies or the ranking of strategies related to sustainable development [107,109].
- Evaluation methods for innovative power technologies, including electric vehicles, renewable fuels, and energy sources such as solar, wind, hydrogen, and hybrid solutions [112,113].
- Energy management approaches for urban transport systems [66,108,111,114–116].
- The rationale for implementing shared mobility systems such as carsharing, bike sharing, and ridesharing [36,109].
- Financing methods for the implementation of sustainable development strategies [108,114].
- The ranking and identification of stakeholders responsible for shaping and implementing these strategies and managing transport decarbonization efforts [107].
- Logistical challenges specific to cold-chain vehicles related to maintaining cargo freshness while simultaneously planning battery charging [115].

The necessary actions should include the following:

- Formulating medium- and long-term green transport development plans, with a clear definition of the status and role of electric vehicles within the urban green transport network [107].
- Increasing investment in research and development related to key battery and electric vehicle technologies in order to improve energy density, charging speed, and battery life [112,113].
- Conducting market and user behavior analyses, along with detailed emissions assessments across the entire life cycle of electric vehicles, from production to end-of-life, to better understand their comprehensive environmental impact [66,107,115].
- Developing digital twin concepts to monitor complex procedures and reduce uncertainty and risk [10,113].
- Expansion of vehicle-to-grid (V2G) services and virtual energy storage systems (VESS) to stabilize the power grid, taking into account the strategic placement of V2G hubs in locations with extended vehicle parking times and their integration with commercial buildings [36,111].
- Expanding charging infrastructure and optimizing its location. Forecasting the optimal placement of charging stations requires a comprehensive approach that significantly reduces range anxiety and ensures equitable access to infrastructure. The use of GIS tools and publicly available spatial data enables efficient identification of suitable locations for new charging stations [13].
- Advancing technological innovations, particularly in the development of sophisticated algorithms such as DHFO and ECNN, as well as artificial intelligence technologies that improve the accuracy of charging station location forecasting and management. For example, wireless power transfer (WPT) technology enables contactless and safe charging, which is particularly important for autonomous systems and smart cities. Further breakthroughs are anticipated in battery technologies, including improvements in energy density, charging speed, and battery lifespan [14,113].

- Design of dynamic tariff systems aimed at maximizing photovoltaic energy self-consumption by EV users, which according to simulations can increase the self-consumption rate by at least 13 percent and reduce charging costs by approximately 25 percent [114].
- Optimization of cold-chain vehicle schedules using advanced algorithms to minimize energy consumption and prevent the degradation of cold-chain quality [115].
- Improvement of energy efficiency in logistics hubs through the optimal selection of management strategies, which helps reduce emissions and operational costs in urban terminals [116].
- Implementation of signal prioritization algorithms in traffic management to reduce energy consumption while maintaining traffic flow continuity in coordinated systems [110].

Considering the above, it is important to emphasize that electric vehicles contribute to reducing air pollution and noise, while also helping to mitigate the urban heat island effect. These benefits lead to a higher quality of life in cities. Additionally, new optimization models and algorithms enable the effective management of electric vehicle fleets.

Author Contributions: Conceptualization, M.J. and R.Ż.; methodology, E.S. and M.I.; formal analysis, E.S., R.Ż., M.I. and M.J.; investigation, M.J. and E.S.; resources, E.S., R.Ż., M.I. and M.J.; writing—original draft preparation, E.S., R.Ż., M.I. and M.J.; writing—review and editing, M.J. and E.S.; visualization, R.Ż.; supervision, M.J.; funding acquisition, M.J. All authors have read and agreed to the published version of the manuscript.

Funding: This research received no external funding.

Acknowledgments: The authors would like to thank the editorial team of the *Energies* journal for their valuable support in organizing the Special Issue "Electric Vehicles for Smart Cities: Trends, Challenges and Opportunities" and for ensuring its high scientific quality, thereby contributing meaningfully to the advancement of knowledge in the field.

Conflicts of Interest: The authors declare no conflict of interest.

References

1. Jelti, F.; Allouhi, A.; Tabet Aoul, K.A. Transition Paths towards a Sustainable Transportation System: A Literature Review. *Sustainability* **2023**, *15*, 15457. [CrossRef]
2. Li, S.; Wang, B.; Zhou, H. Decarbonizing Passenger Transportation in Developing Countries: Lessons and Perspectives1. *Reg. Sci. Urban Econ.* **2024**, *107*, 103977. [CrossRef]
3. Meyer, T. Decarbonizing Road Freight Transportation—A Bibliometric and Network Analysis. *Transp. Res. Part D Transp. Environ.* **2020**, *89*, 102619. [CrossRef]
4. Patil, G.; Pode, G.; Diouf, B.; Pode, R. Sustainable Decarbonization of Road Transport: Policies, Current Status, and Challenges of Electric Vehicles. *Sustainability* **2024**, *16*, 8058. [CrossRef]
5. Noussan, M.; Hafner, M.; Tagliapietra, S. *The Future of Transport Between Digitalization and Decarbonization: Trends, Strategies and Effects on Energy Consumption*; SpringerBriefs in Energy; Springer: Cham, Switzerland, 2020. [CrossRef]
6. Yan, Y.; Fu, Z.; Yue, J. Study on the environmental benefits of truck parking space for trucks waiting to enter freight terminal—A case study of Gangchen Logistics Park. *Arch. Transp.* **2024**, *71*, 7–23. [CrossRef]
7. Kozyra, J.; Łukasik, Z.; Kuśmińska-Fijałkowska, A.; Janota, A. Reduction of CO_2 emission as a result of the use of 48-volt electrical installations in passenger cars. *Arch. Transp.* **2023**, *66*, 7–20. [CrossRef]
8. Saeid Atabaki, M.; Bagheri, M.; Aryanpur, V. Exploring the Role of Electrification and Modal Shift in Decarbonizing the Road Passenger Transport in British Columbia. *Sustain. Energy Technol. Assess.* **2023**, *56*, 103070. [CrossRef]
9. Jaździk-Osmólska, A.M.; Kalbarczyk, M.K.; Trzaskowski, D.; Wojtachnik, R.; Trzaskowska-Dmoch, A. Economic impact of road accidents on costs of employee turnover: A quantitative study from Poland. *Arch. Transp.* **2024**, *72*, 151–165. [CrossRef]
10. Nochta, T.; Wan, L.; Schooling, J.M.; Parlikad, A.K. A Socio-Technical Perspective on Urban Analytics: The Case of City-Scale Digital Twins. *J. Urban Technol.* **2021**, *28*, 263–287. [CrossRef]

11. Bhandary, R.R.; Gallagher, K.S.; Zhang, F. Climate Finance Policy in Practice: A Review of the Evidence. *Clim. Policy* **2021**, *2* 529–545. [CrossRef]
12. Dhulasi, P.S.; Saranya, K.G. Significance of artificial intelligence in the development of sustainable transportation. *Sci. Tempe* **2023**, *14*, 418–425. [CrossRef]
13. Soczówka, P.; Lasota, M.; Franke, P.; Żochowska, R. Method of Determining New Locations for Electric Vehicle Charging Station Using GIS Tools. *Energies* **2024**, *17*, 4546. [CrossRef]
14. Paul, R.; Selvan, M.P. Predicting and Forecasting of Vehicle Charging Station Using ECNN with DHFO Algorithm. *Energies* **2024** *17*, 4308. [CrossRef]
15. Jacyna, M.; Żochowska, R.; Sobota, A.; Wasiak, M. Decision support for choosing a scenario for organizing urban transpor system with share of electric vehicles. *Sci. J. Silesian Univ. Technology. Ser. Transp.* **2022**, *117*, 69–89. [CrossRef]
16. Jacyna-Gołda, I.; Izdebski, M.; Lewczuk, K. The multi-criteria location problem of the municipal plants. In *Communications in Computer and Information Science Management Perspective for Transport Telematics*; Springer: Cham, Switzerland, 2018; pp. 493–505 [CrossRef]
17. Ibrahim, A.; Qader, H.; Latif, N. Use the Power of a Genetic Algorithm to Maximize and Minimize Cases to Solve Capacity Supplying Optimization and Travelling Salesman in Nested Problems. *J. Comput. Commun.* **2023**, *11*, 24–31. [CrossRef]
18. Bendiab, G.; Hameurlaine, A.; Germanos, G.; Kolokotronis, N.; Shiaeles, S. Autonomous Vehicles Security: Challenges and Solutions Using Blockchain and Artificial Intelligence. *IEEE Trans. Intell. Transp. Syst.* **2023**, *24*, 3614–3637. [CrossRef]
19. Khan, S.; Nazir, S.; García-Magariño, I.; Hussain, A. Deep learning-based urban big data fusion in smart cities: Towards traffic monitoring and flow-preserving fusion. *Comput. Electr. Eng.* **2021**, *89*, 106906. [CrossRef]
20. Ullah, Z.; Al-Turjman, F.; Mostarda, L.; Gagliardi, R. Applications of Artificial Intelligence and Machine learning in smart cities. *Comput. Commun.* **2020**, *154*, 313–323. [CrossRef]
21. Izdebski, M.; Jacyna, M. An Efficient Hybrid Algorithm for Energy Expenditure Estimation for Electric Vehicles in Urban Service Enterprises. *Energies* **2021**, *14*, 2004. [CrossRef]
22. Wu, S.; Chen, Y.; Lin, X. Simulation and analysis of optimal energy-saving mode in microfield of underground rail transit. *Arch Transp.* **2024**, *70*, 27–42. [CrossRef]
23. Mrozik, M.; Merkisz-Guranowska, A. Modeling of Material and Energy Inputs in the Life Cycle of a Vehicle. *Arch. Transp.* **2024** *70*, 117–136. [CrossRef]
24. Saleem, M.; Abbas, S.; Ghazal, T.M.; Khan, M.A.; Sahawneh, N.; Ahmad, M. Smart cities: Fusion-based intelligent traffic congestion control system for vehicular networks using machine learning techniques. *Egypt. Inform. J.* **2022**, *23*, 417–426 [CrossRef]
25. Garlík, B. The application of artificial intelligence in the process of optimizing energy consumption in intelligent areas. *Neural Netw. World* **2017**, *27*, 415–446. [CrossRef]
26. Sablin, O.; Bosyi, D.; Kuznetsov, V.; Lewczuk, K.; Kebal, I.; Myamlin, S.S. Efficiency of energy storage control in the electric transport systems. *Arch. Transp.* **2022**, *62*, 105–122. [CrossRef]
27. Jeong, J.; Ghaddar, B.; Zufferey, N.; Nathwani, J. Adaptive robust electric vehicle routing under energy consumption uncertainty. *Transp. Res. Part C Emerg. Technol.* **2024**, *160*, 104529. [CrossRef]
28. Caballero, H.; Sang, Y.; Wang, S. An Energy-Efficient Routing Algorithm for Electric Delivery Trucks Considering Battery Constraints and Traffic Flow. In Proceedings of the 2024 56th North American Power Symposium (NAPS), El Paso, TX, USA, 13–15 October 2024; IEEE: New York, NY, USA, 2024; pp. 1–6. [CrossRef]
29. Duda, J.; Fierek, S.; Karkula, M.; Kisielewski, P.; Puka, R.; Redmer, A.; Skalna, I. Determining lower bound on number of vehicle blocks in multi-depot vehicle scheduling problem with mixed fleet covering electric buses. *Arch. Transp.* **2023**, *65*, 27–38. [CrossRef]
30. Trends in Heavy Electric Vehicles. Available online: https://www.iea.org/reports/global-ev-outlook-2024/trends-in-heavy-electric-vehicles (accessed on 10 July 2025).
31. Global EV Outlook 2025. Expanding Sales in Diverse Markets. Available online: https://iea.blob.core.windows.net/assets/0aa4762f-c1cb-4495-987a-25945d6de5e8/GlobalEVOutlook2025.pdf (accessed on 10 July 2025).
32. How Air Pollution Affects Our Health. Available online: https://www.eea.europa.eu/en/topics/in-depth/air-pollution/eow-it-affects-our-health (accessed on 10 July 2025).
33. Europe's Air Quality Status 2024. Available online: https://www.eea.europa.eu//publications/europes-air-quality-status-2024 (accessed on 10 July 2025).
34. Carbon Intensity of the Power Sector in the European Union in 2023, by Country. Available online: https://www.statista.com/statistics/1291750/carbon-intensity-power-sector-eu-country/ (accessed on 10 July 2025).

35. EV Life Cycle Assessment Calculator. Available online: https://www.iea.org/data-and-statistics/data-tools/ev-life-cycle-assessment-calculator (accessed on 10 July 2025).
36. Comi, A.; Elnour, E. Challenges for Implementing Vehicle-to-Grid Services in Parking Lots: A State of the Art. *Energies* **2024**, *17*, 6240. [CrossRef]
37. Biszko, K.; Oskarbski, J.; Żarski, K. Impact of different car-following models on estimating safety and emissions on signal-controlled intersections using microscopic simulations. *Arch. Transp.* **2024**, *72*, 43–73. [CrossRef]
38. Izdebski, M.; Jacyna-Gołda, I.; Gołda, P. Minimisation of the probability of serious road accidents in the transport of dangerous goods. *Reliab. Eng. Syst. Saf.* **2022**, *217*, 108093. [CrossRef]
39. European Cyclists' Federation. Cycling Facts and Figures. Available online: https://ecf.com/resources/cycling-facts-and-figures (accessed on 10 July 2025).
40. Central London Ultra Low Emission Zone—Ten Month Report. April 2020. Available online: https://www.london.gov.uk/sites/default/files/ulez_ten_month_evaluation_report_23_april_2020.pdf (accessed on 10 July 2025).
41. Kilic, M.; Demirel, T. Fuzzy Decision Systems for Sustainable Public Transport: A Literature Review. *Int. J. Comput. Intell. Syst.* **2025**, *18*, 164. [CrossRef]
42. Enjoy Mobile Freedom with WienMobil. Available online: https://www.wienerlinien.at/web/wl-en/wienmobil-enjoy-mobile-freedom (accessed on 10 July 2025).
43. Link, C.; Heinemann, A.; Gerike, R.; Jonuschat, H.; Maryschka, M. Who uses a mobility card? A case study on the wienmobil card. *Int. J. Transp. Dev. Integr.* **2017**, *1*, 225–234. [CrossRef]
44. Hyperconnected City Case Study. Mobility as a Service in Helsinki. Available online: https://econsultsolutions.com/wp-content/uploads/2019/11/Helsinki-Whim_CaseStudy_MaaS.pdf (accessed on 10 July 2025).
45. Soczówka, P.; Kłos, M.J.; Żochowska, R.; Sobota, A. An analysis of the influence of travel time on access time in public transport. *Sci. J. Silesian Univ. Technology. Ser. Transp.* **2021**, *111*, 137–149. [CrossRef]
46. Karoń, G.; Żochowska, R. Modelling of Expected Traffic Smoothness in Urban Transportation Systems for ITS Solutions. *Arch. Transp.* **2015**, *33*, 33–45. [CrossRef]
47. Musa, A.A.; Malami, S.I.; Alanazi, F.; Ounaies, W.; Alshammari, M.; Haruna, S.I. Sustainable traffic management for smart cities using Internet-of-Things-oriented intelligent transportation systems (ITS): Challenges and recommendations. *Sustainability* **2023**, *15*, 9859. [CrossRef]
48. Pamukcu-Albers, P.; Pauleit, S.; Ambrose-Oji, B. Building green infrastructure to enhance urban resilience to climate change and pandemics. *Landsc. Ecol.* **2021**, *36*, 665–673. [CrossRef]
49. Drobniak, A.; Janiszek, M.; Plac, K. Zielona gospodarka i zielona infrastruktura jako mechanizmy wzmacniania gospodarczo-środowiskowego wymiaru prężności miejskiej. *Współczesne Probl. Ekon.* **2016**, *12*, 9–22. [CrossRef]
50. Moura, P.; Mohan, A.; Lau, S.; Mohammadi, J. Maximizing Decarbonization Benefits of Transportation Electrification in the U.S. *Electricity* **2023**, *4*, 46–61. [CrossRef]
51. Alanazi, F. Electric Vehicles: Benefits, Challenges, and Potential Solutions for Widespread Adaptation. *Appl. Sci.* **2023**, *13*, 6016. [CrossRef]
52. Soleimani, A.; Dolatabadi, S.H.H.; Heidari, M.; Pinnarelli, A.; Khorrami, B.M.; Luo, Y.; Vizza, P.; Brusco, G. Progress in Hydrogen Fuel Cell Vehicles and Up-and-Coming Technologies for Eco-Friendly Transportation: An International Assessment. *Multiscale Multidiscip. Model. Exp. Des.* **2024**, *7*, 3153–3172. [CrossRef]
53. Inderwildi, O.; Kraft, M. (Eds.) *Intelligent Decarbonisation: Can Artificial Intelligence and Cyber-Physical Systems Help Achieve Climate Mitigation Targets?* Springer: Cham, Switzerland, 2022. [CrossRef]
54. Archibald, B.; Harvey, P.; Sevegnani, M. A digital twinning approach to decarbonisation: Research challenges. *arXiv* **2025**, arXiv:2506.23733. [CrossRef]
55. Plantinga, A.; Scholtens, B. The financial impact of fossil fuel divestment. *Clim. Policy* **2021**, *21*, 107–119. [CrossRef]
56. Ekins, P.; Zenghelis, D. The costs and benefits of environmental sustainability. *Sustain. Sci.* **2021**, *16*, 949–965. [CrossRef] [PubMed]
57. Pfoser, S. Policy Measures to Promote Sustainable Freight Transport. In *Decarbonizing Freight Transport: Acceptance and Policy Implications*; Springer: Cham, Switzerland, 2022. [CrossRef]
58. Di Ciommo, F.; Shiftan, Y. Transport equity analysis. *Transp. Rev.* **2017**, *37*, 139–151. [CrossRef]
59. Młodystach, Ł.; Orczyk, M.; Tomaszewski, F. Evaluation of public transport in Poland form the perspective of the deaf and hard of hearing people towards the improvement of mobility. *Arch. Transp.* **2023**, *66*, 61–76. [CrossRef]
60. Gössling, S.; Scott, D.; Hall, C.M. Pathway to zero emissions in global tourism: Opportunities, challenges and policy implications. *J. Sustain. Tour.* **2024**, *32*, 751–772. [CrossRef]
61. Shoaib, M.; Abbas, Z.; Yousaf, M.; Zámečník, R.; Ahmed, J.; Saqib, S. The role of GHRM practices towards organizational commitment: A mediation analysis of green human capital. *Cogent Bus. Manag.* **2020**, *7*, 1870798. [CrossRef]

62. Motowidlak, U. Analiza strategii rozwoju transportu na rzecz redukcji emisji CO_2. *Stud. I Pr. Wydziału Nauk Ekon. I Zarządzania Uniw. Szczecińskiego* **2015**, *42*, 69–78. [CrossRef]
63. Lasota, M.; Zabielska, A.; Jacyna, M.; Gołębiowski, P.; Żochowska, R.; Wasiak, M. Method for Delivery Planning in Urban Areas with Environmental Aspects. *Sustainability* **2024**, *16*, 1571. [CrossRef]
64. He, J.; Li, Z.; Zhang, X.; Wang, H.; Dong, W.; Chang, S.; Ou, X.; Guo, S.; Tian, Z.; Gu, A.; et al. *Institute of Climate Change and Sus China's Long-Term Low-Carbon Development Strategies and Pathways: Comprehensive Report*; Springer: Singapore, 2022; ISBN 978-981-16-2523-7. [CrossRef]
65. Argyriou, I. Bus regulation and the net-zero transition dynamics in Great Britain. *Int. J. Sustain. Transp.* **2025**. [CrossRef]
66. Izdebski, M.; Jacyna, M.; Bogdański, J. Minimisation of the Energy Expenditure of Electric Vehicles in Municipal Service Companies, Taking into Account the Uncertainty of Charging Point Operation. *Energies* **2024**, *17*, 2179. [CrossRef]
67. Liu, R.; Xie, X.; Augusto, V.; Rodriguez, C. Heuristic algorithms for a vehicle routing problem with simultaneous delivery and pickup and time windows in home health care. *Eur. J. Oper. Res.* **2013**, *230*, 475–486. [CrossRef]
68. Krasuski, K.; Lalak, M.; Gołda, P.; Ciećko, A.; Grunwald, G.; Mrozik, M.; Kozuba, J. Analysis of the precision of determination of aircraft coordinates using EGNOS+SDCM solution. *Arch. Transp.* **2023**, *67*, 105–117. [CrossRef]
69. Kumar, S.N.; Panneerselvam, R. Development of an Efficient Genetic Algorithm for the Time Dependent Vehicle Routing Problem with Time Windows. *Am. J. Oper. Res.* **2017**, *7*, 1–25. [CrossRef]
70. Joo, H.; Lim, Y. Ant colony optimized routing strategy for electric vehicles. *J. Adv. Transp.* **2018**, *2018*, 5741982. [CrossRef]
71. Yu, B.; Yang, Z.Z. An ant colony optimization model: The period vehicle routing problem with time windows. *Transp. Res. Part E* **2011**, *47*, 166–181. [CrossRef]
72. Wilck, J.H.; Cavalier, T.M. A genetic algorithm for the split delivery vehicle routing problem. *Am. J. Oper. Res.* **2012**, *2*, 207–216. [CrossRef]
73. Lewczuk, K.; Żak, J.; Pyza, D.; Jacyna-Gołda, I. Vehicle routing in urban area: Environmental and technological determinants. *WIT Trans. Built Environ.* **2013**, *130*, 373–384. [CrossRef]
74. Olayode, I.O.; Tartibu, L.K.; Okwu, M.O. Prediction and modeling of traffic flow of human-driven vehicles at a signalized road intersection using artificial neural network model: A South African road transportation system scenario. *Transp. Eng.* **2021**, *6*, 100095. [CrossRef]
75. Ma, X.; Tao, Z.; Wang, Y.; Yu, H.; Wang, Y. Long short-term memory neural network for traffic speed prediction using remote microwave sensor data. *Transp. Res. Part C Emerg. Technol.* **2015**, *54*, 187–197. [CrossRef]
76. Reis, M.J.C.S. Internet of Things and Artificial Intelligence for Secure and Sustainable Green Mobility: A Multimodal Data Fusion Approach to Enhance Efficiency and Security. *Multimodal Technol. Interact.* **2025**, *9*, 39. [CrossRef]
77. Pielecha, J.; Skobiej, K.; Kurtyka, K. Exhaust Emissions and Energy Consumption Analysis of Conventional, Hybrid, and Electric Vehicles in Real Driving Cycles. *Energies* **2020**, *13*, 6423. [CrossRef]
78. Erickson, L.; Ma, S. Solar-Powered Charging Networks for Electric Vehicles. *Energies* **2021**, *14*, 966. [CrossRef]
79. Merkisz, J.; Sordyl, A.; Chłopek, Z. Non-repeatability of the WLTP vehicle test results. *Arch. Transp.* **2024**, *71*, 25–49. [CrossRef]
80. Veres, M.; Moussa, M. Deep Learning for Intelligent Transportation Systems: A Survey of Emerging Trends. *IEEE Trans. Intell. Transp. Syst.* **2020**, *21*, 3152–3168. [CrossRef]
81. Yuan, T.T.; Da Rocha Neto, W.; Rothenberg, C.E.; Obraczka, K.; Barakat, C.; Turletti, T. Machine learning for next-generation intelligent transportation systems: A survey. *Trans. Emerg. Telecommun. Technol.* **2022**, *33*, e4427. [CrossRef]
82. Abduljabbar, R.; Dia, H.; Liyanage, S.; Bagloee, S.A. Applications of artificial intelligence in transport: An overview. *Sustainability* **2019**, *11*, 189. [CrossRef]
83. Kchaou-Boujelben, M. Charging station location problem: A comprehensive review on models and solution approaches. *Transp. Res. Part C Emerg. Technol.* **2021**, *132*, 103376. [CrossRef]
84. Xue, B.; Kwai-Sang, C.; Zhili, Z. A bi-objective model for location planning of electric vehicle charging stations with GPS trajectory data. *Comput. Ind. Eng.* **2019**, *128*, 591–604. [CrossRef]
85. Dong, J.; Liu, C.; Lin, Z. Charging infrastructure planning for promoting battery electric vehicles: An activity-based approach using multiday travel data. *Transp. Res. Part C Emerg. Technol.* **2014**, *38*, 44–55. [CrossRef]
86. Yi-Wen, C.; Chen-Yang, C.; Shu-Fen, L.; Chung-Hsuan, Y. Location optimization for multiple types of charging stations for electric scooters. *Appl. Soft Comput.* **2018**, *67*, 519–528. [CrossRef]
87. Li, J.; Wang, F.; He, Y. Electric Vehicle Routing Problem with Battery Swapping Considering Energy Consumption and Carbon Emissions. *Sustainability* **2020**, *12*, 10537. [CrossRef]
88. Erdeli'c, T.; Cari'c, T. A Survey on the Electric Vehicle Routing Problem: Variants and Solution Approaches. *J. Adv. Transp.* **2019**, *54*, 1–48. [CrossRef]

89. Mohammed, A.; Fadhel, M.A.; Duhaim, A.M.; Saihood, A.; Sewify, A.; Al-Hamadani, M.N.A.; Albahri, A.S.; Alzubaidi, L.; Ashish Gupta, A.; Sayedali Mirjalili, S.; et al. Comprehensive systematic review of information fusion methods in smart cities and urban environments. *Inf. Fusion* **2024**, *107*, 102317. [CrossRef]
90. Szpilko, D.; Jimenez Naharro, F.; Lăzăroiu, G.; Nica, E.; de la Torre Gallegos, A. Artificial intelligence in the smart city—A literature review. *Eng. Manag. Prod. Serv.* **2023**, *15*, 53–75. [CrossRef]
91. Yigitcanlar, T.; Desouza, K.C.; Butler, L.; Roozkhosh, F. Contributions and risks of artificial intelligence (AI) in building smarter cities: Insights from a systematic review of the literature. *Energies* **2020**, *13*, 1473. [CrossRef]
92. Yun, J.; Lee, D.; Ahn, H.; Park, K.; Yigitcanlar, T. Not deep learning but autonomous learning of open innovation for sustainable artificial intelligence. *Sustainability* **2016**, *8*, 797. [CrossRef]
93. Chui, K.; Lytras, M.; Visvizi, A. Energy sustainability in smart cities: Artificial intelligence, smart monitoring, and optimization of energy consumption. *Energies* **2018**, *11*, 2869. [CrossRef]
94. Guerrero-Ibáñez, J.; Zeadally, S.; Contreras-Castillo, J. Sensor technologies for intelligent transportation systems. *Sensors* **2018**, *18*, 1212. [CrossRef]
95. Iyer, L.S. AI enabled applications towards intelligent transportation. *Transp. Eng.* **2021**, *5*, 100083. [CrossRef]
96. Olayode, O.I.; Tartibu, L.K.; Okwu, M.O. Application of artificial intelligence in traffic control system of non-autonomous vehicles at signalized road intersection. *Proc. CIRP* **2020**, *91*, 194–200. [CrossRef]
97. Niestadt, M.; Debyser, A.; Scordamaglia, D.; Pape, M. Artificial intelligence in transport. European Parliamentary Research Service. *European Parliament*, 19 October 2019. [CrossRef]
98. Akhtar, M.; Moridpour, S. A review of traffic congestion prediction using artificial intelligence. *J. Adv. Transp.* **2021**, *18*, 8878011. [CrossRef]
99. Zia, S.; Naseem, M.; Mala, I.; Mughal, J.A. Smart traffic light system by using Artificial intelligence. *Sindh Univ. Res. J.* **2018**, *50*, 639–646. [CrossRef]
100. Boppuru, P.R.; Kukatlapalli, P.K.; Chowdary, C.R.; Hussain, J. AI-based YOLO V4 Intelligent Traffic Light Control System. *J. Autom. Mob. Robot. Intell. Syst.* **2023**, *16*, 53–61. [CrossRef]
101. Rani, P.; Shaw, K.D. Artificial neural networks approach induced by fuzzy logic for traffic delay estimation. *J. Eng. Technol.* **2018**, *6*, 127–141.
102. Park, S.J.; Hong, S.; Kim, D.; Hussain, I.; Seo, Y. Intelligent in-car health monitoring system for elderly drivers in connected car. In *Advances in Intelligent Systems and Computing;* Springer International Publishing: Cham, Switzerland, 2019; pp. 40–44, ISBN 978-3-319-96073-9. [CrossRef]
103. Garikapati, D.; Shetiya, S.S. Autonomous Vehicles: Evolution of Artificial Intelligence and the Current Industry Landscape. *Big Data Cogn. Comput.* **2024**, *8*, 42. [CrossRef]
104. Fildes, B.; Keall, M.; Bos, N.; Lie, A.; Page, Y.; Pastor, C.; Tingvall, C. Effectiveness of low speed autonomous emergency braking in real-world rear-end crashes. *Accid. Anal. Prev.* **2015**, *81*, 24–29. [CrossRef] [PubMed]
105. Norouzi, A.; Heidarifar, H.; Borhan, H.; Shahbakhti, M.; Koch, C.R. Integrating machine learning and model predictive control for automotive applications: A review and future directions. *Eng. Appl. Artif. Intell.* **2023**, *120*, 105878. [CrossRef]
106. Zhou, H.; Zhou, A.L.J.; Wang, Y.; Wu, W.; Qing, Z.; Peeta, S. Review of learning-based longitudinal motion planning for autonomous vehicles: Research gaps between self-driving and traffic congestion. *Transp. Res. Rec.* **2021**, *2676*, 324–341. [CrossRef]
107. Li, D.; Lau, A.D.; Gong, Y. Electric Vehicles Empowering the Construction of Green Sustainable Transportation Networks in Chinese Cities: Dynamic Evolution, Frontier Trends, and Construction Pathways. *Energies* **2025**, *18*, 1943. [CrossRef]
108. Lewicki, W.; Coban, H.H.; Wróbel, J. Integration of Electric Vehicle Power Supply Systems—Case Study Analysis of the Impact on a Selected Urban Network in Türkiye. *Energies* **2024**, *17*, 3596. [CrossRef]
109. Pokorska, A.; Wiśniewski, T. Electromobility as a Challenge of Modern City Logistics—Indicator Analysis. *Energies* **2024**, *17*, 3167. [CrossRef]
110. Górka, A.; Czerepicki, A.; Krukowicz, T. The Impact of Priority in Coordinated Traffic Lights on Tram Energy Consumption. *Energies* **2024**, *17*, 520. [CrossRef]
111. Brhane, G.Y.; Oh, E.; Son, S.-Y. Virtual Energy Storage System Scheduling for Commercial Buildings with Fixed and Dynamic Energy Storage. *Energies* **2024**, *17*, 3292. [CrossRef]
112. Lemssaddak, S.; Ait Elmahjoub, A.; Tabaa, M.; El-Alami, A.; Zegrari, M. A Novel Multilevel Inverter Topology Generating a 19-Level Output Regulated by the PD-PWM Method. *Energies* **2025**, *18*, 3227. [CrossRef]
113. Razek, A. One Health Ecological Approach to Sustainable Wireless Energy Transfer Aboard Electric Vehicles for Smart Cities. *Energies* **2024**, *17*, 4349. [CrossRef]
114. Etxegarai, G.; Camblong, H.; Ezeiza, A.; Lie, T.T. Design of Three Electric Vehicle Charging Tariff Systems to Improve Photovoltaic Self-Consumption. *Energies* **2024**, *17*, 1806. [CrossRef]

115. Gan, Z.; Dong, P.; Fu, Z.; Ju, Y.; Shen, Y. Optimizing Electric Cold-Chain Vehicle Scheduling for Sustainable Urban Logistics: Novel Framework Balancing Freshness and Vehicle Charging. *Energies* **2025**, *18*, 1705. [CrossRef]
116. Kłodawski, M.; Jachimowski, R.; Chamier-Gliszczyński, N. Analysis of the Overhead Crane Energy Consumption Using Differen Container Loading Strategies in Urban Logistics Hubs. *Energies* **2024**, *17*, 985. [CrossRef]

energies

MDPI

Article

The Impact of Priority in Coordinated Traffic Lights on Tram Energy Consumption

Anna Górka *, Andrzej Czerepicki and Tomasz Krukowicz

Faculty of Transport, Warsaw University of Technology, 75 Koszykowa Street, 00-662 Warsaw, Poland; tomasz.krukowicz@pw.edu.pl (T.K.)
* Correspondence: anna.gorka@pw.edu.pl

Abstract: Traffic signal priority issues have been a research subject for several decades in Poland and worldwide. Traffic control algorithms have evolved considerably during this period and have become increasingly advanced. Most of them operate within coordinated street sequences, which adds to their complexity. Tramway priority affects traffic conditions for other road users, so many aspects must be taken into account when choosing a priority solution. Typically, one of the main criteria for evaluating the effectiveness of priority is reducing travel time for the priority vehicle while ensuring that the travel times of other traffic participants through the intersection are maintained or slightly deteriorated. However, the energy aspects are often overlooked. This publication aims to investigate how local priority for tramways in traffic signals of coordinated streets affects energy consumption for tramway traction needs. The study was conducted using a microscopic modeling method with PTV Vissim software (ver. 2021). The models were built for coordinated sequences with different levels of priority. Real traffic control algorithms with priority were implemented into the model on the sequence of Marymoncka Street and Grochowska Street in Warsaw. Then, by introducing changes to the parameters of the algorithms, their effect on traffic characteristics, including estimated power consumption, was studied. The results obtained from the computer simulation were statistically processed using R software (ver. 4.3.2). The analysis results prove the effectiveness of tramway priority operation, show its impact on electricity consumption, and allow us to determine the limits of its effective application. Thus, they complement the knowledge of the impact of tramway priority on traffic. The research results also have practical value, as they help us to make rational decisions in the process of designing traffic control algorithms at intersections with a multi-criteria approach.

Keywords: public transport; tram priority; energy consumption; coordinated routes; traffic control

1. Introduction

On each permanently inhabited continent, human activities have established road networks and traffic signaling systems [1–6]. Over time and with technological advancement, traffic regulations and control systems have become increasingly complex. Rail transportation has always served as the axis of urban development. Residents perceive tramways as a fast means of public transportation [7]. A tram network's presence positively influences areas' development and raises property values [8]. Introducing priority measures improves public transportation perception and makes trams more attractive to passengers [9]. Increasing travel speed is coupled with a reduced need for rolling stock, as shorter travel times enable the exact timetable to be maintained with fewer vehicles. This indirectly reduces energy consumption for traction by decreasing vehicle acceleration and deceleration cycles.

The issue of public transportation priority is an integral element of sustainable urban mobility strategies and is often included in a city's transportation policy. One of the goals of this policy is to provide priority for selected categories of road users, primarily for public transportation, which can accommodate more passengers in less space.

A significant advantage of implementing priority measures is the reduction in electric energy consumption [10]. Priority in traffic signaling affects the energy loss level by minimizing the likelihood of a tram having to stop or reduce speed other than at designated stops or speed limits. Therefore, it follows that only tram approaches where stops have not been built have an impact on the energy balance. From an energy perspective, ensuring full priority in those cases is desirable. Furthermore, shortened travel times decrease energy expenditure for non-speed-dependent vehicle needs, such as air conditioning, ventilation, and passenger information displays.

Evaluating the impact of tram priority on traffic conditions is a complex issue. Constructing a universal model is challenging because intersections vary significantly in geometry, traffic volume, and control algorithms. The problem becomes more complicated when considering coordinated intersections. Thus, research findings in this area tend to be somewhat general or overly focused on specific, overly detailed, and rarely reproducible examples. Consequently, there is a need to develop a method that objectively assesses the impact of traffic control solutions on coordinated road traffic and electric energy demand.

This article proposes such a method and provides an example of its application to assess the impact of tram priority on traction energy consumption. Many publications focusing on priority only address technical aspects of vehicle detection [11,12] or rapid signal clearance for privileged vehicles [11,13–16]. Other aspects of the impact of priority on road traffic are often neglected. Another group in the literature focuses on benefits of prioritization as improving travel speed by 15% to 30% [17] or enhancing punctuality [18–20] due to reducing variability in travel time by minimizing random waiting times. Helsinki's experience shows a 40% increase in punctuality [21].

Another benefit of tram priority is improving intersectional safety by eliminating the dilemma zone phenomenon. An example of this is when the length of the equivalent of a yellow signal for a tram is insufficient to guarantee a safe stopping of the tram and the extension of the signal is not allowed by the national regulations [22]. The dilemma zone phenomenon can be effectively eliminated through accommodative algorithms that prevent the signal's deactivation for vehicles in the dilemma zone.

Prior research indicates that satisfying the needs of all road users during high traffic volume is challenging. Some studies argue that skillful prioritization does not adversely affect other road users [23,24]. In other works, it is suggested that deteriorating individual traffic conditions is a natural consequence of prioritization [20], or that it serves as the basis for lowering the level of priority [25]. These discrepancies in research results arise from the fact that most analyzed studies considered only simplified two-phase algorithms [26–28], or isolated intersections [29,30] or prioritization was controlled by a simplified algorithm that only extended the green light for the main direction [19], with no examination of the impact on other road users [31].

Many publications focus on optimizing control systems [32] and approaching traffic control using various models, including Petri nets [14], particle swarm optimization algorithms [33], fuzzy logic [5], and genetic algorithms [34], among others. However, these solutions relate to closed global systems. The authors of publications [23,28,35] view traffic signal control as a highly simplified phenomenon, making applying research results to solve the priority problem impractical. Few publications refer to more than two-phase solutions [13,30].

In the literature, there are also concepts of allocating priority only in the case of schedule delays [36], sufficient vehicle occupancy [37], the type of line (ordinary or express), or a sufficiently high stopping cost [3,38]. While current technology allows for the easy implementation of such proposals, they often lack substantive justification. The stability of the timetable is considered more important than accelerating a single crowded vehicle [39].

Publications describing practical priority implementations often include numerical indicators related to the actual effectiveness of these solutions [40]. For instance, it has been shown that implementation of partial traffic signal priority results in a 5% reduction in travel time, while significantly more effective solutions for traffic organization, such as dedicated tram lanes, lead to a 25% reduction in travel time. Experiences from Melbourne show a reduction in road incidents by over ten percent after implementing tram priority at 29 intersections [4].

Commonly used traffic coordination methods consider many important factors in tramway traffic [41]. Passive priority solutions [20] do not consider disruptions in tramway movement and are less effective than active priority solutions. The methods described in [42] consider public transport traffic but decrease the number of priority function calls, reducing the impact of public transport on overall traffic. The authors of the method described in [43,44] took a similar approach. In Poland, few studies have addressed integrated tram coordination, as seen in [45]. Currently, the automatization of processing in this area is unavailable. Results obtained by the authors of the publication [46], based on one of the proposed tram routes, were unsatisfactory.

In summary, tram priority is a current research problem in many countries. Different methods of preferring track vehicles have different impacts on traffic conditions. Commonly used control algorithms do not fully utilize the potential, especially on coordinated intersections. Clear guidance is lacking on how specific levels of priorities affect the electric energy demand for tram vehicles.

This study investigates how various variants of local priority algorithms in traffic signaling affect the electric energy consumption necessary for tram traction.

To this end, work has been conducted towards analyzing

- The impact of the use of priority algorithms on the electricity consumption for tram traction needs;
- The impact of parameterization of priority control algorithms on electricity consumption for tram traction needs;
- The impact of tram traffic volume on electricity consumption on a route controlled by a priority algorithm.

To this end, computer simulation models were created, based on the characteristics of actual tram routes in Warsaw, data on tram traffic intensity, and other means of road transport. These models utilize real priority control algorithms modified for research purposes. Details of the model are presented in Section 2 of this article, while the results of modeling and their interpretation are provided in Sections 3–5, respectively.

2. Materials and Methods

Warsaw's tram network, in general, is characterized by separated tracks: more than 95% of tramway approaches of intersections with traffic lights are segregated from general traffic. Two coordinated routes in Marymoncka Street (Figure 1) and Grochowska Street (Figure 2) were selected for the study, where the tram track occurs beside the roadway and on the axis of the roadway. The geometrical solutions of the junctions in each route include both situations with and without dedicated lanes for turning onto the track. The routes also include pedestrian crossings with traffic lights outside the crossings.

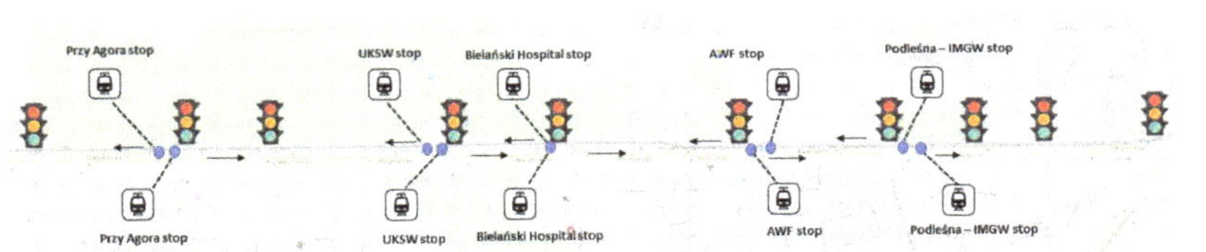

Figure 1. Schematic of the Marymoncka Street route. Based on [47].

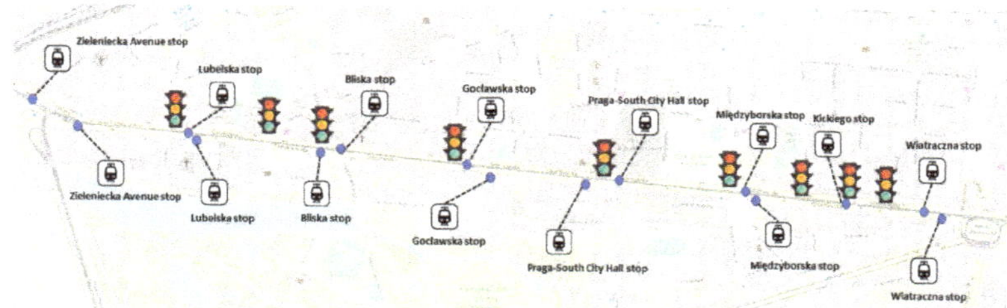

Figure 2. Schematic of the Grochowska Street route. Based on [47].

Modeling tramway traffic is performed using a microscopic model, which allows analysis of the results, taking into account the analysis that considers individual interactions between individual vehicles. Therefore, the simulation was conducted using the PTV Vissim modeling environment [48], based on the Wiedemann car-following model developed in 1974 [49]. The software offers advanced capabilities for implementing complex traffic control algorithms, is highly efficient when working with large intersection networks [50,51] and features a rich user interface. All around the world, Vissim is widely used as a tool for testing theoretical solutions [41], comparing solution variants [1,52], optimizing simple traffic designs [53], and checking traffic control algorithms before their implementation.

Input data for the traffic simulation model include geometric data describing the modeling area, traffic signal algorithms, and vehicle traffic volumes and directions, as well as public transportation parameters. The modeling results encompass vehicle speeds, spatial positions over time, derived measures of traffic conditions, auxiliary variables defined in the modeling algorithms and their values, and diagrams displaying signals (Figure 3).

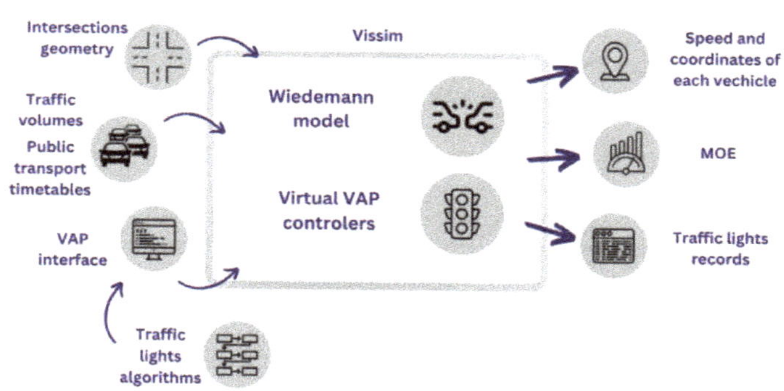

Figure 3. Schematic of the simulation model. Developed using [54,55].

To represent the geometry of the modeling area, an orthophoto map provided by the capital city of Warsaw [56] was used. During the model construction, geometric data for intersections, lane counts, additional turning lanes, pedestrian crossings, curve geometries, detectors, and public transport stop locations, as well as speed limits, were incorporated.

The model in Vissim was prepared based on standard Wiedemann 74 parameters. The speed limit for trams and cars was set to 50 km/h, the desired speed on curves and minimum headway and gap time in collision fields were modified manually accordingly to the observations. The simulation length during the experiment was 11,800 s (3 h of observations + 1000 s for model filling). Traffic volumes and type and directional structure were adopted based on field measurements individually for each intersection approach.

Data describing vehicle traffic were acquired from traffic condition calculations included in traffic signal projects and tram schedules [57]. They encompass traffic volumes, directional traffic structures at various nodes of the model, vehicle types and desired speeds along the route, as well as public transportation schedules.

Real traffic control algorithms functioning at intersections and approved by the city traffic management authority were introduced into the model. These include both fixed-time base algorithms and complete adaptive traffic control algorithms with all logical and temporal conditions.

Unlike the articles presented in the other studies (where the algorithms are very simplified), our study was prepared on the basis of actual algorithms operating in Warsaw at 19 intersections gathered into two arterials. The volume of their documentation does not allow us to present them in full (171 A3 pages). It is worth mentioning here that the algorithms studied are algorithms without an objective function.

After developing road network models, the accuracy of lane alignments, distances between intersections, permissible relationships at individual intersections, and right-of-way rules at collision points were verified. The outcome of this verification is that the road network models for both examined routes adequately represent the geometry of the road network and the applied traffic organization for the research.

The results of computer simulation provide data describing the status of all traffic participants (trams, buses, cars, and pedestrians) moving within the network during the simulation. Files with a *.fzp extension are saved in CSV format with additional metadata in the header. Each row contains a record describing the status of a given vehicle during each simulation step. Another type of output file with a *.ldp extension contains data on the internal states of the traffic control algorithm for each intersection and controlled pedestrian crossings, recorded at 1-s intervals. The file includes values of variables, logical conditions of the algorithm, counters, and markers defined for the operation of tram priority, as needed for the work. Additional auxiliary files with a *.att extension are dictionaries of vehicle types, and data on traffic signal controllers, including the intersection name, cycle length, control type, and geometry of zones designated for analysis for individual intersections and entrances.

The overall research concept is presented in Figure 4.

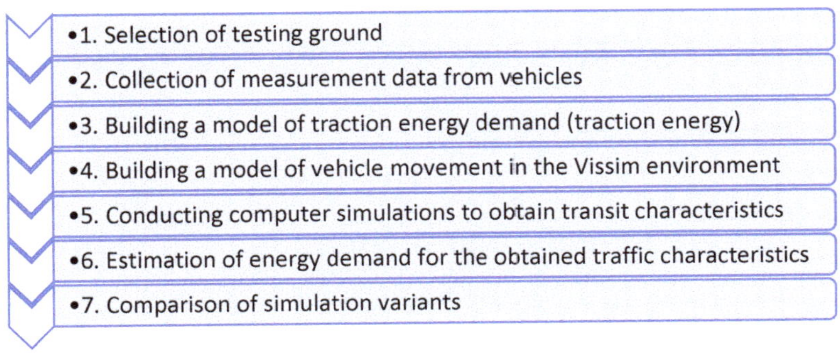

Figure 4. Flow chart of the study.

The data obtained from computer simulations were subjected to analysis using mathematical methods from the field of statistics. The process begins with preprocessing, which involves loading records from files, correcting variable names (as the names generated in the Vissim software are not syntactic names in the R language), and merging data with the dictionary of vehicle types. Subsequently, the sets of output data are grouped by vehicle number and simulation number, and the average speed of each vehicle for a given simulation is calculated.

To analyze the impact of priority on the consumption of electric energy by the tramway traction system, a computer model described in [10] was used. The model was built based on actual measurements of energy consumption by PESA-type trams that operate on Warsaw's routes. This resulted in a highly accurate model, with an error on test sets of <1%.

Measurement data were obtained during a single measurement day in November 2020 collected from all vehicles passing along the route with on-board recorders (modern low-floor trams of the Pesa Jazz model). The owner of the data is Warsaw Tramways Ltd., and they can be requested through public data access. The vehicles moved in real conditions, during a standard workday, according to the trams schedule in typical weather conditions. On-board recorders recorded electricity consumption for the vehicle's traction needs, ignoring the vehicle's own needs (other equipment on the vehicle), whose consumption does not depend on the driving mode.

The authors of the article present the data in an aggregated manner for each time of day. They are shown in Figure 5, which include data on the length of stretches of uninterrupted driving, time, speed and power consumed for sections of the studied route.

The model for emissions calculations was fed with data (positions, speeds, and accelerations) generated from the Vissim model based on vehicle movement diagrams. The differences between the variants were due to the different movement characteristics of the trams in each variant. Energy consumption was calculated according to the basic principles of physics outlined below.

For discrete calculations, we used the formula:

$$E = \sum_{i=1}^{n} P_i \cdot \Delta t_i$$

where

E is electrical energy in Joules (J);

P_i is the Power in the i-th time interval, expressed in Watts (W);

Δt_i is the length of the i-th time interval, expressed in seconds (s);

n is the number of time intervals.

Power was calculated by knowing instantaneous values of voltage U and current I in the time interval, which makes Energy formula look like

$$E = \sum_{i=1}^{n} (U_i \cdot P_i) \cdot \Delta t_i$$

The comparative analysis included the computer simulation scenarios presented in Table 1.

Table 1. Description of computer simulation variants.

Variant	Short Description
A	Fixed time programs
B	Accommodation programs without tram priority
C	Accommodation programs with tram priority (real)
D	Accommodation programs without tram
E	Accommodation programs with lack of tram detection
F	Accommodation with limited cycle regeneration
G	Accommodation with extended cycle regeneration
H	Accommodation with limited tram phase extension
I	Accommodation with extended tram phase extension

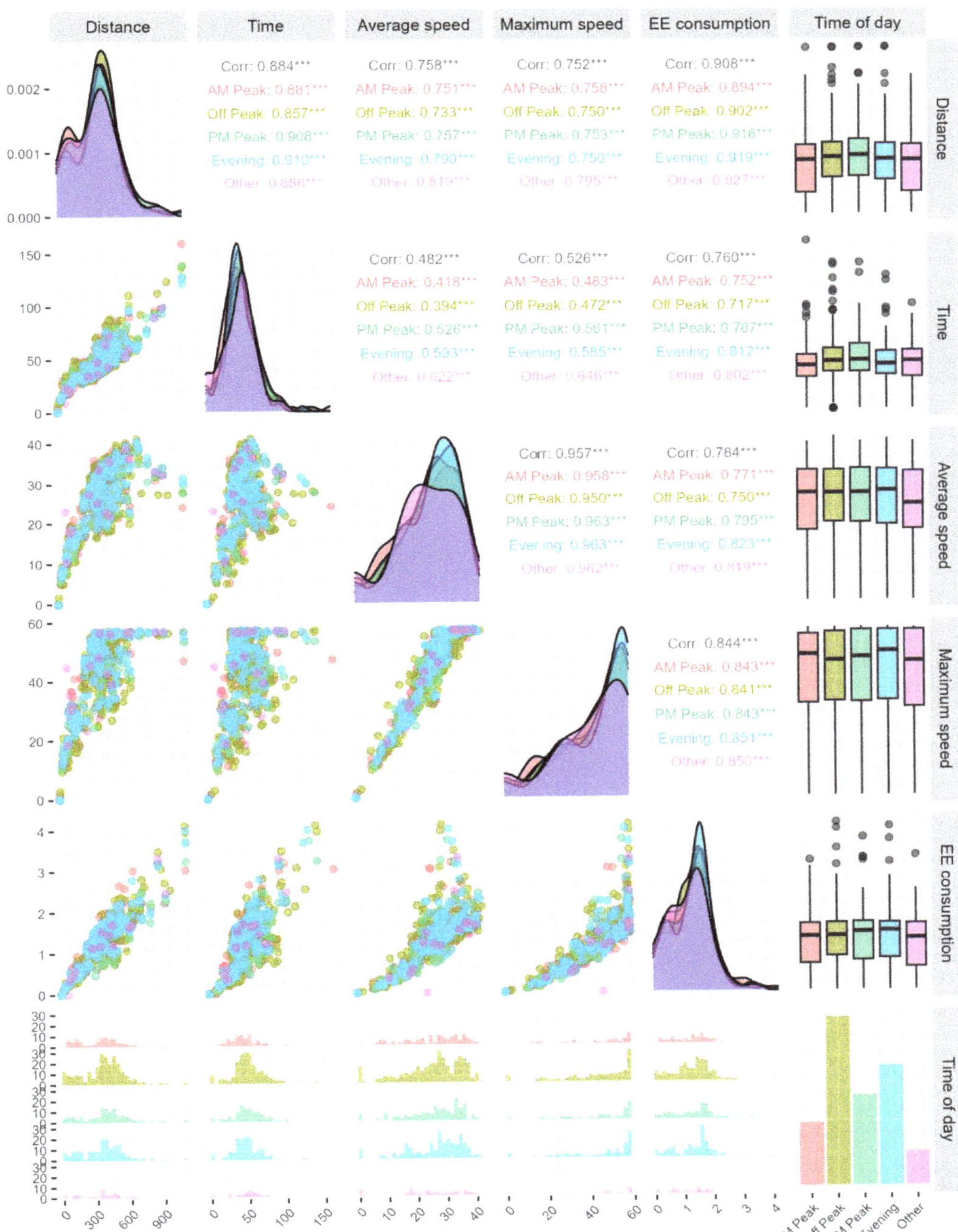

Figure 5. Aggregated data obtained during the measurement, (*** correlation is statistically significant, with a p-value of less than 0.001).

The fixed-time programs (A) and original accommodation programs (C) are faithfully replicated traffic signal design projects for individual intersections that have been introduced into the model in accordance with the approval of the Traffic Management Bureau, the traffic management authority in the city. The research used traffic volumes and signal programs reflecting the morning peak on Grochowska Street and the afternoon peak on Marymoncka Street. All subsequent simulation scenarios were based on the original accommodation projects (C).

Accommodation programs without priority (B) are algorithms in which the priority modules are disabled. They operate with full accommodation, treating trams like other road users.

Scenario (D) implements a simulation variant in which full accommodation algorithms are active, but trams have been removed from the model. This variant allows for determination of traffic performance indicators that are derived only from the movement of other road users and are independent of tram traffic.

Accommodation projects with reduced detection (E) implement a simulation variant in which full-priority module functionality is maintained, but tram detection is limited to a detector placed before the stop line. Such a solution is often used in practice to reduce construction costs by eliminating distant detectors. It may result in detection deficits, causing trams to be detected too late or imprecisely, making it impossible to guarantee tram passage without stopping.

Subsequent simulation scenarios were developed to study the impact of signal priority on other vehicles on the corridor by determining the extent to which algorithms disrupt the basic operation of traffic signals. This disruption is referred to as the "aggressiveness" of priority. To investigate this issue, two basic parameters were modified: the intersection cycle coordination return speed and the potential for priority phase extensions.

The first parameter affecting the aggressiveness of priority is the flexibility of offset synchronization. When granting priority to trams, it is possible to extend the cycle duration beyond the variant adopted in the base traffic signal program. After such an extension, it is necessary to synchronize the offsets between subsequent signals. The accommodation variant with limited cycle regeneration capabilities (F) assumes a 50% reduction in the time parameters of the algorithm compared to the base variant (C), while the accommodation with extensive cycle regeneration capabilities (G) assumes a 150% increase in the time parameters compared to the original variant (150%).

The second parameter affecting the aggressiveness of priority is the maxPT parameter, determining the maximum duration of the tram phase when granting priority. In the slightly aggressive accommodation variant (H), allowable tram phase extensions and reductions were reduced by 50% compared to the original variant (C). The strongly aggressive accommodation variant (I) combines the last two extensions, allowing maximum phase extension or reduction times to be increased to 150% while also increasing the ability to synchronize offsets, as in variant F.

The R language was chosen as the statistical analysis tool, using the RStudio graphical interface (version 2023.12.0).

3. Results

3.1. Experimental Run and Model Verification

The simulation parameters included simulation speed and duration, the number of simulation steps per second (recalculation frequency), and the random number generator seed, which forms the basis for generating parameters for individual road traffic participants. For the experiment, the simulation duration was set to 11,800 s (3 h of observation + 1000 s to fill the model).

As a result of the simulation, 170 files containing vehicle data and algorithm parameters were obtained for Marymoncka Street, and 153 for Grochowska Street, in total 12.84 GB of data.

The correctness of tram traffic was verified by generating vehicles at specified time intervals and observing their movement. Automatic verification was performed for tram stops, subsequent traffic lights, and the time spent at stops. An example of the registration process is shown in Figure 6.

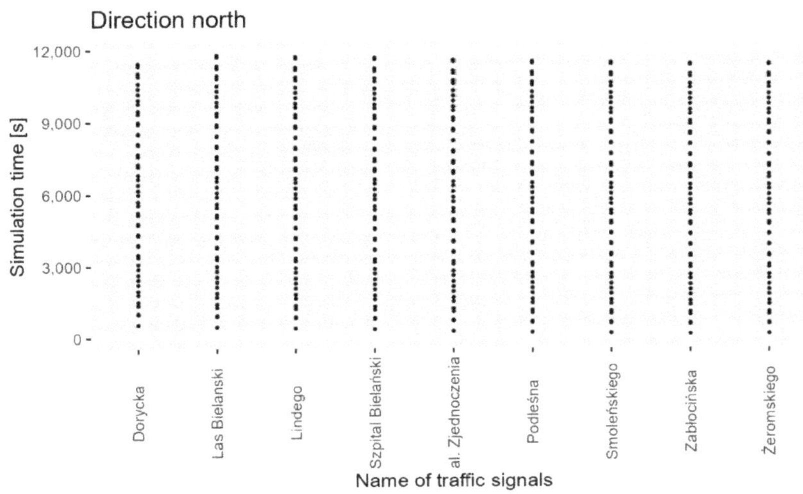

Figure 6. Tram notification moments at each traffic light.

Next, travel times between stops were calculated and compared to the carrier's schedule data [57]. The compatibility of model travel times with the timetables is an indicator of the correct operation of the models because the punctuality of Tramwaje Warszawskie (Warsaw Tramways) is 98% [58]. The comparison is presented in Table 2.

Table 2. Comparison of modelled and scheduled [57] tram travel times.

Route (Stops)	Average Travel Time from Model [min]	Average Travel Time from Timetable, [min]
Cm. Włoski 03—Park Kaskada 03	8.21	8.00
Park Kaskada 04—Cm. Włoski 04	7.93	7.00
Wiatraczna 05—Al. Zieleniecka 05	7.85	8.00
Al. Zieleniecka 06—Wiatraczna 04	7.66	7.00

A parameter describing the movement of passenger cars in the model is the desired speed and traffic intensities. Traffic intensities in the model are consistent with measurements made along the route for the purpose of designing traffic lights. The correctness of the vehicle generator configurations and the generation of vehicles on free-flow sections were verified. Situations in which vehicles were unable to enter the network were eliminated from the models. Running the simulation showed that the model with default settings (following leader driving parameters) adequately replicates reality.

3.2. Impact of the Use of Prioritized Algorithms on Electricity Consumption for Tram Traction Needs

The text mentions Table 3, which presents the results of the estimation of the total electric energy consumption (EE) for tram traction, considering the basic traffic control algorithms on the route, both with and without priority. The tram traffic intensity was set to match real-life conditions and was 16 trams per hour on Marymoncka Street and 17 trams per hour on Grochowska Street.

Table 3. Aggregate EE consumption for baseline variants to simulation.

Street	Variant	Algorithm	Pearson Test p-Value	Normal Distribution	Sum, [Wh]	Avg, [Wh]	Std Dev [Wh]
Grochowska	A	Fixed time programs	0.001	false	472,263	4541	517
	B	Real algorithms	0.132	true	459,313	4416	497
	C	Lack of detection	0.012	false	426,107	4058	501
Marymoncka	A	Fixed time programs	0.863	true	574,615	5804	474
	B	Real algorithms	0.703	true	583,997	5899	448
	C	Lack of detection	0.179	true	465,166	4652	390

Figure 7 illustrates a histogram of electric energy consumption for tram traction, categorized by the number of trams participating in the simulation for both examined routes. The histogram shows how energy consumption varies based on the number of trams involved in the simulation.

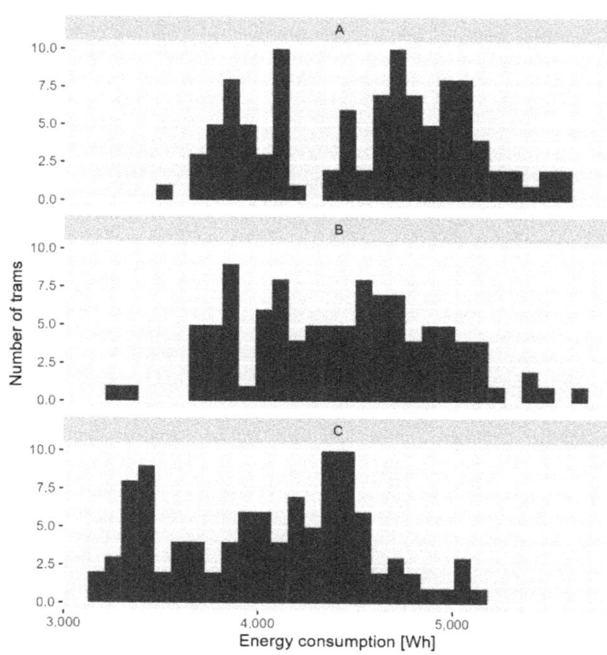

(a)

Figure 7. *Cont.*

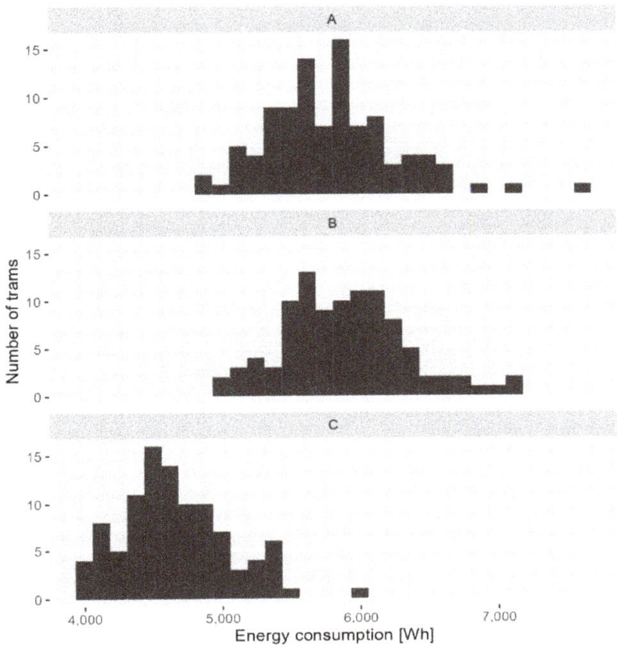

(**b**)

Figure 7. Histogram of electricity consumption for traction under different control variants: (**a**) Grochowska Street; (**b**) Marymoncka Street.

The course of the power taken from the overhead contact line for Grochowska Street and Marymoncka Street is shown in Figures 8 and 9, respectively. The value shown in the diagrams is a weighted average over a period of 120 s.

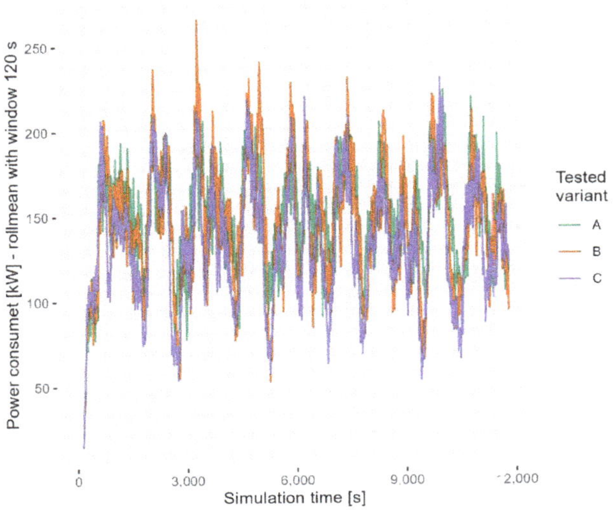

Figure 8. The course of the power taken from the overhead line for Grochowska Street.

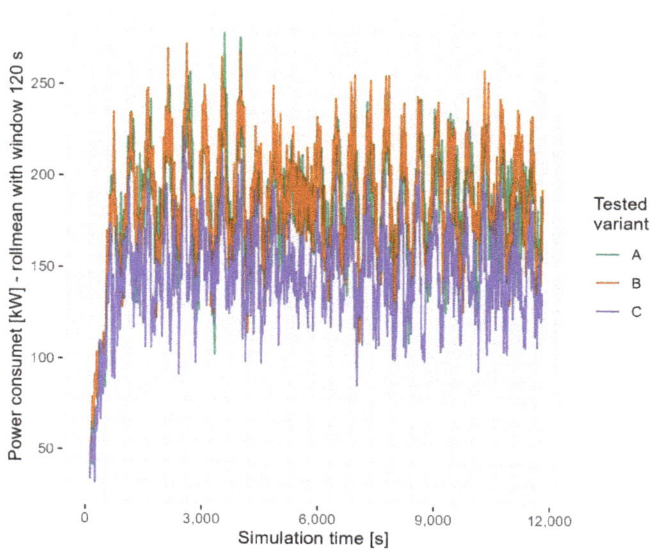

Figure 9. The course of the power taken from the overhead line for Marymoncka Street.

Assessing the normality of the EE consumption distributions using Pearson's χ^2 method showed that the distributions for variants A and C for Grochowska Street were not normal. Therefore, further analyses of these variants were conducted using the Kruskal–Wallis and Dunn tests.

3.3. Influence of Parameterization of Control Algorithms with Priority on Electricity Consumption for Tram Traction Needs

Traffic control algorithms that take tram priority into account can be tuned to individual traction characteristics by means of parameters. A study of the impact of the parameterization of priority control algorithms on traction energy consumption was conducted on both trains using the scenarios described in Table 1. The results of the simulations are presented in Table 4.

Table 4. Summarized electricity consumption for traction for priority control algorithms with different parameters.

Street	Variant	Algorithm	Pearson Test *p*-Value	Standard Deviation	Sum, [Wh]	Avg, [Wh]	Std Dev [Wh]
Grochowska	A	Fixed time programs	<0.001	false	472,263	4241	517
	C	Real algorithms	0.012	false	426,106	4058	501
	E	Lack of detection	0.483	true	462,948	4409	473
	F	Limited cycle regeneration	0.198	true	426,106	4058	484
	G	Extended cycle regeneration	0.052	true	425,385	4051	475
	H	Limited tram phase extension	0.238	true	427,574	4072	462
	I	Extended tram phase extension	0.008	false	424,226	4040	490
Marymoncka	A	Fixed time programs	0.863	true	574,614	5804	473
	C	Real algorithms	0.179	true	465,166	4651	390
	E	Lack of detection	0.059	true	539,566	5395	458
	F	Limited cycle regeneration	0.398	true	463,990	4639	399
	G	Extended cycle regeneration	0.317	true	463,662	4636	381
	H	Limited tram phase extension	0.281	true	484,905	4849	397
	I	Extended tram phase extension	0.00	false	461,625	4616	364

Figure 10 presents the histograms of the EE consumption of the different parameters of control algorithms with priority.

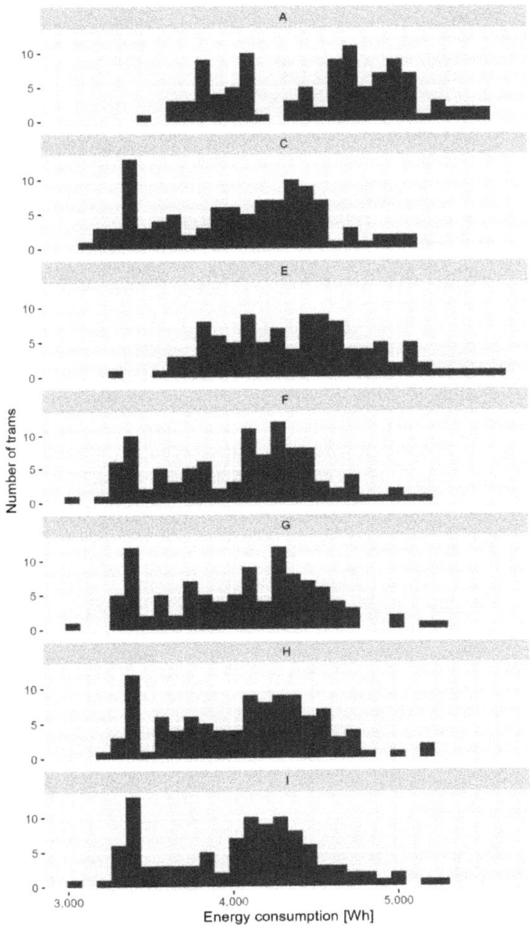

(a)

Figure 10. *Cont.*

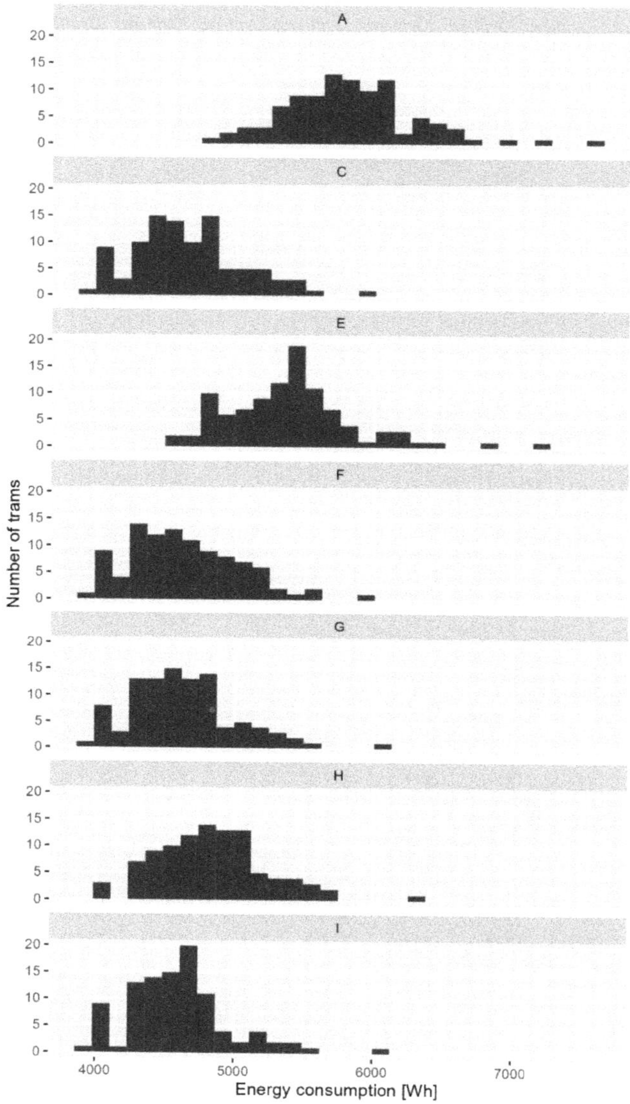

(b)

Figure 10. Histogram of electricity consumption for different parameters of control algorithms with priority: (**a**) Grochowska Street; (**b**) Marymoncka Street.

3.4. *Influence of Tram Traffic Volume on Electricity Consumption on a Route Controlled by a Priority Algorithm*

As the base scenario for further analysis, scenario (C) was selected, which involves real traffic control algorithms with tram priority. The variable parameter in each independently run simulation was the tram traffic intensity, expressed as the number of trams per hour (1/h). The range of frequencies used in this study corresponded to the frequencies observed in the Warsaw tram network, ranging from sporadically operating trams (every 10 min) to the network's capacity limit (36 vehicles per direction during peak hours).

The results of the computer simulations conducted to analyze the impact of tram traffic intensity on electric energy consumption (EE) are presented in Table 5. Modeling was conducted for the algorithms used in scenario C, which includes adaptive traffic signal control with full or high priority for trams.

Table 5. Summary EE consumption for control algorithms with different parameters.

Street	Volume [tram/h]	Pearson Test *p*-Value	Normal Distribution	Sum, [Wh]	Avg, [Wh]	Std Dev [Wh]
Grochowska	6	0.153894	true	161,261	4243	295
	10	0.144151	true	256,706	4140	253
	12	0.001301	False	307,288	4152	263
	15	0.070233	True	398,427	4238	275
	20	0.015014	false	500,465	4170	266
	24	0.057677	true	596,828	4173	268
	30	0.007429	false	732,581	4186	279
	36	0.062859	true	913,610	4210	270
Marymoncka	6	0.179585	true	174,061	4580	377
	10	0.554131	true	291,724	4705	475
	12	0.465158	true	347,878	4701	446
	15	0.042196	false	434,021	4617	384
	20	0.036616	false	557,339	4644	396
	24	0.000018	false	675,019	4720	403
	30	0.013682	false	800,710	4655	390
	36	0.778498	true	1,001,257	4657	355

The total electric energy consumption for tram traction in accordance with the simulation scenarios presented in Table 5 is depicted in Figure 11.

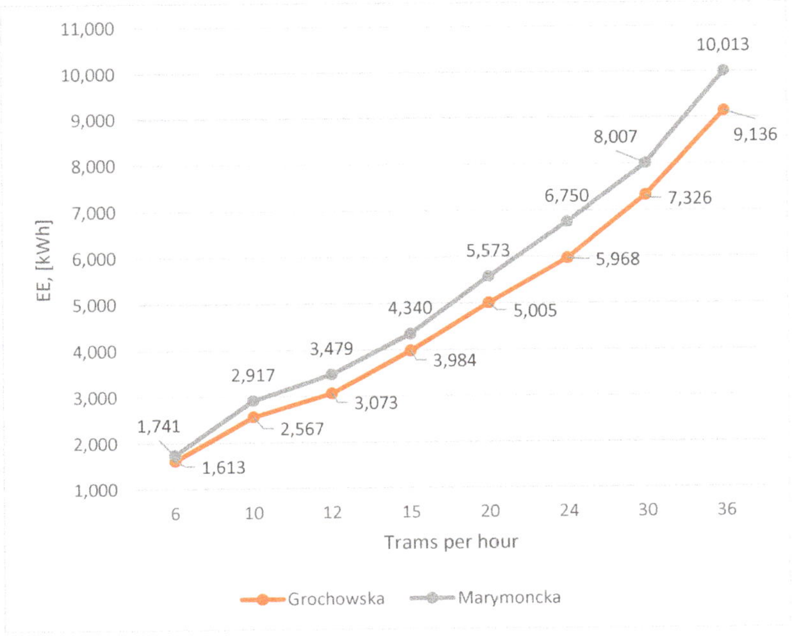

Figure 11. Dependence of EE consumption on tram traffic volume.

4. Discussion

The first stage of the study, the results of which are presented in Section 3.2, aimed to determine whether the use of priority-based algorithms would result in a noticeable effect in terms of reducing the total electric energy consumption for tram traction compared to classical traffic control algorithms without tram priority.

The analysis of statistically significant differences between scenarios with priority (C) and scenarios without priority (A and B) revealed that the application of adaptive control algorithms without priority (B) did not significantly affect the electric energy consumption of trams compared to the baseline scenario (A). The analysis of travel times for these two cases also did not show a statistically significant reduction in travel time.

On the other hand, the use of tram priority-based traffic control algorithms (C) led to an approximate 10% reduction in energy consumption on Grochowska Street and about a 19% reduction on Marymoncka Street compared to fixed-time algorithms (Figure 12). It should be noted that for the Marymoncka Street route, the use of accommodation algorithms resulted in an increase in energy consumption compared to the baseline algorithm.

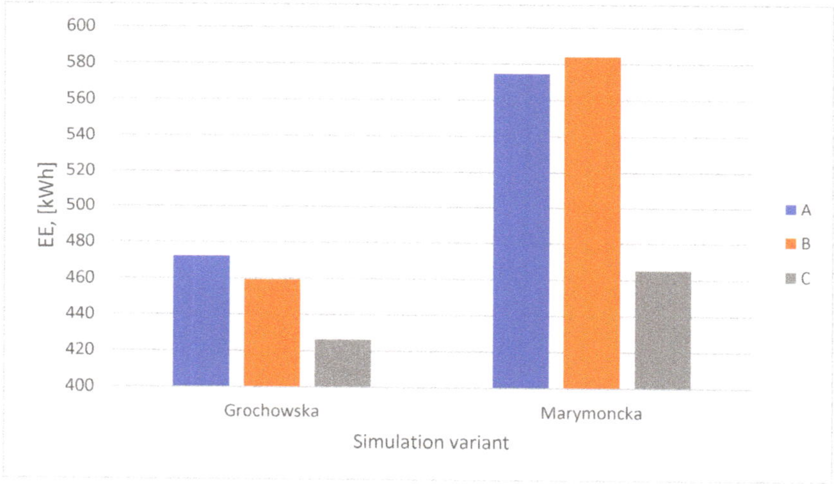

Figure 12. Summary of total traction power consumption for traffic control algorithms with and without priority.

The conclusion from the first stage of the study: the introduction of tram priority on the route contributes to a reduction in the demand for electric energy for tram traction. The subsequent stages of the study aimed to determine the extent of this change and its dependence on the parameterization of traffic control algorithms with priority, as well as the traffic intensity on the routes.

The aim of the second stage of the study, the results of which are presented in Section 3.3, was to answer the following question: how does changing the parameter values of the priority algorithms (the so-called "aggressiveness" of the priority) affect the electricity consumption of the traction? The modifications of the parameters of the priority control algorithms listed in Table 1 were used for the study. A visualization of the simulation results for the different variants is shown in Figure 13.

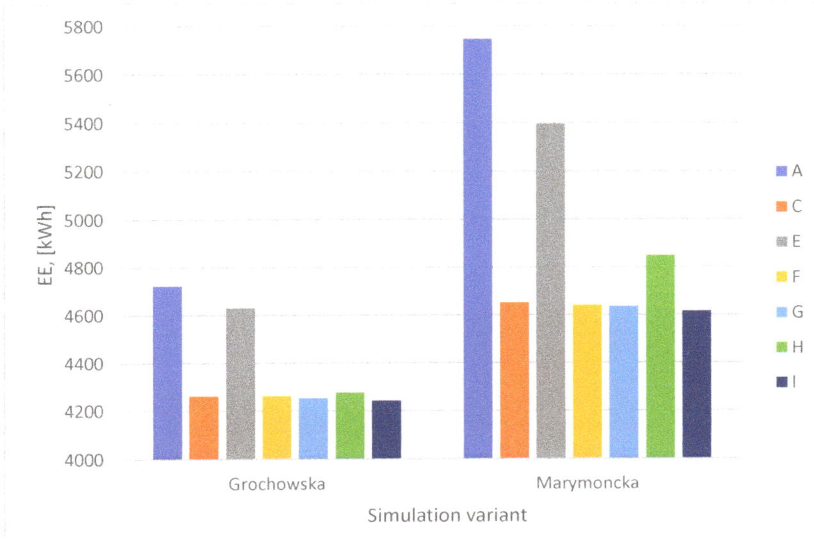

Figure 13. EE consumption for the different parameterization variants of the priority control algorithms.

Based on the analysis of the simulation results, it can be concluded that changing the parameters of traffic control algorithms with tram priority does not significantly impact energy consumption for tram traction. Both fixed-time priority-based algorithms and those with limited detection priority result in higher electric energy consumption (up to ~9% for Grochowska Street and up to ~20% for Marymoncka Street) compared to the scenario with minimal EE consumption (Scenario I—increased maximum times). Therefore, it can be stated that the selection of parameter values for priority-based traffic control algorithms can be made without considering the criterion of energy consumption minimization.

The aim in the next stage of the study, the results of which are presented in Section 3.4, was to analyze the impact of tram traffic intensity on routes controlled by priority-based algorithms on energy consumption for trams. Based on the results presented in Table 5, the average electric energy consumption per tram participating in the simulation was calculated. Figure 14 shows a bar chart illustrating the average energy demand for one tram, depending on the traffic intensity on the route.

The average energy consumption for one tram was 4189 kWh for Grochowska Street and 4659 kWh for Marymoncka Street, respectively. The standard deviation as a percentage of the mean in both cases did not exceed 3%. Based on this, it can be concluded that increasing the traffic intensity on the route does not significantly affect electric energy consumption per unit. A slight downward trend in energy consumption with an increase in the number of trams on the route resulted from the prioritization of larger groups of trams by traffic control algorithms, which allowed for a slight reduction in the number of stops. At the same time, an increase in energy demand was observed as traffic intensity approached the upper limit defined by the route's capacity.

In summary, in this chapter, the computer modeling process and subsequent statistical analysis provided answers to the research questions posed. These answers, in a synthesized form, will be presented in the Conclusions section. It should be emphasized that the article did not delve deeply into other aspects of tram priority and its impact on other road users. Due to the breadth of the topic, this would require a separate article. Nonetheless, some interesting, yet not obvious, conclusions can already be presented.

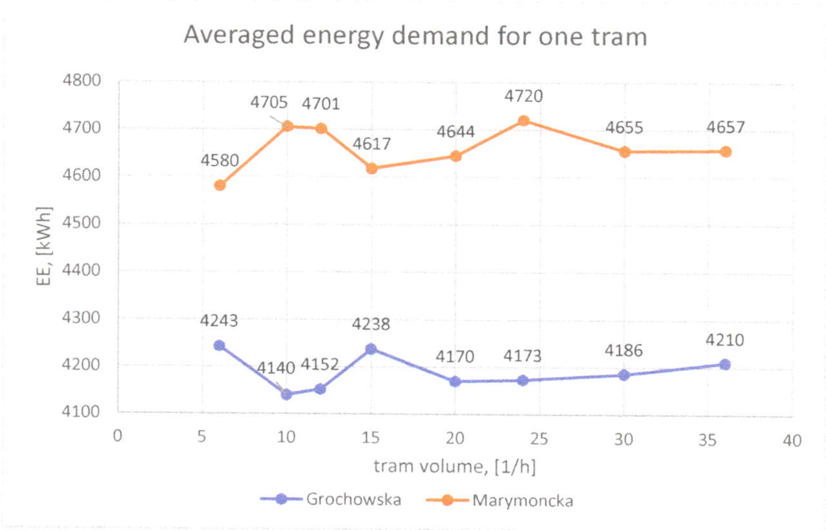

Figure 14. Averaged energy demand for one tram, depending on the volume of traffic on the route.

Firstly, the introduction of priority had a significantly positive impact on tram traffic. Average tram speeds in the studied scenarios increased by approximately 19% and 14%. At the same time, for other vehicles, there was a decrease of just under 3% for Marymoncka Street and an increase of 9% for Grochowska Street. These differences result from the nature of traffic on these routes. Marymoncka Street has lower traffic volume, so the introduction of local priority caused the temporary disruption of coordination for cars, resulting in more stops and a decrease in their average speed. In the case of Grochowska Street, a highly congested route, introducing priority improved the main traffic flow, also for other vehicles, which is related to the appropriate distribution of public transport stops and adjustments to pedestrian phase opening times.

Secondly, during the study of the impact of tram traffic intensity on priority operation, a difference in average speeds of passenger cars was observed. With the introduction of high priority, the speed decreased slightly from 32.57 km/h to 31.57 km/h, while for full priority, a decrease was observed from 30.91 km/h to 28.02 km/h. Therefore, it is not recommended to use full priority on routes with very high tram traffic. This relationship is also confirmed by the time losses, where the standard deviation starts to increase sharply when the tram intensity reaches 20 trams per hour in one direction.

Thirdly, the results of the time loss analysis show that changes in individual priority parameters (unit cycle regeneration range for scenarios F and G and maximum tram phase extension in scenarios H and I) did not significantly affect the previously achieved results. Additional testing of priority operation in cases with detection deficits resulted in a slight decrease in tram speeds, along with a significant increase in time losses. The only statistically significant changes were observed for the implementation of the lack of a detection variant, proving that proper placement of detection devices is a key aspect of implementing traffic signal priority.

5. Comparative Analysis

The energy consumption studies were conducted for railways [59–61] and metro [62–64]. Due to the nature of the movement of these vehicles, the analysis focused on the movement of a single vehicle within the network segment. However, the tram network significantly differs from railway and metro systems. Trams are not separated from other traffic participants. In this case, interactions with other vehicles and pedestrians occur. Even greater interactions occur in the case of trolleybuses, whose movement resembles that of buses, with only the power source being analogous to trams [65].

Energy consumption studies for trams focus on the driving style of a driver [66] and the creation of schedules [67], resulting in energy savings ranging from several to over a dozen percent. These studies assume no influence on infrastructure-related factors. In other works, authors developed energy-optimal driving methods considering only the influence of traffic signals [68], yielding model efficiencies of over 23% energy savings [69]. However, this study presented an analysis for a single intersection, and the traffic control method is not detailed.

The approach proposed in this article is innovative as it involves matching the traffic lights to the way the vehicle is driven, resulting in a reduction in the number of stops. The obtained results align with the outcomes of an experiment conducted by temporarily disabling the traffic signal priority on Marymoncka Street. A recent field study [10] showed a 13% energy increase due to the priority system's deactivation, similar to the 10% and 19% energy reduction obtained in this article depending on the analyzed stretch. The difference in research outcomes may stem from the fact that the applied model does not account for energy recuperation, an assumption often made in energy consumption studies, e.g., in [68].

Other studies focus on energy storage in the vehicle and the optimization of tram movement [70–73]; those topics are not covered in this article. The study [74] indicates 2.5% energy savings, while study [75] shows a 5% reduction, both inferior to the solution proposed in this work, emphasizing the significance of designing traffic signals to reduce energy consumption, potentially yielding greater savings. The prediction error for regression models in energy consumption was approximately 2% [76], comparable to the effects of some studies. Comparative analysis indicates that incorporating tram movement into traffic signals yields excellent results; however, combining such a solution with other energy-saving methods should provide synergistic effects, enabling energy cost reduction and, in the case of trams with energy storage, reducing the required supercapacitor or battery capacity.

The resulting method also has some limitations. The method does not take into account the recuperation of electrical energy. In the case of tramways, the energy extracted during the braking of the vehicle is given back to the power grid, if the grid has the capacity to receive it. So it does not directly affect the distribution of EE to traction. Since the calculation model was built on data that do not take into account recuperation, a comparison of different simulation variants is possible (when calculating differences in EE consumption, recuperation is not taken into account).

The method does not take into account the possibility of the line being served by different types of trams. The introduction of a new type of tram into service will require the construction of its model in the simulation program; among other things, measurement data for this model will be needed.

Undoubtedly, a characteristic feature of the method is that it is based on real data, with which it is necessary to feed the model, which can be considered both a disadvantage (the need for measurements) and an advantage (high accuracy) of the model.

Our research method can be used for any control method that can be connected to Vissim. The program has the ability to be used to control vehicle traffic in a traffic signal controller emulator model. It is also possible to control using an external traffic control model described in Python, so it would be possible to use reinforcement learning control too. With the method we developed, it would be possible to compare the results of the control algorithm used in the study and the control using RL. This is an interesting direction for further research. Recently, there have been many studies published related to traffic signal control using RL, such as [34,77,78]. Most of them are implemented based on SUMO software (version 1.18.0) [79]. SUMO software is less sophisticated when it comes to tools for analyzing traffic data and exporting them. However, it is possible to include in the

Python script used for traffic control (e.g., using RL) data acquisition analogous to that obtained by the authors from Vissim software in .fzp and .ldp files. Then, in the case of the study [34,77,78], after adjusting the data to the same structure, the comparison methods developed by the authors can be used.

The genesis of the method stemmed from the requirements of Polish regulations, which mandate an explicit description of the operation of the control algorithm [80]. A neural network or any other machine learning algorithm does not meet this requirement. The methods used may be used in other countries that have such requirements in regulations or guidelines.

6. Conclusions

The use of traffic control algorithms with tram priority on road networks not only shortens travel times and reduces the number of stops but also lowers the demand for electrical energy. Depending on the type of priority applied, the reduction in energy consumption for the selected analyzed road sections reached a maximum of 10% and 19%. It is worth emphasizing that while the study focused on two typical road sections, in practice, the achieved effect will always depend on the specific configuration of these sections. This is why it is crucial to model traffic during the design of new road solutions or when planning the modernization of existing ones. An interesting finding from the study is the low dependence of energy consumption on the parameters of priority control algorithms. In practice, this can be used to fine-tune a selected priority control algorithm more precisely to improve traffic indicators for other participants. Another conclusion from the research that supports the use of priority control algorithms is the low dependence of average energy consumption on the number of trams on the section, at least until approaching the section's capacity limits.

It is important to emphasize that mathematical models commonly used to describe priority-controlled traffic at individual intersections become highly complex in the case of a network of intersections. Hence, the research used a computer model built using Vissim software. This model exhibits a high degree of fidelity to real-world conditions, as confirmed by the validation results using real-world data.

The innovative aspect of the study presented in this article primarily stems from the fact that it addresses not just one intersection (as shown in the literature review in the introduction) but a network of interconnected intersections. It uses algorithms that are genuinely used in the field and parameterizes them appropriately for computer simulation.

As mentioned in the introduction, this article provided a detailed analysis of just one of the consequences of implementing complex priority control algorithms on coordinated road sections: electrical energy consumption for traction. The direction of further research is to compare the obtained results with changes in the situation for other road users, primarily cars (waiting time at intersections, the number of vehicles in the queue, etc.), and the associated effects, such as emissions of pollutants and delays. Conducting such studies will help to specify guidelines for selecting control algorithms considering a wider range of criteria.

Author Contributions: Conceptualization, A.C., A.G. and T.K.; methodology, A.C. and T.K.; software, A.C., T.K. and A.G.; validation, A.G. and T.K.; formal analysis, A.C.; investigation, T.K. and A.C.; resources, A.G.; data curation, T.K.; writing—original draft preparation, A.G. and A.C.; writing—review and editing, A.G. and T.K.; visualization, A.G., A.C. and T.K.; supervision, A.C.; project administration, A.G.; funding acquisition, A.G. All authors have read and agreed to the published version of the manuscript.

Funding: This research was funded by Warsaw University of Technology founds, Grant RND ILGiT 2023.

Data Availability Statement: The data presented in this study are available on request from the corresponding author. The data are not publicly available due to privacy policy of cooperating companies.

Conflicts of Interest: The authors declare no conflicts of interest.

References

1. Desta, R.; Tesfaye, D.; Tóth, J. Microscopic Traffic Characterization of Light Rail Transit Systems at Level Crossings. *Adv. Civ. Eng.* **2021**, *2021*, 1–11. [CrossRef]
2. Leitner, D.; Meleby, P.; Miao, L. Recent Advances in Traffic Signal Performance Evaluation. *J. Traffic Transp. Eng. Engl. Ed.* **2022**, *9*, 507–531. [CrossRef]
3. Li, J.; Liu, Y.; Yang, H.; Chen, B. Bus Priority Signal Control Considering Delays of Passengers and Pedestrians of Adjacent Intersections. *J. Adv. Transp.* **2020**, *2020*, 1–12. [CrossRef]
4. Naznin, F.; Currie, G.; Sarvi, M.; Logan, D. An Empirical Bayes Safety Evaluation of Tram/Streetcar Signal and Lane Priority Measures in Melbourne. *Traffic Inj. Prev.* **2016**, *17*, 91–97. [CrossRef] [PubMed]
5. de Oliveira, G.G.; Iano, Y.; Vaz, G.C.; Negrete, P.D.M.; Negrete, J.C.M.; Chuma, E.L. Intelligent Mobility: A Proposal for Modeling Traffic Lights Using Fuzzy Logic and IoT for Smart Cities. In *Proceedings of the Soft Computing and its Engineering Applications*; Patel, K.K., Doctor, G., Patel, A., Lingras, P., Eds.; Springer International Publishing: Cham, Switzerland, 2022; Volume 1572, pp. 302–311.
6. Zhang, X.; Blokpoel, R.; Lu, M. Adaptive Green Wave with Speed Advice for Automated Vehicles. In Proceedings of the ITS Europe congress, Eindhoven, The Netherlands, 3–6 June 2019.
7. Ito, H.; Kawazoe, N. Promoting Urban Light Rail Transit in a Compact City Context: The Case of Toyama City, Japan. *Reg. Stud. Reg. Sci.* **2022**, *9*, 776–793. [CrossRef]
8. Gadziński, J.; Radzimski, A. The First Rapid Tram Line in Poland: How Has It Affected Travel Behaviours, Housing Choices and Satisfaction, and Apartment Prices? *J. Transp. Geogr.* **2016**, *54*, 451–463. [CrossRef]
9. Rabay, L.; Meira, L.H.; de Andrade, M.O.; de Oliveira, L.K. A Portrait of the Crisis in the Brazilian Urban Bus System: An Analysis of Factors Influencing the Reduction in Usage. *Case Stud. Transp. Policy* **2021**, *9*, 1879–1887. [CrossRef]
10. Czerepicki, A.; Górka, A.; Szustek, J. Analysis of Trams' Consumption Depending on the Type of Traffic Light Used. *Nauka Teh.* **2019**, *18*, 490–494. [CrossRef]
11. Nellore, K.; Hancke, G.P. Traffic Management for Emergency Vehicle Priority Based on Visual Sensing. *Sensors* **2016**, *16*, 1892. [CrossRef]
12. Sundar, R.; Hebbar, S.; Golla, V. Implementing Intelligent Traffic Control System for Congestion Control, Ambulance Clearance, and Stolen Vehicle Detection. *IEEE Sens. J.* **2015**, *15*, 1109–1113. [CrossRef]
13. Chakraborty, P.S.; Tiwari, A.; Sinha, P.R. Adaptive and Optimised Emergency Vehicle Dispatching Algorithm for Intelligent Traffic Management System. *Procedia Comput. Sci.* **2015**, *57*, 1384–1393. [CrossRef]
14. Huang, Y.-S.; Shiue, J.-Y.; Luo, J. A Traffic Signal Control Policy for Emergency Vehicles Preemption Using Timed Petri Nets. *IFAC Pap.* **2015**, *48*, 2183–2188. [CrossRef]
15. Louati, A.; Elkosantini, S.; Darmoul, S.; Louati, H. Multi-Agent Preemptive Longest Queue First System to Manage the Crossing of Emergency Vehicles at Interrupted Intersections. *Eur. Transp. Res. Rev.* **2018**, *10*, 52. [CrossRef]
16. Parthasarathi, V.; Surya, M.; Akshay, B.; Murali Siva, K.; Vasudevan, S.K. Smart Control of Traffic Signal System Using Image Processing. *Indian J. Sci. Technol.* **2015**, *8*, 1–5. [CrossRef]
17. Madrjas, G.; Szustek, J.; Górka, A.; Guralski, B.; Stasiak, M. Dziesięć lat wdrażania priorytetu tramwajowego w Warszawie.(eng. Ten years of implementing tramway priority in Warsaw). In Proceedings of the Krakowskie Dni Bezpieczeństwa w Miastach, Kraków, Poland, 12–14 October 2022.
18. Dera, P. Stosowanie priorytetów dla pojazdów transportu zbiorowego w szczególności w zakresie wspólnych pasów autobusowo-tramwajowych w Krakowie. (The use of priority for public transport vehicles, especially in terms of shared tram-bus lanes in Kraków). *Transp. Miej. I Reg.* **2013**, *6*, 20–25.
19. Estrada, M.; Mension, J.; Salicrú, M. Operation of Transit Corridors Served by Two Routes: Physical Design, Synchronisation, and Control Strategies. *Transp. Res. Part C Emerg. Technol.* **2021**, *130*, 103283. [CrossRef]
20. Zhou, W.; Bai, Y.; Li, J.; Zhou, Y.; Li, T. Integrated Optimization of Tram Schedule and Signal Priority at Intersections to Minimise Person Delay. *J. Adv. Transp.* **2019**, *2019*, 4802967. [CrossRef]
21. Lehtonen, M.; Kulmala, R. Benefits of Pilot Implementation of Public Transport Signal Priorities and Real-Time Passenger Information. *Transp. Res. Rec.* **2002**, *1799*, 18–25. [CrossRef]
22. Akgüngör, A.P.; Mercan, E.Z. An Analysis of Type I Dilemma Zone at Signalised Intersections. *Sci. J. Silesian Univ. Technol. Ser. Transp.* **2021**, *112*, 5–16. [CrossRef]
23. Carvalho Barbosa, R.; Shoaib Ayub, M.; Lopes Rosa, R.; Zegarra Rodríguez, D.; Wuttisittikulkij, L. Lightweight PVIDNet: A Priority Vehicles Detection Network Model Based on Deep Learning for Intelligent Traffic Lights. *Sensors* **2020**, *20*, 6218. [CrossRef]
24. Górka, A.; Szustek, J. Wdrożenie priorytetu dla tramwajów na rondzie Zgrupowania AK „Radosław" w Warszawie metodą dekompozycji na podskrzyżowania częściowe. (Implementation of the priority for trams on the "Zgrupowanie AK Radosław" roundabout in Warsaw using the partial intersections division method). *Transp. Miej. I Reg.* **2018**, *12*, 17–21.
25. Molecki, A. Dostosowywanie układu drogowego w kontekście funkcjonalności sieci tramwajowe. (Adaptation of the road system in the context of the tram network functionality). *TTS Tech. Transp. Szyn.* **2020**, *5–6*, 29–34.

26. Chentoufi, M.A.; Ellaia, R. A Novel Metaheuristic for Adaptive Signal Timing Optimization Considering Emergency Vehicle Preemption and Tram Priority. *SAE Int. J. Transp. Saf.* **2019**, *7*, 129–151. [CrossRef]
27. Hounsell, N.B.; McLeod, F.N.; Shrestha, B.P. Bus Priority at Traffic Signals: Investigating the Options. In Proceedings of the 12th IEE International Conference on Road Transport Information and Control 2004 (RTIC 2004), London, UK, 20–22 April 2004; Institute of Electrical and Electronics Engineers: Piscataway, NJ, USA; pp. 287–294.
28. Qiu, Y.; Zheng, L.; Du, Y.; Jiang, S. Performance Evaluation of Transit Signal Priorities on Bus Transit Corridor Based on Data Envelopment Analysis. *Int. J. Transp. Sci. Technol.* **2020**, *9*, 322–333. [CrossRef]
29. Li, Y.; Cai, Q.; Xu, Y.; Shi, W.; Chen, Y. Design of Real-Time Actuated Control System for Modern Tram at Arterial Intersections Based on Logic Rules. *Adv. Mech. Eng.* **2018**, *10*, 168781401881542. [CrossRef]
30. Yan, H.; Ou, D.; Chen, Z.; Yang, Y. Research on Tram Detector Location Based on Vehicle–Infrastructure Communication. In *Proceedings of the 3rd International Conference on Electrical and Information Technologies for Rail Transportation (EITRT) 2017*; Springer: Singapore, 2018; pp. 915–925.
31. Koehler, L.A.; Seman, L.O.; Kraus, W.; Camponogara, E. Real-Time Integrated Holding and Priority Control Strategy for Transit Systems. *IEEE Trans. Intell. Transport. Syst.* **2019**, *20*, 3459–3469. [CrossRef]
32. Mirchandani, P.; Head, L. A Real-Time Traffic Signal Control System: Architecture, Algorithms, and Analysis. *Transp. Res. Part C Emerg. Technol.* **2001**, *9*, 415–432. [CrossRef]
33. Celtek, S.A.; Durdu, A.; Ali, M.E.M. Real-Time Traffic Signal Control with Swarm Optimization Methods. *Measurement* **2020**, *166*, 108206. [CrossRef]
34. Gao, J.; Shen, Y.; Liu, J.; Ito, M.; Shiratori, N. Adaptive Traffic Signal Control: Deep Reinforcement Learning Algorithm with Experience Replay and Target Network. *arXiv* **2017**, arXiv:1705.02755. [CrossRef]
35. Suthaputchakun, C.; Sun, Z. A Novel Traffic Light Scheduling Based on TLVC and Vehicles' Priority for Reducing Fuel Consumption and CO2 Emission. *IEEE Syst. J.* **2018**, *12*, 1230–1238. [CrossRef]
36. Li, M.; Wu, G.; Li, Y.; Bu, F.; Zhang, W. Active Signal Priority for Light Rail Transit at Grade Crossings. *Transp. Res. Rec.* **2007**, *2035*, 141–149. [CrossRef]
37. Aleksandrowicz, J.; Stróżek, A. Priorytet dla pojazdów transportu zbiorowego w programie sygnalizacji świetlnej na podstawie liczby pasażerów. (Priority for public transport vehicles in a traffic light control plan based on number of passengers). *Transp. Miej. I Reg.* **2021**, *3*, 14–18.
38. Chow, A.H.F.; Sha, R.; Li, S. Centralised and Decentralised Signal Timing Optimisation Approaches for Network Traffic Control. *Transp. Res. Part C Emerg. Technol.* **2020**, *113*, 108–123. [CrossRef]
39. Molecki, A. Kształtowanie priorytetu tramwajowego w procesie wdrażania ITS we Wrocławiu. (Shaping tram priority in the process of ITS implementation in Wroclaw). *Przegląd Komun.* **2015**, *22*, 20–22.
40. Vujičić, M.; Prester, J. Assessing Service Quality of Public Tram Transport in Zagreb City Using P-Transqual Model. *Zb. Ekon. Fak. U Zagreb* **2019**, *17*, 19–31. [CrossRef]
41. Xu, Z.; Lv, S.; Li, D.; Quan, M. Coordinated Control Method for Trams on Urban Arterial. *IOP Conf. Ser. Earth Environ. Sci.* **2020**, *587*, 012092. [CrossRef]
42. Lin, Y.; Yang, X.; Zou, N. Passive Transit Signal Priority for High Transit Demand: Model Formulation and Strategy Selection. *Transp. Lett.* **2019**, *11*, 119–129. [CrossRef]
43. Jeong, Y.; Kim, Y. Tram Passive Signal Priority Strategy Based on the MAXBAND Model. *KSCE J. Civ. Eng.* **2014**, *18*, 1518–1527. [CrossRef]
44. Bai, Y.; Li, J.; Li, T.; Yang, L.; Lyu, C. Traffic Signal Coordination for Tramlines with Passive Priority Strategy. *Math. Probl. Eng.* **2018**, *2018*, 1–14. [CrossRef]
45. Kaczmarek, M. Zintegrowana Koordynacja Tramwajów i Pojazdów Indywidualnych w Korytarzach Transportowych (Integrated Coordination of Trams and Individual Vehicles in Transportation Corridors). In Proceedings of the VIII Konferencja Naukowo, Techniczna, Rosnówko, 15–17 June 2011; SITK: Poznań, Poland, 2011.
46. Górka, A.; Szustek, J. Związek Priorytetu Dla Tramwajów w Sygnalizacji Świetlnej z Zużyciem Energii Elektrycznej (Relation of Priority for Trams in Traffic Signals to Electricity Consumption). *Plan. Ruchu A Wyzwania Glob. Annały Inżynierii Ruchu I Badań Transp.* **2019**, *3*, 383–394.
47. OpenStreetMap. Available online: https://www.openstreetmap.org (accessed on 10 January 2023).
48. *Vissim 21 User Manual*; PTV Planung Transport Verkehr AG: Karlsruhe, Germany, 2021.
49. Wiedemann, R. *Simulation Des Straßenverkehrsflusses (Traffic Flow Simulations)*; Publication series of the Institute for Traffic, University of Karlsruhe; Institute for Traffic, University of Karlsruhe: Karlsruhe, Germany, 1974.
50. Papageorgiou, G.; Ioannou, P.; Pitsillides, A.; Aphamis, T.; Maimaris, A. Development and Evaluation of Bus Priority Scenarios via Microscopic Simulation Models. *IFAC Proc. Vol.* **2009**, *42*, 434–441. [CrossRef]
51. Ejercito, P.M.; Nebrija, K.G.E.; Feria, R.P.; Lara-Figueroa, L.L. Traffic Simulation Software Review. In Proceedings of the 2017 8th International Conference on Information, Intelligence, Systems & Applications (IISA), Larnaca, Cyprus, 27–30 August 2017; pp. 1–4.
52. Bandi, M.M.; George, V. Microsimulation Modelling in VISSIM on Short-Term and Long-Term Improvements for Mangalore City Road Network. *Transp. Res. Procedia* **2020**, *48*, 2725–2743. [CrossRef]

53. Ahuja, S.; van Vuren, T.; MacDonald, M.; Priest, N. Public Transport Priority Schemes–Comparing Microsimulation with Traditional TRANSYT and LINSIG Models. In Proceedings of the European Transport Conference (Etc) 2003, Strasbourg, France, 8–10 October 2003.
54. Shastry, K. Code Icon–Free PNG & SVG 2187377–Noun Project. Available online: https://thenounproject.com/icon/code-2187377/ (accessed on 14 April 2023).
55. SAM Designs. Performance Icon–Free PNG & SVG 5329663–Noun Project. Available online: https://thenounproject.com/icon/performance-5329663/ (accessed on 14 April 2023).
56. Map of Warsaw. Available online: https://mapa.um.warszawa.pl/mapaApp1/mapa?service=mapa&L=pl (accessed on 10 January 2023).
57. Timetables–Warsaw Public Transport. Available online: https://www.wtp.waw.pl/rozklady-jazdy/ (accessed on 13 April 2023).
58. *Zarząd Transportu Miejskiego. Raport Roczny 2022*; Annual Report; Public Transport Authority: Warszawa, Poland, 2022.
59. Miyatake, M.; Ko, H. Optimization of Train Speed Profile for Minimum Energy Consumption. *IEEJ Trans. Elec Engng.* **2010**, *5*, 263–269. [CrossRef]
60. Sicre, C.; Cucala, P.; Fernández, A.; Jiménez, J.A.; Ribera, I.; Serrano, A. A Method to Optimise Train Energy Consumption Combining Manual Energy Efficient Driving and Scheduling. In Proceedings of the 12th International Conference on Computer System Design and Operation in Railways and Other Transit Systems, Beijing, China, 4 August 2010; pp. 549–560.
61. Liu, J.; Guo, H.; Yu, Y. Research on the Cooperative Train Control Strategy to Reduce Energy Consumption. *IEEE Trans. Intell. Transport. Syst.* **2017**, *18*, 1134–1142. [CrossRef]
62. Domínguez, M.; Fernández, A.; Cucala, A.P.; Lukaszewicz, P. Optimal Design of Metro Automatic Train Operation Speed Profiles for Reducing Energy Consumption. *Proc. Inst. Mech. Eng. Part F J. Rail Rapid Transit* **2011**, *225*, 463–474. [CrossRef]
63. He, D.; Yang, Y.; Chen, Y.; Deng, J.; Shan, S.; Liu, J.; Li, X. An Integrated Optimization Model of Metro Energy Consumption Based on Regenerative Energy and Passenger Transfer. *Appl. Energy* **2020**, *264*, 114770. [CrossRef]
64. Feng, Z.; Chen, W.; Liu, Y.; Chen, H.; Skibniewski, M.J. Long-Term Equilibrium Relationship Analysis and Energy-Saving Measures of Metro Energy Consumption and Its Influencing Factors Based on Cointegration Theory and an ARDL Model. *Energy* **2023**, *263*, 125965. [CrossRef]
65. Mwambeleko, J.J.; Kulworawanichpong, T.; Greyson, K.A. Tram and Trolleybus Net Traction Energy Consumption Comparison. In Proceedings of the 2015 18th International Conference on Electrical Machines and Systems (ICEMS), Pattaya, Thailand, 25–28 October 2015; pp. 2164–2169.
66. Kubin, J.; Ferkova, Z. Influnce of Driving Style of a Tram Driver on the Tram's Energy Consumption. In Proceedings of the 2015 International Conference on Electrical Drives and Power Electronics (EDPE), Tatranska Lomnica, Slovakia, 21–23 September 2015; pp. 417–421.
67. Zhang, T.; Mao, B.; Xu, Q.; Feng, J. Timetable Optimization for a Two-Way Tram Line with an Active Signal Priority Strategy. *IEEE Access* **2019**, *7*, 176896–176911. [CrossRef]
68. He, J.; Li, Y.; Long, S.; Xu, Y.; Chen, J. Energy-Efficient Tram Speed Trajectory Optimization Considering the Influence of the Traffic Light. *Front. Energy Res.* **2022**, *10*, 963275. [CrossRef]
69. Xing, Z.; Zhu, J.; Zhang, Z.; Qin, Y.; Jia, L. Energy Consumption Optimization of Tramway Operation Based on Improved PSO Algorithm. *Energy* **2022**, *258*, 124848. [CrossRef]
70. Yan, Y.; Li, Q.; Huang, W.; Chen, W. Operation Optimization and Control Method Based on Optimal Energy and Hydrogen Consumption for the Fuel Cell/Supercapacitor Hybrid Tram. *IEEE Trans. Ind. Electron.* **2021**, *68*, 1342–1352. [CrossRef]
71. Chen, L.; Wang, D. Energy Consumption Modeling and Energy Saving Analysis of Supercapacitor Tramcar. In Proceedings of the 2018 IEEE International Conference of Intelligent Robotic and Control Engineering (IRCE), Lanzhou, China, 24–27 August 2018; pp. 79–84.
72. Yan, Y.; Li, Q.; Chen, W.; Huang, W.; Liu, J. Hierarchical Management Control Based on Equivalent Fitting Circle and Equivalent Energy Consumption Method for Multiple Fuel Cells Hybrid Power System. *IEEE Trans. Ind. Electron.* **2020**, *67*, 2786–2797. [CrossRef]
73. Moskowitz, J.-P.; Cohuau, J.-L. STEEM: ALSTOM and RATP Experience of Supercapacitors in Tramway Operation. In Proceedings of the 2010 IEEE Vehicle Power and Propulsion Conference, Lille, France, 1–3 September 2010; pp. 1–5.
74. Yan, Y.; Li, Q.; Chen, W.; Su, B.; Liu, J.; Ma, L. Optimal Energy Management and Control in Multimode Equivalent Energy Consumption of Fuel Cell/Supercapacitor of Hybrid Electric Tram. *IEEE Trans. Ind. Electron.* **2019**, *66*, 6065–6076. [CrossRef]
75. Liu, J.; Wu, X.; Li, H.; Qi, L. An Optimal Method of the Energy Consumption for Fuel Cell Hybrid Tram. *Int. J. Hydrog. Energy* **2020**, *45*, 20304–20311. [CrossRef]
76. Zhaolu, S.; Liwei, D. Prediction of Tram Energy Consumption Based on Modified Regression Model. In Proceedings of the 2019 2nd International Conference on Safety Produce Informatization (IICSPI), Chongqing, China, 28–30 November 2019; pp. 188–193.
77. Kolat, M.; Kővári, B.; Bécsi, T.; Aradi, S. Multi-Agent Reinforcement Learning for Traffic Signal Control: A Cooperative Approach. *Sustainability* **2023**, *15*, 3479. [CrossRef]
78. Zeinaly, Z.; Sojoodi, M.; Bolouki, S. A Resilient Intelligent Traffic Signal Control Scheme for Accident Scenario at Intersections via Deep Reinforcement Learning. *Sustainability* **2023**, *15*, 1329. [CrossRef]

79. SUMO User Documentation. Available online: https://eclipse.dev/sumo/ (accessed on 7 January 2024).
80. Rozporządzenie Ministra Infrastruktury z Dnia 3 Lipca 2003 r. w Sprawie Szczegółowych Warunków Technicznych Dla Znaków i Sygnałów Drogowych Oraz Urządzeń Bezpieczeństwa Ruchu Drogowego i Warunków ich Umieszczania na Drogach (t.j. Dz. U z 2019 r. poz. 2311 z późn. zm.). poz. 2311 z późn. zm.). Available online: https://isap.sejm.gov.pl/isap.nsf/DocDetails.xsp?id WDU20190002311 (accessed on 7 January 2024).

energies

MDPI

Article

Analysis of the Overhead Crane Energy Consumption Using Different Container Loading Strategies in Urban Logistics Hubs

Michał Kłodawski [1,*], Roland Jachimowski [1] and Norbert Chamier-Gliszczyński [2]

[1] Faculty of Transport, Warsaw University of Technology, 00-661 Warsaw, Poland; roland.jachimowski@pw.edu.pl

[2] Faculty of Technology and Education, Koszalin University of Technology, 75-453 Koszalin, Poland; norbert.chamier-gliszczynski@tu.koszalin.pl

* Correspondence: michal.klodawski@pw.edu.pl

Abstract: This study addresses the critical gap in the literature regarding the energy efficiency of intermodal terminals in smart cities, mainly focusing on crane operations during train loading processes. Novelty's contribution lies in developing a comprehensive simulation model in FlexSim, where quantitative analysis of crane energy consumption, factoring in container location in the storage yard, rehandling operations, and crane movement strategies were performed. Moreover, the analysis of hoist, trolley, and gantry movements was performed to evaluate their impact on overall container loading process energy efficiency. The findings reveal that the choice of train loading method significantly influences crane energy consumption, thereby affecting the operational costs, environmental footprint, and energy efficiency of the logistics hub in the form of an intermodal terminal. This research provides a methodology for assessing and enhancing the energy efficiency of intermodal terminals and highlights the broader implications for smart city sustainability goals, including reduced greenhouse gas emissions, lower operating costs, and improved transportation infrastructure. The outcomes of this research can possibly support smart city planners and logistics managers in making informed decisions to optimise intermodal terminal operations, contributing to urban areas' sustainable development and economic resilience.

Keywords: crane energy consumption; intermodal terminal; container handling; smart city; intermodal train loading; logistics hubs

1. Introduction

The concept of a smart city is based on the utilisation of advanced information and communication technologies (ICT), large datasets, data analysis tools, and innovative solutions to improve the quality of life for residents, enhance the efficiency of municipal services, and promote sustainable urban development. The goals of the smart city concept are achieved by leveraging modern technologies such as the Internet of Things (IoT), artificial intelligence (AI), data analysis, and blockchain to develop intelligent solutions and services. ICT infrastructure plays a crucial role in realising this concept [1]. As a result, the smart city concept aims to integrate various resources such as transportation, education, health, energy, and public safety to maximise the efficiency of managing these resources. An essential element of this concept is also the pursuit of sustainable development through proper monitoring and management of energy consumption. From the perspective of freight transport, this concept envisions the use of modern technologies to create a more efficient, sustainable, and secure logistics system that contributes to improving the quality of life for residents. Freight transport in a smart city is crucial for the functioning of the entire urban logistics system. It encompasses the service of businesses and residents, who are both senders and receivers of goods, and efficient logistics within the city. From the perspective of freight transport, a smart city utilises advanced technologies such as supply chain management (SCM) systems to optimise logistics processes. Additionally, automation, data analysis, and real-time monitoring enable effective route planning, minimisation of delays, and optimisation of resource utilisation in the form of transportation means or storage space for goods. These systems, in turn, are supported by integrated logistics platforms that connect various supply chain elements, including storage, transportation, and data management. These platforms enable better planning and coordination of logistical activities [2]. Utilising these technologies enables effective planning of the city's freight transport service to minimise the impact of freight transport on urban traffic [3].

Due to limitations on the movement of freight vehicles in cities, arising from their size, permissible total weight, and often emission levels, the efficient execution of freight services for senders and recipients in a smart city requires the use of facilities that will enable the deconsolidation of large cargo flows entering the city and the consolidation of cargoes sent from the city and its immediate surroundings. Examples of such facilities are Urban Consolidation Centres and intermodal terminals. Urban Consolidation Centres are facilities where large streams of goods arriving in the city are collected, grouped, and then distributed to their final destinations within the city. The issue of Urban Consolidation Centres is one of the most frequently addressed topics in the literature related to urban logistics [4]. This is because such facilities can potentially address the environmental and social challenges urban transportation systems pose [5]. The first publications addressing this issue date back to the early 1970s [6]. However, in practice, initiatives involving the construction of such facilities often do not come to fruition [7]. This results from a series of challenges that such concepts must face. Coordinating various suppliers, carriers, and recipients while dealing with diverse delivery requirements can lead to operational difficulties. Stakeholders, including businesses, transport companies, and local authorities, may resist such changes due to concerns about disruptions, costs, or favouring established practices [8].

Therefore, one of the solutions to the cargo logistics problem in a smart city can be intermodal terminals, which, as practice indicates, are often located inside cities at nodes and railway lines. Intermodal terminals also offer storage space, allowing for flexible management of deliveries and pickups. Moreover, with proper storage and transhipment infrastructure and information technology infrastructure, intermodal terminals often become central hubs where various transport operators can collaborate effectively, exchange information, and plan transportation [9].

In addition to the functions performed by intermodal terminals, a significant aspect favouring their use for smart city operations is the rapid development of containerisation. Since its inception in the mid-20th century, containerised transport has enabled a standardised method of freight transport, ensuring the safety and efficiency of goods transportation. In 2012, container throughput at ports worldwide was at the level of 662 million TEUs. 10 years later, it achieved the value of 866 million TEUs. The global container transport market is forecast to grow at a compound annual growth rate (CAGR) of over four percent between 2019 and 2027 [10]. The presented values clearly indicate that with the increase in intermodal transport, there will be a need for the development of both infrastructure and organisational aspects of container terminals, including those located in or near the city, to handle large volumes of cargo both inbound to and outbound from the city and its surroundings [11].

The use of intermodal terminals in urban logistics also aligns with the European Commission's 2011 White Paper on Transport recommendations. According to the goals set forth in this document, it is anticipated that by 2030, 30% of road freight transport over distances exceeding 300 km should be shifted to other modes of transport, such as rail or water transport. By 2050, this percentage should exceed 50%. Therefore, the realisation of deliveries to and from the city via rail-road transport, with operations conducted in intermodal terminals, may become necessary soon [12].

The literature analyses conducted concerning the use of intermodal terminals for cargo handling in the city indicate that so far, there have been few publications on this subject.

The application of a keyword query ("urban" OR "city") AND ("intermodal terminal" OR "container terminal") in the Web of Science database returned only 55 results. Most of these publications address the issue of work organisation in maritime terminals located in large ports, which are usually part of urban agglomerations [13,14].

Among the works on rail-road intermodal terminals located in the city or its vicinity, one can distinguish works concerning the rationalisation of their location [15,16], or studies examining and assessing the investment risk associated with the location of a terminal in a given city area [17]. Some publications investigate the issue of the impact of intermodal terminals on logistic processes in the city [18]. From the perspective of urban logistics, some publications focus on studying the impact of freight traffic serving intermodal terminals on city congestion as well as the impact of using intermodal terminals in cities to reduce air pollution [19,20].

Literature analysis shows that there is essentially a lack of publications on the energy efficiency of intermodal terminals handling cargo in the city. Indeed, in their work [21], the authors address the issue of the energy efficiency of a rail-road intermodal terminal, but they do not present the necessary calculations related to energy consumption. The authors of this work focus on developing a generic algorithm for scheduling crane operations, and the study of the energy efficiency of the intermodal terminal is merely reduced to identifying factors that influence energy consumption in the outbound container handling process.

Therefore, the aim of the article is to calculate and examine the crane energy consumption during the train loading in the intermodal terminal handling freight from the city. These studies were conducted on a simulation model developed in Flexsim in which the storage yard of the intermodal terminal, the loading front, and consequently the process of loading containers onto wagons were replicated. The energy efficiency of an intermodal train loading process is conditioned by the location of containers in the storage yard, the necessity of rehandling containers in the storage yard in order to pick up those designated for loading, as well as the crane operation strategy resulting from the distribution of coupling pins on container wagons (understood as the algorithm for allocation of containers to wagons and the order of their handling by the crane).

As a result of the conducted research, it was possible to answer the question of how the aforementioned factors affect crane energy consumption and consequently the energy efficiency of the entire process. The data for the crane energy consumption calculation was determined based on analyses conducted in work [22] for an RTG crane.

The remaining part of the article is organised as follows: A brief literature review on the optimisation of handling processes in intermodal terminals and related energy consumption by handling equipment is presented in Section 2. In Section 3, the article outlines the description of the research problem addressed. Section 4 conducts simulation experiments and presents the results of the calculations. Section 5 summarises the conducted research.

2. Literature Review

2.1. Intermodal Train Loading

In addition to providing various logistics services such as consolidation, deconsolidation, transportation, forwarding, and customs clearance, an intermodal terminal in a smart city primarily handles container transshipments in the truck-train relationship. Containers arriving at the terminal by rail are transferred to the yard. Some of them are then transported by truck directly to customers, while others undergo deconsolidation, allowing small batches of goods to be delivered to recipients in the city. The process of container dispatch is precisely the opposite. Some containers are delivered directly from senders, while others are formed in the terminal by consolidating smaller shipments sent from the city. The specificity of the intermodal terminal's operation leads to a series of decision-making problems directly affecting cargo handling time in the terminal, transport, and transhipment cycle times, and consequently, the efficiency of process execution and the energy efficiency of the terminal. Among the fundamental decision-making problems determining the efficiency of the terminal, it is necessary to distinguish [23–25]: the distribution of containers in the storage yard, scheduling the work of yard cranes, and loading intermodal trains. The first of these problems involves determining the optimal location for container storage to prevent the need to move other containers to extract a specific one. Scheduling the work of yard cranes at the terminal involves allocating containers to cranes and determining the order in which the crane handles these containers. The loading of intermodal trains is directly related to these issues. The literature on maritime terminals devotes a significant amount of space to issues such as scheduling the work of cranes or arranging containers in the storage yard [26,27]. Unfortunately, the processes occurring in land terminals, although they may appear such as those in maritime terminals, have a different course in land terminals [24]. This is because deliveries to the terminal and shipments are characterised by high variability, mainly influenced by urgent containers that need handling [28]. One of the very first publications that advanced the understanding of the train loading problem by offering a substantial contribution by creating practical models that can be adapted to diverse, specific environments were the papers [29,30].

The issue of loading intermodal trains is a challenging one because, when planning the arrangement of containers on wagons, it is crucial to consider factors such as the configuration of coupling pins on wagons, the gross weights of loaded containers, and the associated permissible axle loads of the wagons, as well as the maximum filling of wagon capacity to minimise the train length and aerodynamic resistance. One of the studies extensively addresses the issue of loading intermodal trains [30]. In addition to the configuration of coupling pins on wagons, the authors also considered weight restrictions for wagons and various types and weights of containers, including the arrangement of their coupling corners and centre of mass. Later in work [31], they added uncertainty in the weight of containers and faults in the wagons. Similarly, the loading of a train was considered in the work [32].

In addition to the configuration of coupling pins on wagons and the proper distribution of the load on wagon axles resulting from the weight of containers, another factor influencing the strategy of intermodal train loading is the train's aerodynamics, as considered in the publication [33]. According to the authors, the train's aerodynamics significantly impact the train's stability in motion and its energy consumption.

The planning of the intermodal train loading is directly related to the optimisation issues of container storage and movement in the yard. In [34], the authors formulated a MILP load planning model to maximise train utilisation while minimising reshuffling costs in the stacking area and unproductive crane movements as the train is loaded. Also, in [35], the authors considered the situation of train loading, where containers were sequentially loaded on the wagons to maximise the priority of containers to be loaded while minimising the number of rehandles in the yard. Similar issues were considered in [21,36].

The literature analysis in this area indicates a lack of publications that comprehensively consider all aspects of loading trains while simultaneously considering the location of containers in the storage yard.

2.2. Crane Energy Efficiency

In the inland terminals serving urban agglomerations and their surroundings, the typical equipment used includes RMG cranes, RTG cranes, or Reachstackers [37,38]. Various energy sources and power loading equipment are adapted to the specific location and operations performed at a given terminal. Many intermodal terminals use electricity supplied by the power grid. Loading machines, such as rail-mounted gantry cranes and sometimes rubber-tired gantry cranes, are connected to a constant power source, enabling efficient and continuous operation. Some loading machines in intermodal terminals are equipped with internal combustion engines, allowing them to be self-powered. This solution is particularly useful in locations where access to electricity may be limited. This type of power is typically used in Reachstackers and rubber-tired gantry cranes (RTG). In RTG cranes powered this way, a diesel engine drives a generator that produces the necessary electricity. The electricity is then rectified, and the DC power is distributed to all electric motors driving the hoist, trolley, and RTG crane's gantry [39].

Contemporary technologies increasingly utilise electric batteries to power loading machines. These environmentally friendly solutions can be charged from both traditional and renewable sources, such as photovoltaic panels [40]. Some loading machines utilise hybrid power systems, combining various energy sources, such as a combustion engine and electric batteries. This allows for the optimal use of resources and reduces greenhouse gas emissions [39].

The choice of an appropriate power source depends on various factors, such as the type of machine, local infrastructure conditions, the availability of energy resources, and priorities related to ecology and sustainable development. The final choice often considers a combination of different technologies, ultimately aiming to optimise energy consumption and minimise the impact of loading equipment in terminals on the environment.

Technological progress in energy generation, storage, distribution, conversion, and consumption significantly influences the energy efficiency of an intermodal terminal [41]. In recent years, scientific publications addressing energy consumption by loading devices have increasingly focused on energy storage methods [42]. This energy can be recovered by the loading device, for example, during the lowering of a container by the gantry crane hoist. Research indicates that implementing an appropriate energy storage system on the crane can reduce fuel consumption, decrease the size of the combustion engine, and simultaneously extend its engine life [43]. These storage systems may include supercapacitors [44], batteries [45], and flywheel systems [46].

Equally crucial for improving the energy efficiency of an intermodal terminal is proper planning and coordinating the work of devices. Extensive studies have been conducted on reducing the energy consumption of equipment operations, especially in automated container terminals. Here, we mention just a few examples. In [47], the authors proposed an optimisation model that minimises the total number of operations performed on containers using machine learning techniques. In [48], the authors delved into the determinants impacting energy efficiency during container loading operations. Their research focused on addressing the challenge of reducing energy consumption in container handling and reshuffling processes. In one study [49], the authors examined the energy consumption patterns of diverse handling equipment within a rail-water intermodal container terminal. Their primary objective was to optimise overall operational efficiency by minimising processing times in the context of multi-device coordination, integrated scheduling, and constrained storage space. Some publications on optimising intermodal terminals' energy efficiency are dedicated to AGV vehicles, their scheduling, and route planning [50–53].

A distinct group of publications related to terminal energy efficiency comprises works in which authors, alongside minimising energy consumption, also pay attention to air pollution. One of the publications that indicates the correlation between energy consumption, CO_2 emissions, and the layout of an intermodal terminal is the study [54]. In the course of this investigation, an estimation was made regarding the energy utilisation and carbon dioxide (CO_2) emissions within container terminals, considering the variations in their layouts. The computation of energy consumption involved the analysis of utility data, in conjunction with the assessment of fuel and electricity usage specific to each container-handling equipment present in the terminal. To gauge CO_2 emissions, a methodology based on the movement characteristics—such as the frequency and distance covered by each container-handling equipment—was employed. Similar considerations were presented in [55–57].

For the purposes of the research conducted in this article, literature on crane energy efficiency was analysed to find data on crane energy consumption. Such data was presented in a paper [58]. Referring to data from the port of Rotterdam, the authors of [58] presented the values of instantaneous energy consumption by the STS crane. Unfortunately, due to the dimensions of the STS crane and the operations it performs, this data could not be used to calculate the energy consumption of the RTG crane at the land intermodal terminal.

Further analysis of the literature revealed that an interesting approach to the analysis of the energy efficiency of an intermodal terminal was presented in [22]. The authors conducted a study examining energy consumption by RTG cranes in a maritime intermodal terminal over eight days. An energy analysis concerning the various motors of the crane has been carried out. The range of percentages of the total recoverable energy has been calculated depending on the workload of the specific day. The data on the instantaneous energy consumption of the RTG crane presented in this work were the starting point for the analyses carried out in our article.

The literature analysis on the optimisation of energy consumption by handling equipment indicates that publications mainly focus on maritime terminals. It is challenging to find publications that address this issue in land terminals located in or near cities, especially considering that transport and handling processes in land terminals differ from those in maritime terminals.

Therefore, this study aimed to analyse the energy efficiency of an intermodal terminal located in a city where the gantry crane is the primary handling equipment. The research focused on energy consumption during the loading of intermodal trains, as this process significantly determines the cycle time of the gantry crane, and minimising it is crucial for the energy efficiency of the intermodal terminal.

3. Problem Description

3.1. Basic Assumptions for Train Loading

The article addresses the issue of loading intermodal trains and its impact on the energy efficiency of a land intermodal terminal located in a city, particularly in the context of energy consumption by the gantry crane. Generally, a land-based intermodal terminal consists of freight tracks, a roadway for trucks and various handling machines, and a storage yard for containers. An example of the layout of the basic functional zones of the terminal is presented in Figure 1.

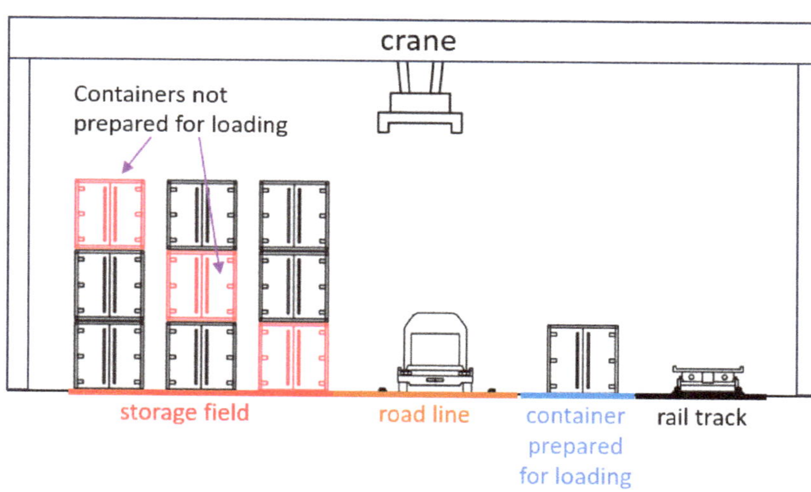

Figure 1. An example of the intermodal terminal layout.

The practice of handling containers in an intermodal terminal indicates that it is reasonable to prepare containers in advance for loading onto trains. Containers prepared for loading onto trains are usually positioned along the loading track, minimising the train handling time at the terminal. Of course, these containers need to be retrieved from the storage yard. Depending on the stacking height of containers, retrieving containers from the storage yard may involve the need to move other containers. This means that to handle one container, the gantry crane will have to perform several cycles, affecting the crane's working time and the amount of energy consumed. The article assumes that containers in the storage yard will be stacked in a maximum of three layers.

Again, the practice of loading intermodal trains suggests that despite preparing containers for train loading, some will still need to be loaded directly from the storage yard. This is justified by the uncertainty regarding the final list of containers leaving the terminal on a given train, as a previously planned container may be replaced by an unplanned one at the last moment. This issue has been considered in the simulation model presented later in the article.

In addition to the arrangement of containers in the terminal planned for loading, a significant influence on the time required for this loading process i the configuration of coupling pins on wagons. Container wagons are generally adapted for transporting containers of various sizes by allowing changes to the configuration of coupling pins that secure the container to the wagon (see Figure 2). Incorrect configuration of coupling pins on a particular wagon results in the inability to load the container and will require manual changes to the pin configuration. The article assumes that the pin configuration on all wagons allows for loading all planned containers without requiring manual pin configuration changes. Additionally, in the simulation studies, the loading of ISO 20′, 30′, and 40′ containers onto wagons with a capacity of 3 TEU (TEU—twenty-foot equivalent unit) was assumed.

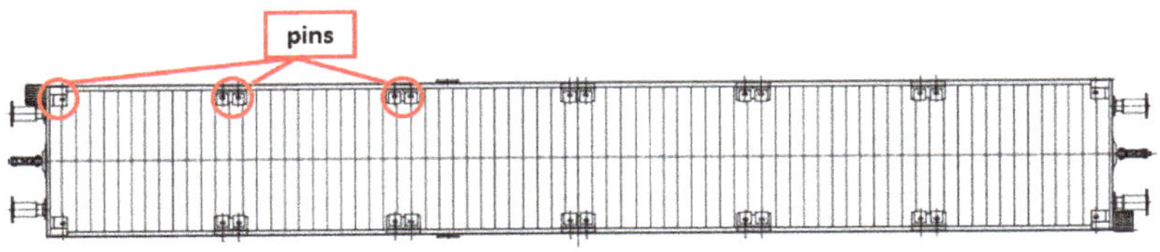

Figure 2. Example of pin configuration of the wagon. (Own work based on [59]).

The factors mentioned above have been considered in further calculations and strongly influence the distances covered by the gantry crane gate, the trolley, and the hoist, resulting in energy consumption during the loading of intermodal trains.

3.2. Decision Model Assumptions

The decision problem of loading a train involves determining the allocation of containers to slots on wagons and the sequence in which containers are loaded onto the wagons. For the study, five strategies were developed based on which containers are loaded onto wagons in the appropriate order:

- L1—priority of the wagons—algorithm chooses the first slot on the first wagon in the head of a train and then searches for the first matching container starting from the head of the yard (search for a container that satisfies pin configuration constraints as well as the wagon boogie maximum payload constraints). Pin and boogie constraints are checked in every strategy.
- L2—priority of the containers—algorithm chooses the first container starting from the head of the yard and then searches for the first matching slot on the wagon starting from the head of the train.
- L3—shortest distance from the wagon to a container—algorithm chooses the first slot on the first wagon in the head of a train and then searches for the closest matching container.
- L4—shortest distance from the container to the wagon—algorithm chooses the first container starting from the head of the yard and then searches for the closest matching slot on the wagon.
- L5—shortest distance from the current container to the wagon and from the current wagon to the container (nearest neighbour algorithm). The algorithm chooses the

first available container and then searches for the closest matching slot on the wagon. In the next step, the algorithm searches for the container closest to the slot from the previous step.

A proper approach to handling containers in the storage yard significantly impacts the working time of the gantry crane in an intermodal terminal. Monitoring the arrangement of containers in the storage yard allows for minimising the crane's movements, thereby reducing energy consumption. Therefore, the article considered two strategies for handling containers retrieved for loading directly from the storage yard:

- S1—containers set aside to extract a specific container were rearranged to their original positions in the storage yard in each loading cycle of the gantry crane.
- S2—containers set aside to extract a specific container were not returned to their original positions by the gantry crane.

As mentioned earlier, the practice of loading containers onto trains suggests preparing these containers in advance along the loading track. The article considered two variants of preparing containers along the track:

- V1—containers prepared along the track were arranged sequentially from the beginning of the loading track, regardless of the location of a specific container in the storage yard.
- V2—containers placed along the track were positioned perpendicular to their initial position in the storage yard.

As a result of compiling strategies for the gantry crane's operation, handling containers in the storage yard, and strategies for preparing them for loading along the track, 20 variants of handling trains and containers at the intermodal terminal were obtained. These variants were further subjected to simulation studies later in the article.

3.3. *Basic Assumptions for Research Methodology*

The choice of research methodology usually depends on the research problem's objectives, available resources (time, financial, technological), and data. Combining several methods could also be practical and provide a more comprehensive picture of the solved problem.

In this case, a specific issue related to analysing the energy consumption of the container train loading process is examined. This problem requires estimating the energy consumption by the crane engines during all of its movements performed througout various methods of train loading. Therefore, conducting measurements in the actual process at the intermodal terminal could require developing and applying a mathematical model that estimates energy consumption during crane operations or utilising simulation modelling tools. In the first case, the research is expensive, time-consuming, and often impossible from the point of view of access to the actual terminal. Mathematical modelling involves applying certain simplifications and averages due to the loading process's dynamics, variability, and randomness (which is not necessary in the case of simulation). Therefore, we decided to combine all three methods. A simulation model of the crane operation was developed. It includes mathematical formulas for estimating energy consumption based on actual measurements (taken from [22]).

The simulation model calculates the speeds and distances covered by specific elements of the gantry crane during the loading process, which were identified for each of the considered variants. This information determined the working time of the gantry crane gate, trolley, and hoist. The identified times in each variant were then used to calculate the energy consumption of these gantry crane components.

4. Simulation Research

Experiments were conducted using an original simulation model. The model was developed using the FlexSim 23.2.3 tool. It is an environment designed for building discrete simulation models and conducting simulation experiments. As it is a commercial and licensed tool, the research presented in this paper was carried out under a license owned by the Warsaw University of Technology, Faculty of Transport. The prepared model mapped the various elements of a container terminal using appropriately parameterised Flexsim library objects. All the processes regarding selecting, preparing, and loading containers onto the train were programmed in Process Flow and FlexScript modules (integral FlexSim modules). The research also considered several parameters and functions that enabled estimating crane energy consumption during individual intermodal train loading strategies.

4.1. Simulation Model and Assumptions

The simulation model consists of several essential elements and areas (see Figure 3):

- RTG gantry crane,
- container storage field, containing containers that wait for loading
- rail-road, on which the train with wagons is placed for loading,
- storage lane, on which containers are laid down in a specific sequence before the loading process
- containers.

The parameters of an RTG gantry crane that were used in simulation experiments are shown in Table 1. Crane serviced 20 ft (1C, TEU), 30 ft (1B), and 40 ft (1A) containers. For each type of container, the different weight categories have been assigned. It was assumed that the maximum permissible weights of individual container types do not exceed 23.2 t (20 ft), 25.4 t (30 ft), and 30.48 (40 ft). In the research conducted, it was assumed that the container's weight would impact the lifting and lowering speed of the hoist. By analysing actual measurements from the publication [22], we noticed that the lifting speed of a 40 tonne container is limited to 26 m/min. In contrast, a container weighing approximately 2 tonnes achieves a lifting speed of around 52 m/min. Therefore, it was assumed that the container's lifting speed (V_{HL}) can be written as a linear function, depending on the weight of the lifted container (W_C).

$$V_{HL} = -0.684 \cdot W_C + 53.368 \quad (1)$$

We also assumed that the lifting and lowering speeds would be the same ($V_{HL} = V_{HD}$).

The container storage field is an area intended for storing containers. It has around 360 containers in capacity. The containers are stored over 485 m in two levels and three rows. At the beginning of the simulation, containers are randomly placed in the container storage field. There are always available containers needed to load the complete train among the stored containers.

Table 1. Crane parameters (own work).

Parameter	Value
Gantry speed [m/s]	0.66
Gantry acceleration [m/s^2]	1
Trolley speed [m/s]	1.16
Trolley acceleration [m/s^2]	1
Hoist lift speed [m/s]	V_{HL}, depending on the container's weight
Hoist drop speed [m/s]	V_{HD}, depending on the container's weight
Hoist acceleration [m/s^2]	1
Lift height [m]	10.5
Container picking up time [s]	5
Container put off time [s]	10

The rail road (on which the train with wagons is placed for loading) is long enough to accommodate around 30 wagons. Each wagon has a total capacity of up to 3 TEU (60 ft). Therefore, the following container combinations can be created on a single wagon: 20 ft + 40 ft, 30 ft + 30 ft, 20 ft + 30 ft, and 20 ft + 20 ft + 20 ft.

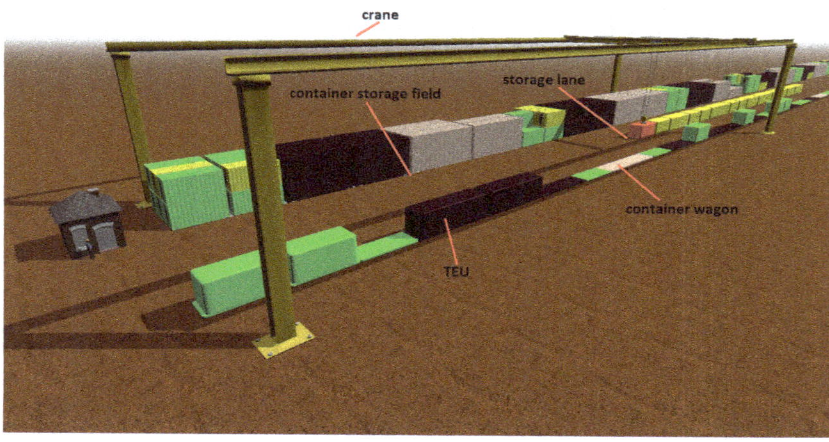

Figure 3. Container loading front. Screen captured from the simulation model in FlexSim.

4.2. Simulation Process

The operating logic of the simulation model can be briefly presented in the following several bullet points:

1. Generating an initial stock of containers in the container storage field.
2. Generating a set of wagons and assigning them types of containers.
3. Checking the availability of containers at the storage field that are necessary for loading. If the availability condition is not met, random containers in the storage field are replaced with those for which there is demand.
4. Carrying out the container preparation process (placing containers along the rail road).
5. Carrying out the train loading process (container transportation between the storage line and wagons on the rail-road).
6. Collecting characteristics and parameters and estimating the energy consumption.
7. Checking if all wagons are loaded. If so, the simulation ends.

The container preparation process for loading occurs when containers are transferred from the container storage field to the storage line (located along the rail road). The purpose of this process is to shorten the overall train loading process. During preparation, there may be a need to conduct the reshuffling process. Reshuffling occurs when the container essential for retrieval (from the storage field) is at the bottom of the stack. In such a situation, containers stacked on top of it need to be lifted and moved to the nearest stack (stack with a height not exceeding the maximum crane's lifting height). The process continues until there is a direct opportunity to lift the needed container. After the reshuffling, containers can be returned to the stack from which they were retrieved. This process will be referred to as additional reshuffling. Its execution is optional and can take place after completing the preparation process for each container.

4.3. Estimation of Energy Consumption

The primary objective of the research is to estimate the energy consumption of the train loading process for various scenarios of crane operation. Process characteristics are collected to calculate the energy consumption during simulation (see Table 2). The total energy consumed by the crane is the sum of the energy consumption by the drives of its structural elements as well as energy losses (E_{CL}), namely:

$$E_C = E_{CX} + E_{CY} + E_{CZ} + E_{CL} \quad (2)$$

Actual energy consumption measurements by the RTG crane, published in [22], have been considered. The publication's authors conducted measurements for eight days of the container handling system's operation. Varying workloads characterised the days, and the different weights had containers. The obtained results are summarised in Table 3. The publication presents the energy consumption values for the drives of the crane's structural elements and the amounts of energy that can be recovered during crane operation.

Table 2. Characteristics collected from the simulation.

Characteristics	Unit	Description
L_X	[m]	Total distance travelled by gantry
L_Y	[m]	Total distance travelled by trolley
L_Z (L_{ZDL}, L_{ZDE}, L_{ZUL}, L_{ZUE})	[m]	Total distance travelled by hoist (down loaded, down empty, up loaded, up empty)
T_X	[s]	Total time of gantry working
T_Y	[s]	Total time of trolley working
T_Z (T_{ZDL}, T_{ZDE}, T_{ZUL}, T_{ZUE})	[s]	Total time of hoist working (down loaded, down empty, up loaded, up empty)
T_{CC}	[s]	Total time of container pick-up
T_{CD}	[s]	Total time of container picking-off
T_P	[s]	Total time of the whole process
T_{PP}	[s]	Total time of the container preparation process
T_{PD}	[s]	Total time of the container reshuffling process
T_{PC}	[s]	Total time of the additional container reshuffling process
E_{CX}	[kWh]	Total energy consumed by gantry operations
E_{CY}	[kWh]	Total energy consumed by trolley operations
E_{CZ}	[kWh]	Total energy consumed by hoist operations
E_C	[kWh]	Total energy consumed during the whole process
E_{RX}	[kWh]	Total energy recovered from gantry operations
E_{RZ}	[kWh]	Total energy recovered from hoist operations
E_R	[kWh]	Total energy recovered during the whole process
E_{CPP} (E_{RPP})	[kWh]	Total energy consumed during the preparation process (Total energy recovered during the preparation process)
E_{CPD} (E_{RPD})	[kWh]	Total energy consumed during the reshuffling process (Total energy recovered during the reshuffling process)
E_{CPC} (E_{RPC})	[kWh]	Total energy consumed during the additional reshuffling process (Total energy recovered during the additional reshuffling process)

Table 3. Energy consumption results for the RTG crane [22].

Day	Working Time	Total		Gantry		Hoist		Trolley and Losses
		Energy Consumption	Energy Recovery Share	Energy Consumption	Energy Recovery Share	Energy Consumption	Energy Recovery Share	Energy Consumption
	[h]	[kWh]	[%]	[kWh]	[%]	[kWh]	[%]	[kWh]
1	10.3	711.2	57.1	220.47	5.06	440.94	89.00	49.784
2	11.1	748.36	50.1	231.99	4.09	463.99	86.50	52.38
3	11.9	785.53	57.7	243.51	4.47	487.03	83.80	54.99
4	12.7	822.7	51.9	255.04	4.30	510.07	84.10	57.59
5	6.7	439.9	50.8	136.37	2.68	272.74	84.30	30.79
6	7.3	499.2	54.0	154.75	2.80	309.50	72.60	34.94
7	1.4	112.8	49.8	34.97	3.54	69.94	67.10	7.89
8	2.1	160.8	30.2	49.85	2.87	99.69	84.20	11.25

Based on the data presented in Table 1, the energy consumption curve for the crane was plotted (Figure 4). Subsequently, the formula for the cumulative energy consumption function after a specified crane work time was determined (Equation (3)):

$$E_C = \frac{64.34 \cdot T_P}{3600} + 24.37 \quad (3)$$

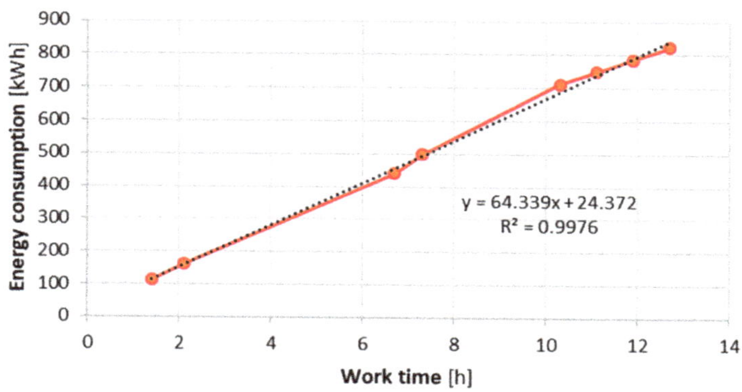

Figure 4. Average energy consumption of the crane over time.

As a result of the simulation studies, estimated distances (L_X, L_Y, L_Z) and working times for individual structural elements of the crane (T_X, T_Y, T_Z) can be determined. Consequently, to estimate the total energy consumed by the crane, the relationship (2) can be utilised. According to [22], the energy consumed by the gantry (S_{GEC}), hoist (S_{HEC}), and trolley (S_{TEC}) constitutes approximately 31%, 62%, and 7% of the total energy consumed by the crane, respectively. It is assumed, therefore, that individual energy consumption can be described using the relationships (4)–(6):

$$E_{CX} = S_{GEC} \cdot \left(\frac{64.34 \cdot T_X}{3600} + 24.37 \right) \quad (4)$$

$$E_{CY} = S_{TEC} \cdot \left(\frac{64.34 \cdot T_Y}{3600} + 24.37 \right) \quad (5)$$

$$E_{CZ} = S_{HEC} \cdot \left(64.34 \cdot \left(\frac{T_{ZUL} + T_{ZUE}}{3600} \right) + 24.37 \right) \quad (6)$$

The total time of hoist working (T_Z) used in Formula (6) is divided into the time of hoist lifting loaded (T_{ZUL}) and hoist lifting empty (T_{ZUE}). The time T_{ZUL} depends on the hoist lifting speed (V_{HL}), and the weight of the lifted container (W_C). Thus, the weight of the containers influences the hoist energy consumption.

During the operation of the crane, energy recovery is also possible. According to the research results presented in [22] and Table 2, the percentage share of energy recovery from the gantry's work (S_{GER}) and hoist (S_{HER}) reaches values of approximately 3.7% and 81.5%, respectively. In the case of the hoist's operation, energy is recovered only during lowering with a load (T_{ZDE}) and without a load (T_{ZDE}). Therefore, the following relationships were adopted in the simulation studies to estimate the total amount of recovered energy (7), the amount of energy recovered during the gantry's operation (8), and the hoist's operation (9):

$$E_R = E_{RX} + E_{RZ} \quad (7)$$

$$E_{RX} = S_{GER} \cdot S_{GEC} \cdot \left(\frac{64.34 \cdot T_X}{3600} + 24.37 \right) \quad (8)$$

$$E_{RZ} = S_{HER} \cdot S_{HEC} \cdot \left(64.34 \cdot \left(\frac{T_{ZDL} + T_{ZDE}}{3600}\right) + 24.37\right) \quad (9)$$

To summarise, the estimation of energy consumption in the simulation model is based on:

- basic stacker crane parameters (e.g., speed, lifting height) implemented in the model—see Table 1;
- distances between the places of collection and storage of containers, estimated by the simulation model at various stages of the simulation;
- operating times of individual engines of the crane's structural elements (gantry, hoist, trolley), estimated in the model based on the distances covered and the parameters of the stacker crane (e.g., travel and lifting speed);
- individual data on energy consumption by stacker crane motors measured and presented by authors of publication [22]—see Table 3;
- functional dependencies (2)–(9) developed based on research from a previous publication [22].

4.4. Simulation Scenarios

In order to evaluate different approaches to the train loading process, 20 different simulation scenarios were formulated. Each scenario was characterised by one of the five analysed crane operation strategies, an additional reshuffling process, and the selected container preparation method. The data for each simulation scenario is presented in Table 4.

Table 4. Simulation scenarios.

Scenario No.	Crane Operation Strategy	Additional Reshuffling	Container Preparing Method	Scenario No.	Crane Operation Strategy	Additional Reshuffle	Container Preparing Method
Scenario 1	1	No	1	Scenario 11	1	No	2
Scenario 2	2	No	1	Scenario 12	2	No	2
Scenario 3	3	No	1	Scenario 13	3	No	2
Scenario 4	4	No	1	Scenario 14	4	No	2
Scenario 5	5	No	1	Scenario 15	5	No	2
Scenario 6	1	Yes	1	Scenario 16	1	1	2
Scenario 7	2	Yes	1	Scenario 17	2	1	2
Scenario 8	3	Yes	1	Scenario 18	3	1	2
Scenario 9	4	Yes	1	Scenario 19	4	1	2
Scenario 10	5	Yes	1	Scenario 20	5	1	2

Reshuffling is a process in which containers placed on top of the container needed for retrieval must be moved to adjacent stacks to make the target container accessible. An example of this process is illustrated in Figure 5.

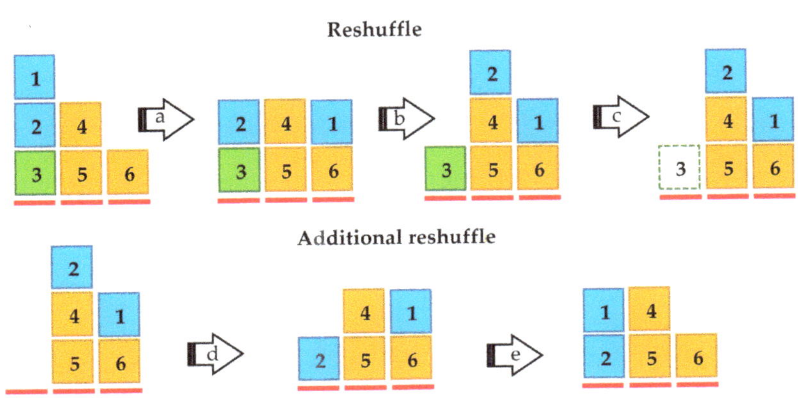

Figure 5. Reshuffle (a, b, c) and additional reshuffle processes (d, e).

In order to retrieve container number 3, it is necessary to shift containers 1 and 2. In the subsequent stages of the process, container 1 is moved onto container number 6, and then container 2 onto container 4 (or onto 1). This sequence enables access to container number 3, which can be easily retrieved. The simulation model performs this reshuffling process (if necessary) during the containers' loading preparation. After their completion, the containers are moved to the storage line, awaiting further loading stages. At this stage, the shifted containers (numbers 1 and 2) can either remain in their new locations for further processing or return to the stack from which they were initially taken (an additional reshuffle). The additional reshuffle process involves moving the containers previously relocated in the reshuffling process (container numbers 1 and 2 in Figure 5) back to the stack from which they were initially retrieved.

Both methods of preparing containers for loading, as analysed in the scenarios, are presented in Figure 6. They differ in the way containers are arranged in the storage line. In the first case (Figure 6a), containers taken from the storage field are placed sequentially along the rail road, starting from the beginning. In the second case (Figure 6b), containers taken from the storage field are transported to the storage line at the nearest available location relative to the pickup point.

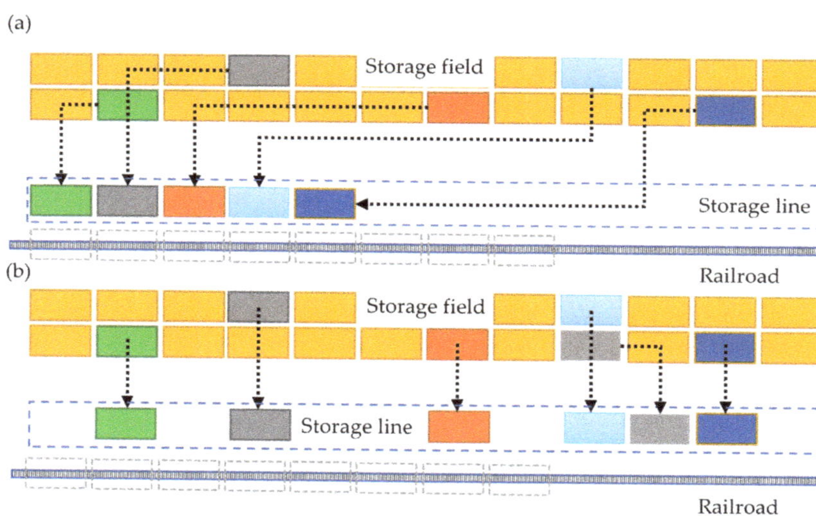

Figure 6. Container preparation methods—first method (**a**) and second one (**b**).

For the study, each of the analysed variants underwent ten simulation experiments. In each experiment, the distribution of containers stored in the storage field, as well as the distribution of pins and wagons in the serviced trains, was independently randomised. As a result, some quantities were estimated, such as: distances covered by individual structural elements of the crane (gantry, hoist, trolley), drive times (gantry, hoist, trolley), energy consumed by the crane's drives (gantry, hoist, trolley), energy recovered during the crane's drive operation (gantry, hoist), total train loading time, time for individual stages of the loading process (preparation, cleaning, loading), and energy consumed and recovered during these stages. The average values of these quantities were also calculated. The results are summarised in Table A1 in Appendix A.

4.5. Simulation Experiments and Results

In the area of conducted research, the most significant criteria for evaluation were considered to be the total energy consumption by the crane and the total loading time of the train. These criteria directly influence the cost of the train loading process and its efficiency. Therefore, in the first instance, the total amounts of consumed and recovered energy and the duration of the train loading process were compared for each scenario (Figure 7 and Table 5). The difference between the total operation time and the process time (Table 5) lies in the fact that in the former case, the time of operation for all drives of the container handling system is summed. In contrast, in the latter, the time between the start and end of the train loading process is considered (including the time for preparing containers for loading).

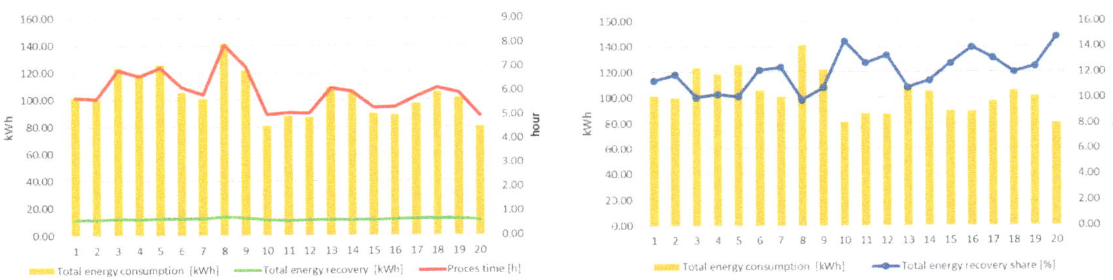

Figure 7. Total energy consumption and recovery, total process time for each scenario.

Table 5. Total average distance, total average time and energy consumption, and total average energy recovery for each scenario.

Scenario	Total Distance [m]	Total Operations Time [h]	Total Energy Consumption [kWh]	Total Energy Recovery [kWh]	Total Energy Recovery Share [%]	Process Time [h]
1	13,504.37	5.51	101.43	11.36	11.35	5.72
2	13,389.64	5.46	100.01	11.51	11.82	5.66
3	16,213.42	6.64	123.71	12.29	10.05	6.86
4	15,628.74	6.40	118.82	12.20	10.27	6.62
5	16,479.89	6.76	125.95	12.40	10.11	6.96
6	14,290.71	5.85	105.45	12.55	12.16	6.14
7	13,642.30	5.58	100.75	12.42	12.39	5.83
8	18,508.80	7.64	141.14	13.76	9.86	7.92
9	16,215.41	6.66	121.82	13.07	10.79	6.98
10	11,368.79	4.62	80.94	11.64	14.45	5.02
11	11,984.83	4.87	87.96	11.11	12.73	5.10
12	11,955.56	4.85	87.41	11.35	13.33	5.08
13	14,421.63	5.89	108.90	11.73	10.83	6.12
14	14,041.98	5.73	104.99	11.75	11.33	5.96
15	12,260.34	4.99	89.84	11.32	12.69	5.30
16	12,343.95	5.03	88.94	12.14	13.96	5.33
17	13,352.03	5.46	97.34	12.50	13.12	5.76
18	14,265.51	5.84	105.78	12.50	12.01	6.12
19	13,771.59	5.63	101.12	12.42	12.45	5.92
20	11,347.31	4.61	80.21	11.75	14.75	4.98

The total energy consumption in the analysed variants ranged from 80.21 kWh to 141.14 kWh. It means that the increase in energy consumption between the best scenario (Scenario 20) and the worst scenario (Scenario 8) was approximately 75.96%. A similar ranking of scenarios occurred when comparing the duration of the process (Table 6). In this case, the process time ranged from 4.98 h to 7.92 h, resulting in a 59.04% increase in the train loading time. For both criteria (energy consumption and time), the two best scenarios were Scenario 20 and Scenario 10. These scenarios used the fifth strategy for the crane's operation and additional reshuffling procedures.

Table 6. The scenario sequence complies with the energy consumption and process time criteria.

Criterion	Scenarios from the Best to the Worst																			
Energy consumption	20	10	12	11	16	15	17	2	7	19	1	14	6	18	13	4	9	3	5	8
Process time	20	10	12	11	15	16	2	1	17	7	19	14	18	13	6	4	3	5	9	8

Simulation results obtained from all scenarios are presented in Figure 8. For all of them, with the increase in total energy consumption:

- the percentage of gantry working time increases,
- the percentage of energy consumed by the gantry increases,
- the percentage of energy consumed by the hoist and trolley decreases,
- the percentage of energy recovered from total energy consumption decreases.

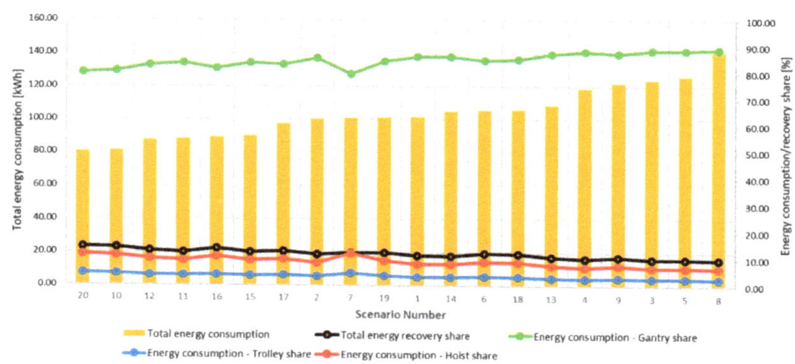

(**a**)

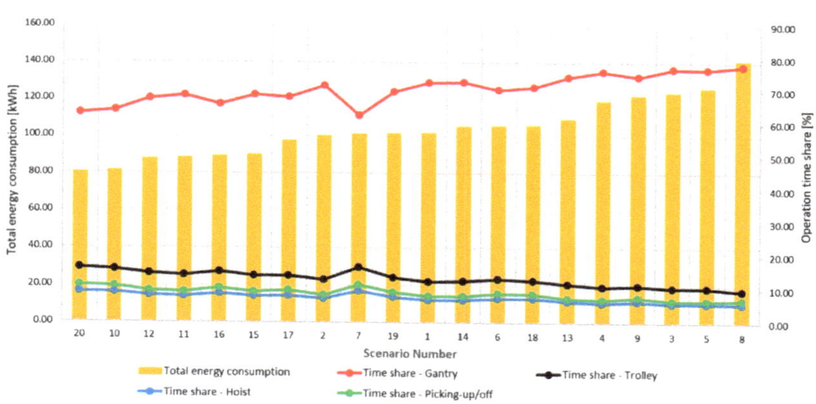

(**b**)

Figure 8. Energy (**a**) and time (**b**) consumption share—scenarios sequence according to increasing total energy consumption.

This indicates that energy consumed by the gantry has the most significant impact on overall energy consumption. The scenarios with more extended gantry operations (i.e., covering greater distances along the loading track) are more energy-intensive. Even though the hoist drives have the highest individual energy consumption, the share of energy consumed by the gantry is approximately 6.8 to 13 times greater than that of the hoist, accounting for 79% to 89% of the energy consumed by the crane (Figure 8a). The reduction in the share of energy consumed by the hoist (in relation to the total energy) also leads to a decrease in the amount of recovered energy (the majority of recovered energy comes from the hoist's operation). The changes in the share of energy consumption and time during different operations in different scenarios are presented in Table 7.

Table 7. Scenario sequence for the energy consumption and process time criteria.

Scenario	20	10	12	11	16	15	17	2	7	19	1	14	6	18	13	4	9	3	5	8
Energy consumption share [%]																				
Gantry	80.2	80.8	82.9	83.8	81.8	83.9	83.3	85.5	79.6	84.4	86.2	86.3	84.8	85.3	87.3	88.3	87.3	88.7	88.6	89.1
Hoist	11.9	11.5	10.3	9.7	11.0	9.7	10.1	8.7	12.3	9.4	8.2	8.2	9.1	8.9	7.6	7.0	7.6	6.8	6.9	6.6
Trolley	4.7	4.6	4.1	4.0	4.3	3.9	4.0	3.6	4.8	3.7	3.4	3.4	3.6	3.5	3.1	2.9	3.0	2.8	2.8	2.6
Picking up/off	3.2	3.1	2.7	2.6	2.9	2.6	2.7	2.3	3.3	2.5	2.2	2.2	2.5	2.4	2.0	1.8	2.1	1.8	1.8	1.8
Energy recovery share [%]																				
Total	14.7	14.4	13.3	12.7	14.0	12.7	13.1	11.8	12.4	12.4	11.4	11.3	12.2	12.0	10.8	10.3	10.8	10.1	10.1	9.9
Time consumption share [%]																				
Gantry	72.6	71.5	76.8	76.0	76.8	70.4	62.4	77.7	74.5	64.0	68.5	67.5	74.3	72.6	68.7	65.7	68.0	71.1	69.6	63.1
Hoist	6.9	7.2	5.9	6.1	5.9	7.5	9.6	5.7	6.5	9.1	7.9	8.3	6.5	6.9	7.9	8.7	8.1	7.4	7.7	9.3
Trolley	12.5	13.0	10.6	10.9	10.5	13.2	16.6	9.8	11.3	16.0	14.4	14.7	11.8	12.5	14.2	15.3	14.2	12.8	13.6	16.5
Picking up/off	8.1	8.3	6.7	7.0	6.8	9.0	11.4	6.8	7.8	10.9	9.3	9.5	7.5	8.0	9.2	10.3	9.7	8.7	9.1	11.2

For each type of crane operation strategy, the average values of various characteristics obtained from all simulation experiments were estimated (Table 8). Thus, it was possible to identify the best strategy. The results indicate that the best strategy in terms of energy consumption is Strategy No. 5. It had the shortest total distance covered by all structural elements of the crane (gantry, trolley, hoist) and the shortest train loading process time. Each subsequent (worse) strategy was characterised by an increase in the total distances covered, an increase in the loading process time, and an increase in the energy consumption by the gantry in relation to the hoist and trolley. It is also worth noting that the differences in the results obtained for the first three strategies (strategy No. 5, 1, and 2) are small. The increases in energy consumption between strategies No. 5 and 2 are only 2.28%, and the increase in loading time is 0.38%. Given that the average values are comparable, it can be assumed that they are equivalent in terms of their attractiveness and applicability. However, Strategies No. 4 and 3 proved to be significantly worse. In this case, the increase in energy consumption (compared to Strategy No. 5) was 27.22%, and the increase in the total time of the loading process was 21.56%.

Table 8. Average values of characteristics for each crane operation strategy.

Crane Operation Strategy	Total Distance [m]	Total Energy [kWh]		Process Time [h]	Energy Consumption Share [%]			
		Consumption	Recovery		Gantry	Trolley	Hoist	Picking Up/Off
5	12,864.08	94.23	11.78	5.56	83.36	4.00	9.98	2.67
1	13,030.96	95.94	11.79	5.57	84.15	3.83	9.49	2.53
2	13,084.88	96.38	11.94	5.58	84.44	3.73	9.36	2.48
4	14,914.43	111.69	12.36	6.37	86.56	3.24	8.06	2.14
3	15,852.34	119.88	12.57	6.76	87.59	2.98	7.46	1.97

The analysis of the results indicates that the container preparation method can significantly impact the train loading time and the energy consumed during the process. The application of the second container preparation method in various scenarios (with the same crane operation strategy and additional reshuffling method—see Table 9) allowed for a reduction in the loading process time by an average of approximately 12% and a decrease in energy consumption by an average of about 14% (Table 9).

Table 9. Average benefits from using the second container preparation method.

Crane Operation Strategy	Additional Reshuffling	Energy Consumption Reduction [%] Comparing Scenarios	Energy Consumption Reduction [%] Average	Time Consumption Reduction [%] Comparing Scenarios	Time Consumption Reduction [%] Average
1	No	13.3	14	10.9	12
	Yes	15.6		13.2	
2	No	12.6		10.3	
	Yes	3.4		1.3	
3	No	12.0		10.8	
	Yes	25.1		22.7	
4	No	11.6		9.9	
	Yes	17.0		15.2	
5	No	28.7		23.8	
	Yes	0.9		0.8	

Additional reshuffling is a process that, in most cases, prolongs the train loading time and increases energy consumption. This is because it involves additional crane movements, returning to the location of container retrieval from the storage field, and performing additional operations on the relocated containers. This is reflected in almost all the analysed scenarios. An exception is Strategy 5 (Scenarios 10 and 20), for which the total loading time and overall energy consumption in variants with additional reshuffling were reduced. This is related to the specificity of this strategy. After loading a container onto the train, the crane searches for the next container to retrieve closest to that location. In the case of additional reshuffling, the gantry's work cycle did not end at the place where the container was deposited but at the point of its retrieval. This disrupted the standard sequence of containers being retrieved for loading. In this case, the disruption led to positive results, shortening the distance covered by the gantry, the loading time, and the resulting energy consumption.

A comparison of the results for the best and worst scenarios showed potential benefits that a logistics hub could achieve. These benefits include, among others, a reduction in the intermodal train loading time by approximately 37% (2.94 h) and a decrease in electricity consumption by about 43% (60.49 kWh).

The reduction in loading time enables a 37% decrease in the crane's engagement time. During this period, the crane can be used for other tasks in the storage yard or to prepare containers to load another train. It leads to an increase in the logistics hub's efficiency and the number of trains and containers served daily. The hub's handling potential (with one track) could be increased from 2 to 3 trains per day.

The reduction in energy consumption leads to financial savings. Assuming the cost of 1 kWh is approximately €0.57, annual savings can reach up to €75,023.17 (energy savings at 131.62 MWh, serving three trains per day, operating 20 days per month).

Another crucial aspect is the carbon footprint generated and the amount of CO_2 emissions released into the environment. According to [60], the carbon intensity of electricity per kilowatt-hour in Poland is 633 g of CO_2 equivalents per kilowatt-hour. Therefore, reducing energy consumption through proper train loading organisations can decrease CO_2 emissions by approximately 27,771.74 kg of CO_2 annually.

Based on simulation results from all scenarios, the shares of average energy consumption (Figure 9a) and time consumption (Figure 9b) were determined. Gantry has the largest share in energy consumption (85%), followed by hoists (9%), trolleys (4%), and containers picking up and picking off (2%). The shares of individual operation duration and motor work were as follows: gantry (71%), trolley (13%), hoist (7%), and containers picking up and picking off (9%).

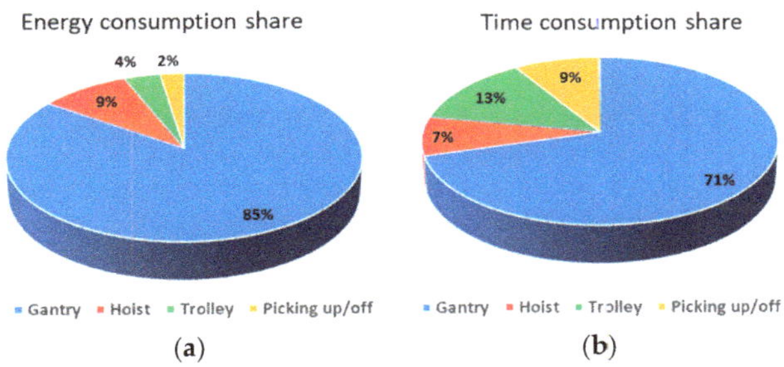

Figure 9. Shares of average energy (**a**) and time (**b**) consumption for all scenarios.

The values of energy consumption by individual stacker crane drives that we used for research came from the publication [22]. Therefore, we compared the results presented in Figure 9a with the shares of energy consumption by the RTG crane from that publication. In [22], hoist operations had the largest share of average energy consumption (61%), followed by gantry operations (31%), and finally, trolley operations, losses, and others (7%). Thus, these results are significantly different from those obtained from our research. It is related to the operating characteristics of the stacker crane at various types of intermodal terminals.

In marine terminals (such as those analysed in [22]), containers are taken from the ship and moved to the yard or train (or vice versa). There is a significant difference in height between the places of collecting and depositing the containers, even up to several dozen metres (40–60 m). Thus, hoist processes take a long time and consume up to 61% of energy. The gantry (31%) and trolley (7%) have a significantly smaller share of energy consumption.

In the case of land intermodal terminals, the differences in height between the places where containers are picked up and put away are significantly smaller. In such situations, the most common and time-consuming movements are those along the railroad (71%), followed by trolleys and hoists (Figure 9b). Regarding energy consumption, the gantry has the largest share, followed by the hoist and trolley (Figure 9a).

It can be noticed that the share of individual crane movements is related to the type of intermodal terminal. For this reason, the type of terminal will also determine crane energy consumption. The results of our research indicate that due to the enormous workload of the hoist, the unit hourly energy consumption by cranes in marine terminals is almost four times higher than in the case of land terminals. In marine terminals, overhead cranes' average hourly energy consumption is approximately 69.8 kWh (see Table 3 and [22]), while in the land terminal, it is approximately 17.23 kWh.

5. Conclusions

The aim of the article was to calculate and examine the crane energy consumption during the train loading in the intermodal terminal handling freight from the smart city.

Based on the literature research, it was found that there is a lack of publications on the use of intermodal terminals for smart city freight handling. There is also a lack of publications on the study of the energy efficiency of such facilities. Therefore, taking into account the smart city's requirement to minimise energy consumption of transport processes, our original contribution to the article was the analysis of the energy efficiency of the intermodal train loading process performed by a crane, including the location of containers in the storage yard, container rehandling in the storage yard, as well as the crane operation strategy. This made it possible to determine the energy consumption generated by hoist, trolley, and gantry moves. As a result of the conducted research, it was possible to answer the question of how the aforementioned factors affect crane energy consumption and consequently the energy efficiency of the entire process.

The research in the paper unequivocally demonstrated that selecting an intermodal train loading scenario impacts the crane's working time and energy consumption. Consequently, choosing an appropriate method for intermodal train loading can determine the efficiency of the entire logistics hub, its operational costs, and its environmental and energy efficiency. Improving the energy efficiency of a logistic hub in the form of an intermodal terminal can bring many benefits to a smart city. It primarily means reduced energy consumption, contributing to the mitigation of greenhouse gas emissions, and thereby supporting sustainable development goals.

Furthermore, enhancing the energy efficiency of the logistic hub will also positively impact the overall transportation infrastructure in the smart city, facilitating a smoother and more sustainable flow of goods. It, in turn, will contribute to reducing traffic congestion, lowering emissions, and ultimately minimising external transportation costs.

The simulation model presented in the paper is a crucial tool for researching cranes' energy consumption in intermodal terminals. However, analyses of its functionality have revealed certain limitations that require further development. Specifically, this involves the need to supplement the model with actual measurements of energy consumption by various crane engines (currently, mathematical formulas yielding approximate values are utilised). Additionally, it is essential to consider randomness and variability in the processes carried out in the model and the needs of its users. There are plans to develop and implement advanced optimisation algorithms (based on AI, neural networks, and machine learning) in the model, enabling productive and energy-efficient management of loading intermodal trains and handling the container storage yard. Working on these aspects is crucial for further refining the model and maximising the benefits of its application for smart cities' needs.

The findings from this study have the potential to guide smart city planners and logistics managers towards strategic enhancements in intermodal terminal operations, fostering sustainable urban development, and bolstering the economic stability of city environments.

Author Contributions: Conceptualisation, M.K. and R.J.; methodology, M.K., N.C.-G. and R.J.; validation, M.K. and N.C.-G.; formal analysis, M.K. and R.J.; investigation, M.K.; resources, R.J.; data curation, M.K.; writing—original draft preparation, M.K., N.C.-G. and R.J.; writing—review, R.J.; visualisation, M.K.; supervision, M.K., R.J. and N.C.-G.; project administration, R.J. and N.C.-G. All authors have read and agreed to the published version of the manuscript.

Funding: This research received no external funding.

Data Availability Statement: Data are contained within the article.

Acknowledgments: This research is a result of a project funded by the National Science Centre, Poland (project no. 2022/04/Y/ST8/00134). Project title: Energy optimal urban logistics as a service (E-Laas). Project implemented as part of the call ERA-NET Cofund Urban Accessibility and Connectivity (EN-UAC China Call) organised by JPI Urban Europe and the National Natural Science Foundation of China (NSFC) funded from the European Union's Horizon 2020 research and innovation programme under grant agreement No 875022.

Conflicts of Interest: The authors declare no conflicts of interest.

Appendix A

Table A1. Simulation results—mean values from the results of all scenario replications.

Scenario	Type of Operation/Motor Usage	Distance [m]	Time of Operation [h]	Time Share [%]	Energy Consumption [kWh]	Energy Share [%]	Energy Recovery [kWh]	Energy Recovery Share [%]	Process Time [h]
1	Gantry	9561.43	4.024	72.56	87.75	86.24	3.25	3.7	0.000
1	Trolley	2834.32	0.679	12.51	3.35	3.37	0	0	0.000
1	Hoist	1108.62	0.373	6.86	8.17	8.21	8.11	99.36	0.000
1	Picking up/Picking off/idle	0	0.438	8.07	2.16	2.17	0	0	0.000
1	Total	13,504.37	5.514	100	101.43	100	11.36	11.35	5.719
2	Gantry	9383.35	3.949	71.5	86.1	85.52	3.19	3.7	0.000
2	Trolley	2873.81	0.688	13.02	3.4	3.55	0	0	0.000
2	Hoist	1132.48	0.381	7.18	8.34	8.67	8.33	99.79	0.000
2	Picking up/Picking off/idle	0	0.440	8.3	2.17	2.26	0	0	0.000
2	Total	13,389.64	5.458	100	100.01	100	11.51	11.82	5.659
3	Gantry	12,184.76	5.128	76.83	109.97	88.69	4.07	3.7	0.000
3	Trolley	2890.12	0.692	10.59	3.36	2.77	0	0	0.000
3	Hoist	1138.54	0.383	5.85	8.24	6.78	8.22	99.83	0.000
3	Picking up/Picking off/idle	0	0.440	6.74	2.14	1.76	0	0	0.000
3	Total	16,213.42	6.644	100	123.71	100	12.29	10.05	6.864
4	Gantry	11,567.47	4.868	76.03	104.88	88.26	3.88	3.7	0.000
4	Trolley	2909.58	0.697	10.89	3.39	2.86	0	0	0.000
4	Hoist	1151.69	0.387	6.05	8.37	7.05	8.32	99.38	0.000
4	Picking up/Picking off/idle	0	0.450	7.02	2.19	1.84	0	0	0.000
4	Total	15,628.74	6.402	100	118.82	100	12.2	10.27	6.615
5	Gantry	12,455.95	5.242	76.8	112.17	88.58	4.15	3.7	0.000
5	Trolley	2878.91	0.689	10.54	3.34	2.76	0	0	0.000
5	Hoist	1145.03	0.385	5.9	8.3	6.88	8.25	99.36	0.000
5	Picking up/Picking off/idle	0	0.442	6.77	2.15	1.78	0	0	0.000
5	Total	16,479.89	6.759	100	125.95	100	12.4	10.11	6.962
6	Gantry	9874.69	4.156	70.35	89.97	84.84	3.33	3.7	0.000
6	Trolley	3148.51	0.754	13.23	3.7	3.63	0	0	0.000
6	Hoist	1267.51	0.426	7.45	9.27	9.08	9.22	99.45	0.000
6	Picking up/Picking off/idle	0	0.511	8.96	2.51	2.46	0	0	0.000
6	Total	14,290.71	5.847	100	105.45	100	12.55	12.16	6.140
7	Gantry	9315.05	3.920	70.12	85.4	84.67	3.16	3.7	0.000
7	Trolley	3062.56	0.733	13.21	3.61	3.61	0	0	0.000
7	Hoist	1264.69	0.425	7.66	9.28	9.27	9.26	99.83	0.000
7	Picking up/Picking off/idle	0	0.500	9.01	2.46	2.46	0	0	0.000
7	Total	13,642.3	5.580	100	100.75	100	12.42	12.39	5.831
8	Gantry	14,148.45	5.955	77.65	126.04	89.1	4.66	3.7	0.000
8	Trolley	3084.46	0.739	9.8	3.54	2.55	0	0	0.000
8	Hoist	1275.88	0.429	5.7	9.11	6.57	9.09	99.86	0.000
8	Picking up/Picking off/idle	0	0.515	6.84	2.47	1.78	0	0	0.000
8	Total	18,508.8	7.638	100	141.14	100	13.76	9.86	7.923
9	Gantry	11,809.2	4.970	74.46	106.51	87.33	3.94	3.7	0.000
9	Trolley	3135.87	0.751	11.32	3.64	3	0	0	0.000
9	Hoist	1270.35	0.427	6.45	9.18	7.6	9.13	99.43	0.000
9	Picking up/Picking off/idle	0	0.515	7.77	2.5	2.06	0	0	0.000
9	Total	16,215.41	6.663	100	121.82	100	13.07	10.79	6.982
10	Gantry	7044.24	2.965	63.97	65.47	80.77	2.42	3.7	0.000
10	Trolley	3079.02	0.737	16.02	3.68	4.58	0	0	0.000
10	Hoist	1245.53	0.419	9.08	9.28	11.53	9.22	99.43	0.000
10	Picking up/Picking off/idle	0	0.504	10.94	2.52	3.13	0	0	0.000
10	Total	11,368.79	4.624	100	80.94	100	11.64	14.45	5.022
11	Gantry	7959.5	3.350	68.47	73.84	83.76	2.73	3.7	0.000
11	Trolley	2892.61	0.693	14.37	3.45	3.98	0	0	0.000
11	Hoist	1132.72	0.381	7.88	8.43	9.69	8.38	99.37	0.000
11	Picking up/Picking off/idle	0	0.448	9.28	2.24	2.57	0	0	0.000
11	Total	11,984.83	4.872	100	87.96	100	11.11	12.73	5.096
12	Gantry	7885.07	3.319	67.47	73.05	82.94	2.7	3.7	0.000
12	Trolley	2905.68	0.696	14.74	3.47	4.12	0	0	0.000
12	Hoist	1164.82	0.391	8.29	8.66	10.28	8.64	99.83	0.000
12	Picking up/Picking off/idle	0	0.449	9.5	2.24	2.66	0	0	0.000
12	Total	11,955.56	4.854	100	87.41	100	11.35	13.33	5.077
13	Gantry	10,427.73	4.389	74.34	95.17	87.29	3.52	3.7	0.000
13	Trolley	2869.03	0.687	11.75	3.37	3.12	0	0	0.000
13	Hoist	1124.87	0.378	6.46	8.22	7.61	8.21	99.83	0.000
13	Picking up/Picking off/idle	0	0.436	7.45	2.14	1.98	0	0	0.000
13	Total	14,421.63	5.890	100	108.9	100	11.73	10.83	6.125

Table A1. *Cont.*

Scenario	Type of Operation/Motor Usage	Distance [m]	Time of Operation [h]	Time Share [%]	Energy Consumption [kWh]	Energy Share [%]	Energy Recovery [kWh]	Energy Recovery Share [%]	Process Time [h]
14	Gantry	9937.25	4.182	72.62	90.86	86.28	3.36	3.7	0.000
14	Trolley	2953.69	0.707	12.53	3.48	3.38	0	0	0.000
14	Hoist	1151.04	0.387	6.85	8.44	8.19	8.39	99.35	0.000
14	Picking up/Picking off/idle	0	0.452	8	2.22	2.15	0	0	0.000
14	Total	14,041.98	5.728	100	104.99	100	11.75	11.33	5.963
15	Gantry	8162.84	3.436	68.71	75.49	83.88	2.79	3.7	0.000
15	Trolley	2939.12	0.704	14.2	3.5	3.92	0	0	0.000
15	Hoist	1158.39	0.389	7.86	8.59	9.65	8.53	99.39	0.000
15	Picking up/Picking off/idle	0	0.457	9.24	2.27	2.55	0	0	0.000
15	Total	12,260.34	4.986	100	89.84	100	11.32	12.69	5.303
16	Gantry	7929.61	3.337	65.67	73.21	81.75	2.71	3.7	0.000
16	Trolley	3136.25	0.751	15.26	3.73	4.33	0	0	0.000
16	Hoist	1278.09	0.430	8.73	9.48	10.99	9.43	99.43	0.000
16	Picking up/Picking off/idle	0	0.509	10.34	2.53	2.93	0	0	0.000
16	Total	12,343.95	5.027	100	88.94	100	12.14	13.96	5.328
17	Gantry	8893.32	3.743	67.95	81.53	83.28	3.02	3.7	0.000
17	Trolley	3166.66	0.758	14.18	3.74	3.95	0	0	0.000
17	Hoist	1292.05	0.435	8.13	9.5	10.05	9.48	99.85	0.000
17	Picking up/Picking off/idle	0	0.521	9.74	2.57	2.72	0	0	0.000
17	Total	13,352.03	5.457	100	97.34	100	12.5	13.12	5.757
18	Gantry	9930.75	4.180	71.13	90.56	85.28	3.35	3.7	0.000
18	Trolley	3079.6	0.737	12.84	3.62	3.49	0	0	0.000
18	Hoist	1255.15	0.422	7.36	9.17	8.87	9.15	99.81	0.000
18	Picking up/Picking off/idle	0	0.498	8.67	2.44	2.36	0	0	0.000
18	Total	14,265.51	5.837	100	105.78	100	12.5	12.01	6.123
19	Gantry	9355.43	3.937	69.59	85.61	84.37	3.17	3.7	0.000
19	Trolley	3148.74	0.754	13.58	3.71	3.73	0	0	0.000
19	Hoist	1267.41	0.426	7.69	9.31	9.39	9.25	99.38	0.000
19	Picking up/Picking off/idle	0	0.507	9.15	2.5	2.52	0	0	0.000
19	Total	13,771.59	5.625	100	101.12	100	12.42	12.45	5.919
20	Gantry	6931.09	2.917	63.07	64.46	80.19	2.39	3.7	0.000
20	Trolley	3153.06	0.755	16.47	3.77	4.74	0	0	0.000
20	Hoist	1263.16	0.425	9.27	9.42	11.85	9.37	99.4	0.000
20	Picking up/Picking off/idle	0	0.513	11.19	2.56	3.22	0	0	0.000
20	Total	11,347.31	4.610	100	80.21	100	11.75	14.75	4.982

References

1. Gracias, J.S.; Parnell, G.S.; Specking, E.; Pohl, E.A.; Buchanan, R. Smart Cities—A Structured Literature Review. *Smart Cities* **2023**, *6*, 1719–1743. [CrossRef]
2. Pečar, M.; Papa, G. Transportation problems and their potential solutions in smart cities. In Proceedings of the 2017 International Conference on Smart Systems and Technologies (SST), Osijek, Croatia, 18–20 October 2017; IEEE: Piscataway, NJ, USA, 2017; pp. 195–199.
3. Potdar, V.; Batool, S.; Krishna, A. Risks and challenges of adopting electric vehicles in smart cities. In *Smart Cities: Development and Governance Frameworks*; Springer: Cham, Switzerland, 2018; pp. 207–240.
4. Lagorio, A.; Pinto, R.; Golini, R. Research in urban logistics: A systematic literature review. *Int. J. Phys. Distrib. Logist. Manag.* **2016**, *46*, 908–931. [CrossRef]
5. Allen, J.; Browne, M.; Woodburn, A.; Leonardi, J. The role of urban consolidation centres in sustainable freight transport. *Transp. Rev.* **2012**, *32*, 473–490. [CrossRef]
6. Paddeu, D.; Fancello, G.; Fadda, P. An experimental customer satisfaction index to evaluate the performance of city logistics services. *Transport* **2017**, *32*, 262–271. [CrossRef]
7. Nordtømme, M.E.; Bjerkan, K.Y.; Sund, A.B. Barriers to urban freight policy implementation: The case of urban consolidation center in Oslo. *Transp. Policy* **2015**, *44*, 179–186. [CrossRef]
8. Allen, J.; Browne, M.; Woodburn, A.; Leonardi, J. A review of urban consolidation centres in the supply chain based on a case study approach. *Supply Chain Forum Int. J.* **2014**, *15*, 100–112. [CrossRef]
9. Gronalt, M.; Schultze, R.C.; Posset, M. Intermodal transport—Basics, structure, and planning approaches. In *Sustainable Transportation and Smart Logistics*; Elsevier: Amsterdam, The Netherlands, 2019; pp. 123–149.
10. Statista Container Shipping. Available online: https://www.statista.com/topics/1367/container-shipping/ (accessed on 23 December 2023).
11. Škultéty, F.; Beňová, D.; Gnap, J. City logistics as an imperative smart city mechanism: Scrutiny of clustered EU27 capitals. *Sustainability* **2021**, *13*, 3641. [CrossRef]

12. European Commission (EC). *Roadmap to a Single European Transport Area—Towards a Competitive and Resource Efficient Transport System*; White Paper; European Commission: Brussels, Belgium, 28 March 2011. Available online: https://eur-lex.europa.eu/legal-content/EN/TXT/PDF/?uri=CELEX:52011DC0144&from=EN (accessed on 23 December 2023).
13. Buchari, E.; Putranto, D.D.A.; Rahman, A. Analysis of model loading and unloading time of ships at Boom Baru Port, Palembang, Indonesia. *MATEC Web Conf.* **2017**, *138*, 07012. [CrossRef]
14. Koo, P.H.; Lee, W.S.; Jang, A.D.W. Fleet sizing and vehicle routing for container transportation in a static environment. *OR Spectr.* **2004**, *26*, 193–209. [CrossRef]
15. Martinov, S. Increasing the accuracy of evaluation in selecting a location for establishing an intermodal terminal. *MATEC Web Conf.* **2018**, *234*, 06004. [CrossRef]
16. Lee, E. Spatial analysis for an intermodal terminal to support agricultural logistics: A case study in the upper great plains. *Manag. Res. Rev.* **2015**, *38*, 299–319. [CrossRef]
17. Raicu, R.; Raicu, Ş.; Popa, M.; Costescu, D. On the evaluation of urban logistics intermodal terminal projects. *Procedia-Soc. Behav. Sci.* **2012**, *39*, 726–738. [CrossRef]
18. Görçün, Ö.F. Analysis of the Effects of Intermodal Terminals for the Solutions of Urban Logistics Problems in Istan-bul City. In *Dynamics in Logistics: Proceedings of the 4th International Conference LDIC*; Springer International Publishing: Berlin/Heidelberg, Germany, 2016; pp. 685–691.
19. Grubisic, N.; Krljan, T.; Maglić, L.; Vilke, S. The microsimulation model for assessing the impact of inbound traffic flows for container terminals located near city centers. *Sustainability* **2020**, *12*, 9478. [CrossRef]
20. Brnjac, N.; Badanjak, D.; Babi, D. Reducing congestion in urban transport by using intermodal transport solutions. *WIT Trans. Built Environ.* **2007**, *96*, 45–52.
21. Wang, L.; Zhu, X. Container Loading Optimization in Rail–Truck Intermodal Terminals Considering Energy Consumption. *Sustainability* **2019**, *11*, 2383. [CrossRef]
22. Papaioannou, V.; Pietrosanti, S.; Holderbaum, W.; Becerra, V.M.; Mayer, R. Analysis of energy usage for RTG cranes. *Energy* **2017**, *125*, 337–344. [CrossRef]
23. Jacyna, M.; Jachimowski, R.; Szczepański, E.; Izdebski, M. Road vehicle sequencing problem in a rail-road intermodal terminal–simulation research. *Bull. Pol. Acad. Sci. Tech. Sci.* **2020**, *68*, 135–1148.
24. Boysen, N.; Fliedner, M.; Jaehn, F.; Pesch, E. A survey on container processing in railway yards. *Transp. Sci.* **2013**, *47*, 312–329. [CrossRef]
25. Nehring, K.; Jachimowski, R. Modelling of container train handling in the land intermodal terminal. In *Carpathian Logistics Congress 2019*; Akademia Górniczo-Hutnicza im. Stanisława Staszica w Krakowie: Kraków, Poland, 2020.
26. Steenken, D.; Voß, S.; Stahlbock, R. Container Terminal Operation and Operations Research—A Classification and Literature Review. *OR Spectr.* **2004**, *26*, 3–49.
27. Stahlbock, R.; Voß, S. Operations research at container terminals: A literature update. *OR Spectr* **2008**, *30*, 1–52. [CrossRef]
28. Heggen, H.; Braekers, K.; Caris, A. Optimising train load planning: Review and decision support for train planners. In *Computational Logistics*; Springer: Berlin/Heidelberg, Germany, 2016; pp. 193–208.
29. Corry, P.; Kozan, E. Optimised loading patterns for intermodal trains. *OR Spectr.* **2008**, *30*, 721–750. [CrossRef]
30. Bruns, F.; Knust, S. Optimised load planning of trains in intermodal transportation. *OR Spectr.* **2012**, *34*, 511–533. [CrossRef]
31. Bruns, F.; Goerigk, M.; Knust, S.; Schöbel, A. Robust load planning of trains in intermodal transportation. *OR Spectr.* **2014**, *36*, 631–668. [CrossRef]
32. Heggen, H.; Braekers, K.; Caris, A. A multi-objective approach for intermodal train load planning. *OR Spectr.* **2018**, *40*, 341–366. [CrossRef]
33. Li, C.; Burton, D.; Kost, M.; Sherida, J.; Thompson, M.C. Flow Topology of a Container Train Wagon Subjected to Varying Local Loading Configurations. *J. Wind Eng. Ind. Aerodyn.* **2017**, *169*, 12–29. [CrossRef]
34. Ambrosino, D.; Siri, S. Models for train load planning problems in a container terminal. In *Computer-based Modelling and Optimisation in Transportation*; Springer: Berlin/Heidelberg, Germany, 2014; Volume 262, pp. 15–25.
35. Ambrosino, D.; Caballini, C.; Siri, S. A mathematical model to evaluate different train loading and stacking policies in a container terminal. *Marit. Econ. Logist.* **2013**, *15*, 292–308. [CrossRef]
36. Li, X.; Otto, A.; Pesch, E. Solving the single crane scheduling problem at rail transshipment yards. *Discret. Appl. Math.* **2019**, *264*, 134–147. [CrossRef]
37. Jacyna, M.; Pyza, D.; Jachimowski, R. *Intermodal Terminal Designing—Case Study*; Warsaw University of Technology: Warsaw, Poland, 2019.
38. Carboni, A.; Deflorio, F. Simulation of rail-road terminal operations and traffic control strategies in critical scenarios. *Transp. Res. Procedia* **2020**, *45*, 325–332. [CrossRef]
39. Konecranes Rubber Tired Gantry Cranes. Available online: www.konecranes.com/port-equipment-services/container-handling-equipment/rubber-tired-gantry-cranes (accessed on 21 December 2023).
40. Kalmar. Kalmar Hands over Its First Fully Electric Straddle Carrier. Available online: https://www.kalmarglobal.com/news--insights/press_releases/2023/kalmar-hands-over-its-first-fully/ (accessed on 21 December 2023).
41. Parise, G.; Parise, L.; Malerba, A.; Pepe, F.M.; Honorati, A.; Chavdarian, P.B. Comprehensive peak-shaving solutions for port cranes. *IEEE Trans. Ind. Appl.* **2017**, *53*, 1799–1806. [CrossRef]

42. Mahlia, T.M.I.; Saktisahdan, T.J.; Jannifar, A.; Hasan, M.H.; Matseelar, H.S.C. A review of available methods and development on energy storage; technology update. *Renew. Sustain. Energy Rev.* **2014**, *33*, 532–545. [CrossRef]
43. Flynn, M.M.; McMullen, P.; Solis, O. High-speed flywheel and motor drive operation for energy recovery in a mobile gantry crane In Proceedings of the APEC 07-Twenty-Second Annual IEEE Applied Power Electronics Conference and Exposition, Anaheim CA, USA, 25 February–1 March 2007; p. 11511157.
44. Kim, S.; Sul, S. Control of rubber tyred gantry crane with energy storage based on supercapacitor. *IEEE Trans. Power Electron* **2006**, *21*, 1420–1427. [CrossRef]
45. Baalbergen, F.; Bauer, P.; Ferreira, J.A. Energy storage and power management for typical 4Q-load. *IEEE Trans. Ind. Electron.* **2009** *56*, 1485–1498. [CrossRef]
46. Pietrosanti, S.; Alasali, F.; Holderbaum, W. Power management system for RTG crane using fuzzy logic controller. *Sustain. Energy Technol. Assess.* **2020**, *37*, 100639. [CrossRef]
47. Zhang, C.; Guan, H.; Yuan, Y.; Chen, W. Machine learning-driven algorithms for the container relocation problem. *Transp. Res Part B Methodol.* **2020**, *139*, 102–131. [CrossRef]
48. Li, Y.; Chu, F.; Zheng, F.; Liu, M. A bi-objective optimisation for integrated berth allocation and quay crane assignment with preventive maintenance activities. *IEEE Trans. Intell. Transp. Syst.* **2022**, *23*, 2938–2955. [CrossRef]
49. Yang, Y.; Zhu, X.; Ali, H. Multiple Equipment Integrated Scheduling and Storage Space Allocation in Rail-Water Intermodal Container Terminals Considering Energy Efficiency. *Transp. Res. Rec.* **2019**, *2673*, 199–209. [CrossRef]
50. Zhong, M.; Yang, Y.; Zhou, Y.; Postolache, O. Adaptive auto-tuning mathematical approaches for integrated optimisation of automated container terminal. *Math. Probl. Eng.* **2019**, *2019*, 7641670. [CrossRef]
51. Chen, X.; He, S.; Zhang, Y.; Tong, L.; Zhou, X. Yard crane and AGV scheduling in automated container terminal: A multi-robot task allocation framework. *Transp. Res. Part C Emerg. Technol.* **2020**, *114*, 241–271. [CrossRef]
52. Jonker, T.; Duinkerken, M.B.; Yorke-Smith, S.; de Waal, A.; Negenborn, R.R. Coordinated optimisation of equipment operations in a container terminal. *Flex. Serv. Manuf. J.* **2021**, *33*, 281–311. [CrossRef]
53. Hu, Y.; Dong, L.; Xu, L. Multi-AGV dispatching and routing problem based on a three-stage decomposition method. *Math. Biosci. Eng.* **2020**, *17*, 5150–5172. [CrossRef]
54. Budiyanto, M.A.; Huzaifi, M.H.; Sirait, S.J.; Prayoga, P.H.N. Evaluation of CO_2 emissions and energy use with different container terminal layouts. *Sci. Rep.* **2021**, *11*, 5476. [CrossRef]
55. Okşaş, O. Carbon emission strategies for container handling equipment using the activity-based method: A case study of Ambarlı container port in Turkiye. *Mar. Policy* **2023**, *149*, 105480. [CrossRef]
56. Linder, A.J. CO_2 restrictions and cargo throughput limitations at California ports: A closer look at AB 32 and port-to-port shipping. *Public Work. Manag. Policy* **2010**, *14*, 374–391. [CrossRef]
57. Garg, A.; Shukla, P.R.; Kankal, B.; Mahapatra, D. CO_2 emission in India: Trends and management at sectoral, sub-regional and plant levels. *Carbon Manag.* **2017**, *8*, 111–123. [CrossRef]
58. Geerlings, H.; Heij, R.; van Duin, R. Opportunities for peak shaving the energy demand of ship-to-shore quay cranes at container terminals. *J. Shipp. Trade* **2018**, *3*, 3. [CrossRef]
59. PKP Cargo. Rail Wagons Catalogue. Available online: www.pkpcargo.com/wp-content/uploads/2023/10/pkpcargo_katalogwagonow_3008_19.pdf/ (accessed on 28 December 2023).
60. Ember—Yearly Electricity Data (2023); Ember—European Electricity Review (2022); Energy Institute—Statistical Review of World Energy (2023). Available online: https://ourworldindata.org/grapher/carbon-intensity-electricity (accessed on 30 December 2023).

energies MDPI

Article

Design of Three Electric Vehicle Charging Tariff Systems to Improve Photovoltaic Self-Consumption

Garazi Etxegarai [1,2,*], Haritza Camblong [1,3], Aitzol Ezeiza [1] and Tek Tjing Lie [3]

1 Department of Systems Engineering & Control, Faculty of Engineering of Gipuzkoa, University of the Basque Country (UPV/EHU), Europa Plaza 1, E-20018 Donostia, Spain; aritza.camblong@ehu.eus (H.C.); aitzol.ezeiza@ehu.eus (A.E.)
2 ESTIA Institute of Technology, University of Bordeaux, 64210 Bidart, France
3 Department of Electrical and Electronic Engineering, Auckland University of Technology, Auckland 1010, New Zealand; tek.lie@aut.ac.nz
* Correspondence: garazi.etxegarai@ehu.eus

Abstract: Electric vehicles (EVs) are emerging as one of the pillars for achieving climate neutrality. They represent both a threat and an opportunity for the operation of the network. Used as flexible loads, they can favor the self-consumption of photovoltaic (PV) energy. This paper presents three EV charging tariff systems (TSs) based on the self-consumption of excess PV energy. The TS objectives are to increase the self-consumption rate (SCR) and thus indirectly decrease the charging cost of the EV users. Two of the proposed TSs correspond to an indirect control of EV charging. The third TS is a hybrid system where the charging power is controlled. The TS is designed using a series of rules that consider the momentary PV surplus and the charging power of each EV. The influence of the TS is simulated by considering real data from a PV collective self-consumption project in the Basque Country (Spain). The TS simulations performed with 6 months of data show a 13.1% increase in the SCR when applying the third TS, reaching an average of 93.09% for the SCR. In addition, the cost of EV charging is reduced by 25%.

Keywords: electric vehicles; charging system control; pricing schemes; PV energy; collective self-consumption; self-consumption rate

1. Introduction

The transport sector accounts for a large contribution to greenhouse gas emissions. In this context, electric vehicles (EVs) are emerging as one of the solutions that will help the European Union (EU) achieve its climate neutrality targets [1]. Thus, the EU has taken the measure of banning the sale of new petrol and diesel cars from 2035 [2,3]. In the first quarter of 2023, over 2.3 million EVs were sold worldwide, 25% more than in the same period the precedent year [4]. To cope with this growth, EV chargers will have to be installed every 60 km by 2026 [5].

According to [6], uncontrolled deployment of EVs would increase peak electricity demand by 35% to 51%. Without proper control, the high penetration of EVs will widen the gap between peak and off-peak loads on the grid [7]. This fact could overload distribution lines and transformers, leading to higher grid losses and reduced equipment lifetime [8].

However, coordinating the charging of EVs could also offer significant flexibility [9]. One solution involves using renewable energy sources (RES) for EV charging. By scheduling charging during times when RES is available, more RES is integrated into the grid and the gap between peak and off-peak power is reduced [9]. This way, local solar energy self-consumption holds the potential to emerge as a multifaceted solution.

When using EVs as flexible loads to help reduce grid operation issues, it is essential to manage the charging pattern in a controlled manner. EV control systems that intend to ensure an acceptable charging profile can be classified into two main types: direct and indirect control [10]. Direct control involves acting on the load patterns directly by controlling the charging power. One of the drawbacks of direct control systems is the high communication and computational infrastructure required. On the other hand, indirect control is based on encouraging EV users to adapt their charging patterns. The most common method of influence is through charging tariffs. There are several charging tariff systems, such as real-time pricing (RTP), critical peak pricing (CPP), and time-of-use (ToU) charging. In the RTP solution, prices for each time slot are announced shortly before the interval starts. Although this scheme is very efficient due to its dynamic nature, it requires a large information and communication technology (ICT) infrastructure. Furthermore, it requires substantial user participation [11]. In the case of CPP, extremely high prices are charged for occasions where peak demand is very high [12]. This type of pricing can be associated with others, whereas in ToU strategies, different fixed windows in the day are related to different fixed rates [13]. ToU charging system is the most widely used because of the simplicity of implementation [10,14].

Several papers have studied how EVs can benefit the electricity grid [15]. In [16], pricing mechanisms to incentivize EV users to shift their charging to fill the off-peak zones are proposed. The authors highlight that by using ToU-based charging, EV users would choose to charge the vehicle at the beginning of the cheapest zone, causing a new consumption peak. For this reason, the authors propose two charging mechanisms: one non-cooperative and the other cooperative, which consider the charging schedules of the vehicles that have already arrived. In the non-cooperative scenario, each EV schedules its own charging without cooperating with the other EVs, while in the cooperative scenario, all EVs are controlled by an aggregator. In [17], a day-ahead dispatch strategy is proposed for EVs considering the carbon quota. In addition, this strategy can provide peak shaving and valley-filling services.

Other works have studied the synergy between photovoltaic (PV) energy and EV charging. In [18,19], an indirect control approach presents a dynamic pricing system based on the Stackelberg game for an EV charging station associated with a PV system. The Stackelberg equilibrium seeks a win–win situation, reducing the cost for users and increasing the profit of the charging station. In [19], the charging station sets an appropriate selling price to maximise its profit. Then, the Stackelberg equilibrium is solved considering the user's criteria, and the prices are determined by the charging station.

There are also studies examining the role of EVs in increasing the PV self-consumption rate (SCR). Studies in [20,21] present improvements in PV self-consumption and self-sufficiency using batteries and EVs without applying any control. For instance, in [20], households with PV generation exhibited a SCR of 26% without storage, 59% with an EV, and 31% with batteries sized for that case study. Conversely, works [22,23] implement a direct control with the objective of maximising the SCR. In [22], direct control is used to define the charging pattern of EVs considering them as a flexible charging source, and in the case of vehicle to grid (V2G) as a storage device. In this work, three charging algorithms are proposed. The first algorithm uses real-time information, the second incorporates V2G technology, and the third is an optimisation algorithm using predictions for both demand and production, taking into account V2G technology. The results present a SCR of 49% for the uncontrolled case, 62% with the first algorithm, 79% with the second, and 87% with the third. The study conducted in [23] presents a combination of smart metering and smart charging that helps local energy communities increase self-consumption. A case with four consumers and a prosumer is studied. In the scenario without EVs, one-fifth of the PV energy is consumed, while with smart metering and EVs, the SCR is increased by 45%. In [24], a distributed and centralised smart charging scheme for EVs in residential buildings with PV systems is presented. The aim of smart charging is to minimise net load variability, thereby increasing self-consumption and reducing consumption peaks. In the centralised charging scheme, a central unit determines the charging time and power for a fleet of EVs, whereas, in the distributed charging approach, charging decisions are made at the user level. The proposed smart charging schemes consider EV energy demand, arrival and departure times, and predictions of building consumption and PV production.

Considering the state-of-the-art study conducted, there are no studies that assess the improvement of PV SCR through the use of indirect control systems in EV charging.

The work presented in this paper proposes three charging tariff schemes based on temporary PV surpluses in a real PV collective self-consumption (CSC) project in Aduna (Basque Autonomous Community, Spain). All three tariff systems (TS) share the same objective: to influence user behaviour in shifting the timing of EV charging to increase the consumption of local PV energy, and thus indirectly decrease the charging cost of the EV users.

The following hypotheses were considered:

I. EV users prefer to charge their vehicles when prices are lower. Therefore, if reduced prices are offered during PV surplus hours, users will adjust their charging times.
II. The resolution in the control of EV charging power considered in this article is ideal.

The two main contributions of the research study are as follows:

- The design of indirect control EV charging is based on temporary PV surplus, with the main objective of increasing the SCR of a real PV CSC project.
- A detailed description of the design of three TSs for EV charging, which can be easily replicated and adjusted to any case.

It is also worth highlighting that these TSs would promote the development of EVs, which is the aim of the Aduna town council, the owner of the PV panels and EV chargers.

The rest of this document is organised as follows. Section 2 presents the case study and the data used. In Section 3, the proposed three TSs are explained. Numerical results and discussion are provided in Section 4. Finally, the last section concludes this article.

2. Case Study

The three TSs proposed in this work were simulated with real historical data collected from the PV CSC project. This project takes place in the municipality of Aduna, located in the Basque Country, Spain. There are eight consumption points associated with the CSC project. One of them is a public EV charging point. For administrative reasons, the PV panels with a capacity of 62.4 kWp were not yet installed during this study.

The mentioned data can be divided into three groups: the data related to the eight consumption points, the data on the public EV charging station obtained from the Charge and Parking application, and the data related to PV power generation.

The consumption of the seven points (all except the EV charger) is considered to be the basic consumption for this case study. This consumption is used to obtain the PV surplus by calculating the difference between PV production and basic consumption. All proposed TSs are based on the PV surpluses. In this document, these surpluses are directly calculated by subtracting the basic consumption from the PV generation. However, when implementing the TSs in real time, the energy management system (EMS) will use the day-ahead forecasts of both PV production and consumption (as carried out in [25,26]).

The consumption historical data were recorded hourly and were available for the last 3 years. However, since the public charging point was not installed until November 2021, the simulations considered data obtained between November 2021 and April 2022, the same period as for the EV charging data.

On the other hand, the sampling time was reduced from 1 h to 10 min (the sample time corresponding to PV generation), with the objective of producing more accurate simulations. In this way, the EV charging time was more closely adjusted to the real charging time. To perform the change in the sampling time, the value of the power consumed over one hour was used in the six intervals that constitute one hour.

The public EV charging station is a 22 kW three-phase charger with two connectors. Regarding the EV charging data, the following information was obtained using the Charge and Parking application: the time at which an EV was connected and disconnected, the

EV brand and model, and the total energy consumed for each charge. By knowing the EV brand and model, it was possible to check and record the maximum charging power for each EV. It should be noted that not all EV models charge at the same power. Among the 23 EV models registered in the Charge and Parking application, the most common maximum charging power values were 3.7 kW, 7.2 kW, and 11 kW. These maximum power values were used in the design of TS1 and 2.

Regarding PV data, they were obtained using local historical solar irradiance data for the simulation period, from the Basque Meteorological Agency, Euskalmet [27], with a sampling interval of 10 min. PV production was estimated by multiplying the irradiation by the peak power of the PV panels and applying a correction factor to account for efficiency.

3. Proposed Pricing Methods

The following section describes the three proposed TSs to influence the EV user's charging pattern.

Although the three TS algorithms share the same objective, their rules are different. TS1 and 2 are indirect controls. TS1 is the least complex: when the PV surplus exceeds 7 kW, a lower price is offered to EV users to encourage charging at that time. On the other hand, TS2 is more personalised and is offered when the PV surplus exceeds different levels corresponding to the different charging power of EVs. Finally, TS3 is based on a hybrid control. In this TS, a cheaper charging price is always offered when there is a PV surplus, regardless of the amount, and the EV on-board battery management system (BMS) and the charger modify the charging power depending on the surplus at any given time, without consuming energy from the grid.

The three proposed TSs offer a cheaper price than the market price to encourage charging at the most opportune times. However, when there is no surplus, the market price is offered.

3.1. *Market Tariff Used as Base Tariff*

The market tariff used as the base tariff for the three TSs is one of the GoiEner cooperatives [28], specifically the 3.0 TDVE tariff, which is tailored for public charging points. The 3.0 TDVE tariff has different prices per hourly slot, with modifications each month, as described in Figure 1. There are six periods with different energy and power rates used in the various hourly slots. Regarding the GoiEner's self-consumption compensation price, it is 0.0839 EUR/kWh. The compensation price is the price at which the PV panel owner is compensated for injecting locally generated PV energy into the grid that has not been instantaneously consumed. Charging and compensation prices are reviewed every trimester.

Energy

Period	P1	P2	P3	P4	P5	P6
€/kWh	0.281990	0.244388	0.213676	0.211412	0.184719	0.154102

Power and energy periods

Hour	January	February	March	April	May	June	July	August	September	October	November	December	Saturdays, Sundays and holidays
00:00 - 01:00	P6	P6	P6	P6	P6	P6	P6	P6	P6	P6	P6	P6	P6
01:00 - 02:00	P6	P6	P6	P6	P6	P6	P6	P6	P6	P6	P6	P6	P6
02:00 - 03:00	P6	P6	P6	P6	P6	P6	P6	P6	P6	P6	P6	P6	P6
03:00 - 04:00	P6	P6	P6	P6	P6	P6	P6	P6	P6	P6	P6	P6	P6
04:00 - 05:00	P6	P6	P6	P6	P6	P6	P6	P6	P6	P6	P6	P6	P6
05:00 - 06:00	P6	P6	P6	P6	P6	P6	P6	P6	P6	P6	P6	P6	P6
06:00 - 07:00	P6	P6	P6	P6	P6	P6	P6	P6	P6	P6	P6	P6	P6
07:00 - 08:00	P6	P6	P6	P6	P6	P6	P6	P6	P6	P6	P6	P6	P6
08:00 - 09:00	P2	P2	P3	P5	P5	P4	P2	P4	P4	P5	P3	P2	P6
09:00 - 10:00	P1	P1	P2	P4	P4	P3	P1	P3	P3	P4	P2	P1	P6
10:00 - 11:00	P1	P1	P2	P4	P4	P3	P1	P3	P3	P4	P2	P1	P6
11:00 - 12:00	P1	P1	P2	P4	P4	P3	P1	P3	P3	P4	P2	P1	P6
12:00 - 13:00	P1	P1	P2	P4	P4	P3	P1	P3	P3	P4	P2	P1	P6
13:00 - 14:00	P1	P1	P2	P4	P4	P3	P1	P3	P3	P4	P2	P1	P6
14:00 - 15:00	P2	P2	P3	P5	P5	P4	P2	P4	P4	P5	P3	P2	P6
15:00 - 16:00	P2	P2	P3	P5	P5	P4	P2	P4	P4	P5	P3	P2	P6
16:00 - 17:00	P2	P2	P3	P5	P5	P4	P2	P4	P4	P5	P3	P2	P6
17:00 - 18:00	P2	P2	P3	P5	P5	P4	P2	P4	P4	P5	P3	P2	P6
18:00 - 19:00	P1	P1	P2	P4	P4	P3	P1	P3	P3	P4	P2	P1	P6
19:00 - 20:00	P1	P1	P2	P4	P4	P3	P1	P3	P3	P4	P2	P1	P6
20:00 - 21:00	P1	P1	P2	P4	P4	P3	P1	P3	P3	P4	P2	P1	P6
21:00 - 22:00	P1	P1	P2	P4	P4	P3	P1	P3	P3	P4	P2	P1	P6
22:00 - 23:00	P2	P2	P3	P5	P5	P4	P2	P4	P4	P5	P3	P2	P6
23:00 - 24:00	P2	P2	P3	P5	P5	P4	P2	P4	P4	P5	P3	P2	P6

Figure 1. Different power and energy prices per hourly slot of the GoiEner's 3.0 TDVE tariff.

To conclude, the three proposed TSs offer the compensation price (0.0839 EUR/kWh) when the specific conditions of each TS are fulfilled. When the conditions are not met, the charging price is the market price for the corresponding time slot and month.

3.2. *Tariff System 1*

The first TS is the simplest one. Whenever the PV surplus exceeds 7 kW, the reduced price is offered. There are two main reasons why the threshold is set at 7 kW. On the one hand, if the threshold is very low, by offering the reduced price, EVs would consume more from the grid and less from the surplus, as the users would charge their EVs when the surplus is lower. On the other hand, the average maximum charging power of all registered EVs is around 7 kW.

The main disadvantage of this TS is that there are occasions when the surplus is less than 7 kW and is not used.

To calculate SCRs that would be obtained with the implementation of TS1, a simulation was carried out based on the historical data presented in Section 2. The simulator was coded in MATLAB® software 9.13.0.2166757 (R2022b) Update 4. The simulation was performed for each day where EVs were charged during the above-mentioned six months. The EVs considered in the simulation on a given day were processed in the order of original arrival at the charging point. Additionally, since charging system 1 is an indirect control, once the EV had started to charge in the simulation (when the reduced price was offered), the EV was charged at its maximum power until it was fully charged, even though after a while the reduced price was no longer offered. This process was considered to be the most realistic.

The flow chart in Figure 2 describes the simulation process for one day under the influence of TS1. The diagram can be divided into three phases. The first phase is composed of the two main loops, which ensure that all EVs and all sampling periods of the day are analysed. It starts by detecting how many EVs were charged that day (variable *Num_EV_Charge*). The simulation starts by analysing the first EV of the day, so the index y takes the value 1 ($y = 1$). When the charge of the first EV is simulated, index y is incremented to 2 to consider the next EV of the day. When y is greater than *Num_EV_Charge*, it means that all EVs have been simulated and therefore the simulation of the EV charging pattern ends, leading to Figure 7's flow chart (linked to connector A) where the charging cost is calculated.

Next, the first sampling period ($t = 1$) is analysed. Given that in the simulation the sample time is 10 min, the first sampling period, $t = 1$, starts at 00:10. The index t is incremented every new period until it reaches a value of *Num_samples*, 144 (00:00). When t is greater than *Num_samples*, the simulation ends, leading to Figure 7's flow chart.

The second phase analyses whether the conditions are met to offer the reduced price according to TS1. For this purpose, the difference between the consumption and the PV production at time t ($diff(t)$) is calculated. If $diff(t)$ exceeds 7 kW, the reduced price is offered, and therefore $EV(y)$ (y^{th} EV) charging is carried out. On the contrary, if the difference does not exceed 7 kW, the variable t is incremented, and the difference is recalculated and compared with the threshold of 7 kW. Whenever $diff(t)$ is greater than 7 kW, it is checked if there is free space at the charging point. If the variable $EV\_Connected$ is less than 2, this means that there is space available, and the simulation of the $EV(y)$ charging pattern starts.

In the third phase, the charging pattern of the vehicles is simulated. First, index t_y is created. This index represents how many time intervals the $EV(y)$ has been connected to. It is necessary to (i) know how long $EV(y)$ has needed to complete its charge and (ii) consider that a connector of the charging point has been occupied during the time interval t_y. Index t_y increases each time $EV(y)$ spends a sample period charging. The variable EV_rem_en oversees how much energy the $EV(y)$ must consume to complete its charge. These data are known from the registers of the Charge and Parking application. Whenever EV_rem_en is greater than 0, it means that the $EV(y)$ has not yet completed its charge. The variable EV_char_pat records the $EV(y)$ charge pattern, i.e., the energy consumed in each *t* interval. During an interval, an EV charges the maximum amount of energy. This value is obtained by multiplying the maximum charging power of the EV by the sampling time. In the flow chart, this maximum energy is defined as *EV_max_en_SP*. Whenever the variable *EV_rem_en* is greater than *EV_max_en_SP*, the charge pattern for that interval is the maximum energy that can be charged. After recording in the variable *EV_char_pat* how much energy has been consumed for that interval, *EV_rem_en* is updated, as shown in Equation (1). In addition, the fact that a connector has been occupied by that interval is also recorded (Equation (2)).

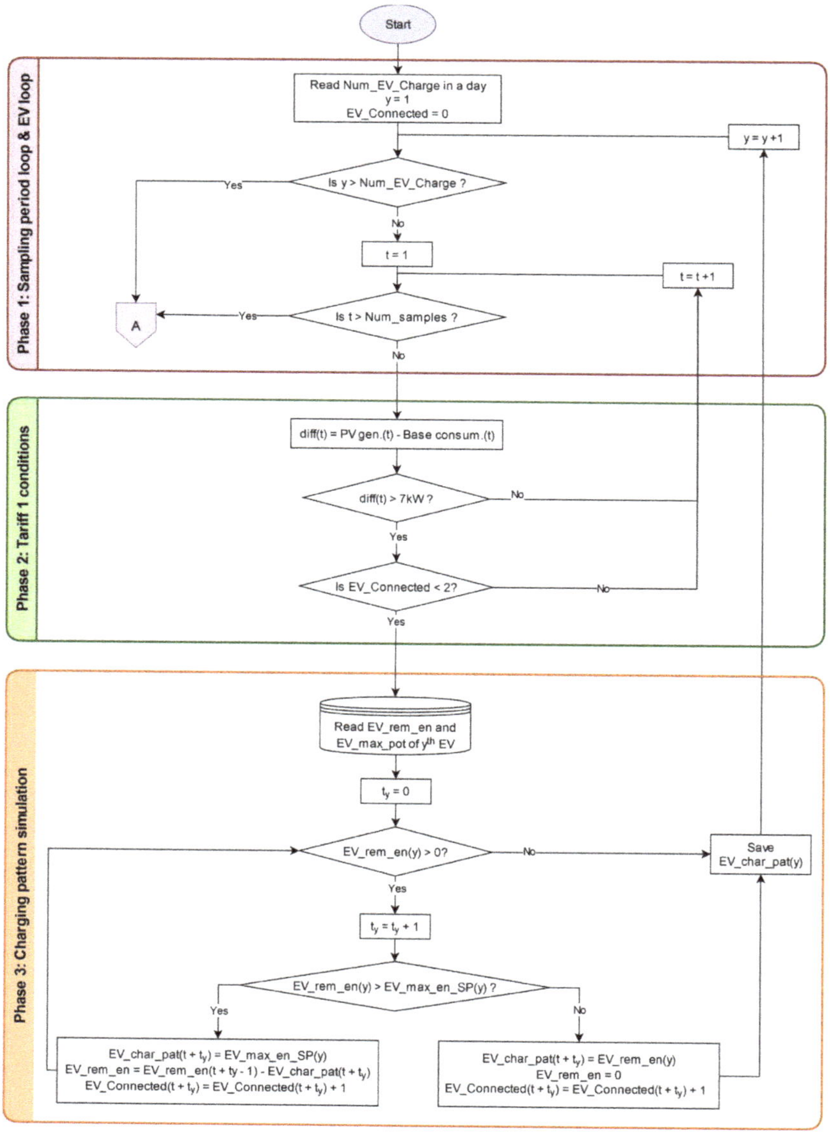

Figure 2. Flow chart related to TS1.

$$EV\_rem\_en = EV\_rem\_en(t + t_y - 1) - EV\_char\_pat(t + t_y), \quad (1)$$

$$EV\_Connected(t + t_y) = EV\_Connected(t + t_y) + 1. \quad (2)$$

Afterwards, EV_rem_en is checked again to ensure that it is still greater than 0. t_y is increased by one, and $EV(y)$ continues to be charged. In the last interval before the end of the charge, EV_rem_en is lower than the maximum power it can charge in one interval. Then, the charge pattern for that interval is the amount of remaining energy (EV_rem_en). Next, the variable EV_rem_en becomes 0, and the variable $EV\_Connected$ registers one last time that the connector is occupied.

Finally, as the charging of $EV(y)$ is completed, its charging pattern is saved, and the variable y is incremented. The simulation starts again from the first phase analysing the charge of the next EV of the day. Although the charging of a single EV is simulated at each round of the flow chart, when starting again from the beginning of the flow chart with the index $y = 2$, the index t is updated to 1. Continuing with the aforementioned steps, it is checked whether the conditions for charging the vehicle are met and whether there is free space. This means that, although only one EV is charged in each round of the flow chart, two EVs can be charged in the same time interval t. The simulation continues to analyse all EVs of the day until y is greater than Num_EV_Charge.

3.3. *Tariff System 2*

The second TS is similar to the first one. However, a more customised approach is considered. As described in Section 2, each EV has a maximum charging power. Three power levels are mainly found: one at 3.4 kW, another one at around 7 kW, and the last one at 11 kW. In TS2, the reduced price is offered when the PV surplus exceeds one of these 3 values. Three different tariffs are published, customised to the maximum charging power of each EV. The aim is to make the most of all the time slots where there is sufficient PV surplus to charge each EV. Thus, EVs that consume less are charged when there is less surplus, and EVs that consume more are charged when there is more surplus, thereby allowing them to consume less energy from the grid.

This charging scheme is also indirect. As with TS1, the simulation of the charging patterns was processed in the recorded historical order of EV charging. In addition, once the EV connected, its charging continued at its maximum power until it was fully charged.

Figure 3 shows the flow chart that describes the simulation process of the charging for one day under the influence of TS2. Phases 1 and 3 of the flow charts are the same as those of Figure 2. The difference between TS1 and 2 lies in the conditions under which the reduced price is offered in the second phase.

In this phase, first, the maximum charging power (EV_max_pot) of the yth EV is detected. Depending on the value of the maximum charging power of the EV, a different tariff is offered. Hence, once the maximum charging power of the EV is known, it is classified into one of these three groups: charging power (a) lower than 7 kW, (b) between 7 and 11 kW, and (c) greater than or equal to 11 kW. After classifying the yth EV within the corresponding group, the difference between the consumption and the PV production at time t ($diff(t)$) is calculated. If the variable $diff(t)$ exceeds the threshold of the corresponding group, y^{th} EV charging is carried out. For group a), for EVs with maximum charging power under 7 kW, $diff(t)$ must exceed 3 kW. In group b), for EVs with charging power between 7 kW and 11 kW, $diff(t)$ must exceed 7 kW. And finally, in group c), for EV maximum charging power greater or equal to 11 kW, $diff(t)$ must exceed 11 kW. On the contrary, if the difference does not exceed the corresponding threshold, t is incremented, and the difference is recalculated and compared with the corresponding threshold for each group. Whenever $diff(t)$ is greater than the threshold, the availability of free space at the charging point is checked. If the variable $EV\_Connected$ is less than 2, this means that there is space available, and the simulation of the $EV(y)$ charging pattern starts.

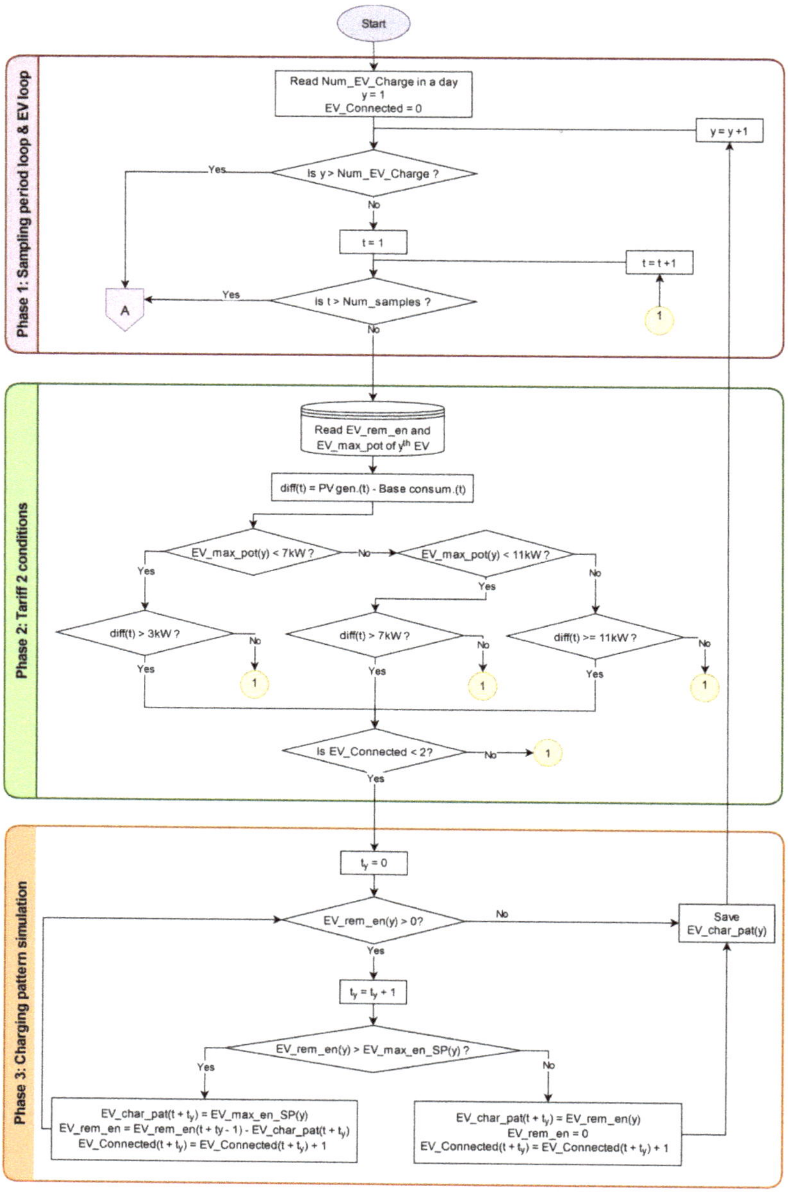

Figure 3. Flow chart related to TS2.

3.4. *Tariff System 3*

TS3 operates as a hybrid configuration. In addition to offering reduced prices to motivate the user to switch the charging to a more convenient time, it also acts on the charging power of EVs. Therefore, it avoids consuming energy from the grid and maximises the use of PV surplus. However, in cases where there is no more PV surplus for the rest of the day, only those EVs that are still connected finish their charge by consuming energy from the grid. This process was considered to be more realistic.

The control of the EV charging power is carried out by the public charger that communicates with the EV batteries' BMS.

Figure 4 shows the flow chart related to TS3. The three phases are different from the diagrams in Figures 2 and 3. This is due to the hybrid nature of the system, in which the charging power is controlled.

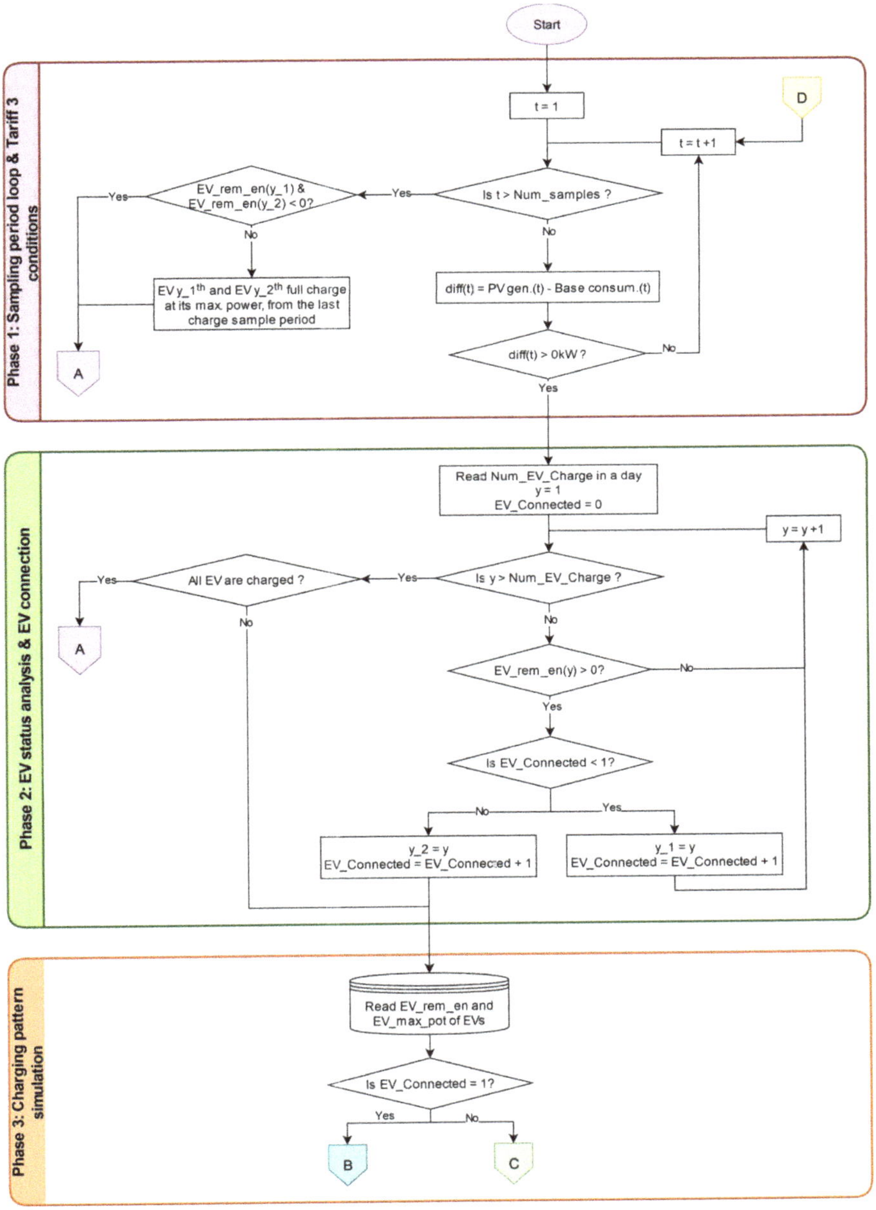

Figure 4. Flow chart related to TS3. The letters A-C are connectors linking the existing flowchart to the flowchart in Figures 5–7 respectively.

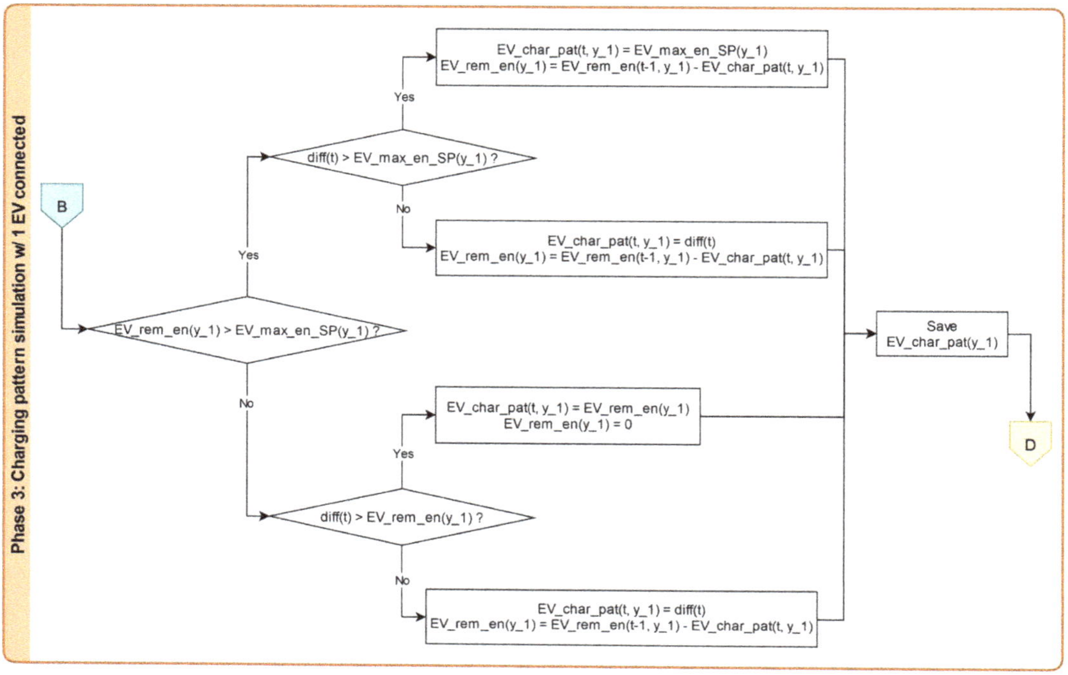

Figure 5. Continuation of Figure 4 flow chart, EV charging when only one EV is connected. The letters B and D refer to the connectors that links the existing diagram to the flowchart in Figure 4.

The initial phase is responsible for the analysis of the PV surplus for all sampling periods within a day. Starting with the first period (t = 1), the difference between PV production and baseline consumption is calculated. If there is a surplus, i.e., $diff(t) > 0$, the simulation for EV charging can begin. Conversely, if no surplus exists, t is incremented, and the difference is recalculated. Finally, when t exceeds *Num_samples* (144), it means that EV charges are completed. At that point, to align with real-world scenarios, if an EV was connected during a surplus period but did not complete its charge, it continues to charge at its maximum power (*EV_max_pot*), consuming electricity from the grid, starting from the last period at which it was charged. The end of Figure 4's flow chart is achieved when the connection point A is reached. At this point, the simulation of EV charging related to TS3 is finished.

In the second phase, EVs are connected. First, EV data for the day to be simulated are read. The algorithm begins by analysing the status of the first EV ($y = 1$). If the yth EV is not yet charged, i.e., if *EV_rem_en* > 0, it is connected. On the other hand, if the yth EV is already charged, y is updated to 2 to assess the status of the next EV of the day. Before EV connection, it is verified that there is an available slot. If no EV is connected (*EV_Connected* < 1), the first EV connects on one connector, index y_1 obtains the value of yth EV ($y\_1 = y$). Conversely, if *EV_Connected* equals 1, the second EV is connected on the last connector, and index y_2 obtains the value of the yth EV ($y\_2 = y$), and the charging simulation begins. Finally, when y is greater than *Num_EV_Charge*, it is checked that all connected EVs are fully charged. If there is an EV connected, it starts being charged.

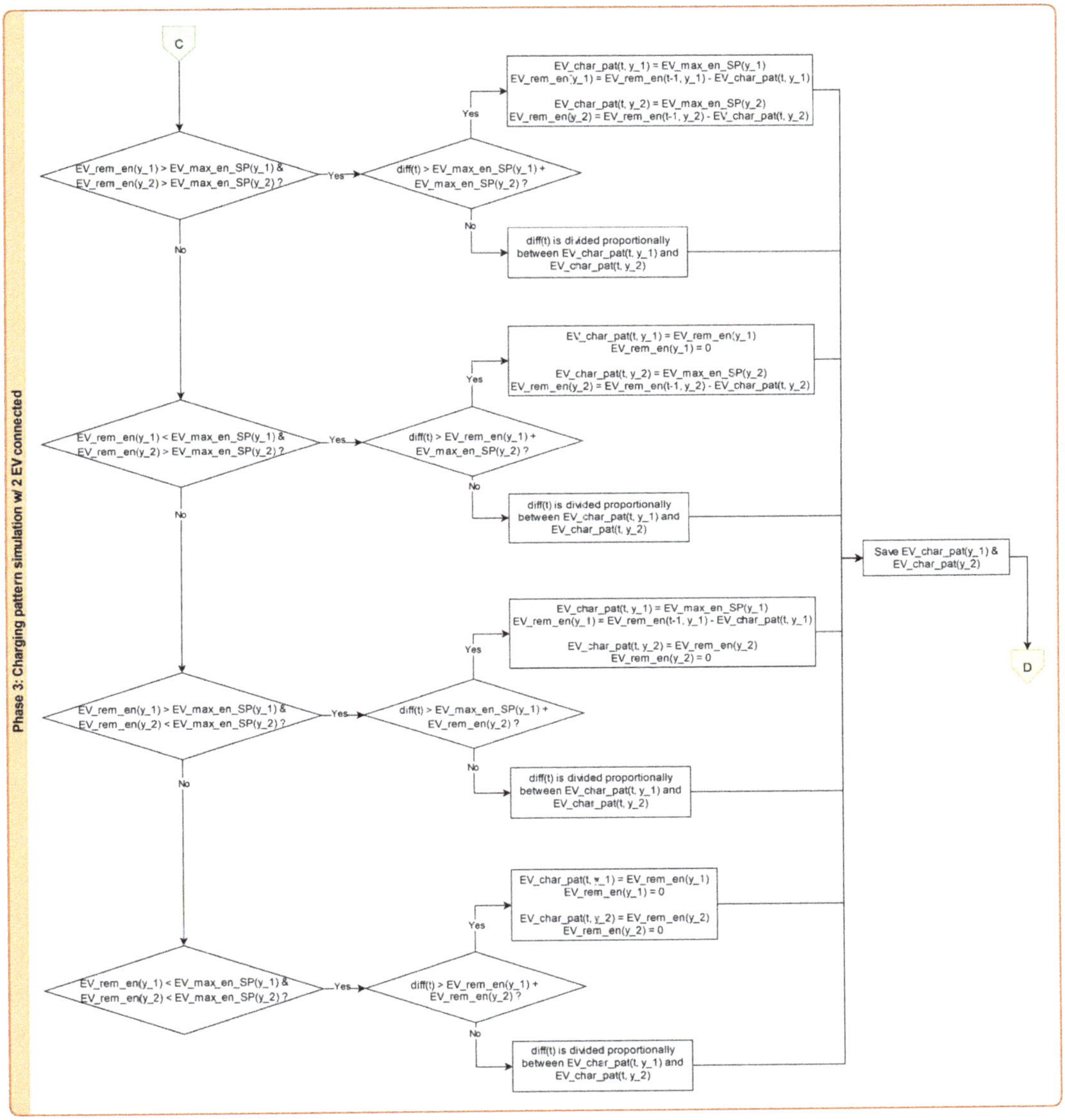

Figure 6. Continuation of Figure 4 flow chart, EV charging when two EVs are connected. The letters C and D refer to the connectors that links the existing diagram to the flowchart in Figure 4.

The third phase oversees the simulation of the EV charging pattern. The charging pattern of the vehicles is not the same if there is a single EV connected or two. When only one EV is connected, all the surplus of the period t is used to charge the single EV, whereas when two EVs are connected, the PV surplus is divided between the two EVs. This is why two different situations are distinguished, leading to a B connector or a C connector.

Figure 5 describes EV charging when only one EV is connected. Firstly, two possible situations are analysed: whether (EV_rem_en) is greater than $EV\_max\_en\_SP$, or the opposite. If the yth EV still has much energy left to complete its charge and there is enough PV energy, the charging is performed at the maximum EV power. Otherwise, y^{th} EV charging power varies to match the momentary surplus, i.e., *diff*(t). This is the main property that characterises the third TS. Similarly, in this case, the yth EV is at the end of its charge and must consume less than $EV\_max\_en\_SP$, and it is checked if there is enough PV surplus. Then, as before, in the case there is sufficient surplus, the EV is fully charged. On the other hand, if *diff*(t) is not enough to finish the charge, the EV modifies its charging power to adjust to the PV surplus. Once the charge has been completed for the sampling period t, the simulation starts again from the first phase (connector D), analysing the situation of the next sampling period (t + 1).

The other possible scenario is described in Figure 6, where the flow chart for charging two EVs at the same time is presented. In this case, with two EVs, four possible situations are considered.

1. Both EVs still have energy remaining to be charged to their maximum power;
2. The EV on connector 1 (y_1) is almost charged and its remaining energy is less than $EV\_max\_en\_SP$, while the EV on the second connector (y_2) can still be charged to maximum power;
3. In contrast to case two, the EV at connector 1 (y_1) still has a considerable amount of charge left, and the EV at connector 2 (y_2) is at the end of its charge;
4. Both EVs are at the end of their charge and consume less than $EV\_max\_en\_SP$.

Once the situation of the two connected EVs is determined, it is analysed whether the PV surplus is sufficient to charge the EVs under the conditions determined in the previous step. If there is enough PV energy, both EVs are charged with the energy they need for each situation. On the other hand, the charging power of the EVs is modified, and the surplus is divided between the two EVs without consuming energy from the grid. Not all EVs are charged at the same power, so the amount of the surplus is considered and divided proportionally. At the end of the interval t, the flow chart starts again from the first phase until variable t is greater than *Num_samples*.

3.5. Charging Cost Calculation

The calculation is carried out at the end of the flow charts for the three charging systems (connector A). The flow chart in Figure 7 determines how much energy has been consumed from the grid and the PV surplus at each interval.

In each interval, the charge pattern of each EV is analysed. If the PV surplus of that sampling interval is greater than the charge of all EVs for that sampling interval, it means that the consumption corresponding to the EV charge pattern for the sampling interval t is completely consumed from the PV surplus. Otherwise, the PV surplus is divided in proportion to the maximum charging power of each EV. The remaining energy to complete the charging pattern is obtained from the grid (*Grid_con*).

This way, the EV charging pattern for the day is divided between the energy obtained from the PV surplus (*Surplus_con*) and the energy obtained from the grid (*Grid_con*).

The energy consumed from the surplus is paid at the reduced price, i.e., the compensation price. Conversely, energy consumed from the grid is paid according to GoiEner's 3.0TDVE tariff, described in Section 3.1, as described in the flow chart in Figure 8.

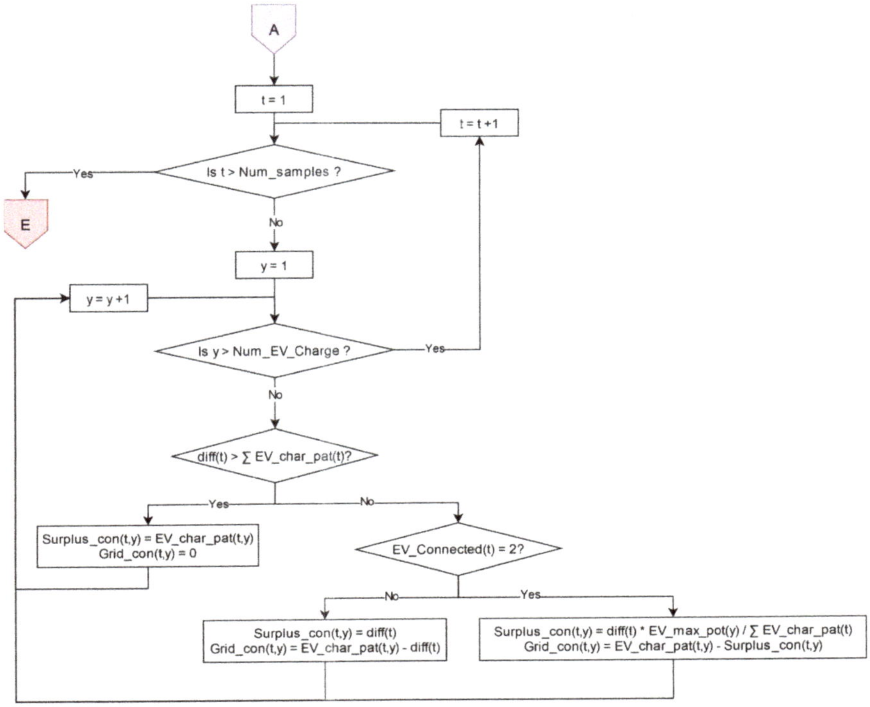

Figure 7. Division of EV consumption between PV surplus and grid. The letters A and E are connectors that link the existing flowchart to the other flowcharts above.

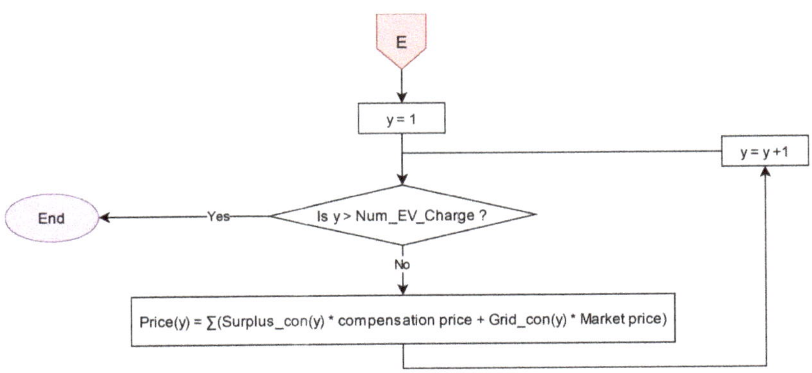

Figure 8. Flow chart to calculate EV charging price.

3.6. Self-Consumption Rate Calculation

Since the main objective of the three designed TSs is to maximise the consumption of locally produced energy and to decrease the consumption of the grid, the metric used to compare the different TSs is the SCR (Equation (3)).

The SCR is defined as the amount of PV energy consumed at each time step, corresponding to each hour in Spain.

$$SCR = \frac{\sum_1^{24} E_{PVSC}(k)}{\sum_1^{24} E_{PV}(k)}, \tag{3}$$

where $E_{PVSC}(k)$ represents the self-consumed energy in each calculation section period k, 1 h in Spain. $E_{PV}(k)$ represents the PV energy produced in each considered k period.

4. Results and Discussion

The following section presents the results obtained from simulations of EV charging patterns under the influence of the three proposed TSs and their impact on the SCR.

4.1. Influence of the Three Tariff Systems for One Day in Detail

4.1.1. Day Example to Understand the Phenomenon

Firstly, 19 March 2022 is chosen as an example to observe the effect of the three proposed tariffs on the SCR.

Figure 9 shows the basic consumption and the EV consumption recorded in the Charge and Parking manager. The purple area reflects the consumption of the seven points associated with the CSC. The yellow curve represents the PV production, whereas the white area refers to the unconsumed PV surplus. The blue area represents the original consumption of the EV charging registered for that day. Two EVs were charged on 19 March 2022, with 6.6 kW and 11 kW of maximum charging power.

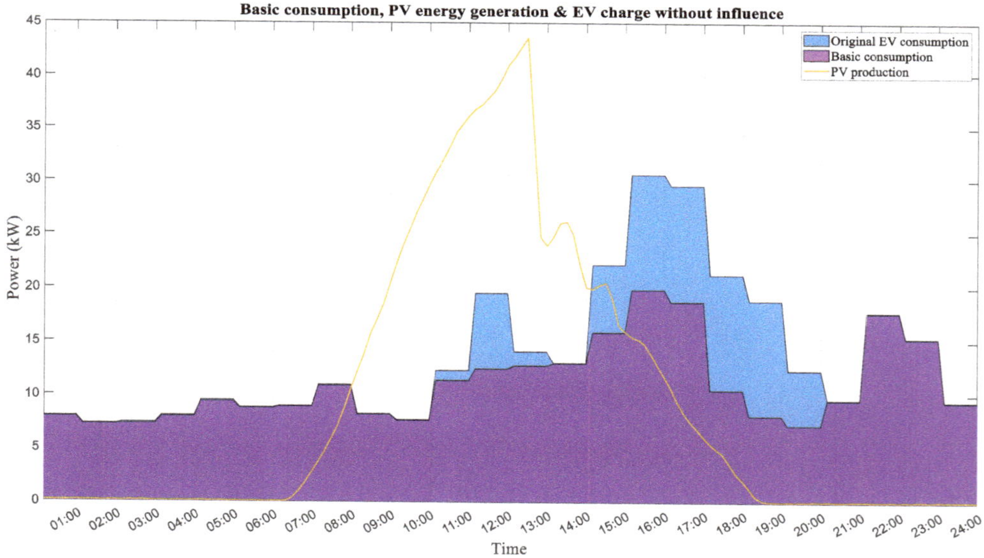

Figure 9. Basic consumption, PV production, and EV consumption for 19 March 2022.

Figures 10 and 11 describe the basic consumption and EV consumption under the influence of TS1. Compared with Figure 9, it can be observed that the charge of the second EV has been shifted to the hours when PV surpluses occur.

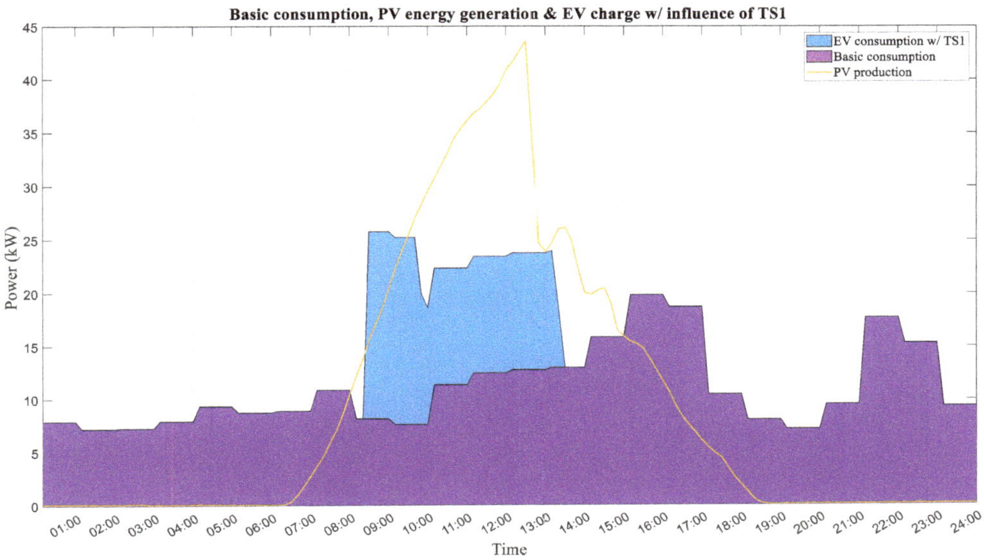

Figure 10. Basic consumption, PV production, and EV consumption under TS1 influence for 19 March 2022.

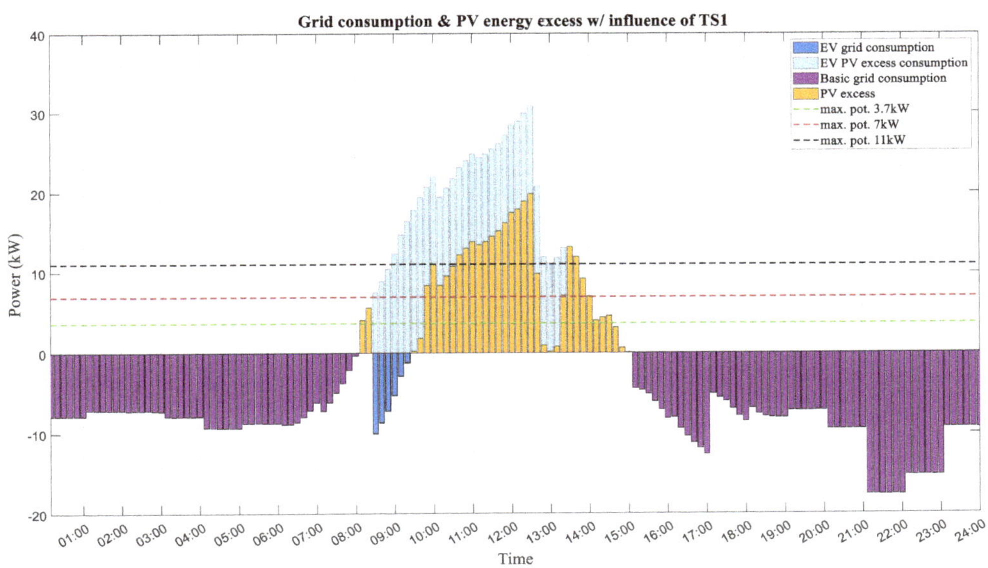

Figure 11. Grid consumption, PV energy surplus, and EV consumption under TS1 influence for 19 March 2022.

A more detailed analysis is provided in Figure 11. Negative consumption denotes energy consumed from the grid. The light blue bars indicate consumption from PV surplus, while the dark blue ones in the negative area are the consumption from the grid. Furthermore, three horizontal lines are included, marking the threshold corresponding to 3.7 kW, 7 kW, and 11 kW. In TS1, the reduced price is offered when the PV surplus exceeds 7 kW. Therefore, the charging of both EVs is not carried out until 08:30. It should be noted that, although Figure 11 shows that there are more PV surpluses, a part is consumed from the grid. This is due to the fact that TS1 is an indirect control where the charging power is not controlled.

Figures 12 and 13 are related to TS2. TS2 offers a reduced price depending on the maximum charging power of each EV. In this case, the first EV with a charging power of less than 7 kW benefits from the reduced tariff when the PV surplus exceeds 3.7 kW. The second EV, with a charging power of 11 kW, benefits from the reduced tariff when the surplus exceeds 11 kW. With TS2, Figure 12 illustrates how the blue area has expanded further into the yellow curve (PV surplus). Nonetheless, a small part is still consumed from the grid.

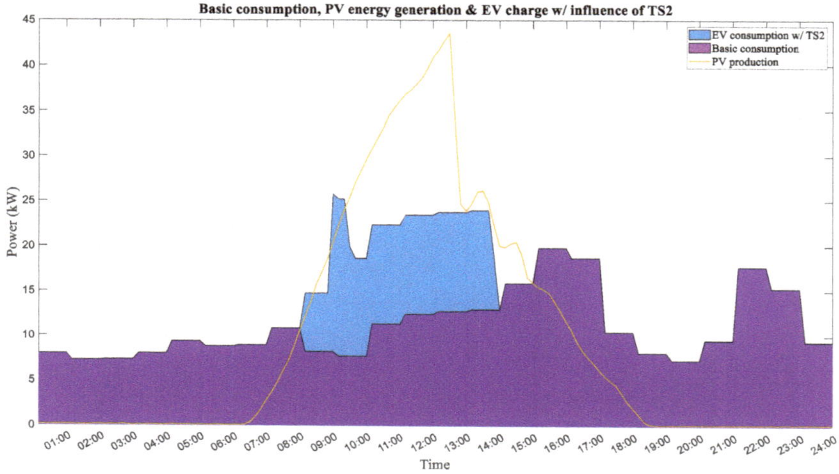

Figure 12. Basic consumption, PV production, and EV consumption under TS2 influence for 19 March 2022.

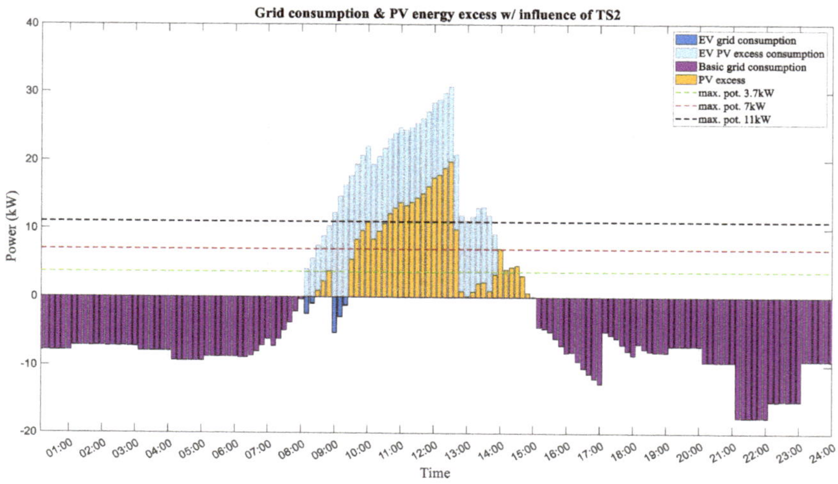

Figure 13. Grid consumption, PV energy surplus, and EV consumption under TS2 influence for 19 March 2022.

Figure 13 shows that by offering tariffs customised to each EV's charging power, the charging time of each EV is better distributed depending on its power and surplus.

Finally, Figures 14 and 15 present the consumption under the influence of TS3, where, whenever surpluses occur, the reduced charging price is offered. In addition, this system operates with direct control, modifying the charging power of the EVs. As illustrated in Figure 14, the EV charging consumption (blue area) is perfectly aligned with the yellow curve (PV surplus), without consuming from the grid. The same result can be seen in Figure 15, where the EVs start charging from the first instant where PV surpluses occur, and by modifying EV charging power, no electricity is consumed from the grid.

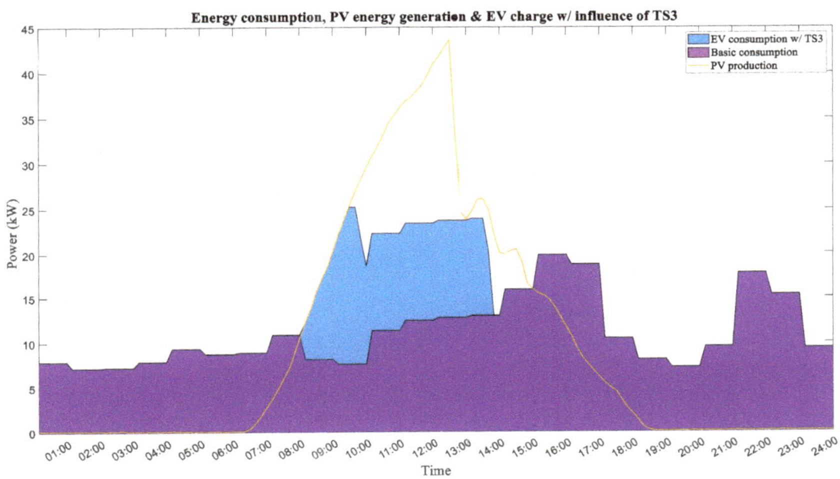

Figure 14. Basic consumption. PV production and EV charging consumption under TS3 influence for 19 March 2022.

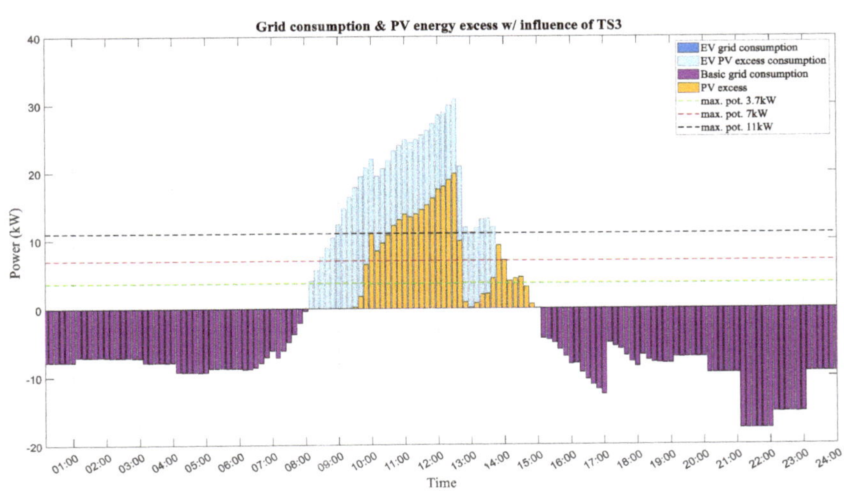

Figure 15. Grid consumption, PV energy surplus, and EV consumption under TS3 influence for 19 March 2022.

It is appropriate to mention that the simulations performed are made under the assumption that human behaviour is ideal. That is, all EV users would be willing to modify their vehicle charging schedule. However, after conducting a survey among the members of the CSC project, the opinion of the users and their likelihood to modify their charging schedule will be considered.

Finally, Table 1 compiles the SCR values of all the cases and the cost of the EV charges. Regarding the SCR values, firstly, it should be noted that on 19 March 2022, a considerable PV surplus was recorded, reaching 30 kW. Furthermore, only two EVs were registered at the charging point that day. Considering this fact, it was clearly impossible to achieve a 100% SCR. The remaining PV surplus, i.e., that not consumed by the EV, could be consumed, at least partially, by the remaining flexible loads of the other seven consumption points of the CSC. Future work will analyse how the SCR could be further increased.

Table 1. SCR and charge cost values for 19 March 2022.

	SCR	Charging Price	
		EV1	EV2
Basic consumption	51.13%	-	-
Original EV charge	56.54%	EUR 3.19	EUR 20.34
EV charge with TS1	76.32%	EUR 2.18	EUR 11.38
EV charge with TS2	77.96%	EUR 1.99	EUR 10.99
EV charge with TS3	78.90%	EUR 1.77	EUR 10.72

4.1.2. Different Characteristic Days Analysis

Two other days with different characteristics are analysed. Figure 16 compiles the charging patterns under the influence of the three tariffs on 9 February 2022, where a single EV with a power of 11 kW was charged. The EV is charged with TS1 as the surplus exceeds the 7 kW threshold at some times. Regarding TS2, it is not applied to this EV since the surplus does not reach 11 kW. Finally, with TS3, the EV is charged as soon as PV surplus energy is available, without the need to consume from the grid. As far as SCR is concerned, with TS1, the SCR increases from 90.49% to 93.27%, while with TS3, it increases to 94.18%.

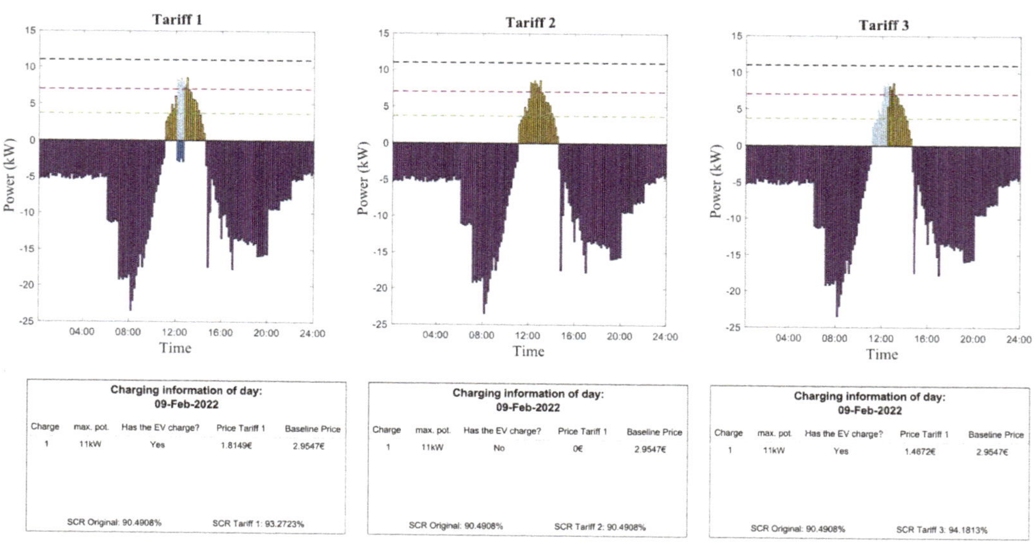

Figure 16. Charging pattern and charging information under the influence of the three proposed tariffs for 9 February 2022.

Figure 17 examines the situation for 6 April 2022, where there are almost no PV surpluses, and the basic SCR is 98.66%. That day, as the surpluses do not reach the 7 kW threshold required for TS1, this TS remains unapplied. Under TS2, a reduced price is offered to the first two EVs that have a charging power of 6.6 kW. The third EV with a power of 7.4 kW is not charged. Indeed, as in TS1, the surplus does not reach 7 kW. With TS2, the SCR increases to 99.25%. Lastly, with TS3, all surpluses are consumed, reaching a SCR of 100%.

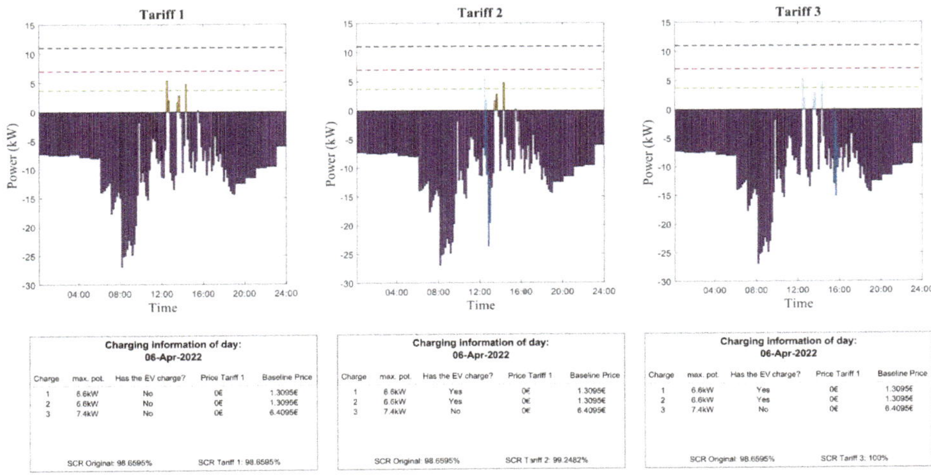

Figure 17. Charging pattern and charging information under the influence of the three proposed tariffs for 6 April 2022.

4.2. Analysis of the Effect of the Three TSs on the SCR over One and Six Months

Figure 18 presents the SCR for all days where charging occurs during the month of March 2022, as long as the basic SCR is below 100% (i.e., there are PV surpluses). SCR values are depicted for the basic consumption, the original charging recorded by the Charge and Parking manager, and the three designed TSs. TS3 always obtains the highest SCR values. On the other hand, most of the time, TS2 obtains higher SCRs than TS1. In addition, it is noticeable that on days with a high basic SCR, the increase in SCR is not as significant as on the other days.

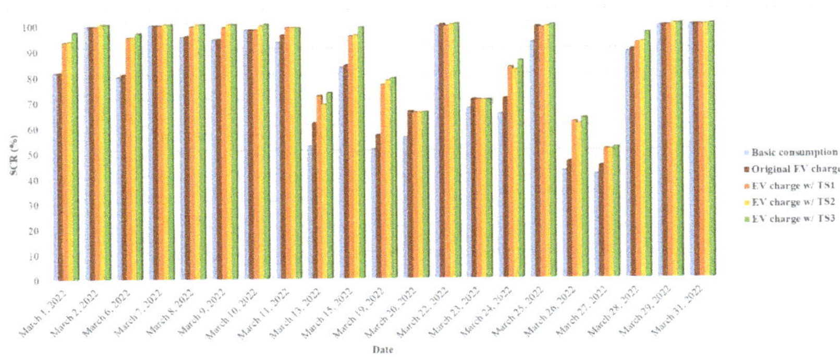

Figure 18. SCR for basic consumption, original charge, and three TS.

Table 2 presents the average SCR values for each month and the overall average for all simulated days over the 6 months. The overall average basic SCR is 83.82%. The SCR obtained with the original charge of the EVs is 86.54%. The average SCR stands at 90.86% with TS1, 91.04% with TS2, and 93.09% with TS3. It can be concluded that these TSs clearly show an improvement in self-consumption, achieving an increase in SCR of 8.8% when applying TS3, for instance.

Table 2. Average SCR values and SCR increase for each month.

Month	SCR Values (%)					SCR Increase/Basic Consumption			
	Basic	Original Charge	TS1	TS2	TS3	Original Charge	TS1	TS2	TS3
November	92.09	94.20	95.07	95.54	98.01	1.2%	1.8%	4.3%	1.2%
December	89.90	92.04	94.10	94.10	96.71	2.4%	2.4%	5.3%	2.4%
January	84.42	87.58	95.27	96.51	97.73	10.4%	11.8%	13.3%	10.4%
February	84.41	87.51	91.16	90.96	92.98	5.6%	5.3%	7.6%	5.6%
March	79.87	82.43	87.90	88.07	89.32	8.4%	8.5%	10.3%	8.4%
April	72.25	75.47	81.66	81.08	83.77	9.0%	8.3%	11.9%	9.0%
Mean	**83.82**	**86.54**	**90.86**	**91.04**	**93.09**	**6.2%**	**6.4%**	**8.8%**	**6.2%**

4.3. Comparison of the SCR Increase in the Three TSs with the Original Charge

Figure 19 presents four pie charts divided into ranges based on the SCR values. For instance, the yellow range shows all days when the SCR is between 90% and 100%. Figure 19a displays SCR values related to the original charge registered by the Charge and Parking manager. Results obtained with TS1, 2, and 3 are shown in Figure 19b–d, respectively. It is noticeable that in 21.7% of the days, SCR values below 70% are obtained with uncontrolled charging. Conversely, with TS1, these values are reduced to 10.8%, 12% with TS2, and 10.8% with TS3. On the other hand, when examining higher SCR, without TS, a SCR higher than 90% is obtained only 48.6% of the time, while these data increase to 60.8% with TS1, 62.2% with TS2, and 68.9% with TS3.

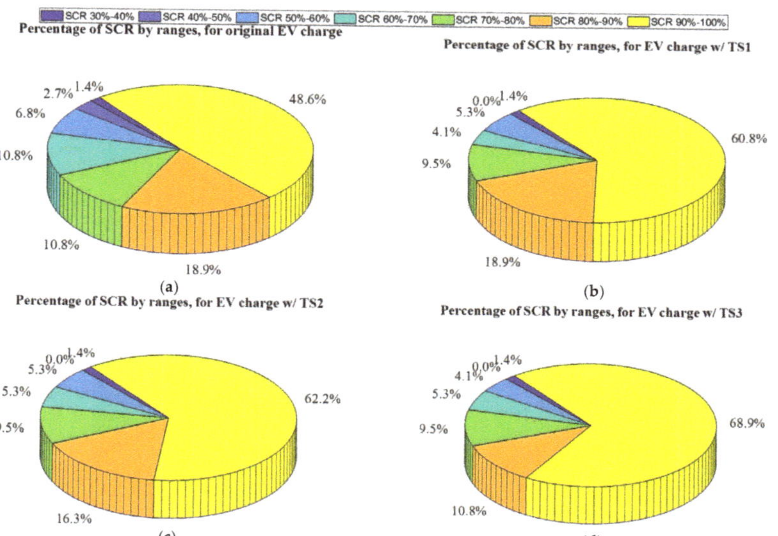

Figure 19. (**a**) Percentage of SCRs by ranges for the original charge; (**b**) percentage of SCRs by ranges with the TS1; (**c**) percentage of SCRs by ranges with the TS2; (**d**) percentage of SCRs by ranges with the TS3.

4.4. Analysis of the Economic Savings of the Three TS

Table 3 compiles the average percentage savings for each month from the EV user's point of view. With TS3, more energy is consumed from PV surplus and less from the grid. Therefore, it is also the TS that provides the most economic savings, achieving savings of 32%. Finally, the values of the global average over the six months indicate that both TS1 and TS2 achieve the same savings of 22%. As for TS3, an average of 25% savings is obtained for each EV charge.

Table 3. Monthly average savings for the three proposed tariffs.

Month	Mean Savings w/TS1	Mean Savings w/TS2	Mean Savings w/TS3
November 2021	10%	13%	11%
December 2021	14%	24%	26%
January 2022	31%	26%	27%
February 2022	29%	23%	25%
March 2022	30%	26%	32%
April 2022	21%	21%	30%
Mean	**22%**	**22%**	**25%**

5. Conclusions and Future Works

The research study presented in this paper analyses three TSs for EV charging based on PV surplus within a CSC project. Contrary to most articles found in the scientific literature, the main objective of this work is to increase the SCR rather than to increase the economic benefits of a charging station owner. In addition to the increase in the SCR, EV users also benefit from a reduction in the charging cost. As three different tariffs are designed and analysed, stakeholders can assess which case suits them best, besides having easy replicability.

Overall, the behaviour of the TSs can be divided into two cases: when PV surpluses are low and when they are high. When they are low, there are often scenarios where TS1 and TS2 are not applied, as the surpluses do not reach the set threshold. TS3 is always offered. However, it is usually necessary to consume from the grid to complete the charging of the EVs. In these cases, the SCR is usually very high, reaching 100%. Regarding days where there is a significant amount of surplus, generally, all three TSs are offered. With a large surplus of PV, the SCR tends to be lower, but the increase in the SCR thanks to the TS is highlighted.

As for which TS is the best, all of them have their advantages and disadvantages. TS1 is the simplest to implement. Additionally, if the surplus exceeds the 7 kW threshold, all EVs are charged. As a drawback, TS1 may lead to underutilisation of PV surplus. Considering the 6-month simulations, TS1 results in a 6.2% increase in SCR compared to the original charge, while the cost of EV charge is reduced by 22%.

Concerning TS2, its main advantage lies in its significant potential on days with substantial PV surplus. However, if there are no high surplus levels, it could result in not offering the tariff to all EVs. Overall, with TS2, the SCR increase is 6.4%, and the EV charging cost is 22% lower.

Finally, TS3 proves to be the most beneficial system regarding the SCR and the economic savings of the users, with a SCR increase of 8.8% and a reduction of the cost of 25%. Its advantages lie in offering a reduced price regardless of surplus levels and controlling the charging power, thereby taking advantage of all the potential for self-consumption. On the other hand, its implementation complexity and related costs stand as a disadvantage.

It is important to note that direct control of EV charging is not always possible, especially when charging points are public. Therefore, although it has been shown that direct control leads to higher SCR increases, it is also essential to analyse ways in which higher SCR can be achieved through indirect control of EV charging. In addition, indirect controls are less costly to implement.

This study is carried out as part of a CSC project, involving eight consumption points and PV production. To conduct the calculations presented in this article, historical recorded data were used. However, in the future, the real-time EMS will use predictions of both PV consumption and production to calculate the PV surplus.

Regarding other future studies, different actions are planned. On the one hand, it is important to better represent the uncertainty of human behaviour. For this purpose, a survey of CSC members conducted by the retailer will be considered. These members will be asked about their willingness to change the EV charging time. Additionally, an analysis of the scientific literature will be carried out to determine how EV users react to charging tariffs. According to the results of these studies, probability factors that depend on the level of change in the charging time will be considered in the simulation. Finally, the confidence interval of the PV production and the consumption predictions will be considered.

Author Contributions: Conceptualisation, G.E. and H.C.; methodology, G.E. and H.C.; software, G.E.; validation, H.C., A.E. and T.T.L.; formal analysis, G.E.; investigation, G.E.; data curation, G.E.; writing—original draft preparation, G.E.; writing—review and editing, G.E., H.C., A.E., and T.T.L.; visualisation, G.E.; supervision, H.C.; project administration, H.C. All authors have read and agreed to the published version of the manuscript.

Funding: This research received no external funding.

Data Availability Statement: Data related to this study will be available upon request.

Acknowledgments: We would like to thank the renewable energy generation and consumption cooperative GoiEner for their assistance in providing us with information about the CSC project of Aduna and for the historical data collected.

Conflicts of Interest: The authors declare no conflicts of interest.

References

1. CORDIS EU Research Results. Enabling a Successful Transition towards Electric-Powered Road Transport. Available online: https://cordis.europa.eu/article/id/443730-enabling-a-successful-transition-towards-electric-powered-road-transport (accessed on 17 November 2023).
2. European Parliament. EU Ban on the Sale of New Petrol and Diesel Cars from 2035 Explained. Available online: https://www.europarl.europa.eu/news/en/headlines/economy/20221019STO44572/eu-ban-on-sale-of-new-petrol-and-diesel-cars-from-2035-explained (accessed on 17 November 2023).
3. European Parliament. Reducing Car Emissions: New CO_2 Targets for Cars and Vans Explained. Available online: https://www.europarl.europa.eu/news/en/headlines/society/20180920STO14027/reducing-car-emissions-new-co2-targets-for-cars-and-vans-explained (accessed on 17 November 2023).
4. IEA, International Energy Agency. Global EV Outlook 2023. Available online: https://www.iea.org/reports/global-ev-outlook-2023 (accessed on 17 November 2023).
5. European Parliament. Car-Recharging Stations Should Be Available Every 60 km, Say MEPs. Available online: https://www.europarl.europa.eu/news/en/press-room/20221014IPR43206/car-recharging-stations-should-be-available-every-60-km-say-meps (accessed on 17 November 2023).
6. Mangipinto, A.; Lombardi, F.; Sanvito, F.D.; Pavičević, M.; Quoilin, S.; Colombo, E. Impact of mass-scale deployment of electric vehicles and benefits of smart charging across all European countries. *Appl. Energy* **2022**, *312*, 118676. [CrossRef]
7. Su, J.; Lie, T.T.; Zamora, R. Modelling of large-scale electric vehicles charging demand: A New Zealand case study. *Electr. Power Syst. Res.* **2019**, *167*, 171–182. [CrossRef]
8. Salah, F.; Ilg, J.P.; Flath, C.M.; Basse, H.; van Dinther, C. Impact of electric vehicles on distribution substations: A Swiss case study. *Appl. Energy* **2015**, *137*, 88–96. [CrossRef]
9. Guthoff, F.; Klempp, N.; Hufendiek, K. Quantification of the Flexibility Potential through Smart Charging of Battery Electric Vehicles and the Effects on the Future Electricity Supply System in Germany. *Energies* **2021**, *14*, 2383. [CrossRef]
10. Venkatraman, A.; Hug, G.; Schaffner, C.; Vayá, M.G. Optimal Design of Time-of-Use Tariffs using Bilevel Optimization. In Proceedings of the IEEE PES Innovative Smart Grid Technologies Conference Europe (ISGT-Europe), Novi Sad, Serbia, 10–12 October 2022; pp. 1–5. [CrossRef]
11. Vardakas, J.S.; Zorba, N.; Verikoukis, C.V. A Survey on Demand Response Programs in Smart Grids: Pricing Methods and Optimization Algorithms. *IEEE Commun. Surv. Tutor.* **2015**, *17*, 152–178. [CrossRef]
12. Herter, K. Residential implementation of critical-peak pricing of electricity. *Energy Policy* **2007**, *35*, 2121–2130. [CrossRef]
13. Li, R.; Wang, Z.; Gu, C.; Li, F.; Wu, H. A novel time-of-use tariff design based on Gaussian Mixture Model. *Appl. Energy* **2016**, *162*, 1530–1536. [CrossRef]

14. Rahman, A.; Aziz, T.; Masood, N.-A.; Deeba, S.R. A time of use tariff scheme for demand side management of residential energy consumers in Bangladesh. *Energy Rep.* **2021**, *7*, 3189–3198. [CrossRef]
15. Kandpal, B.; Pareek, P.; Verma, A. A robust day-ahead scheduling strategy for EV charging stations in unbalanced distribution grid. *Energy* **2022**, *249*, 123737. [CrossRef]
16. Hu, Z.; Zhan, K.; Zhang, H.; Song, Y. Pricing mechanisms design for guiding electric vehicle charging to fill load valley. *Appl. Energy* **2016**, *178*, 155–163. [CrossRef]
17. Zhang, L.; Yin, Q.; Zhu, W.; Lyu, L.; Jiang, L.; Koh, L.; Cai, G. Research on the orderly charging and discharging mechanism of electric vehicles considering travel characteristics and carbon quota. *IEEE Trans. Transp. Electrif.* **2023**. [CrossRef]
18. Tao, J.; Huang, D.; Li, D.; Yang, X.; Ling, C. Pricing strategy and charging management for PV-assisted electric vehicle charging station. In Proceedings of the 2018 13th IEEE Conference on Industrial Electronics and Applications (ICIEA), Wuhan, China, 31 May–2 June 2018; pp. 577–581. [CrossRef]
19. Dai, Y.; Qi, Y.; Li, L.; Wang, B.; Gao, H. A dynamic pricing scheme for electric vehicle in photovoltaic charging station based on Stackelberg game considering user satisfaction. *Comput. Ind. Eng.* **2021**, *154*, 107117. [CrossRef]
20. Gudmunds, D.; Nyholm, E.; Taljegard, M.; Odenberger, M. Self-consumption and self-sufficiency for household solar producers when introducing an electric vehicle. *Renew. Energy* **2020**, *148*, 1200–1215. [CrossRef]
21. Roselli, C.; Sasso, M. Integration between electric vehicle charging and PV system to increase self-consumption of an office application. *Energy Convers. Manag.* **2016**, *130*, 130–140. [CrossRef]
22. van der Kam, M.; van Sark, W. Smart charging of electric vehicles with photovoltaic power and vehicle-to-grid technology in a microgrid; a case study. *Appl. Energy* **2015**, *152*, 20–30. [CrossRef]
23. Barone, G.; Brusco, G.; Menniti, D.; Pinnarelli, A.; Polizzi, G.; Sorrentino, N.; Vizza, P.; Burgio, A. How Smart Metering and Smart Charging may Help a Local Energy Community in Collective Self-Consumption in Presence of Electric Vehicles. *Energies* **2020**, *13*, 4163. [CrossRef]
24. Fachrizal, R.; Munkhammar, J. Improved Photovoltaic Self-Consumption in Residential Buildings with Distributed and Centralized Smart Charging of Electric Vehicles. *Energies* **2020**, *13*, 1153. [CrossRef]
25. Zapirain, I.; Etxegarai, G.; Hernández, J.; Boussaada, Z.; Aginako, N.; Camblong, H. Short-term electricity consumption forecasting with NARX, LSTM, and SVR for a single building: Small data set approach. *Energy Sources Part A Recovery Util. Environ. Eff.* **2022**, *44*, 6898–6908. [CrossRef]
26. Etxegarai, G.; Zapirain, I.; Camblong, H.; Ugartemendia, J.; Hernandez, J.O. Curea. Photovoltaic Energy Production Forecasting in a Short Term Horizon: Comparison between Analytical and Machine Learning Models. *Appl. Sci.* **2022**, *12*, 12171. [CrossRef]
27. Euskalmet. Available online: https://www.euskalmet.euskadi.eus/inicio/ (accessed on 17 November 2023).
28. GoiEner. Renewable Energy Generation and Consumption Cooperative. Available online: https://www.goiener.com/es/ (accessed on 17 November 2023).

energies

MDPI

Article

Minimisation of the Energy Expenditure of Electric Vehicles in Municipal Service Companies, Taking into Account the Uncertainty of Charging Point Operation

Mariusz Izdebski, Marianna Jacyna * and Jerzy Bogdański

Faculty of Transport, Warsaw University of Technology, 00-662 Warsaw, Poland; mariusz.izdebski@pw.edu.pl (M.I.); jerzy.bogdanski@pw.edu.pl (J.B.)

* Correspondence: marianna.jacyna@pw.edu.pl

Abstract: This article presents an original method for minimising the energy expenditure of electric vehicles used in municipal service undertakings, taking into account the uncertainty in the functioning of their charging points. The uncertainty of the charging points' operation was presented as the probability of the occurrence of an emergency situation hindering a point's operation, e.g., a breakdown or lack of energy supply. The problem is how to calculate the driving routes of electric vehicles so that they will arrive at charging points at times at which there is a minimal probability of breakdowns. The second aspect of this problem to be solved is that the designated routes are supposed to ensure the minimum energy expenditure that is needed for the vehicles to complete the tasks assigned. The developed method is based on two heuristic algorithms, i.e., the ant algorithm and genetic algorithms. These algorithms work in a hybrid combination, i.e., the ant algorithm generates the initial population for the genetic algorithm. An important element of this method is the decision-making model for defining the driving routes of electric vehicles with various restrictions, e.g., their battery capacity or the permissible risk of charging point breakdown along the routes of the vehicles. The criterion function of the model was defined as the minimisation of the energy expenditure needed by the vehicles to perform their transport tasks. The method was verified against real-life data, and its effectiveness was confirmed. The authors presented a method of calibrating the developed optimisation algorithms. Theoretical distributions of the probability of charging point failure were determined based on the Statistica 13 program, while a graphical implementation of the method was carried out using the PTV Visum 23 software.

Keywords: electric vehicles; energy expenditure; ant algorithm; genetic algorithm

1. Introduction

The use of ecological means of transport characterized by low costs and levels of emissions of harmful exhaust gases, such as electric and hybrid cars, is of key importance for developing various modes of transport. Electric mobility seems to be one of the best solutions, enabling transport development without endangering future generations and the environment [1]. Many studies indicate the efficiency of electric cars [2] as they have zero pollutant emissions, which is a huge advantage when using them in populated cities [3]. Due to restrictions on the entry of conventional vehicles into city centres, service companies are analysing the possibility of introducing electric vehicles. Compared to conventional vehicles, electric vehicles require constant recharging due to their limited battery capacity. This involves developing effective vehicle routes that include loading and unloading points and battery charging stations [4].

The research carried out in this article focuses on the subject of calculating the energy expenditure of electric vehicles performing transport tasks in municipal service companies, such as municipal utilities [5] or courier companies [6]. These types of issues are very challenging to analyse due to the complexity of the tasks—customer service in densely built-up areas and the occurrence of breakdowns or lack of energy supply at charging points. The transport task for these types of companies was defined in two ways as part of the research presented herein. In the first case, it is the driving route of the electric vehicle from the base to the first loading point, the route between the loading points carried out until the entire capacity of the vehicle is used, and the driving route to the collective unloading point. In the second way of defining the transport task, the electric vehicle starts the route from the first loading point directly, after leaving the collective unloading point.

The main objective of our research was to develop a method for calculating the energy expenditure of electric vehicles that takes into account the uncertainty [7,8] of charging point operation. The developed method consists of a decision-making model calculating the energy expenditure of vehicles over their routes and optimisation algorithms minimising this consumption. The uncertainty of charging point operation was presented as the probability of a random situation hindering the functioning of this point, e.g., a failure occurring or a lack of energy supply. The probability of emergencies hindering the operation of the charging point, e.g., a failure, was determined based on a defined random variable, which is the moment of the occurrence of an undesirable event in the operation of the charging station. This probability is known from the perspective of the operation of the charging station. Taking into account the fact that the research undertaken aims to minimize the energy expenditure of electric vehicles, which depends on the designated driving route of these vehicles, the uncertainty of the operation of charging stations is seen from the point of view of electric vehicles performing transport tasks, not from the point of view of the operation of the station itself. For the electric vehicles, the moment a vehicle appears at the charging station determines the probability of the station failing. The moments when the vehicles appear at stations are not known until the driving routes of these vehicles are determined. Without designated vehicle driving routes, it is impossible to determine the probability of adverse events at each charging point seen from the perspective of an electric vehicle, not the station itself. Therefore, the authors use the concept of the uncertainty of charging point operation. The uncertainty of station operation seen from the vehicle's point of view is a measurable value and is defined by the probability of an emergency occurring at the station. This probability can be determined after determining the vehicle's driving routes.

Bearing in mind the complexity of the problem of calculating energy expenditure and taking into account the uncertainty of charging point operation, the authors developed an original decision-making model and solved it using artificial intelligence algorithms. In the driving route model, the routes of electric vehicles are calculated in such a way that they appear at charging points at times when the probability of a point breakdown is minimal. These routes are supposed to entail the minimum energy expenditure required by vehicles to carry out the transport tasks assigned. In order to control the level of uncertainty in the functioning of the charging point, the decision model was developed to limit the permissible level of the risk of charging point breakdown along vehicle driving routes. A function minimising the energy expenditure of all electric vehicles executing transport tasks was defined as the criterion function. In addition to the typical limitations encountered in this type of model, e.g., the time for the vehicles to complete tasks, the model takes into account the limitations typical of electric vehicles, e.g., the maximum distance covered without charging the battery.

Two optimisation algorithms were used to minimize the energy expenditure of electric vehicles, i.e., the ant algorithm and the genetic algorithm. These algorithms display different modes of operation, so they were mutually used for the verification of our results. The specificity of the genetic algorithm operation requires generating initial solutions, which are modified in further iterations of this algorithm. In classic approaches, initial solutions are generated randomly. In this study, a combination of the ant and genetic algorithm was employed, which created a hybrid genetic algorithm. The initial population for the genetic algorithm was generated by the ant algorithm. Genetic or ant algorithms are often used and recommended for calculating vehicle driving routes [9–11]. Their main advantage is the short amount of time needed to generate solutions.

The developed method was verified against real-life data, and its effectiveness was confirmed. The article presents a method of calibrating the developed optimisation algorithms. The theoretical distributions of the probability of charging point breakdown were determined based on the Statistica 13 program, whereas a graphical implementation of the method was carried out using the PTV Visum software.

The article proposes a new approach to determining the driving routes of electric vehicles in urban service enterprises, taking into account the uncertainty of the operation of vehicle battery charging stations. Determining driving routes for electric vehicles is a complex optimization problem based on the VRP (vehicle routing problem) with pick-up and delivery [12]. In this problem, vehicles visit individual customers, pick up or deliver the load, and return to the collection point. The authors focused on determining the energy expenditure of electric vehicles by using a hybrid genetic algorithm. A literature analysis has shown that using electric vehicles in urban service enterprises, e.g., courier companies or municipal enterprises, is still a big challenge for researchers.

The research in this work enabled the development of a new decision-making model defining the energy expenditure of electric vehicles performing transport tasks in the context of the operating uncertainty of battery charging stations and the development of a practical algorithm minimizing this expenditure.

This study is divided into five sections. The first section describes the research problem and the purpose of the research. The second section indicates the research gap and our premise to address this subject. The decision model calculating the energy expenditure of electric vehicles is presented in the third section, while the optimisation algorithms employed to minimize the expenditure are presented in the fourth section. The verification of the developed method against real-life data is described in the fifth section.

2. Analysis of the Literature

Electric vehicles play a vital role in transportation companies due to the development of sustainable transport systems [2,13,14]. In these companies, minimising the energy expenditure of electric vehicles involves calculating the shortest driving routes for vehicles executing transport tasks. In the work of [15], the authors determined the energy expenditure in municipal enterprises based on a decision-making model in which the criterion function was defined as the amount of energy needed to overcome rolling forces and aerodynamic resistance. An ant algorithm and a genetic algorithm were used to calculate the shortest driving routes of vehicles. A verification of the algorithms confirmed their high level of effectiveness.

The problem of minimising the energy expenditure of electric vehicles belongs to the group of problems described in the literature as electric vehicle routing problems (E-VRPs) [16–18]. This problem is a version of the vehicle routing problem (VRP) extended to issues related to electric vehicles. The classic vehicle routing problem consists in designating optimal transportation routes for a strictly defined number of vehicles, which are tasked with serving a set of customers located at various points, while adhering to a series of constraints, such as vehicle capacity and task completion time [19]. Various types of this problem occur, such as the capacitated VRP (CVRP), which incorporates vehicle carrying capacity constraints, the VRP with time windows (VRPTW), and the VRP with pickup and delivery (VRPPD). The vehicle routing problem is an NP-hard optimisation problem. Heuristic algorithms, primarily genetic [20] or ant algorithms, are commonly applied to solve it [21,22].

The electric vehicle routing problem, just like the classic VRP, has various variants and draws upon the travelling salesman problem [23,24]. Setting routes for electric vehicles depends on the location and number of available charging points. In the course of this study, mathematical models were developed to set vehicle routes, taking into account the location of both individual charging points [25] and multiple ones [26–28]. The time windows for customer service by electric vehicles have been described in the works of [29–31].

The problem of minimising the energy expenditure of electric vehicles is a complex decision-making problem, and therefore, artificial intelligence algorithms are used to solve it. In the work of [32], a novel energy management strategy based on machine learning was proposed for a hybrid electric bus. On the other hand, in the work of [33], a battery charging strategy was defined while meeting the operational constraints of vehicles by the consideration of dynamic programming and a genetic algorithm. The particle swarm optimisation algorithm for energy management in plug-in hybrid electric vehicles (PHEVs) was presented in the publication of [34]. Their verification against real-life data confirmed its high level of effectiveness. A comprehensive analysis of artificial intelligence algorithms in managing the efficiency of battery consumption in electric vehicles was presented in the work of [35]. Machine learning algorithms based on a feed forward neural network (FFNN) for diagnosing defects in electric vehicle batteries were presented in the work of [36]. A neuro-fuzzy inference system for estimating the state of charge of lithium-ion batteries was presented in the work of [37]. Advanced research on optimal energy management in electric vehicles using machine learning techniques was presented in the publications of [38–42]. In the works of [43,44], the classic electric vehicle routing problem was solved using the simulated annealing algorithm. The ant algorithm was applied to the same problem in the paper of [45], while the genetic algorithm was used in the paper of [46,47].

After a literature analysis, it can be concluded that energy expenditure is minimized in two ways. The first approach involves optimal energy management and its appropriate selection suited to the prevailing conditions along the route. In the second case, minimal driving routes for vehicles are set in order to reduce the vehicle's involvement in performing transportation tasks. Considering that the energy expenditure of electric vehicles in service companies depends on the routes travelled by these vehicles, and that heuristic algorithms are recommended and commonly used for vehicle routing problems, two algorithms were applied to minimize expenditure: the ant algorithm and the genetic algorithm. These algorithms are effective optimisation algorithms that are commonly applied in complex decision-making problems. Our literature analysis also confirmed the lack of publications regarding the coverage of regions by electric vehicles with the consideration of the reliability of vehicle charging points.

3. Decision-Making Model Determining the Energy Expenditure of Electric Vehicles

3.1. Assumptions

An essential aspect in determining the energy expenditure of electric vehicles in municipal service companies, considering the uncertainty of charging point operation, is to define routes characterized by the minimum distance travelled to accomplish all transportation tasks. Additionally, an electric vehicle performs battery-charging operations when the probability of a charging point breakdown or unavailability due to energy supply shortages is minimal. It was also assumed that the minimisation of the distance travelled along vehicle routes is proportional to the minimisation of energy expenditure. The minimisation of the energy expenditure of electric vehicles in municipal service companies was adopted as the quality assessment of the solution to the decision-making model for setting the minimum driving routes for electric vehicles. The faulty process of the operation of vehicle charging points and their unavailability on a given workday were presented using theoretical distribution models for the probability of the uncertainty of charging point operation. The random variable representing the probability distribution of the uncertainty of charging point operation was defined as the moment of point malfunctions, power supply interruptions, or any other random situations affecting its correct operation. Therefore, the routes of electric vehicles should be set in such a way that the probability of vehicles appearing at charging points during the time of their unavailability is below the permissible level of risk. In the model, it was assumed that there are no delays over sections of the transport network due to traffic congestion, and electric vehicles move along the designated lanes of the road. Additional assumptions of the model include the following [15]:

- The volume of the cargo in the given loading places is not greater than the capacity of the vehicle.
- A battery charging station can be located either at transportation bases or at particular places of a transport network.
- The criterion function determining the energy consumption takes into account the energy required to overcome air resistance and rolling resistance arising from road surface resistance. It was assumed that the vehicle travels over route sections in a uniform motion, and the forces generating vehicle acceleration were disregarded. It was assumed that the vehicle travels over a flat surface, and the resistance forces arising from overcoming inclined route sections were disregarded;
- The electric vehicle departs from the transportation base with a 100% charged battery;
- The transportation task was defined in two ways: (1) the departure of a vehicle from the base to the loading points and, after filling up its entire capacity, the transportation of its cargo to the unloading point; or (2) the departure of the vehicle from the unloading point, which loads its cargo at the loading points and returns to the unloading point;
- In a single transportation task, it was assumed that the electric vehicle may charge its battery only once due to the vehicle's capacity limitation;
- The trip from the unloading point to the transportation base was not included in the model. The route of this trip does not contain loading points and is not determined.

3.2. Input Data of the Decision-Making Model and Decision Variables

In the model, there are several types of data, e.g., sets and parameters. The sets include the identification of characteristic types of points, such as the starting and ending points of vehicle work, intersections, charging points, and vehicle types, as well as the tasks being performed. A summary of the sets is presented in Table 1. In addition, the parameters characteristic of the transportation network sections were defined, e.g., the distances between points in the network, time required for loading, unloading, and travel, and those taking into account the risk of damage to the vehicle charging points. A summary of the parameters is presented in Table 2.

Table 1. The sets used in the decision model.

The Elements of Sets	The Description
b	Transport base number, $b \epsilon \boldsymbol{B}$
pz	Loading point number, $pz \epsilon \boldsymbol{PZ}$
p	Intermediate point number, e.g., intersection, $p \epsilon \boldsymbol{P}$
s	Charging point number, $s \epsilon \boldsymbol{S}$
pw	Unloading point number, $pw \epsilon \boldsymbol{PW}$
v	Electric vehicle number, $v \epsilon \boldsymbol{V}$
z	Transport task number, $z \epsilon \boldsymbol{Z}$

Table 2. The parameters used in the decision model.

The Parameters	The Description
tz^{pz}	Loading time at the loading point
$t1^{v,b,p}$	Driving time between the base and an intermediate point
$t2^{v,p,p'}$	Driving time between intermediate points
$t3^{v,p,pw}$	Driving time between an intermediate point and unloading point
$t4^{v,pw,p}$	Driving time between an unloading point and intermediate point
$s1^{b,p}$	Distance between the base and an intermediate point
$s2^{p,p'}$	Distance between intermediate points
$s3^{p,pw}$	Distance between an intermediate point and unloading point
$s4^{pw,p}$	Distance between an unloading point and intermediate point

Table 2. *Cont.*

The Parameters	The Description
c^v	Vehicle capacity
$q1^{pz,b,p}$	Cargo volume at a loading point between the base and intermediate point
$q2^{pz,p,p'}$	Cargo volume at a loading point between intermediate points
$q3^{pz,p,pw}$	Cargo volume at a loading point between an intermediate point and unloading point
$q4^{pz,pw,p}$	Cargo volume at a loading point between an unloading point and intermediate point
T^{max}	Maximum time to complete a task by a vehicle
ΔT	Operating time of the unloading point
d^v	Permissible distance after which the electric vehicle heads to the charging point
$t^{v,s}$	Charging time for the electric vehicle at a charging point
$pE^{s,t}$	Probability of the uncertainty of charging point operation, where $E^{s,t}$ is a random variable determining the moment t of a charging point breakdown, etc.
R	Permissible risk of uncertainty in the operation of the charging point along the entire vehicle route
$m1^{v,b,p}$	Vehicle weight between the base and intermediate point
$m2^{v,p,p'}$	Vehicle weight between intermediate points
$m3^{v,p,pw}$	Vehicle weight between an intermediate point and unloading point
$m4^{v,pw,p}$	Vehicle weight between an unloading point and intermediate point
ρ	Air density $\rho = 1.225\ \text{kg/m}^3$
μ	The coefficient of road friction $\mu = 2.02 \times 10^{-5}$
cx	The drag coefficient cx = 0.2
A	The vehicle frontal area A = 20 m^2

The energy expenditure depends on the length of the route covered by a vehicle. It is necessary to set vehicles' routes in such a way as to minimize their expenditure. The energy expenditure is determined by the decision variables determining the route of electric vehicles, which take the following form:

- $x1^{v,b,p,z}$—a transport link carried out by an electric vehicle between the base and an intermediate point in a given transportation task, if $x1^{v,b,p,z} = 1$ is a transport link;
- $x2^{v,p,p',z}$—a transport link carried out by an electric vehicle between intermediate points in a given transport task, if $x2^{v,p,p',z} = 1$ is a transport link;
- $x3^{v,p,pw,z}$—a transport link carried out by an electric vehicle between an intermediate point and unloading point in a given transport task, if $x3^{v,p,pw,z} = 1$ is a transport link;
- $x4^{v,pw,p,z}$—a transport link carried out by an electric vehicle between an unloading point and intermediate point in a given transportation task, if $x4^{v,pw,p,z} = 1$ is a transport link.

In addition, a decision variable representing the interpretation of the choice of a given charging point by the electric vehicle was introduced: if $x^{v,s,z} = 1$, then a given charging point is used by a given vehicle for a given task.

3.3. Model Limitations

The following limitations were taken into account in the decision model:

- not exceeding the maximum time of task completion by an electric vehicle. The task type: bases—loading points—unloading points is shown in Formula (1). The task type: unloading points—loading points—unloading points is shown in Formula (2) (***LPP***—a set of transport links between intermediate points, and n—the number of loading points in the task).

$$\forall\ v\epsilon \boldsymbol{V},\ b\epsilon \boldsymbol{B},\ p\epsilon \boldsymbol{P},\ pz\epsilon \boldsymbol{PZ},\ z\epsilon \boldsymbol{Z},\ s\epsilon \boldsymbol{S},\ pw\epsilon \boldsymbol{PW}$$

$$x1^{v,b,p,z}\cdot t1^{v,b,p} + \sum_{(p',p'')\in \boldsymbol{LPP}} x2^{v,p',p'',z}\cdot t2^{v,p',p''} + x3^{v,p,pw,z}\cdot t3^{v,p,pw} + x^{v,s,z}\cdot t^{v,s} + n\cdot tz^{pz} \le T^{max} \quad (1)$$

$$x4^{v,pw,p,z} \cdot t4^{v,pw,p} + \sum_{(p',p'')\in \boldsymbol{LPP}} x2^{v,p',p'',z} \cdot t2^{v,p',p''} + x3^{v,p,pw,z} \cdot t3^{v,p,pw} + x^{v,s,z} \cdot t^{v,s} + n \cdot tz^{pz} \le T^{max} \tag{2}$$

- not exceeding the operating time of the unloading point by a single vehicle ($T1^{z,v}$—the completion time of the type one task calculated from the left part of the inequality (1); $T2^{z,v}$—the completion time of the type two task calculated from the left part of the inequality (2)):

$$\forall\, v \epsilon \boldsymbol{V}$$

$$\sum_{z\in \boldsymbol{Z}} T1^{z,v} + \sum_{z\in \boldsymbol{Z}} T2^{z,v} \le \Delta T \tag{3}$$

- not exceeding the permissible risk related to the uncertainty of the charging point's operation (damage to the charging point, lack of energy supply, etc.) along each electric vehicle route:

$$\forall\, v\ \epsilon \boldsymbol{V}$$

$$\prod_{s\in \boldsymbol{S}} \prod_{z\in \boldsymbol{Z}} x^{v,s,z} \cdot pE^{s,t} \le R \tag{4}$$

where t is the time of arrival of the electric vehicle at a given charging point.

- not exceeding the maximum distance after which the electric vehicle should be routed to the charging point in a type one task (5) and in a type two task (6):

$$\forall\, v\epsilon \boldsymbol{V},\ b\epsilon \boldsymbol{B},\ p\epsilon \boldsymbol{P},\ pz\epsilon \boldsymbol{PZ},\ z\epsilon \boldsymbol{Z},\ pw\epsilon \boldsymbol{PW}$$

$$x1^{v,b,p,z} \cdot s1^{b,p} + \sum_{(p',p'')\in \boldsymbol{LPP}} x2^{v,p',p'',z} \cdot s2^{p,p'} + x3^{v,p,pw,z} \cdot s3^{p,pw} \le d^{v} \tag{5}$$

$$x4^{v,pw,p,z} \cdot s4^{pw,p} + \sum_{(p',p'')\in \boldsymbol{LPP}} x2^{v,p',p'',z} \cdot s2^{p,p'} + x3^{v,p,pw,z} \cdot s3^{p,pw} \le d^{v} \tag{6}$$

- volumetric limitation of the vehicle when collecting cargo from the loading points in a type one task (7) and in a type two task (8):

$$\forall\, v\epsilon \boldsymbol{V},\ b\epsilon \boldsymbol{B},\ p\epsilon \boldsymbol{P},\ pz\epsilon \boldsymbol{PZ},\ z\epsilon \boldsymbol{Z},\ pw\epsilon \boldsymbol{PW}$$

$$x1^{v,b,p,z} \cdot q1^{pz,b,p} + \sum_{(p',p'')\in \boldsymbol{LPP}} x2^{v,p',p'',z} \cdot q2^{pz,p',p''} + x3^{v,p,pw,z} \cdot q3^{pz,p,pw} \le c^{v} \tag{7}$$

$$x4^{v,pw,p,z} \cdot q4^{pz,pw,p} + \sum_{(p',p'')\in \boldsymbol{LPP}} x2^{v,p',p'',z} \cdot q2^{pz,p',p''} + x3^{v,p,pw,z} \cdot q3^{pz,p,pw} \le c^{v} \tag{8}$$

3.4. Criterion Function

The criterion function minimises the energy expenditure of all vehicles carrying out transportation tasks. This function takes into account the energy expenditure needed to overcome road resistance forces and aerodynamic forces. The components of the formula include the following (9): (10)—the expenditure for the sections between the base and the intermediate point, where ***LBP***—a set of transport links between the base and the intermediate point; (11)—the expenditure for the sections between the intermediate points; (12)—the expenditure for the sections between the intermediate point and the unloading point, where ***LPW***—a set of transport links between the intermediate point and the unloading point; and (13)—the expenditure for the sections between the unloading point and the intermediate point, where ***LWP***—a set of transport links between the unloading point and the intermediate point:

$$F = F1^{v,b,p,z} + F2^{v,p,p',z} + F3^{v,p,pw,z} + F4^{v,pw,p,z} \rightarrow min \tag{9}$$

$$F1^{v,b,p,z} = \sum_{v\in \mathbf{V}} \sum_{(b,p)\in \mathbf{LBP}} \sum_{z\in \mathbf{Z}} x1^{v,b,p,z} \cdot s1^{b,p} \cdot \left(F_{air}{}^{v,b,p,z} + F_r{}^{v,b,p,z} \right) \tag{10}$$

$$F2^{v,p,p',z} = \sum_{v\in \mathbf{V}} \sum_{(p,p')\in \mathbf{LPP}} \sum_{z\in \mathbf{Z}} x2^{v,p,p',z} \cdot s2^{p,p'} \cdot \left(F_{air}{}^{v,p,p'z} + F_r{}^{v,p,p',z} \right) \tag{11}$$

$$F3^{v,p,pw,z} = \sum_{v\in \mathbf{V}} \sum_{(p,pw)\in \mathbf{LPW}} \sum_{z\in \mathbf{Z}} x3^{v,p,pw,z} \cdot s3^{p,pw} \cdot \left(F_{air}{}^{v,p,pw,z} + F_r{}^{v,p,pw,z} \right) \tag{12}$$

$$F4^{v,pw,p,z} = \sum_{v\in \mathbf{V}} \sum_{(pw,p)\in \mathbf{LWP}} \sum_{z\in \mathbf{Z}} x4^{v,pw,p,z} \cdot s4^{pw,p} \cdot \left(F_{air}{}^{v,pw,p,z} + F_r{}^{v,pw,p,z} \right) \tag{13}$$

where:

- The air resistance force is as follows:

$$F_{air}{}^{v,b,p,z} = \frac{1}{2} \cdot \rho \cdot cx \cdot A \cdot \left(\frac{s1^{b,p}}{t1^{v,b,p}} \right)^2 \tag{14}$$

- The rolling resistance force is as follows:

$$F_r{}^{v,b,p,z} = \mu \cdot m1^{v,b,p} \cdot g \cdot \cos\alpha \tag{15}$$

4. Algorithms Minimising the Energy Expenditure of Electric Vehicles

4.1. Ant Algorithm

Bearing in mind the fact that the energy expenditure of electric vehicles depends on the length of the driving routes of these vehicles, it is important to determine the vehicle driving routes that are characterised by the minimum distance travelled for cargo pickup. In addition, these routes should comply with the limitations of the developed decision-making model. The ant algorithm uses the so-called artificial ants to set the driving routes of electric vehicles. An ant creates its own route (Figure 1), which is the driving route of all the vehicles completing transportation tasks.

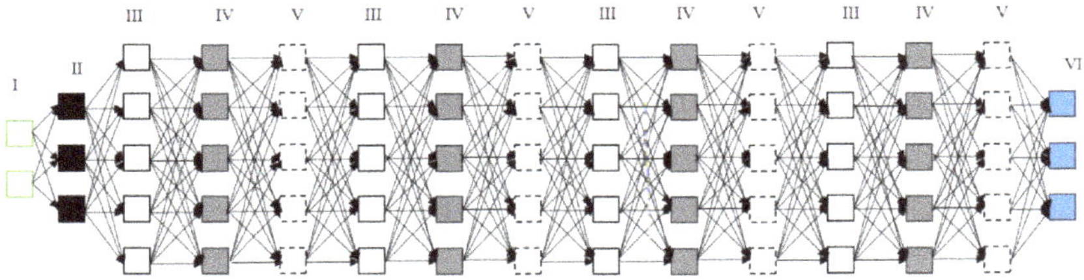

Figure 1. The elements of the ant's route.

Six types of layers were distinguished in the ant route as follows: layer I represents the transport base from which the electric vehicles depart to the initial loading points; layer II represents the points determining the electric vehicles that are assigned to tasks; layer III represents the intermediate points of the ant's route, i.e., intersections; layer IV represents electric vehicle charging points; layer V represents loading points; and layer VI represents collective unloading points. The number of intermediate points along an ant's route, the number of points interpreted as a charging point, and the number of unloading points are known only after the ant has created the entire route. The frequency of unloading points along the ant's route depends on the size of the collected cargo and the capacity of the electric vehicles. On the other hand, the frequency of the occurrence of points interpreted as a charging point depends on the length of the route generated by the ant. The points from layer I in the ant's route appear when the ant leaves the transport base, and the number of loading points is known. The ant returns to the base when all loading points have been visited, or when the operating time of the unloading point is exceeded (restriction (3)). In the event that all loading points are not visited, the ant sets off from the base again to visit these points. The process of selecting further route points by the ant is repeated until all the loading points have been visited. In any case, when the ant leaves the base, the moment of departure of the ant from the base is always determined as the moment (time) of commencing operation in the service company. The departure of the ant from the base represents the departure of another vehicle to perform tasks. An example of an ant's route is shown in Figure 2. Along this route, the ant has set driving routes for two electric vehicles.

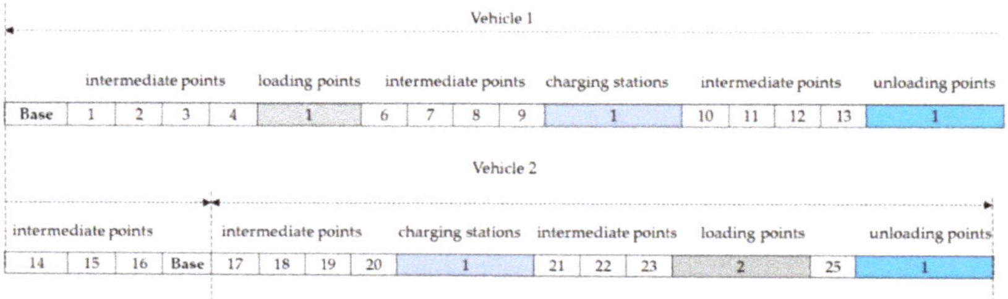

Figure 2. An example of an ant's route.

The starting point of each ant's route is the transport base from which the vehicles depart to the loading points. The ant proceeds to the following route points in accordance with the established probability.

$$PR^{mr}{}_{ab}(t) = \begin{cases} \frac{[\tau_{ab}(t)]^{\alpha}\cdot[\eta_{ab}(t)]^{\beta}}{\sum_{l\in\Omega^{mr}}[\tau_{al}(t)]^{\alpha}\cdot[\eta_{al}(t)]^{\beta}} & b\in\Omega^{mr} \\ 0 & b\notin\Omega^{mr} \end{cases} \tag{16}$$

where:

$\tau_{ab}(t)$—the intensity of the pheromone trace between the a-th point of the ant's route and the b-th point in the t-iteration of the algorithm implemented by the mr ant;

$\eta_{\text{ab}}(t)$—heuristic information:

$$\eta_{a,b}(t) = \frac{1}{d1(a,b) + d2(a,pz)} \tag{17}$$

where:

$d1(a,b)$—the distance between the selected points of the ant's route;

$d2(a,pz)$—the distance between the selected point of the ant's route and the loading point;

α, β—the influence of pheromones and heuristic data on the behaviour of the ants;

Ω^{mr}—a set of all the elements of the ant's route;

l—potential points of the ant's route that are taken into account when choosing the next point of the ant's route.

The heuristic information takes into account both the distance between the route points $d1(a,b)$ and the distance between the route point and the loading point $d2(a,pz)$. The parameter $d2(a,pz)$ was introduced in order to guide the ant to a given loading point to avoid ant accidentally wandering through all the points of its route. The distance $d2(a,pz)$ is the Euclidean distance calculated based on the location coordinates of the midpoints in the ant's route and each loading point. In the event that the ant proceeds to a charging point, base, or unloading point, the Euclidean distances are determined between these points and the intermediate points of the ant's route.

Each ant in the population determines its individual route. After completing these routes, the pheromone trace along them is updated. At the beginning, it is assumed that the trace on the transport links between the route points is equally strong. In the next iterations, the pheromone trace is calculated according to the following formula:

$$\tau_{a,b}(t+1) = (1-\rho)\tau_{a,b}(t) + \sum_{mr\epsilon \boldsymbol{MR}} \Delta\tau_{a,b}{}^{mr}(t) \tag{18}$$

The first component of formula (18) determines the pheromone volatilisation rate, while the second one determines the pheromone amplification and takes on a specific value when the segment (a,b) is used by the ant; otherwise, it is 0, i.e.:

$$\Delta\tau_{a,b}{}^{mr}(t) = \begin{cases} \frac{1}{L^{mr}(t)} - K1^{mr}(t) - K2^{mr}(t) \\ 0 \end{cases} \tag{19}$$

where:

$L^{mr}(t)$—the length of the entire route determined by the ant in a given iteration;

$K1^{mr}(t)$—a penalty for exceeding the maximum time of task implementation by an electric vehicle in the designated routes (1) and (2). It is assumed that this penalty is half of the pheromone accumulated along the route;

$K2^{mr}(t)$—a penalty for exceeding the permissible risk related to the uncertainty of charging point operation by the electric vehicle along the designated route (4). It is assumed that this penalty is half of the pheromone accumulated along the route;

The ant algorithm is an iterative algorithm and works until a fixed number of iterations has been reached. The algorithm diagram is shown in Figure 3. The number of ants in the population and the number of iterations are determined at the beginning of the algorithm's implementation. The operation of the ant algorithm is presented in the following steps:

- Step 1. An ant in the population randomly selects the transport base from which it starts its route, and a vehicle number is assigned to the ant;
- Step 2. The selection of the loading point located nearest to the transport base. The movement of the ant to the intermediate points is shown in Formula (16). After the ant arrives at the loading point, it picks up the load and selects the next closest loading point. The ant collects its cargo until the vehicle capacity is exceeded (limitation (7)). After exceeding this capacity, the ant goes to the nearest unloading point. During each passage of the ant to the subsequent points, it is checked whether the distance limit is exceeded, after which the electric vehicle should be directed to the charging point (5);
- Step 3. After unloading, the ant goes to the next closest loading point, taking into account the volumetric limit (8). The cargo collection process is carried out until the electric vehicle (3) exceeds the maximum time for completing tasks or until all loading points have been visited. After exceeding this limit, the ant returns to the base and then step 1 is implemented again. The ant is assigned to another electric vehicle of a given capacity. The departure of each ant from the base takes place the moment the service company starts work;
- Step 4. Repetition of steps 1–3 to complete all loading points;
- Step 5. Another ant from the population sets its own route, repeating steps 1–4 until the routes have been completed by all the ants in the population;
- Step 6. Updating the pheromone according to Formula (18);
- Step 7. Repeat steps 1–6 until the stop condition is reached;
- Step 8. The route of the ant with the highest pheromone value among all the routes generated in the population is the final solution.

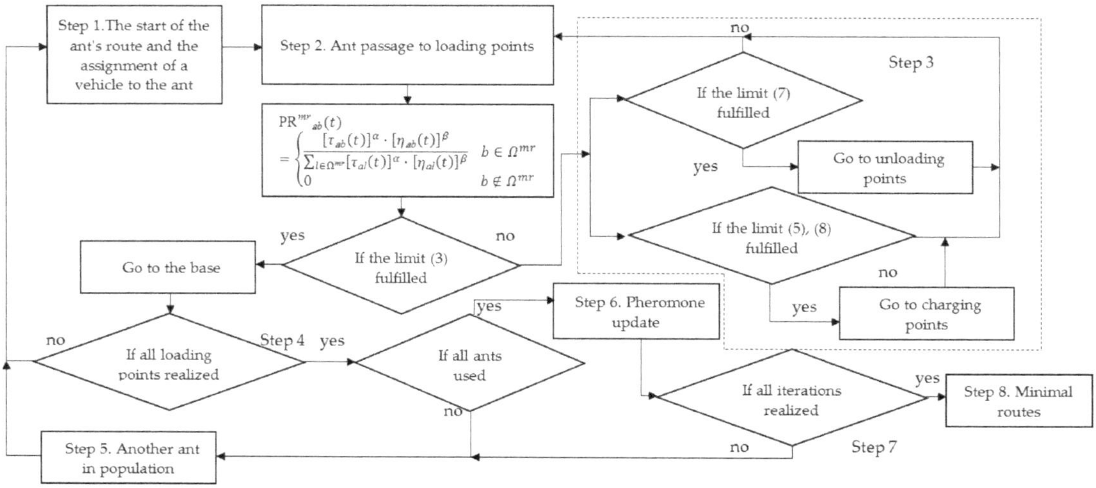

Figure 3. Diagram of the ant algorithm's operation.

4.2. Hybrid Genetic Algorithm

The first step in the operation of the genetic algorithm is to generate the initial population of the algorithm. The initial population consists of chromosomes representing the driving routes of electric vehicles. The initial population was established by the ant algorithm. The following stages were distinguished in the genetic algorithm: chromosome evaluation, reproduction, and crossbreeding. Chromosome mutation is not carried out due to its negative impact on the correctness of the generated solutions. The evaluation of the chromosome is carried out on the basis of the adaptation function. The best chromosomes with the greatest adaptation function are multiplied in the reproduction process. Reproduction generates new populations according to the principle of the roulette method [15,48], in which linear scaling was used to counteract the premature convergence of the algorithm in initial iterations. The algorithm for selecting chromosomes for crossbreeding randomly pairs two chromosomes and randomly selects chromosome cut-points. The adaptation function for the k-th M chromosome (t, k) can be presented as follows (where $K = \{1, \ldots, k, \ldots, K\}$—a set of chromosomes in the population, and t—iteration of the algorithm):

$$F(k,t) = \frac{1}{L(k,t)} - K1(k,t) - K2(k,t) \rightarrow \max \tag{20}$$

where:

$L(k,t)$—the length of the entire chromosome route in a given iteration;

$K1(k,t)$—the penalty for exceeding the maximum time for the implementation of the task by an electric vehicle in chromosomes (1) and (2). It is assumed that the penalty for exceeding this is that the generated value of the adaptation function for the chromosome is reduced by half;

$K2(k,t)$—the penalty for exceeding the acceptable risk associated with the uncertainty of charging point operation by an electric vehicle in chromosome (4); it is assumed that the penalty for exceeding this is that the generated value of the adaptation function for the chromosome is reduced by half.

The process of chromosome crossbreeding is shown in Figure 4 and involves the exchange of gene sequences between two chromosome cut-points. In both chromosomes, the cut-points must be identical.

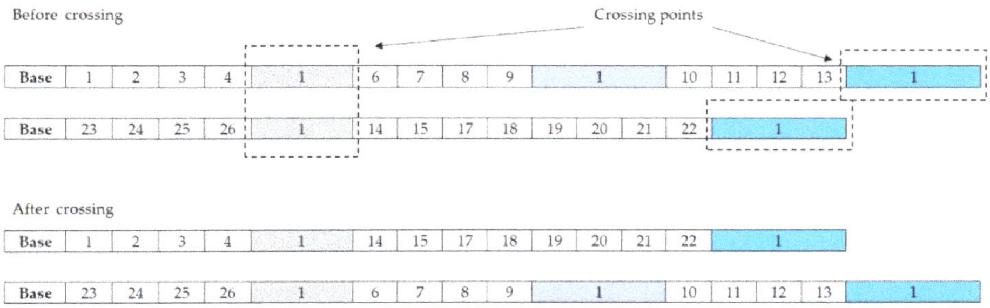

Figure 4. Chromosome crossbreeding.

The hybrid genetic algorithm is an iterative algorithm, and its solution is improved in subsequent iterations. The principle of the algorithm is presented in the following steps:

- Step 1. Generation of the initial population by the ant algorithm;
- Step 2. Calculation of the adaptation function for each chromosome;
- Step 3. Reproduction;
- Step 4. Crossbreeding;
- Step 5. Repeat steps 2–4 until the stop condition is reached;
- Step 6. The chromosome with the highest value of adaptation function in the population is the final solution.

5. The Case Study

5.1. Input Data

A service company carries out the collection of municipal waste from a given region. It was assumed that the waste loading time at each loading point is 5 min, and the unloading time at the unloading points was also set at 5 min. The company owns five DAF CF electric vehicles with a gross permissible weight of 28 tons, load-carrying capacity of 3 tons, battery capacity of 170 kWh, maximum driving distance without charging of 70 km in urban conditions, and charging time of 1 h. The permissible duration of a single task combined with the battery charging time is 3 h. The maximum operating time of the unloading point was set at 4 h. The service company starts work at 8:00 a.m. The accepted risk related to the uncertainty of charging point operation along each designated route was set at 0.40. The distances between the route points, driving times, and location coordinates of all points were generated using the PTV Visum software. The volume of cargo at the loading points is shown in Table 3.

Table 3. The size of the waste.

No.	Size (kg)	No.	Size (kg)	No.	Size (kg)
1	400	7	400	13	400
2	500	8	400	14	300
3	550	9	550	15	550
4	150	10	500	16	400
5	600	11	650	17	600
6	1450	12	450	18	450

The location of the base, waste collection points, unloading point, charging point, and linear infrastructure of the transport network are shown in Figure 5.

Figure 5. Transport network of a service company.

5.2. Distributions of the Probability of Uncertainty in Charging Point Operation

The location of the battery charging points is shown in Figure 6. Theoretical distributions of the probability of uncertainty in charging point operation were determined for these points. The time of the occurrence of malfunctions at each point is shown in Table 4. The time of damages, breakdowns, or other events that disrupt the operation of the charging point are presented in a format totalling the number of minutes, e.g., 8:20 a.m., that is 500 min, 8 times 60 min plus 20 min.

Table 4. Charging point breakdown moments (minutes).

Point 1	Point 2	Point 3	Point 4
521/516	680/707	670/693	720/751
526/550	683/711	671/694	721/752
534/551	687/717	672/695	745/760
535/552	690/723	680/696	731/761
537/556	691/-	681/697	733/726
538/557	692/-	686/698	734/727
541/558	694/-	683/702	741/736
544/561	696/-	684/703	748/737
545/563	697/-	685/704	747/738
548/567	703/-	690/711	744/739
549/572	704/-	691/-	745/-
527/539	706/-	692/-	746/-

Bearing in mind that the shapes of the histograms determining the frequency of charging point failures are similar to normal distributions, the Chi-squared test was used to determine the type of distribution. The samples are small, so in addition to the Chi-squared test, the Kolmogorov–Smirnov test was also employed. The use of the Kolmogorov–Smirnov test is only an additional step to help support the final decision regarding the type of distribution. This test detects only large deviations from the assumed distributions. The null hypothesis referring to the assumed type of distribution is rejected when the calculated value of the statistics belongs to the critical area determined by the assumed level of significance $\alpha = 0.05$ (when $p < \alpha$, p—the probability determined in the tests). The values of the Chi-squared and Kolmogorov–Smirnov concordance tests and the parameters of the distributions investigated are presented in Table 5. Adjustments of the theoretical distributions for the random variable of the time when disruptions in the operation of the charging point occur are shown in Figure 6.

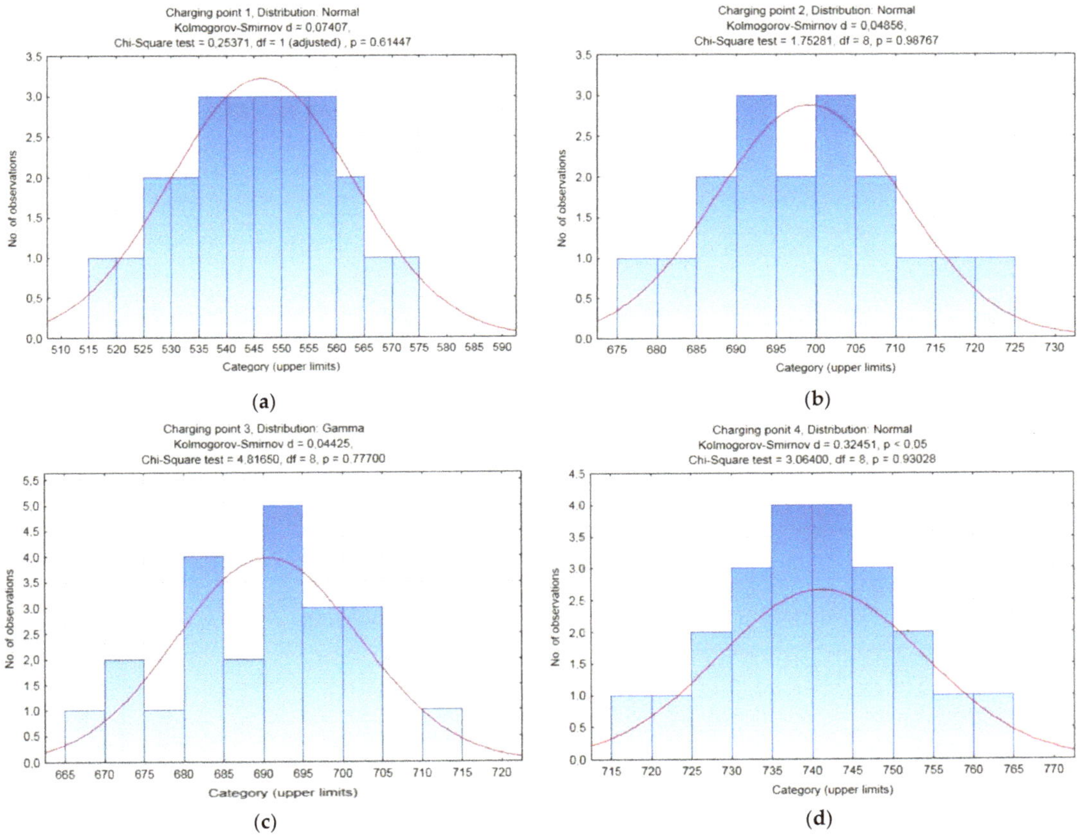

Figure 6. Theoretical distributions of the uncertainty of charging point operation, (**a**) point 1, (**b**) point 2, (**c**) point 3, and (**d**) point 4.

Table 5. Concordance tests and distribution parameters.

Charging Points	Statistic	Chi-Square Test Probability	Statistic	K-S Test Probability	Parameters of Distribution	Distribution
1	0.253	0.61	0.07	-	$\mu = 546.40$; $s^2 = 279.63$	normal
2	1.752	0.98	0.04	-	$\mu = 699.11$; $s^2 = 138.48$	normal
3	4.816	0.77	0.04	-	$\theta = 0.16$; $k = 4111.173$	gamma
4	3.064	0.93	0.32	-	$\mu = 741.11$; $s^2 = 162.23$	normal

5.3. Calibration and Verification of Algorithms

The calibration of the algorithm consists of finding the settings for the algorithm parameters that generate the best solution. The following values were adopted for the parameter α: 1, 3, 5, 10, and 20; parameter β: 0.5, 1, and 5; and parameter ρ: 0.2, 0.4, 0.6, and 0.8 [48] in order to calibrate the ant algorithm. For this number of specific parameters, 60 possible test settings were created: (5 (parameter α) × 3 (parameter β) × 4 (parameter ρ) = 60). The combinations of the ant algorithm settings that were tested are shown in Table 6. Table 7 presents a cumulative summary of the ant algorithm results for each test, i.e., the value of the energy expenditure for all vehicles (kWh), the number of vehicles performing tasks (V), and the breakdown probability for the charging points determined collectively along all routes (P). The number of iterations of the algorithm was set at 200 repetitions and the population size was set at 50 ants in an experimental way.

Table 6. Test settings of the ant algorithm parameters.

Test	α	β	ρ	Test	α	β	ρ	Test	α	β	ρ
1	1	0.5	0.2	21	1	1	0.2	41	1	5	0.2
2	1	0.5	0.4	22	1	1	0.4	42	1	5	0.4
3	1	0.5	0.6	23	1	1	0.6	43	1	5	0.6
4	1	0.5	0.8	24	1	1	0.8	44	1	5	0.8
5	3	0.5	0.2	25	3	1	0.2	45	3	5	0.2
6	3	0.5	0.4	26	3	1	0.4	46	3	5	0.4
7	3	0.5	0.6	27	3	1	0.6	47	3	5	0.6
8	3	0.5	0.8	28	3	1	0.8	48	3	5	0.8
9	5	0.5	0.2	29	5	1	0.2	49	5	5	0.2
10	5	0.5	0.4	30	5	1	0.4	50	5	5	0.4
11	5	0.5	0.6	31	5	1	0.6	51	5	5	0.6
12	5	0.5	0.8	32	5	1	0.8	52	5	5	0.8
13	10	0.5	0.2	33	10	1	0.2	53	10	5	0.2
14	10	0.5	0.4	34	10	1	0.4	54	10	5	0.4
15	10	0.5	0.6	35	10	1	0.6	55	10	5	0.6
16	10	0.5	0.8	36	10	1	0.8	56	10	5	0.8
17	20	0.5	0.2	37	20	1	0.2	57	20	5	0.2
18	20	0.5	0.4	38	20	1	0.4	58	20	5	0.4
19	20	0.5	0.6	39	20	1	0.6	59	20	5	0.6
20	20	0.5	0.8	40	20	1	0.8	60	20	5	0.8

Table 7. Expenditure, number of vehicles and breakdown probability for test settings.

Test	(kWh)	(V)	(P)	Test	(kWh)	(V)	(P)	Test	(kWh)	(V)	(P)
1	489	2	0.12	21	523	2	0.04	41	398	2	0.06
2	477	2	0.20	22	500	2	0.05	42	467	2	0.09
3	643	2	0.12	23	380	2	0.04	43	395	2	0.10
4	386	2	0.12	24	410	2	0.07	44	465	2	0.09
5	399	2	0.23	25	541	2	0.08	45	460	2	0.11
6	603	2	0.14	26	534	2	0.11	46	499	2	0.15
7	678	2	0.21	27	555	2	0.12	47	512	2	0.13
8	612	2	0.22	28	532	2	0.13	48	532	2	0.16
9	600	2	0.14	29	567	2	0.11	49	612	2	0.18
10	632	2	0.17	30	543	2	0.14	50	611	2	0.19
11	611	2	0.21	31	653	2	0.12	51	600	2	0.20
12	636	2	0.20	32	632	2	0.15	52	623	2	0.22
13	611	3	0.19	33	612	3	0.22	53	614	3	0.24
14	676	3	0.14	34	600	3	0.21	54	632	3	0.21
15	688	3	0.23	35	632	3	0.23	55	655	3	0.20
16	704	3	0.14	36	668	3	0.12	56	663	3	0.22
17	743	3	0.12	37	704	3	0.14	57	776	3	0.32
18	789	3	0.43	38	777	3	0.32	58	843	3	0.31
19	801	3	0.33	39	754	3	0.14	59	883	3	0.21
20	898	3	0.22	40	756	3	0.32	60	804	3	0.25

The smallest value of energy expenditure reaching 380 (kWh) was generated for test number 23 with the following parameters for the algorithm: $\alpha = 1$, $\beta = 1$, and $\rho = 0.6$. The method of operation of the algorithm for test 23 and an example featuring test 6 are presented in Figure 7. A graphical presentation of the driving routes of the two vehicles generated in test 23 is presented in Figure 8. The energy expenditure for the first vehicle was 200 (kWh), the route travelled was 82 km, the breakdown uncertainty en route was 0.2, there was a single charging stop at charging point 4, and the loading points were as follows: 14, 9, 7, 3, 1, 4, 18, 16, 15, 12, 10, 13, and 17. The energy expenditure for the second vehicle was 180 (kWh), the route travelled was 75 km, the breakdown uncertainty en route was 0.2, there was a single charging stop at charging point 3, and the loading points were as follows: 8, 5, 2, 6, and 11.

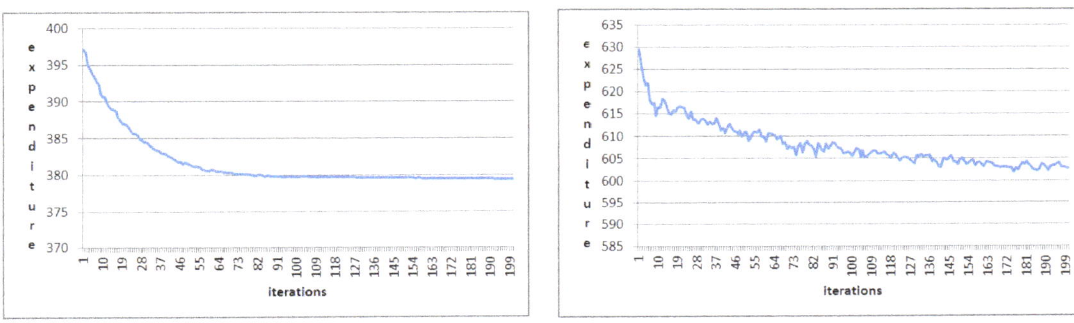

(**a**) (**b**)

Figure 7. Operation of the ant algorithm: (**a**) test 23; (**b**) test 6.

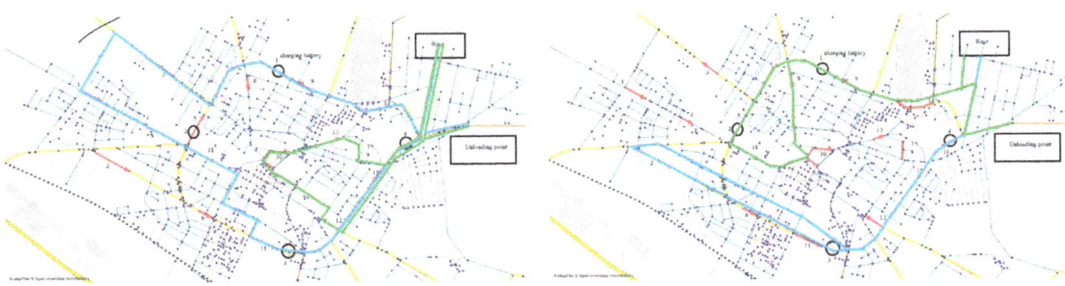

(**a**) (**b**)

Figure 8. The driving route of the vehicles created by the ant algorithm: (**a**) the first vehicle; (**b**) the second vehicle.

The initial population for the calibration of the genetic algorithm was generated using the ant algorithm based on the set of parameters from test 23. The crossbreeding operator, using $p = 0.2$, 0.4, 0.6, 0.8, and 1, is a parameter of the genetic algorithm [48]. Experimentally, the number of algorithm iterations was set to 200 repetitions and the population size was set to 50 chromosomes. The lowest value of energy expenditure for the parameter $p = 0.8$ amounted to 362 (kWh), and the uncertainty of charging point operation along the route of two vehicles was 0.04. The method of operation of the algorithm is presented in Figure 9. A visual presentation of the driving routes of two vehicles with the minimum energy expenditure is presented in Figure 10. The energy expenditure for the first vehicle was 180 (kWh), the distance travelled was 75 km, the breakdown uncertainty en route was 0.22, there was a single charging stop at charging point 2, and the loading points were as follows: 16, 15, 12, 10, 13, 9, 7, 3, 1, 4, and 5. The energy expenditure for the second vehicle was

182 (kWh), the distance travelled was 76 km, the breakdown uncertainty en route was 0.21, there was a single charging stop at charging point 4, and the loading points were as follows: 2, 6, 11, 8, and 18.

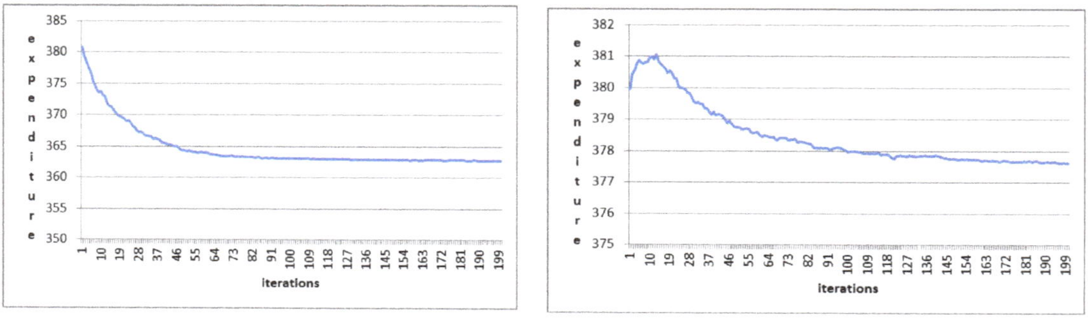

Figure 9. Operation of the hybrid genetic algorithm: (**a**) p = 0.8; (**b**) p = 0.2.

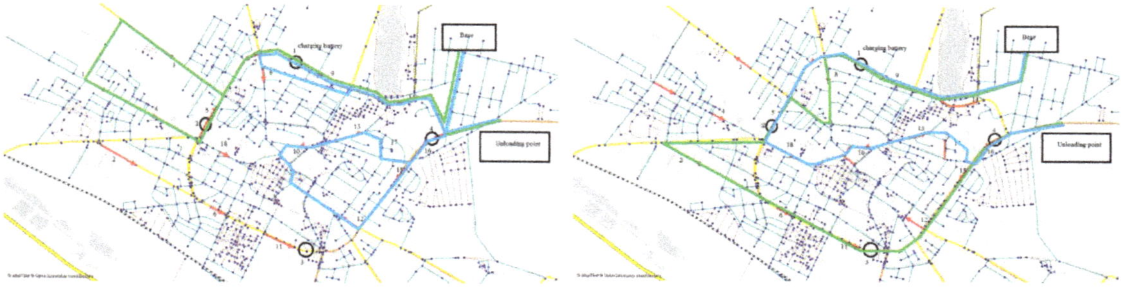

Figure 10. The driving route of the vehicles created by the hybrid genetic algorithm: (**a**) the first vehicle; (**b**) the second vehicle.

The ant and genetic algorithms are probabilistic algorithms, so each time they are run, they can generate different solutions. In order to verify the accuracy of the ant algorithm (A) and the hybrid genetic algorithm (G), 60 solutions were generated. Subsequently, the energy expenditure of the vehicles (E) generated by these algorithms and the probability associated with the uncertainty of charging point operation along these routes (P) were compared. Although the driving routes were determined in accordance with a limitation on the permissible risk related to the uncertainty of charging point operation (4), in the test column of Table 8, the time of the arrival of the vehicles at the loading points was compared with the time of the charging points' breakdowns presented in Table 2. If the time of the vehicle's arrival at a given point is inconsistent with the time of a charging point malfunction, the status OK is entered; in the case of both events being aligned in time, the point at which this alignment occurred is indicated. The average time required to generate a solution with the ant algorithm is seven minutes, while the hybrid genetic algorithm takes four minutes. The algorithms were run according to the best set of parameters established in the calibration process, i.e., $\alpha = 1$, $\beta = 1$, $\rho = 0.6$, and $p = 0.8$.

Table 8. Verification of the ant and hybrid genetic algorithms.

No.	A/G (E)	A/G (P)	A/G Test	No.	A/G (E)	A/G (P)	A/G Test	No.	A/G (E)	A/G (P)	A/G Test
1	380/364	0.09/0.08	OK/OK	21	377/365	0.08/0.12	OK/OK	41	398/373	0.07/0.12	OK/OK
2	370/353	0.10/0.12	OK/OK	22	380/360	0.11/0.11	OK/OK	42	467/380	0.10/0.11	OK/OK
3	391/372	0.12/0.13	OK/OK	23	380/376	0.12/0.10	OK/OK	43	395/377	0.12/0.13	OK/OK
4	386/370	0.07/0.06	OK/OK	24	410/370	0.06/0.08	OK/OK	44	365/376	0.16/0.14	OK/OK
5	399/369	0.12/0.15	OK/OK	25	441/373	0.12/0.06	OK/OK	45	360/350	0.12/0.13	OK/OK
6	403/380	0.14/0.12	OK/OK	26	434/367	0.14/0.09	OK/OK	46	399/371	0.14/0.05	OK/OK
7	378/377	0.15/0.09	OK/OK	27	455/360	0.13/0.05	OK/OK	47	412/345	0.12/0.06	OK/OK
8	412/393	0.11/0.13	OK/OK	28	432/345	0.12/0.12	OK/OK	48	432/355	0.12/0.08	OK/OK
9	400/380	0.06/0.11	OK/OK	29	377/350	0.07/0.15	OK/OK	49	412/345	0.11/0.10	OK/OK
10	392/380	0.07/0.14	OK/OK	30	443/355	0.07/0.05	OK/OK	50	411/360	0.10/0.11	OK/OK
11	380/360	0.11/0.13	OK/OK	31	412/367	0.05/0.04	OK/OK	51	400/355	0.14/0.13	OK/OK
12	436/379	0.12/0.08	OK/OK	32	432/369	0.12/0.06	OK/OK	52	423/370	0.13/0.11	OK/OK
13	411/366	0.13/0.05	OK/OK	33	412/370	0.09/0.05	OK/OK	53	414/345	0.12/0.08	OK/OK
14	376/350	0.11/0.13	OK/OK	34	398/369	0.10/0.12	OK/OK	54	432/367	0.06/0.11	OK/OK
15	388/365	0.12/0.14	OK/OK	35	389/350	0.11/0.14	OK/OK	55	365/345	0.06/0.12	OK/OK
16	404/360	0.08/0.16	OK/OK	36	368/340	0.07/0.11	OK/OK	56	363/344	0.19/0.13	OK/OK
17	443/367	0.04/0.14	OK/OK	37	404/350	0.05/0.14	OK/OK	57	376/333	0.10/0.14	OK/OK
18	389/359	0.05/0.16	OK/OK	38	377/345	0.05/0.12	OK/OK	58	443/356	0.11/0.11	OK/OK
19	401/380	0.05/0.08	OK/OK	39	354/340	0.05/0.06	OK/OK	59	383/345	0.07/0.08	OK/OK
20	380/372	0.10/0.09	OK/OK	40	456/347	0.14/0.08	OK/OK	60	404/380	0.12/0.07	OK/OK

Table 8 shows that the hybrid connection in which the ant algorithm generates the initial population for the genetic algorithm is an effective combination of the algorithms.

6. Discussion

The quality of the results obtained using the algorithms developed depends on the calibration process. In the calibration process, 60 test settings of the ant algorithm parameters and five crossbreeding parameter settings in the hybrid genetic algorithm were used. The space that allows for the search for an optimal solution is limited by the number of verified test combinations of these parameters. In order to expand this space, it is additionally necessary to check other parameter settings $\alpha = 1.5; 2$, whether $p = 0.5; 0.7$.

The best solution generated by the ant algorithm was determined for the small parameter values, e.g., $\alpha = 1$, $\beta = 1$, and $\rho = 0.6$. Maintaining a balance between the influence of the pheromone (α) and the heuristic information is important in the calibration of the ant algorithm (β). A large discrepancy between these two parameters means that the ant algorithm ceases to learn from the experiences of other ants and takes the form of a random search algorithm. The crossbreeding parameter plays a key role in the hybrid genetic algorithm. In the case of a small value of this parameter, the population of chromosomes is not diverse enough to generate new solutions. At a fast pace, the algorithm converges to the local optimum. In the case of a larger value of the crossing parameter, there is a strong level of competition between chromosomes in the population, and thus, there is an increase in the space allowing for the search for an optimal solution. The premature convergence of the ant algorithm in the initial phase of iterations is blocked by the continuous updating of the pheromone in each iteration of the algorithm.

The combination of the ant and genetic algorithms confirmed the effectiveness of the hybrid connection. In each of the tests conducted, the hybrid genetic algorithm found a better solution than the ant algorithm. The use of the genetic algorithm on its own is difficult due to the issue of generating an initial population. In the classic genetic algorithm, this population is randomly generated. The issue of determining vehicle driving routes requires designating a specific sequence of route points, which is difficult to achieve in a random manner. The use of a hybrid connection made it possible to implement a genetic algorithm and check its effectiveness in determining the energy expenditure of electric vehicles.

7. Conclusions

The aim of the conducted research was to develop a method that minimizes the energy expenditure of electric vehicles used in municipal service companies, taking into account the uncertainty of the operation of their charging points. Our approach to determining the routes of electric vehicles presented in this article shows that the use of appropriate algorithms reduces the energy expenditure of electric vehicles with the consideration of malfunctions in the operation of charging points. Our verification of the developed algorithms, i.e., the ant and hybrid genetic algorithms, confirmed their effectiveness. These algorithms can be an effective tool for assessing energy expenditure in service companies. The advantage of the presented algorithms is they are able to generate solutions in a short amount of time, which emphasizes their practical application. Our case study and its results provide valuable information on the feasibility of using ant and genetic algorithms to solve complex problems related to the routing of electric vehicles.

It should be emphasized that classic forms of the developed algorithms were implemented. For further research, their subsequent variations should be used, e.g., the use of ant hill systems, or research on other forms of selection in the genetic algorithm.

The decision-making model developed in this article can serve as a base model for testing other optimisation algorithms. The model developed was presented in a single-criterion approach. In further research, it should be expanded with other functions of the criterion, minimising the time of task implementation or its cost. Further research may also focus on combining the developed method with the charging problem of electric vehicles by optimal energy management for active distribution networks (ADNs) at electric vehicle charging stations [49].

For the proper operation of the heuristic algorithms used in this article, it is important to carry out the process of parameter calibration.

The developed hybrid combination of algorithms can be used to solve complex optimization problems in road transport and other transport branches, e.g., determining aircraft routes in air or rail transport, allocating a means of transport for certain tasks, or solving scheduling issues.

Author Contributions: Conceptualization, M.J. and M.I.; methodology, M.J.; software, J.B.; validation, M.J., M.I. and J.B.; formal analysis, M.J.; investigation, M.I.; resources, J.B.; data curation, J.B.; writing—original draft preparation, M.J.; writing—review and editing, M.I.; visualization, J.B.; supervision, M.J.; project administration, M.I.; funding acquisition, M.J. All authors have read and agreed to the published version of the manuscript.

Funding: This research received no external funding.

Data Availability Statement: Data are contained within the article.

Acknowledgments: This research was funded by the National Science Centre, Poland (project no. 2022/04/Y/ST8/00134; project title: Energy optimal urban logistics as a service (E-LaaS)). The project was implemented as part of the call ERA-NET Cofund Urban Accessibility and Connectivity (EN-UAC China Call) organized by JPI Urban Europe and the National Natural Science Foundation of China (NSFC) funded by the European Union's Horizon 2020 research and innovation programme under grant agreement No. 875022.

Conflicts of Interest: The authors declare no conflicts of interest.

References

1. Cieśla, M.; Sobota, A.; Jacyna, M. Multi-Criteria Decision Making Process in Metropolitan Transport Means Selection Based on the Sharing Mobility Idea. *Sustainability* **2020**, *12*, 7231. [CrossRef]
2. Pielecha, J.; Skobiej, K.; Kurtyka, K. Exhaust Emissions and Energy Consumption Analysis of Conventional, Hybrid, and Electric Vehicles in Real Driving Cycles. *Energies* **2020**, *13*, 6423. [CrossRef]
3. Zhou, G.; Hu, W.; Huang, W. Are Consumers Willing to Pay More for Sustainable Products? A Study of Eco-Labeled Tuna Steak. *Sustainability* **2016**, *8*, 494. [CrossRef]
4. Erickson, L.; Ma, S. Solar-Powered Charging Networks for Electric Vehicles. *Energies* **2021**, *14*, 966. [CrossRef]

5. Jiang, Y.; Leng, B.; Xi, J. Assessing the social cost of municipal solid waste management in Beijing: A systematic life cycle analysis. *Waste Manag.* **2024**, *173*, 62–74. [CrossRef] [PubMed]
6. Yıldız, B. Package routing problem with registered couriers and stochastic demand. *Transp. Res. Part E Logist. Transp. Rev.* **2021**, *147*, 102248. [CrossRef]
7. Jafar-Zanjani, H.; Zandieh, M.; Sharifi, M. Robust and resilient joint periodic maintenance planning and scheduling in a multi-factory network under uncertainty: A case study. *Reliab. Eng. Syst. Saf.* **2022**, *217*, 108113. [CrossRef]
8. Jacyna, M.; Semenov, I. Models of vehicle service system supply under information uncertainty. *Eksploat. I Niezawodn.—Maint. Reliab.* **2020**, *22*, 694–704. [CrossRef]
9. Kumar, S.N.; Panneerselvam, R. Development of an Efficient Genetic Algorithm for the Time Dependent Vehicle Routing Problem with Time Windows. *Am. J. Oper. Res.* **2017**, *7*, 1–25. [CrossRef]
10. Ibrahim, A.; Qader, H.; Latif, N. Use the Power of a Genetic Algorithm to Maximize and Minimize Cases to Solve Capacity Supplying Optimization and Travelling Salesman in Nested Problems. *J. Comput. Commun.* **2023**, *11*, 24–31. [CrossRef]
11. Adebayo, K.; Aderibigbe, F.; Dele-Rotimi, A. On the Development of a Hybridized Ant Colony Optimization (HACO) Algorithm. *Am. J. Comput. Math.* **2019**, *9*, 358–372. [CrossRef]
12. Liu, R.; Xie, X.; Augusto, V.; Rodriguez, C. Heuristic algorithms for a vehicle routing problem with simultaneous delivery and pickup and time windows in home health care. *Eur. J. Oper. Res.* **2013**, *230*, 475–486. [CrossRef]
13. Li, J.; Wang, F.; He, Y. Electric Vehicle Routing Problem with Battery Swapping Considering Energy Consumption and Carbon Emissions. *Sustainability* **2020**, *12*, 10537. [CrossRef]
14. Quak, H.; Nesterova, N.; van Rooijen, T.; Dong, Y. Zero Emission City Logistics: Current Practices in Freight Electromobility and Feasibility in the near Future. *Transp. Res. Procedia* **2016**, *14*, 1506–1515. [CrossRef]
15. Izdebski, M.; Jacyna, M. An Efficient Hybrid Algorithm for Energy Expenditure Estimation for Electric Vehicles in Urban Service Enterprises. *Energies* **2021**, *14*, 2004. [CrossRef]
16. Erdelić, T.; Carić, T. A Survey on the Electric Vehicle Routing Problem: Variants and Solution Approaches. *J. Adv. Transp.* **2019**, *54*, 1–48. [CrossRef]
17. Juan, A.; Mendez, C.; Faulin, J.; de Armas, J.; Grasman, S. Electric vehicles in logistics and transportation: A survey on emerging environmental, strategic, and operational challenges. *Energies* **2016**, *9*, 86. [CrossRef]
18. Keskin, M.; Çatay, B. Partial recharge strategies for the electric vehicle routing problem with time windows. *Transp. Res. Part C Emerg. Technol.* **2016**, *65*, 111–127. [CrossRef]
19. Vidal, T.; Crainic, T.G.; Gendreau, M.; Prins, C. Heuristics for multi-attribute vehicle routing problems: A survey and synthesis. *Eur. J. Oper. Res.* **2013**, *231*, 1–21. [CrossRef]
20. Agrawal, A.K.; Yadav, S.; Gupta, A.A.; Pandey, S. A genetic algorithm model for optimizing vehicle routing problems with perishable products under time-window and quality requirements. *Decis. Anal. J.* **2022**, *5*, 100139. [CrossRef]
21. Phuc, K.; Nguyen, P.; Thao, N.L.P. Ant Colony Optimization for Multiple Pickup and Multiple Delivery Vehicle Routing Problem with Time Window and Heterogeneous Fleets. *Logistics* **2021**, *5*, 28. [CrossRef]
22. Reed, M.; Yiannakou, A.; Evering, R. An ant colony algorithm for the multi-compartment vehicle routing problem. *Appl. Soft Comput.* **2014**, *15*, 169–176. [CrossRef]
23. Doppstadt, C.; Koberstein, A.; Vigo, D. The hybrid electric vehicle—Traveling salesman problem. *Eur. J. Oper. Res.* **2016**, *253*, 825–842. [CrossRef]
24. Roberti, R.; Wen, M. The electric traveling salesman problem with time windows. *Transp. Res. Part E Logist. Transp. Rev.* **2016**, *89*, 32–52. [CrossRef]
25. Barco, J.; Guerra, A.; Muñoz, L.; Quijano, N. Optimal routing and scheduling of charge for electric vehicles: A case study. *Math. Probl. Eng.* **2017**, *2017*, 8509783. [CrossRef]
26. Erdelić, T.; Carić, T.; Erdelić, M.; Tišljarić, L. Electric vehicle routing problem with single or multiple recharges. *Transp. Res. Procedia* **2019**, *40*, 217–224. [CrossRef]
27. Hiermann, G.; Puchinger, J.; Ropke, S.; Hartl, R.F. The electric fleet size and mix vehicle routing problem with time windows and recharging stations. *Eur. J. Oper. Res.* **2016**, *252*, 995–1018. [CrossRef]
28. Zhang, S.; Gajpal, Y.; Appadoo, S.S.; Abdulkader, M.M.S. Electric vehicle routing problem with recharging stations for minimizing energy consumption. *Int. J. Prod. Econ.* **2018**, *203*, 404–413. [CrossRef]
29. Macrina, G.; Pugliese, L.D.P.; Guerriero, F.; Laporte, G. The green mixed fleet vehicle routing problem with partial battery recharging and time windows. *Comput. Oper. Res.* **2019**, *101*, 183–199. [CrossRef]
30. Bruglieri, M.; Pezzella, F.; Pisacane, O.; Suraci, S. A variable neighborhood search branching for the electric vehicle routing problem with time windows. *Electron. Notes Discret. Math.* **2015**, *47*, 221–228. [CrossRef]
31. Desaulniers, G.; Errico, F.; Irnich, S.; Schneider, M. Exact algorithms for electric vehicle-routing problems with time windows. *Oper. Res.* **2016**, *64*, 1388–1405. [CrossRef]
32. Wu, J.; Wei, Z.; Liu, K.; Quan, Z.; Li, Y. Battery-involved Energy Management for Hybrid Electric Bus Based on Expert-assistance Deep Deterministic Policy Gradient Algorithm. *IEEE Trans. Veh. Technol.* **2020**, *69*, 12786–12796. [CrossRef]
33. Basma, H.; Mansour, C.; Haddad, M.; Nemer, M.; Stabat, P. A novel method for co-optimizing battery sizing and charging strategy of battery electric bus fleets: An application to the city of Paris. *Energy* **2023**, *285*, 129459. [CrossRef]

34. Abd-Elhaleem, S.; Shoeib, W.; Sobaih, A. Improved Power Management Under Uncertain Driving Conditions for Plug-In Hybrid Electric Vehicles via Intelligent Controller. *IEEE Trans. Intell. Transp. Syst.* **2023**, *99*, 1–15. [CrossRef]
35. Manoj, D.; Josh, F.T. A Comprehensive Review on Optimization and Artificial Intelligence Algorithms for Effective Battery Management in EVs. *Int. J. Electr. Electron. Eng. Telecommun.* **2023**, *12*, 334–341. [CrossRef]
36. Caylak, O.; Yaman, A.; Baumeier, B.R. Evolutionary approach to constructing a deep feedforward neural network for prediction of electronic coupling elements in molecular materials. *J. Chem. Theory Comput.* **2019**, *15*, 1777–1784. [CrossRef] [PubMed]
37. Chaoufi, I.; Abdelkhalek, O.; Gasbaoui, B. State of charge estimation of lithium-ion batteries using adaptive neuro fuzzy inference system. *IAES Int. J. Artif. Intell.* **2022**, *11*, 473–484. [CrossRef]
38. Hu, X.; Liu, T.; Qi, X.; Barth, M. Reinforcement learning for hybrid and plug-in hybrid electric vehicle energy management: Recent advances and prospects. *IEEE Ind. Electron. Mag.* **2019**, *13*, 16–25. [CrossRef]
39. Samanta, A.; Chowdhuri, S.; Williamson, S.S. Machine learning-based data-driven fault detection/diagnosis of lithium-ion battery: A critical review. *Electronics* **2021**, *10*, 1309. [CrossRef]
40. Lucas, A.; Barranco, R.; Refa, N. EV idle time estimation on charging infrastructure, comparing supervised machine learning regressions. *Energies* **2019**, *12*, 269. [CrossRef]
41. Li, Y.; Wang, Z.; Xu, W.; Gao, W.; Xu, Y.; Xiao, F. Modeling and energy dynamic control for a ZEH via hybrid model-based deep reinforcement learning. *Energy* **2023**, *277*, 127627. [CrossRef]
42. Sun, X.; Fu, J.; Yang, H.; Xie, M.; Liu, J. An energy management strategy for plug-in hybrid electric vehicles based on deep learning and improved model predictive control. *Energy* **2023**, *269*, 126772. [CrossRef]
43. Normasari, N.M.; Yu, V.F.; Bachtiyar, C. Sukoyo A simulated annealing heuristic for the capacitated green vehicle routing problem. *Math. Probl. Eng.* **2019**, *2019*, 2358258. [CrossRef]
44. Vincent, F.Y.; Redi, A.A.N.P.; Hidayat, Y.A.; Wibowo, O.J. A simulated annealing heuristic for the hybrid vehicle routing problem. *Appl. Soft Comput.* **2017**, *53*, 119–132. [CrossRef]
45. Joo, H.; Lim, Y. Ant colony optimized routing strategy for electric vehicles. *J. Adv. Transp.* **2018**, *2018*, 5741982. [CrossRef]
46. Alesiani, F.; Maslekar, N. Optimization of Charging Stops for Fleet of Electric Vehicles: A Genetic Approach. *IEEE Intell. Transp. Syst. Mag.* **2014**, *6*, 10–21. [CrossRef]
47. Gołda, P.; Zawisza, T.; Izdebski, M. Evaluation of efficiency and reliability of airport processes using simulation tools. *Eksploat. I Niezawodn.—Maint. Reliab.* **2021**, *23*, 659–669. [CrossRef]
48. Izdebski, M.; Jacyna-Gołda, I.; Gołda, P. Minimisation of the probability of serious road accidents in the transport of dangerous goods. *Reliab. Eng. Syst. Saf.* **2022**, *217*, 108093. [CrossRef]
49. Zening, L.; Su, S.; Xiaolong, J.; Mingchao, X.; Qifang, C.; Koji, Y. Stochastic and distributed optimal energy management of active distribution network with integrated office buildings. *CSEE J. Power Energy Syst.* **2024**, *10*, 504–517. [CrossRef]

energies

Article

Electromobility as a Challenge of Modern City Logistics—Indicator Analysis

Agnieszka Pokorska and Tomasz Wiśniewski *

Institute of Management, University of Szczecin, Cukrowa 8 Street, 71-004 Szczecin, Poland; agnieszka.pokorska@usz.edu.pl
* Correspondence: tomasz.wisniewski1@usz.edu.pl

Abstract: Electromobility is the challenge of modern cities. The following article explores this issue. The article presents the authors' assessment of the impact of sustainable urban mobility indicators on the development of electromobility strategies in cities. The purpose of the paper is to describe which indicators of sustainable urban mobility may influence the development of electromobility strategies in cities by increasing the number of not only low-emission cars (electric and hybrid) and low-emission buses but also public bicycles. The hypothesis posited in the work is: sustainable urban mobility is an important factor in the development of electromobility in cities. Our literature analysis, expert research, and statistical analysis allowed us to prepare a possibly comprehensive set of indicators for sustainable urban mobility that have a potential impact on the development of electromobility strategies. We ran a thematic analysis of systematic literature reviews and collected around 50 specific types of indicators and measures of sustainable mobility. After consulting the indicator database with experts, we expanded a list of 21 indicators of sustainable mobility. Next, using regression analysis, it was determined which of the sustainable urban mobility indicators is the most important for the development of electromobility strategies in the three identified transport systems.

Keywords: electromobility; city logistics; sustainable mobility; electric vehicles; indicator analysis

1. Introduction

The global climate and environmental crisis are leading countries and their cities to strive for a low-emission and ecological transformation of society and the economy [1–6] while also using modern technologies for this purpose [7]. Therefore, low-emission transport is considered key to the promotion of low-emission development, with the reduction and neutrality of CO_2 emissions being one of the development goals of this sector [8]. A city developing in a sustainable manner strives to maintain appropriate relations between economic growth, care for the environment, and the quality of life of its inhabitants. All components of the functioning of cities, such as education, culture, transport, housing, urban infrastructure, as well as the natural environment, are subject to constant pressure due to the needs and growing expectations of residents, as well as the increasing requirements resulting from national regulations and international standards [9]. Cities on the path to sustainable development will have to face various barriers and emerging challenges. One way to achieve sustainability, although fraught with many risks, is the electromobility strategy. The choice of research topic was determined by its topicality. The analyses conducted so far and the results published have focused only on issues related to sustainable development or urban mobility; it has been noticed that there is no connection between sustainable urban mobility and the electromobility strategy in cities measured using indicators [10].

A fundamental aspect of sustainable urban mobility is the continuous creation of formal and technical frameworks for economically, socially, and ecologically balanced processes in the city, and one of the actions is to increase the share of electric vehicles in the transport services of the city, economy, and residents. The ongoing energy transformation on a global scale, caused by the development of new technologies, environmental and climate protection policies, as well as the desire to reduce import dependence on fossil fuels, has led to the transformation of economic structure, affecting subsequent industries by increasing energy efficiency and reducing external costs while at the same time improving quality of life. In the short term, countries can continue to implement, enforce, and tighten regulatory measures such as CO_2 emissions and fuel consumption standards and electric vehicle mandates. Taxing gasoline and diesel oil at rates reflecting their impact on the environment and human health may increase state budget revenues, reduce the negative impact on the environment, and consequently accelerate the development of electromobility. Urban electromobility is a strategic priority of the European Union, not only in ecological but also economic aspects. Polish and EU regulations introduce high environmental standards for cities, which affects the planning of transport development and its electrification. The result of these activities, especially in the period of an unstable and dynamic global situation, is a gradual decrease in dependence on oil imports and a decrease in air pollution. However, the pluralism of local economic, social, political, and technical conditions makes it impossible to establish a common standard for implementing sustainable mobility in cities, posing numerous challenges and development barriers. In the face of a number of barriers and challenges facing city authorities, it is necessary to identify actions that will support the development of electromobility in urban areas; one of them may be the creation of a set of indicators for the development of electromobility strategies in cities.

2. Materials and Methods

2.1. Research Procedure for Empirical Analysis

Criticism of the research problem in light of the achievements of science to date have been made continuously, throughout the writing of this work and the collection of scientific materials. The actual analysis and research covered cities in Poland in 2018–2022. The criterion for selecting cities was the population as of 31 December 2022 (Table 1) because the largest urban agglomerations face communication problems, congestion, and environmental pollution, and their greatest challenge is the pursuit of sustainable mobility and the development of electromobility strategies.

Table 1. Number of inhabitants of cities in Poland.

City	Number of Inhabitants
Warszawa	1,792,718
Kraków	780,796
Łódź	667,923
Wrocław	641,201
Poznań	530,464
Gdańsk	470,633
Szczecin	396,472
Bydgoszcz	341,692
Lublin	337,788
Białystok	296,401

Source: Own elaboration based on [11].

The primary research included correspondence carried out with the communication departments of city halls, selected in a targeted selection of 10 agglomerations in Poland, and public transport departments. The collected data concern the number of hybrid and electric passenger cars, the number of low-emission buses, and the price of monthly tickets (for buses) in the years 2018–2022 in a given city. In parallel, based on literature research, a database of sustainable urban mobility indicators was prepared and sent to experts in the form of an expert research questionnaire. The prepared expert research questionnaire concerned the separation of sustainable urban mobility indicators from the indicated database, including those that influence the development of electromobility strategies in cities. The questionnaire was sent to 15 experts in the field of electromobility, both practitioners—car manufacturers, mayors of the surveyed cities, and electricity distributors in Poland—and theoreticians publishing in the field of electromobility, in order to have a broad view of the studied problem. The experts' task was to identify variables from the database of sustainable urban mobility indicators that, according to them, have a significant impact on the possibility of developing an electromobility strategy.

The analysis of the expert research questionnaire was the basis for the selection of sustainable urban mobility indicators for further statistical analysis in order to determine their significance for the development of the electromobility strategy. Statistical research, conducted after obtaining the required secondary data from cities, made it possible to identify connections between sustainable urban mobility and the electromobility strategy and to verify the main research hypothesis: sustainable urban mobility is an important factor in the development of electromobility in cities. The steps presented allowed for the selection of research methods adequate to the obtained data. To achieve the assumed goals, it was decided to choose the linear regression method, which from the created set of sustainable urban mobility indicators identified the key factors for the development of electromobility strategies in cities.

2.2. Identifying Indicators of Sustainable Urban Mobility

A search of the literature on the subject, as well as strategic documents of cities, provided knowledge about a wide set of indicators and measures of sustainable transport and urban mobility that have been accepted by international organizations in order to assess the implemented policy (Table 2).

Table 2. Indicators and measures of sustainable urban mobility (literature review).

Indicator	Author	Title	Source	Subject of Study
OECD indicators for the integration of environmental concerns into transport policies	Working Group on the State of the Environment [12]	Indicators for the Integration of Environmental Concerns into Transport Policies	ENV/EPOC/SE 1/FINAL, 1999	The metrics are grouped into three groups reflecting the transport sector and its impact on the environment
TERM (Transport and Environment Reporting Mechanism)	European Commission; European Environment Agency; Eurostat [13]	Transitions Towards a More Sustainable Mobility System. TERM 2016: Transport Indicators Tracking Progress Towards Environmental Targets in Europe	EEA Report No. 34/2016, Luxemburg, 2016	A set of indicators to monitor the progress of the common environment and transport policy
Community Measure	Transportation Research Board [14]	A Guidebook for Developing a Transit Performance—Measurement System	Washington, 2003	It measures the proportion of public subsidies for the development and operation of particular types of transport
Sustainable Development Goal indicators	United Nations [15]	Transforming our World: The 2030 Agenda for Sustainable Development	A/Res/70/1, 2015	A set of measures and indicators for monitoring the implementation of the 2030 Agenda for Sustainable Development

Table 2. *Cont.*

Indicator	Author	Title	Source	Subject of Study
SMI—Sampling Mobility Index	Frei, F. [16]	Sampling mobility index: Case study in Assis–Brazil	Transportation Research Part A: Policy and Practice 40(9), 2006	A model for assessing mobility in medium-sized cities in Brazil
IGEMUS—Strategic Management Index for Sustainable Urban Mobility	Costa, P.B.; Morais Neto, G.C.; Bertolde, A.I. [17]	Urban Mobility Indexes: A Brief Review of the Literature	Transportation Research Procedia, 25, 2017	A model for strategic management for sustainable urban mobility
Mobility Impact Index (MII)	Travisi, C.M.; Camagni, R.; Nijkamp, P. [18]	Impacts of Urban Sprawl and Commuting: A Modelling Study for Italy	Journal of Transport Geography, 18(3), 2010	Analysis of the impact of mobility on the environment
I_SUM	Lima, J.P.; da Silva Lima, R.; da Silva, A.N.R. [19]	Evaluation and Selection of Alternatives for the Promotion of Sustainable Urban Mobility	Procedia—Social and Behavioral Sciences, 162, 2014	Assessment of medium-sized centers in terms of sustainable mobility
	Silva, A.N.R.; Costa Silva, M.; Ramos, R.A.R. [20]	Development and Application of I_SUM—an Index of Sustainable Urban Mobility	89th Transportation Research Board Annual Meeting, 2010	
Measures of the PROPOLIS project	Lautso K. [21]	The SPARTACUS System for Defining and Analysing Sustainable Urban Land Use and Transport Policies	Springer, Berlin, Heidelberg, 2003	Environmental, social, and economic measures of transport
	Lautso, K.; Spiekermann, K.; Wegener, M.; Sheppard, I.; Steadman, Ph.; Martino, A.; Domingo, R.; Gayda, S. [22]	Planning and Research of Policies for Land Use and Transport for Increasing Urban Sustainability [PROPOLIS]	Brussels, 2004	
Measures and indicators that should be taken into account in the process of planning and managing transport, in particular, public transport	Litman T. [23]	Well Measured: Developing Indicators for Comprehensive and Sustainable Transport Planning	Victoria Transport Policy Institute Canada, 2007	Measures and indicators for the planning and management of transport, in particular, public transport
	Litman T. [24]	Litman T. [2008], Sustainable Transportation Indicators: A Recommended Research Program for Developing Sustainable Transportation Indicators and Data, Transportation Research Board Annual Meeting.	Litman T. [2008]/Sustainable Transportation Indicators: A Recommended Research Program for Developing Sustainable Transportation Indicators and Data, Transportation Research Board Annual Meeting	

Table 2. *Cont.*

Indicator	Author	Title	Source	Subject of Study
	Litman T. [2016]. Well Measured: Developing Indicators for Sustainable and Livable Transport Planning, Victoria Transport Policy Institute Canada 2016 [25]	Litman T. [2016], Well Measured: Developing Indicators for Sustainable and Livable Transport Planning, Victoria Transport Policy Institute Canada 2016	Litman T. [2016], Well Measured: Developing Indicators for Sustainable and Livable Transport Planning, Victoria Transport Policy Institute Canada 2016	
Urban Mobility Index	Van Audenhove, F.J.; Korniichuk, O.; Dauby, L.; Pourbaix J [26]	The Future of Urban Mobility 2.0. Imperatives to Shape Extended Mobility Ecosystems of Tomorrow	A.D. Little, 2014	Determining which transport solutions are closest to becoming emission-free

Source: Own elaboration based on [12–26].

In order to shorten and refine the set for further analysis, a procedure for selecting indicators was developed (Figure 1). First (stage 1), repetitions and unmeasurable indicators, difficult to determine and those that were not specified by the authors of the studies, were eliminated. A detailed diagram of the stages of selecting indicators for analysis is presented in Figure 1.

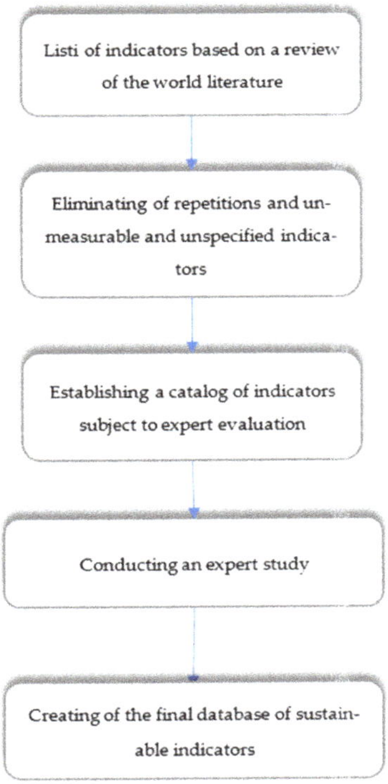

Figure 1. Stages of indicator selection. Source: Own elaboration.

The indicators from stage 1 were sent in the form of a questionnaire to the experts. The experts' task was to select and assess from the indicated database which indicators of sustainable urban mobility influence the development of the electromobility strategy. The catalog of sustainable urban mobility indicators, created on the basis of experts' responses, is presented in Table 3.

Table 3. Catalog of sustainable urban mobility indicators influencing the development of electromobility strategies, along with an explanation.

Indicator	Explanation
Registered unemployment rate (%)	Share of the registered unemployed in the professionally active civilian population, i.e., excluding employees of budgetary units conducting activities in the field of national defense and public security in the population of a given city.
GDP per capita	The ratio of the value of gross domestic product at current prices to the population of a given city.
Average monthly salary per inhabitant (PLN)	The ratio of the sum of gross personal salaries, fees paid to certain groups of employees for work resulting from employment contracts, payments for participation in profit or balance surplus in cooperatives, and additional annual remuneration for employees of public sector units to the average number of employees in a given period, after eliminating people performing home work and those employed abroad, to the population of a given city.
Disposable income per capita (PLN)	The sum of current household income from individual sources, reduced by advances on personal income tax paid by the payer on behalf of the taxpayer, taxes on income from property, taxes paid by self-employed persons, including representatives of free professions and persons using farms and individuals in agriculture, and social and health insurance contributions, depending on the population of a given city.
The ratio of the population of an urban agglomeration to its urbanized area	Population by actual place of residence per 1 km^2.
The ratio of the cost of a 100 km trip by private means of transport to the price of a monthly ticket for public transport in the agglomeration	The ratio of the average cost of traveling by passenger car, taking into account average fuel consumption and fuel price per 100 km, to the price of a monthly ticket (for buses) in a given city.
Average annual concentration value PM2.5	Suspended dust with a diameter of no more than 2.5 μm, the main source of which is road transport.
Average annual concentration value PM10	PM10 suspended dust is a mixture of solid particles and liquid droplets with a diameter not exceeding 10 μm of that remaining in the air. The main source of airborne emissions is road transport, especially from diesel vehicles without particulate filters.
Average annual concentration value NOx	The average annual value of nitrogen oxide concentration is the most harmful component of car exhaust gases. Most nitrogen oxides are produced in diesel engines.
Percentage of residents exposed to excessive noise (%)	Percentage of residents exposed to noise levels harmful to health, above 70 dB. Noise is sounds with a frequency above 16 Hz, undesirable and bothersome for humans, which has a negative impact on the hearing organ and other senses and often disrupts the proper functioning of the entire body. Traditionally powered passenger cars emit 74–85 dB of noise and buses emit 86–92 dB.

Table 3. *Cont.*

Indicator	Explanation
Number of bus stops per 1 inhabitant	The number of stopping places for public transport vehicles, marked with appropriate road signs.
Dedicated bus lane in km in relation to the city area	The length of the part of the road designated and separated from the road by longitudinal road signs for buses and electric cars in relation to the area of a given city.
The level of congestion	The overall congestion level expressed as a percentage represents additional travel time.
Price of energy per capita	The average sales price of electricity on the competitive market includes energy sales (sales volume and value of energy sold) carried out by producers and trading companies in competitive segments of the national wholesale electricity market in relation to the number of inhabitants of a given city.
Number of charging stations in relation to the city area	The number of prepared stations equipped with appropriately designed chargers powered by electricity from the local power grid to which owners connect their electric vehicles to charge their batteries and are able to continue using the energy stored in them per square kilometer of a given city.
Number of low-emission buses per capita	The number of buses refueled with other types of fuel, including mixtures (fuel–oil), compressed natural gas (CNG), liquefied natural gas (LNG), hydrogen, biodiesel, ethanol, gasoline and electricity, diesel oil and energy electricity, and electricity, in relation to the number of inhabitants of a given city.
Number of electric cars per capita	The number of energy-powered cars per number of inhabitants of a given city.
Number of hybrid cars per capita	The number of hybrid cars per number of inhabitants of a given city.
Cycle paths per area per 10 thousand population	The length of roads or parts thereof intended for bicycle traffic, marked with appropriate road signs.
Number of public bicycles per capita	The ratio of the number of public bikes operating in the bike-sharing system in a given city to the number of its inhabitants.
Public transport lines per 1000 inhabitants	The ratio of the number of available public transport lines in a given city to the number of inhabitants.

Source: Own elaboration based on expert research.

The next step is a statistical analysis of the indicators selected in the expert study, taking into account the division into groups of transport subsystems indicated by the experts: low-emission cars (electric and hybrid), low-emission buses, and public bicycles, which are part of the cities' electromobility strategy. Measures and indicators regarding sustainable development, sustainable transport and urban mobility used to monitor the goals set by global, international, and national strategies focus primarily on the negative impact on the natural environment; they do not or only slightly take social and economic impact into account. They lack a correlation between these aspects and a holistic view, which means that a comprehensive measurement of sustainable urban mobility is not possible based on the available indicators.

3. Results

3.1. Modeling Assumptions

In accordance with the nomenclature presented in foreign literature, it is postulated that one of the possible measures of the level of electromobility is the number of electric and hybrid cars registered in a given area. Bartłomiejczyk and Kołacz [27], Połom [28], and

Fernández [29] note that the introduction of electric cars is the main sign of the development of electromobility and sustainable transport in cities. The main goal of implementing the electromobility strategy in cities is to support the broadly understood electromobility policy and the sustainable development of transport and mobility, as well as ecological forms of transport. The strategy's activities are intended to contribute to the creation of a sustainable transport system in cities, based on the development of a public transport system and alternative forms of communication within the city. The electromobility strategy aims to gradually replace the driving of conventional vehicles by residents with electric and emission-free ones. The electromobility strategy seeks to create favorable conditions for the growth of electromobility in cities by modifying current infrastructure and undertaking various initiatives to improve its effectiveness and support the expansion of the low-emission vehicle market for both private and public transportation. This strategy envisions a coherent infrastructure that facilitates the movement of low-emission vehicles, improves transport connections, and includes related enhancements like bicycle paths, parking spaces, bus stops, and bus lanes. The strategy's primary goals are to reduce the negative environmental impact of transport, enhance travel comfort within the city, protect residents' health, and improve overall quality of life. As a result of implementing this strategy, a partial reduction in emissions of harmful substances and noise is expected, leading to long-term improvements in quality of life, including residents' health and the natural environment. The positive impact of the implementation of the strategy will be the result of the development of zero-emission transport in the city (zero-emission buses, a system of bicycle paths, and balancing the mobility of some residents in the form of giving up individual means of transport in favor of public transport or bicycles). The literature review, considerations, and expert research allowed us to clarify the scope of the empirical study. Therefore, the electromobility strategy examined in this work applies to the market of low-emission cars (electric and hybrid), low-emission buses, and public bicycles. The study omitted other elements of micromobility in cities, such as scooters, because these are relatively new solutions; their origin in Polish cities can be said to be in 2019. The urban electromobility strategy examined in this work is therefore based on the following transport subsystems:

- Low-emission public transport (buses),
- Electric and hybrid cars,
- Micromobility means of transport (public bicycles).

The time range of 2018–2022 was adopted for the study because a period of 5 years allows for the observation of potential changes over time and is accepted as sufficient at the statistical level.

3.2. *Factors for Implementing Electromobility in Cities*

When analyzing real phenomena and processes, even in relatively simple situations, we cannot fully explain them. Therefore, when describing the interdependencies between them, we usually use simplified models of real interdependencies. Thus, under the term model, we can understand a useful form of presentation of empirical data. When approaching the process of building a model, it is necessary to make a certain compromise between oversimplifying reality and wanting to include overly detailed data.

A frequently used statistical analysis to find the relationship between explanatory and explanatory variables is linear regression. It takes into account the issue of modeling the interdependence of the studied characteristics and involves estimating some data on the basis of others [30].

In this article, regression analysis is used to explain the effect of changes in independent variables (x_1 to x_{21}) on dependent variables (y_1, y_2, and y_3).

The dependent variables are:

- y_1—number of electric and hybrid cars per capita
- y_2—number of public bicycles per capita
- y_3—number of public transportation vehicles per capita

The independent variables are:

- x_1—the ratio of the total number of cars in car-sharing systems in an urban agglomeration area to the population of that area
- x_2—energy price per capita
- x_3—dedicated bus lane in km in relation to the area of the city
- x_4—number of charging stations in relation to the area of the city
- x_5—number of bus stops per capita
- x_6—bicycle lanes in relation to city area
- x_7—level of congestion
- x_8—noise and vibration levels
- x_9—public transportation lines per 1000 residents
- x_{10}—average annual value of PM2.5 concentration
- x_{11}—average annual value of PM10 concentration
- x_{12}—emissions related to transportation activities NOx
- x_{13}—ratio of the cost of a 100 km trip by private transport to the price of a monthly ticket for a public transport trip in the agglomeration area
- x_{14}—price of gasoline
- x_{15}—price of diesel fuel
- x_{16}—registered unemployment rate
- x_{17}—disposable income per capita
- x_{18}—monthly salary per capita
- x_{19}—GDP per capita
- x_{20}—ratio of the population of an urban agglomeration to its urbanized area
- x_{21}—year

The study began with the construction of a comprehensive model containing the entire set of independent variables affecting the phenomenon under consideration. The components of the built, very comprehensive model were then tested in order to identify less extensive submodels that explain the phenomenon under study. Finally, among these potential submodels, the simplest one was selected, which, on a saving basis, is treated as the best one describing the phenomenon under study. To simplify it, the backward stepwise regression method was used. Simple models are preferred for practical reasons: they are easier to retest, as well as to understand and describe. The data used, presented earlier, were city data on the number of electric and hybrid cars, low-emission buses, public bicycles, and sustainable urban mobility indicators, in such a way that the dependent variables y_1, y_2, and y_3 and all independent variables x_1 to x_{21} are data from each city in successive years, that is, first the 10 values (because there are so many cities) for 2018, then 10 for 2022, and so on. Then, the variable "year" is added, which will have a value of 1 in 2019, 2 in 2020, and so on.

Three analyses were conducted, separately for each of the dependent variables (y_1, y_2, and y_3). For the number of electric and hybrid cars per capita (y_1), a stepwise backward regression was performed and left seven independent variables after 15 steps:

- x_2—energy price per capita
- x_3—dedicated bus lane in km in relation to city area
- x_4—number of charging stations in relation to city area
- x_{10}—average annual value of PM2.5 concentration
- x_{15}—price of diesel fuel
- x_{18}—monthly salary per capita
- x_{21}—year (number)

Table 4 shows significance tests for the number of low-emission cars with effect sizes, *p*-value, and *t*-test. The calculations were performed in Statistica software (Statsoft v.13).

Table 4. Significance tests, effect sizes, and powers for the variable y_1—number of electric and hybrid cars per capita.

Variable—x_i	F	Model Coefficient b_i	p-Value	t-Test
x_2—energy price per capita	28.42	−13.8301	<0.001	−5.33
x_4—number of charging stations in relation to city area	11.54	0.0022	0.002	3.40
x_{10}—average annual value of PM2.5 concentration	13.44	−0.0002	<0.001	−3.67
x_{15}—price of diesel fuel	9.63	0.0012	0.004	3.10
x_3—dedicated bus lane in km in relation to city area	8.36	0.0041	0.006	2.89
x_{18}—monthly salary per capita	6.47	0.0002	0.015	−2.54
x_{21}—year (number)	30.77	0.0023	<0.001	5.55

Source: Own elaboration.

From Table 4, it can be observed that all seven parameters of the variables left after the steps in the backward regression are statistically significant—the p-values are clearly less than 0.05 (p-value is the probability of obtaining test results at least as extreme as the result actually observed, under the assumption that the null hypothesis is correct). In addition, a significance analysis of the structural parameters was carried out.

To test the significance of structural parameters $b_0, b_1, \ldots, b_7$, a t-test was used, in which the t-statistic has a Student's t-test distribution with n-k-1 degrees of freedom.

Null and alternative hypotheses:

H_0: $b_i = 0$ *(no linear dependence).*

H_1: $b_i \neq 0$ *(there is a linear relationship).*

The critical region is two-sided with a critical value, which we read from the t-test distribution tables for a fixed significance level a and n-k-1 degrees of freedom. If the t-value is in the critical region (calculated t-value > t-value from the table), we must reject H_0 in favor of H_1. Otherwise, there is no basis for rejecting H_0. For the parameters in Table 4, all t values are in the critical region, so the relationship is statistically significant.

Another analysis was performed for the second dependent variable, y_2—number of public bicycles (per number of residents). Table 5 shows significance tests with effect sizes. In this case, backward stepwise regression was also used, and after 14 steps, eight independent variables remained:

- x_1—the ratio of the total number of cars in car-sharing systems in a metropolitan area to the population of that area
- x_2—energy price per capita
- x_6—bicycle lanes in relation to city area
- x_{10}—average annual value of PM2.5 concentration
- x_{12}—emissions related to transportation activities NOx
- x_{18}—monthly salary per capita
- x_{19}—GDP per capita
- x_{21}—year (number)

From Table 5, it can be observed that all of the parameters of the variables, after 14 steps of backward regression, are statistically significant—the p-values are clearly less than 0.05. To check the significance of the structural parameters in Table 5, the values of the t-test are also shown, in each case, they are in the critical region, so statistical significance was confirmed.

Table 5. Significance tests, effect sizes, and powers for the variable y_2—number of public bicycles per capita.

Variable—x_i	F	Model Coefficient b_i	p-Value	t-Test
x_1—the ratio	4.47	−0.55061	0.041	−2.11
x_2—energy price per capita	4.65	2.30561	0.038	2.16
x_{10}—average annual value of PM2.5 concentration	4.65	0.00688	0.038	−2.16
x_6—bicycle lanes in relation to city area	21.94	0.00088	<0.001	4.68
x_{12}—emissions related to transportation activities NOx	9.12	0.00009	0.005	3.02
x_{18}—monthly salary per capita	13.58	−0.00001	0.001	−3.69
x_{19}—GDP per capita	25.84	−0.00002	<0.001	5.08
x_{21}—year (number)	10.74	0.0031	0.007	3.14

Source: Own elaboration.

Similarly, the next analysis concerned the regression model for the third dependent variable, y_3—the number of low-emission buses per capita. Table 6 shows significance tests with effect sizes. This time, after 11 steps of backward regression, 11 independent variables were left:

- x_2—energy price per capita
- x_3—dedicated bus lane in km in relation to the area of the city
- x_5—number of bus stops per capita
- x_6—bicycle lanes in relation to city area
- x_9—public transportation lines per 1000 residents
- x_{11}—average annual value of PM10 concentration
- x_{14}—price of gasoline
- x_{15}—price of diesel fuel
- x_{16}—registered unemployment rate
- x_{18}—monthly salary per capita
- x_{21}—year (number)

Table 6. Significance tests, effect sizes, and powers for the variable y_3—the number of low-emission buses per capita.

Variable—x_i	F	Model Coefficient b_i	p-Value	t-Test
x_2—energy price per capita	10.75	0.77447	0.002	3.28
x_3—dedicated bus lane in km in relation to the area of the city	72.94	0.00413	<0.001	−8.54
x_5—number of bus stops per capita	54.99	−0.42413	<0.001	−7.42
x_6—bicycle lanes in relation to city area	22.84	−0.00015	<0.001	−4.78
x_9—public transportation lines per 1000 residents	32.53	0.00046	<0.001	5.70
x_{11}—average annual value of PM10 concentration	9.57	−0.00013	0.004	−3.09
x_{14}—price of gasoline	21.77	0.00047	<0.001	4.67
x_{15}—price of diesel fuel	25.96	0.00061	<0.001	−5.10
x_{16}—registered unemployment rate	144.70	−0.00021	<0.001	−12.03
x_{18}—monthly salary per capita	21.24	0.00002	<0.001	−4.61
x_{21}—year (number)	19.58	0.00019	<0.001	4.43

Source: Own elaboration.

From Table 6, it can be observed that all of the parameters of the variables, after 11 steps of backward regression, are statistically significant—the p-values are clearly less than 0.05. To check the significance of the structural parameters, Table 6 also shows the values of the t-test; in each case, they are in the critical region, so statistical significance is confirmed.

In addition, the model should be verified to see if it is acceptable [31]. A number of indicators were calculated (Table 7) to verify the presented regression model:

- Coefficient of determination, R^2—indicating what proportion of the total variability of the dependent variable—is explained by the variability of the independent variable, $0 \leq R^2 \leq 1$
- Coefficient of variation—$V_{s\varepsilon}$ is a normalized measure of the dispersion of the probability or frequency distribution. The model is better the lower the value of this coefficient is.

Table 7. Measures of regression model goodness of fit.

Model	R^2	$V_{s\varepsilon}$
y_1—number of electric and hybrid cars per capita	0.9074	12.48%
y_2—number of public bicycles per capita	0.8979	13.14%
y_3—number of public transportation vehicles per capita	0.9613	7.27%

Source: Own elaboration.

Table 7 presents the regression model fit measures for all of the regression models presented earlier. In the case of the model describing the dependence of the number of electric and hybrid cars on indicators of sustainable urban mobility, the parameter R^2 shows that 90.74% of changes in the number of electric and hybrid cars are explained by changes in the set of independent variables. For the number of public bicycles, it is very similar, with the model explaining 89.79% of the changes. For the number of public transportation vehicles, meanwhile, the found set of explanatory variables explains 96.13% of the changes. Looking at the values of the coefficient of variation parameter, we obtain the part of the standard deviation of the random component (in percent) in the average value of the dependent variables at 12.48%, 13.14%, and 7.27%, which is an acceptable level.

Summarizing the results in Table 7 and the value of the *t*-test from Tables 4–6, we can conclude that the regression models are statistically significant. The best fit is shown by the model for the number of public transport vehicles (especially the $V_{s\varepsilon}$ index). However, all models show an appropriate level of acceptability.

4. Discussion

By analyzing the results obtained in the linear regression, conclusions can be drawn regarding the significance of indicators for each of the electromobility subsystems indicated for the purposes of the article.

Tables 8–10 present a set of sustainable urban mobility indicators that have the greatest impact on the development of electromobility strategies on the studied transport subsystems in Polish cities.

Table 8. Indicators of sustainable urban mobility affecting the development of the public bicycle transport subsystem.

Sustainable Development Orders	Indicator
Social	The ratio of the total number of cars in car-sharing systems in the urban agglomeration to the number of inhabitants of this area
	Bicycle paths in relation to the city area
Economic	Monthly salary per inhabitant
	GDP per capita
	Energy price per capita
Environment	Emissions related to transport activities NOx
	Average annual value of PM2.5 concentration

Source: Own elaboration.

Table 9. Sustainable urban mobility indicators influencing the development of the low-emission car transport subsystem.

Sustainable Development Orders	Indicator
Social	Number of charging stations in relation to the city area
	Dedicated bus lane in km in relation to the city area
Economic	Monthly salary per inhabitant
	Diesel price
	Energy price per capita
Environment	Average annual concentration value PM2.5

Source: Own elaboration.

Table 10. Indicators of sustainable urban mobility affecting the development of the low-emission bus transport subsystem.

Sustainable Development Orders	Indicator
Social	Public transport lines per 1000 inhabitants
	Dedicated bus lane in km in relation to the city area
	Number of bus stops per 1 inhabitant
	Bicycle paths in relation to the city area
Economic	Diesel price
	Energy price per capita
	Gasoline price
	Registered unemployment rate
	Monthly salary per inhabitant
Environment	Average annual concentration value PM10

Source: Own elaboration.

The most important indicator for the development of the public bicycle subsystem seems to be the length of the infrastructure dedicated to it, i.e., bicycle paths in a given city. Moreover, the number of cars operating in the car-sharing system is important as it may indicate high ecological awareness of residents, just like the variable related to emissions from road transport.

The development of the low-emission car subsystem is dominated by indicators from the economic governance category, which indicate the wealth of residents and the price of the least ecological fuel—diesel oil and the energy necessary to "charge" an electric car. The length of bus lanes, on which electric car drivers can move without restrictions, is also important. Moreover, from the driver's point of view, the availability and density of charging stations are extremely important.

For the development of the low-emission bus subsystem, from the point of view of society, the availability of public transport infrastructure, measured as the number of communication lines, the number of bus stops per inhabitant, and the length of the bus lane, plays an important role. Economic issues are also important, including fuel prices for all types of cars and the financial situation of residents. As in the case of the other two transport subsystems, the environmental indicator seems to be an important development variable of the electromobility strategy.

By analyzing the research results, conclusions can be drawn regarding the attributes necessary to develop an electromobility strategy. Looking at the electromobility strategy from a comprehensive perspective, we can see a balance in the hierarchy of sustainable urban mobility orders. In three groups, the indicator taking into account the price of energy and monthly wages, as well as taking into account emissions related to transport, is repeated. Therefore, these are the most important and priority ones for the development of electromobility strategies in cities. Therefore, five indicators are necessary to build the ideal, desired sustainable mobility management model in terms of the development of electromobility strategies in urban areas. The first one refers to the price of energy, which is extremely important when making decisions among residents and city authorities about the electrification of private vehicles and public transport. The second one refers to the financial situation of city residents, which is important from the point of view of the still high price of purchasing an electric car compared to a traditional one. Next, they concern care for the natural environment, which is a natural consequence of the operation of low-emission vehicles. Therefore, they can be considered the main determinants of the development of electromobility strategies in cities. The general approach, taking into account the level of factors and indicators, of the concept of electromobility strategy is shown in Figure 2.

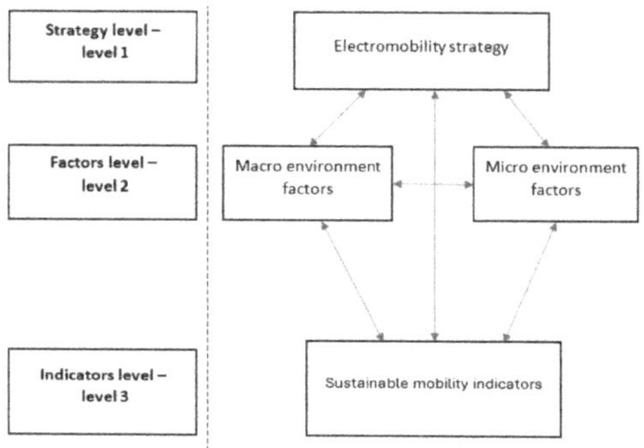

Figure 2. Electromobility strategy in the city—overview. Source: Own elaboration.

A comprehensive approach to activities and factors from the perspective of the macro- and micro-environment and a set of indicators for sustainable urban mobility located in the area of three orders, economic, social, and environmental, influencing the development of electromobility strategies in the city, create a concept of the development of electromobility strategies taking into account sustainable mobility, are presented in Figure 3.

Coherence in the development of electromobility strategies in cities is a complex and multidimensional process. It is characterized by a multitude of entities influencing the development of electromobility (subject complexity), as well as a multifaceted multiplicity of factors that build it (factor complexity). The effectiveness of cooperation between stakeholders, determinants occurring in the micro-environment and macro-environment, and specific indicators of sustainable urban mobility requires cooperation on three levels simultaneously (strategy, factors, and indicators) so that the electromobility strategy can develop.

In the structure of the proposed concept, both micro and macro approaches have been identified at the level of factors at level II. In the first perspective, the development of electromobility strategies is influenced by administrative, legal, and infrastructural determinants. At the local level, all types of legislative and tax support seem to be extremely important, as they will provide details and development of national and international documents. Solutions reflected in the education system, mainly at the level of universities, will also play an important role. Supporting the development of car sharing at the local level will have a positive impact on the development of electromobility strategies in two ways. Firstly, the vehicles used in this solution should be mainly electric and hybrid cars. Secondly, this service reduces the external costs of transport by limiting the number of cars used. This includes as many specific infrastructure solutions as possible in a given city, making it easier for residents to use electric cars, i.e.,

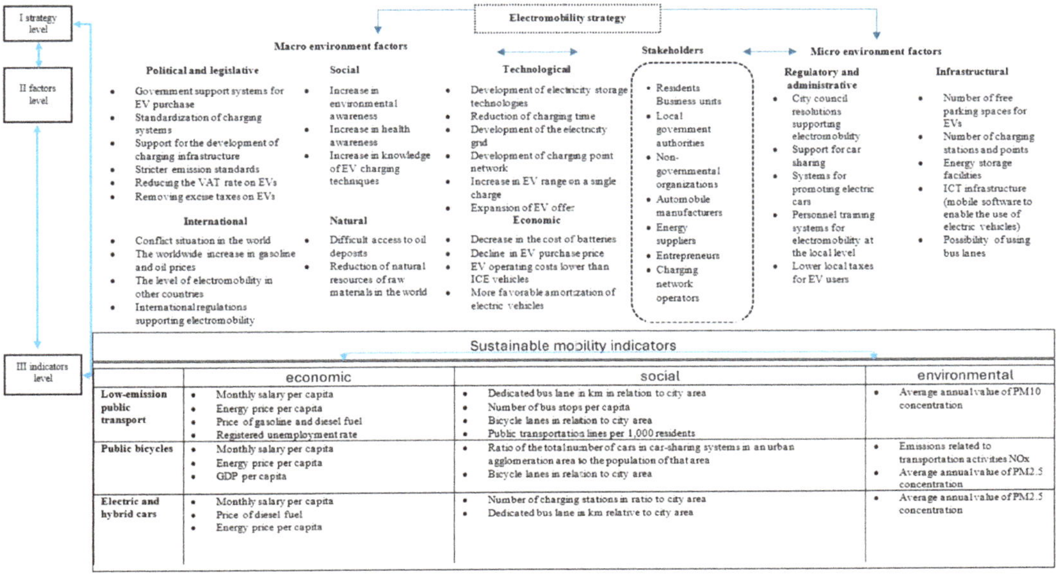

Figure 3. Electromobility strategy in the city—a detailed approach taking into account macro- and micro-environmental factors and indicators of sustainable urban mobility. Source: Own elaboration.

- The number of free parking spaces,
- The number of public stations and charging points,
- The availability of mobile programming enabling the use of electric vehicles,
- With the ability to use bus lanes combined with appropriate promotion tailored to the target group for electric vehicles, they will facilitate the development of electromobility strategies.

Macro-environmental factors significantly influence the development of the examined strategy. In the proposed concept, these factors are categorized into six groups that interconnect. Consistency in political actions at the international, national, and local levels is crucial for implementing solutions and facilities for the electromobility market and enforcing adopted standards and limits. Additionally, when integrating electromobility into urban areas, it is essential to clearly identify the target group for each proposal. Only such action will enable the actual development of the market, in accordance with the assumed forecasts included in the Electromobility Development Plan [32]. The most important task of this plan is to achieve 1 million electric cars in Poland by 2025. Fulfillment of this assumption would allow specific environmental benefits related to the reduction in transport pollutant emissions in cities in Poland to be obtained, which would have a positive impact on the health and quality of life of residents, as well as reducing the country's energy dependence by reducing the demand for liquid fuels and, consequently, inducing a decrease in the volume of crude oil imports, which given the current political and economic situation around the world seems to be a very important, if not the most important, determinant. At the beginning of 2023, there were 32,555 electric cars registered in Poland and 31,146 models with hybrid drive [33] and this is only 3.25% of the target plan of 1 million electric cars.

5. Conclusions

Recently, several global phenomena, including the COVID-19 pandemic, rising inflation, and armed conflicts, have disrupted daily life, diminished residents' comfort, limited access to various services, and impacted the operations of local government institutions. The pandemic also altered how residents perceive urban spaces, which should now be designed to encourage outdoor activity, as people's health awareness has significantly increased. The significant importance attached to health is combined with the permanently increasing ecological awareness of residents and higher standards set for local authorities in the area of the urban environment, such as air quality and air pollution. Education at the national and local level in the field of electromobility, about its usefulness and impact on important values for society, as well as about technical aspects of operation, available solutions, and facilities, also seems to be an important aspect. Coherence in the development of electromobility strategies in cities is a complex and multidimensional process. It is characterized by a multitude of entities influencing the development of electromobility (subject complexity), as well as a multifaceted multiplicity of factors that build it (factor complexity). The effectiveness of stakeholder cooperation, determinants occurring in the micro-environment and macro-environment, and specific indicators dedicated to the three transport subsystems so that the electromobility strategy can develop. The main limitation of the work was the completion of the linear regression for the year 2022 due to the lack of full available data for 2023, as well as the omission from the study of other means of urban micromobility, such as scooters, which will be included in subsequent studies.

Author Contributions: Conceptualization, A.P. and T.W.; methodology, A.P. and T.W.; software, A.P. and T.W.; validation, A.P. and T.W.; formal analysis, A.P. and T.W.; investigation, A.P. and T.W.; resources, A.P. and T.W.; data curation, A.P. and T.W.; writing—original draft preparation, A.P. and T.W.; writing—review and editing, A.P. and T.W.; visualization, A.P. and T.W. All authors have read and agreed to the published version of the manuscript.

Funding: This research was co-financed by the Minister of Science under the "Regional Excellence Initiative".

Data Availability Statement: The data presented in this study are available on request from the corresponding author.

Conflicts of Interest: The authors declare no conflicts of interest.

References

1. Ulewicz, R.; Siwiec, R.; Pacana, A. Sustainable Vehicle Design Considering Quality Level and Life Cycle Environmental Assessment (LCA). *Energies* **2023**, *16*, 8122. [CrossRef]
2. Xiao, M.; Peng, X. Decomposition of Carbon Emission Influencing Factors and Research on Emission Reduction Performance of Energy Consumption in China. *Front. Environ. Sci.* **2023**, *10*, 1096650. [CrossRef]
3. Khanam, T.; Rahman, A.; Xu, X.; Mola-Yudego, B.; Moula, M.M.E.; Pelkonen, P. Assessing the Awareness and Willingness of European Experts to Reduce Their Carbon Footprint in Everyday Consumption. *Environ. Impact Assess. Rev.* **2022**, *97*, 106889. [CrossRef]
4. Pacana, A.; Czerwińska, K. Indicator Analysis of the Technological Position of a Manufacturing Company. *Prod. Eng. Arch.* **2023**, *29*, 162–167. [CrossRef]
5. Wolniak, R. European Union Smart Mobility–Aspects Connected with Bike Road System's Extension and Dissemination. *Smart Cities* **2023**, *6*, 1009–1042. [CrossRef]
6. Kuzior, A.; Sira, M.; Brozek, P. Using Blockchain and Artificial Intelligence in Energy Management as a Tool to Achieve Energy Efficiency. *Virtual Econ.* **2022**, *5*, 69–90. [CrossRef] [PubMed]
7. Jing, Q.-L.; Liu, H.-Z.; Yu, W.-Q.; He, X. The Impact of Public Transportation on Carbon Emissions—From the Perspective of Energy Consumption. *Sustainability* **2022**, *14*, 6248. [CrossRef]
8. Alves, C.; Ferrão, P.M.C.; Silva, A.J.; Reis, L.G.; Freitas, M.; Rodrigues, L.B.; Alves, D.E. Ecodesign of Automotive Components Making Use of Natural Jute Fiber Composites. *J. Clean. Prod.* **2010**, *18*, 313–327. [CrossRef]
9. Moro, A.; Lonza, L. Electricity Carbon Intensity in European Member States: Impacts on GHG Emissions of Electric Vehicles. *Transp. Res. D Transp. Environ.* **2018**, *64*, 5–14. [CrossRef] [PubMed]
10. Kuzior, A.; Postrzednik-Lotko, K.A.; Postrzednik, S. Limiting of Carbon Dioxide Emissions through Rational Management of Pro-Ecological Activities in the Context of CSR Assumptions. *Energies* **2022**, *15*, 1825. [CrossRef]
11. Główny Urząd Statystyczny. Available online: https://stat.gov.pl/obszary-tematyczne/ludnosc/ludnosc/powierzchnia-i-ludnosc-w-przekroju-terytorialnym-w-2020-roku,7,17.html (accessed on 5 January 2023).
12. OECD. Indicators for the Integration of Environmental Concerns into Transport Policies. Available online: https://one.oecd.org/document/ENV/EPOC/SE(98)1/FINAL/en/pdf (accessed on 10 November 2022).

13. European Environment Agency. Transitions Towards a More Sustainable Mobility System. TERM 2016: Transport Indicators Tracking Progress towards Environmental Targets in Europe. Available online: https://www.eea.europa.eu/publications/term-report-2016 (accessed on 7 November 2021).
14. Transportation Research Board. A Guidebook for Developing a Transit Performance—Measurement System. Available online: https://onlinepubs.trb.org/onlinepubs/tcrp/tcrp_report_88/guidebook.pdf (accessed on 7 November 2021).
15. United Nations. Transforming Our World: The 2030 Agenda for Sustainable Development. Available online: https://sdgs.un.org/2030agenda (accessed on 7 November 2021).
16. Frei, F. Sampling mobility index: Case study in Assis–Brazil. *Transp. Res. Part A Policy Pract.* **2006**, *40*, 792–799. [CrossRef]
17. Costa, P.B.; Morais Neto, G.C.; Bertolde, A.I. Urban Mobility Indexes: A Brief Review of the Literature. *Transp. Res. Procedia* **2017**, *25*, 67. [CrossRef]
18. Travisi, C.M.; Camagni, R.; Nijkamp, P. Impacts of Urban Sprawl and Commuting: A Modelling Study for Italy. *J. Transp. Geogr.* **2010**, *18*, 3. [CrossRef]
19. Lima, J.P.; da Silva Lima, R.; da Silva, A.N.R. Evaluation and Selection of Alternatives for the Promotion of Sustainable Urban Mobility. *Procedia–Soc. Behav. Sci.* **2014**, *162*, 408–418. [CrossRef]
20. Silva, A.N.R.; Costa Silva, M.; Ramos, R.A.R. Development and Application of I_SUM—An Index of Sustainable Urban Mobility. In Proceedings of the 89th Transportation Research Board Annual Meeting, Washington, DC, USA, 10–14 January 2010.
21. Lautso, K. *The SPARTACUS System for Defining and Analysing Sustainable Urban Land Use and Transport Policies*; Springer: Berlin/Heidelberg, Germany, 2003.
22. Lautso, K.; Spiekermann, K.; Wegener, M.; Sheppard, I.; Steadman, P.; Martino, A.; Domingo, R.; Gayda, S. Evaluating Urban Sustainability Using Land-Use Transport Interaction Models. *Eur. J. Transp. Infrastruct. Res.* **2004**, *4*, 56–87.
23. Litman, T. *Well Measured: Developing Indicators for Comprehensive and Sustainable Transport Planning*; Victoria Transport Policy Institute: Victoria, BC, Canada, 2007.
24. Litman, T. Sustainable Transportation Indicators: A Recommended Research Program for Developing Sustainable Transportation Indicators and Data. In Proceedings of the Transportation Research Board Annual Meeting, Washington, DC, USA, 13–17 January 2008.
25. Litman, T. *Well Measured: Developing Indicators for Sustainable and Livable Transport Planning*; Victoria Transport Policy Institute: Victoria, BC, Canada, 2016.
26. Van Audenhove, F.J.; Korniichuk, O.; Dauby, L.; Pourbaix, J. *The Future of Urban Mobility 2.0. Imperatives to Shape Extended Mobility Ecosystems of Tomorrow*; Arthur D. Little: Boston, MA, USA, 2014; pp. 10–25.
27. Bartłomiejczyk, M.; Kołacz, R. The reduction of auxiliaries power demand: The challenge for electromobility in public transportation. *J. Clean. Prod.* **2020**, *252*, 22. [CrossRef]
28. Połom, E. E-revolution in post-communist country? A critical review of electric public transport development in Poland. *Energy Res. Soc. Sci.* **2021**, *80*, 9. [CrossRef]
29. Fernández, R.A. Stochastic analysis of future scenarios for battery electric vehicle deployment and the upgrade of the electricity generation system in Spain. *J. Clean. Prod.* **2021**, *316*, 21. [CrossRef]
30. Faraway, J.J. *Practical Regression and Anova Using R*; University of Bath: Bath, UK, 2002; pp. 15–77.
31. Shmueli, G. To Explain or to Predict? *Stat. Sci.* **2010**, *25*, 289–310. [CrossRef]
32. Biuletyn Informacji Publicznej Ministerstwa Energii. Plan Rozwoju Elektromobilności w Polsce. Available online: https://www.nik.gov.pl/plik/id,23045,vp,25751.pdf (accessed on 7 November 2021).
33. Bechmark. Available online: https://www.benchmark.pl/aktualnosci/samochody-elektryczne-w-polsce-2021.html (accessed on 22 January 2023).

energies

MDPI

Article

Virtual Energy Storage System Scheduling for Commercial Buildings with Fixed and Dynamic Energy Storage

Grmay Yordanos Brhane [1], Eunsung Oh [2,*] and Sung-Yong Son [2,*]

[1] Department of Next Generation Energy System Convergence Based-on Techno-Economics, College of IT Convergence, Global Campus, Gachon University, Seongnam-si 13120, Gyeonggi-do, Republic of Korea; jordibrhane@gachon.ac.kr

[2] Department of Electrical Engineering, College of IT Convergence, Global Campus, Gachon University, Seongnam-si 13120, Gyeonggi-do, Republic of Korea

* Correspondence: esoh@gachon.ac.kr (E.O.); xtra@gachon.ac.kr (S.-Y.S.); Tel.: +82-31-750-5347 (S.-Y.S.)

Abstract: This study presents a virtual energy storage system (VESS) scheduling method that strategically integrates fixed and dynamic energy storage (ES) solutions to optimize energy management in commercial buildings. Fixed ES, such as batteries, provides stable flexibility but is expensive and can be inefficiently operated. In contrast, dynamic ES can be utilized as needed but requires validation of their flexibility. By combining fixed ES with dynamic ES utilizing vehicle-to-grid (V2G) capabilities, this approach enhances grid stability and manages energy costs more effectively. Empirical validation using real-world data from Korea demonstrates significant improvements in total net benefits by reducing energy costs, which are crucial for the economic sustainability of commercial energy use. Additionally, the analysis of Pearson's linear correlation coefficient with demand identifies where benefits occur in the scheduling process. The integrated system reduces the need for costly upgrades to the utility grid, suggesting a strategic advantage for large-scale adoption. This study establishes a framework for the broader implementation of such integrated systems, highlighting the potential for substantial improvements in energy efficiency, reduced carbon emissions, and enhanced grid reliability.

Keywords: battery; building; commercial building; energy efficiency; energy storage; vehicle-to-grid; virtual energy storage system

1. Introduction

1.1. Motivation

The commercial sector is a major consumer of energy, representing a significant portion of the global energy demand. According to the International Energy Agency (IEA), buildings were responsible for approximately 30% of global final energy consumption and 26% of global energy-related emissions in 2022 [1]. This substantial impact highlights the critical need to advance energy efficiency in commercial buildings as a strategic response to economic and environmental challenges.

With increasing energy costs and environmental concerns becoming more pressing, there is an urgent need to enhance energy efficiency and foster sustainable practices in commercial buildings [2]. Addressing inefficiencies in energy consumption and adopting energy storage technologies are critical steps toward reducing carbon emissions and advancing sustainable energy initiatives. The integration of energy-efficient systems and energy storage solutions in buildings is a key strategy for environmental conservation [3].

Energy storage systems (ESSs) are essential for optimizing energy consumption in buildings. By storing surplus energy during periods of low demand and releasing it during peak times, these systems mitigate the burden on electrical grids and improve energy efficiency within the premises [4]. However, the high upfront costs of ESSs present a significant barrier to their adoption in commercial settings. Overcoming this financial challenge is crucial for the broader uptake of these technologies [5].

The concept of shared energy storage models offers a promising solution to this challenge. Virtual energy storage systems (VESSs) allow multiple users or entities to share communal energy storage facilities. This collaborative model maximizes the efficiency of storage, diminishes the cost burden on individual participants, and enhances the overall effectiveness of stored energy utilization [6]. By leveraging the shared approach of a VESS, businesses and consumers can access energy storage solutions in a more economically feasible manner.

1.2. Prior Works

The widespread adoption of renewable energy sources in power grids presents several challenges for power system operators. The integration of energy storage systems (ESSs) has emerged as a promising solution [7]. ESSs are utilized in various applications, including interfacing within energy Internet ecosystems [8], smoothing output power in microgrids [9], and enhancing stability for virtual synchronous generators [10]. ESSs also enhance building energy reliability, reduce costs, support grid stability through frequency and voltage regulation, optimize renewable energy use, lower carbon footprints, and provide backup power during outages [5]. However, the initial capital investment required to procure and install ESSs can be substantial, particularly for large-scale systems capable of providing considerable energy storage capacity. This upfront cost burden poses a financial challenge to building owners and operators [11].

VESSs have emerged as a promising approach to address the cost barriers associated with ESS implementation in buildings. VESSs enable multiple users within a building or community to collectively use a single energy storage unit, thereby minimizing the initial capital expenditure [12]. Li et al. [13] developed a cost–benefit analysis framework for shared energy storage systems, emphasizing the economic advantages of collective energy management in reducing both operational and capital costs. Zhu and Ouahada [14] introduced a novel algorithm for the dynamic management of shared energy storage, significantly improving reliability and performance by adapting to real-time energy demands. Jasim et al. [15] proposed a demand response strategy for buildings equipped with VESSs, optimizing energy use and reducing peak loads on the grid, thus enhancing overall grid stability.

Furthermore, understanding and incorporating user behavior into the design and operation of VESSs has also been explored. Mediwaththe et al. [16] developed a user-centric model for VESSs that accounts for individual user preferences and behaviors to optimize energy usage and enhance system efficiency. This approach not only improves user satisfaction but also contributes to more balanced and efficient energy consumption patterns. Gailani and Crosbie [17] analyzed the impact of different regulatory scenarios on the adoption and effectiveness of VESSs, providing insights into how policy adjustments could foster wider implementation.

The integration of vehicle-to-grid (V2G) technology offers a compelling opportunity to enhance energy storage capabilities within buildings. V2G enables electric vehicles (EVs) to serve as mobile energy storage units, storing excess energy from the grid and feeding it back during peak demand periods [18]. One key advantage of V2G is its ability to leverage existing infrastructure—EVs and their batteries—for energy storage, minimizing the need for additional dedicated storage systems. This approach reduces upfront capital costs and maximizes resource utilization, aligning with circular economy principles by repurposing existing resources [19].

V2G also presents unique opportunities for revenue generation and cost offsetting through grid services and energy trading. EV owners can participate in demand response programs or provide ancillary services to the grid, earning incentives or payments for their contributions. In buildings, V2G-enabled EV fleets can be aggregated to provide significant storage capacity and demand flexibility, enhancing the economic viability of energy storage investments and potentially generating additional revenue streams for building owners and operators [20].

Furthermore, V2G integration promotes energy resilience and grid stability by diversifying storage options and enhancing grid flexibility. Building energy management systems can optimize energy use and mitigate grid disturbances by utilizing distributed energy resources such as EVs, solar panels, and stationary batteries. V2G-enabled buildings can function as microgrids, capable of islanding from the main grid during emergencies or blackouts, providing critical backup power with stored energy from EVs [21].

The integration of EVs and ESSs in households holds significant promise for reducing home energy costs and enhancing overall energy efficiency. Homeowners can optimize energy usage, store excess renewable energy, and leverage the flexibility of EV batteries to manage electricity demand more effectively [22]. The combination of ESSs and EVs offers a synergistic solution for energy reduction and optimization. Integrating ESSs and EVs into energy management systems provides numerous advantages, including grid stability, demand response capabilities, and enhanced renewable energy integration. By leveraging EV battery storage capacity and bidirectional energy flow flexibility, this integration enables efficient energy management, cost reduction, and environmental sustainability. Thus, the integration of ESSs and EVs is foundational to modern energy management practices, paving the way for a more resilient, efficient, and sustainable energy future [23].

However, existing research on the combination of ESSs and EVs in buildings has primarily focused on utilizing individual energy storage systems and EVs separately. There is a notable lack of studies exploring the integration of EVs as a source of virtual energy storage in conjunction with fixed shared energy storage systems. To address this gap, our study proposes a novel approach for scheduling energy in commercial buildings by integrating fixed ESSs and EVs. By leveraging the complementary strengths of fixed and mobile storage systems, this combined approach aims to optimize energy management strategies, thereby enhancing the efficiency and effectiveness of commercial building operations.

1.3. Contribution

The main contributions of this study can be summarized as follows:

- Integration of fixed and dynamic ESSs: This study proposes an approach for integrating both fixed and dynamic ESSs within a VESS scheduling framework. Unlike traditional methods that typically focus on either fixed or dynamic storage independently, this study demonstrates the enhanced efficiency and efficacy of combining these systems. By integrating fixed batteries and dynamic V2G capabilities, the proposed model optimizes energy usage within commercial buildings and enhances grid stability and energy cost management. This dual approach allows greater flexibility and resilience in energy management, adapting to both predictable energy demands and peak load fluctuations more effectively than standalone systems.
- Experimental results and discussions regarding a real dataset: The use of real-world data from Korea in the simulation process validates the practical applicability of the proposed scheduling method and enhances its relevance and utility for real-world scenarios. This methodological rigor ensures that the scheduling strategy is not only theoretically sound but also empirically effective, providing a replicable model for other researchers and practitioners in the field. These findings highlight the potential for significant reductions in energy costs and peak demand charges, which are crucial for the economic sustainability of energy use in commercial settings. Moreover, the demonstrated capability of the combined VESS in reducing the need for extensive and costly upgrades to utility grids presents a compelling case for policymakers to consider supportive policies and incentives for integrated energy storage solutions. This could lead to the broader adoption of such systems, ultimately contributing to enhanced energy efficiency, reduced carbon emissions, and greater energy independence on a larger scale.

2. System Description

In this study, we explore the deployment of a VESS in a commercial building comprising multiple units. The VESS integrates both fixed and dynamic ES solutions to enhance the energy efficiency of the participating units, markedly reducing electricity costs. Fixed ES relies on stationary batteries that offer reliable energy reserves. Concurrently, dynamic ES harnesses energy from parked EVs, contributing to the adaptability of the system. The interplay between the participant units and the service provider is managed through clearly defined contractual agreements that stipulate the responsibilities and limits of all parties involved. As depicted in Figure 1, the VESS for commercial buildings is structured around four principal components: the building energy service provider (BESP), participant units, fixed ES, and EV charging stations.

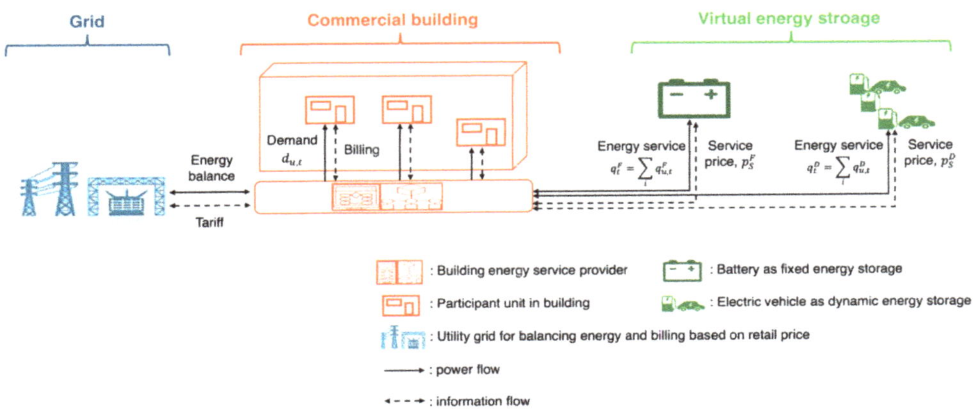

Figure 1. Constitution of a virtual energy storage system (VESS) for a commercial building comprised of multiple units.

In Figure 1, The components marked with green, orange, and blue icons correspond to the virtual energy storage, commercial building, and grid areas, respectively. Additionally, solid lines represent power flow, while dashed lines indicate information flow.

2.1. Building Energy Service Provider

The BESP fulfills a critical function within the VESS, orchestrating the energy flows between the participating units and energy producers. It employs advanced decision-making algorithms to determine the scheduling of the participants in line with the strategic objectives and operational constraints of the VESS. The primary aim is to ensure that energy distribution across building units is conducted smoothly, efficiently, and in an optimized manner.

The BESP operates on a service-oriented basis and offers access to a fixed ES capacity, which includes a fleet of stationary batteries. These batteries provide a reliable and regulated energy supply, particularly during peak demand periods or those related to electricity tariffs. By commercializing access to this fixed ES capacity, the BESP contracts service prices with the participating units, establishing a mutually beneficial economic relationship that incentivizes energy conservation and efficiency.

Although fixed ES implementation costs are shouldered by the BESP, they are considered marginal in the context of a long-term implementation scenario, as envisioned in this study. Such costs are expected to be offset by long-term operational savings and potential revenue generated through the service model.

In contrast, the dynamic ES, which encompasses the energy storage potential of EVs, is managed individually by EV owners. This component introduces a flexible and adaptive energy reserve that complements the fixed ES. The BESP facilitates the integration of this dynamic ES into the energy portfolio of the building but does not directly incur costs or receive benefits. However, it plays a pivotal role in coordinating the availability of EVs for energy storage and ensuring that this dynamic resource is effectively utilized within the VESS framework.

2.2. Participant Unit

Participant units stand to gain economically by engaging with a VESS, primarily through a reduction in their electricity bills, compared to scenarios where no ESS is utilized. By adopting an electricity tariff model, such as Time-of-Use (ToU), the electricity bill for a participant unit u, $B_u(\cdot)$, is calculated as follows:

$$B_u(\cdot) = p_d \max_{t\in\mathcal{T}} d_{u,t} + \sum_{t\in\mathcal{T}} p_{e,t} d_{u,t}, \tag{1}$$

where $\mathcal{T} = \{1,\cdots,t,\cdots,T\}$ represents the scheduling period, such as one month, $d_{u,t}$ indicates the demand for unit u at time t, $\max_{t\in\mathcal{T}} d_{u,t}$ denotes the peak demand among the scheduling period $\mathcal{T}$, p_d denotes the demand price for the billing period, and $p_{e,t}$ indicates the energy price at time t.

When unit u employs fixed ES, denoted by $e^F_{u,t}$, and dynamic ES, denoted by $e^D_{u,t}$, the electricity bill of unit u utilizing fixed and dynamic ES, $B_u(e^F_{u,t}, e^D_{u,t})$, is modified as

$$B_u(e^F_{u,t}, e^D_{u,t}) = p_d \max_{t\in\mathcal{T}} (d_{u,t} - e^F_{u,t} - e^D_{u,t}) + \sum_{t\in\mathcal{T}} p_{e,t} (d_{u,t} - e^F_{u,t} - e^D_{u,t}). \tag{2}$$

Participant units incur costs through service fees associated with the utilization of both fixed and dynamic ES. The cumulative cost for a participant unit u over a given scheduling period, $C_u(E^F_u, e^D_{u,t})$, is represented by

$$C_u(E^F_u, e^D_{u,t}) = \sum_{t\in\mathcal{T}} p^F_s E^F_u + \sum_{t\in\mathcal{T}} p^D_{s,t} |e^D_{u,t}|, \tag{3}$$

where E^F_u denotes the allocated fixed ES capacity for unit u, p^F_s denotes the per-unit service price for the fixed ES, and $p^D_{s,t}$ denotes the per-unit service prices for the dynamic ES unit at time t.

Equation (3) outlines that the fixed ES capacity, which is the battery capacity, is assigned to unit u for the entire scheduling period. In contrast, the dynamic ES capacity is billed at each decision point within that period. This differentiation is caused by the predictable nature of fixed ES capacity as opposed to the variable availability of dynamic ES capacity, which depends on the presence of EVs at any given time.

Considering the costs associated with the usage of ES, the net benefit for a participant unit u utilizing the VESS, $N_u(e^F_{u,t}, e^D_{u,t})$, is computed as

$$N_u(e^F_{u,t}, e^D_{u,t}) = B_u(\cdot) - B_u(e^F_{u,t}, e^D_{u,t}) - C_u(E^F_u, e^D_{u,t}) \ . \tag{4}$$

$B_u(\cdot)$ represents the electricity bill of unit u without employing ES, $B_u(e^F_{u,t}, e^D_{u,t})$ refers to the modified electricity bill of unit u accounting for the use of both fixed and dynamic ES, and $C_u(e^F_{u,t}, e^D_{u,t})$ denotes the cost incurred for the services of fixed and dynamic ES. The net benefit of unit u utilizing the VESS, $N_u(e^F_{u,t}, e^D_{u,t})$, is calculated as the gains obtained from using the VESS minus the costs of usage.

2.3. Fixed Energy Storage

Fixed ES refers to a designated ES capacity that is permanently installed within a building and utilized collectively by various units. This often encompasses conventional battery storage systems that store a specific quantity of energy that is then dispatched based on the proposed scheduling method.

The cost-efficiency of fixed ES is directly related to its capacity. As the capacity of fixed ES expands, the per-unit cost generally declines owing to the benefits of economies of scale. Let E^F denote the capacity of the fixed ES. The unit service price for employing fixed ES throughout the scheduling duration associated with the ESS is mathematically formulated as [24]

$$p_s^F = T\left[\alpha_1 \exp\left(-\alpha_2 E^F\right) + \alpha_3\right], \tag{5}$$

where α_1, α_2, and α_3 are parameters intrinsic to the cost characteristics of fixed ES. In Equation (5), the exponential term reflects the levelized cost of storage (LCOE) [25], which decreases as the capacity increases, whereas the multiplicative factor T represents the duration of service provision.

2.4. Dynamic Energy Storage

In this study, dynamic ES refers to the use of EVs as flexible energy storage systems capable of rapidly responding to fluctuations in energy demand and supply. This dynamic utilization involves tapping the batteries of electric vehicles to store and release energy.

EV owners gain financial benefits from participating in the VESS. They can monetize the energy stored in their vehicle batteries by discharging it during periods of high demand, thereby obtaining a higher price for the energy provided. Consequently, the service price for dynamic ES is directly linked to the EV charging tariff. For this analysis, it is assumed that the EV charging tariff, denoted as $p_{s,t}^D$, remains constant throughout the scheduling period and is represented as [26]

$$p_{s,t}^D = p_{EV}. \tag{6}$$

The model can be easily extended to accommodate variable pricing over time, allowing for adjustments based on temporal changes in energy demand and tariff rates.

3. Virtual Energy Storage Scheduling Method

3.1. Objective Function

The principal objective of VESS scheduling for a commercial building is to maximize social welfare, which is defined here as the total net benefit accrued by all participating units. The objective function of the VESS scheduling problem, $\mathcal{O}(e_{u,t}^F, e_{u,t}^D)$, that quantifies this overarching goal is formulated as follows:

$$\mathcal{O}\left(e_{u,t}^F, e_{u,t}^D\right) = \sum_{u \in \mathcal{U}} N_u\left(e_{u,t}^F, e_{u,t}^D\right), \tag{7}$$

where $N_u(e_{u,t}^F, e_{u,t}^D)$ represents the net benefit for each participant unit u, derived from the use of both fixed (F) and dynamic (D) energy storage options at time t, and $\mathcal{U} = \{1, \cdots, u, \cdots, U\}$ is the set of all participant units within the building.

This function of Equation (7) aims to optimize energy usage across all units by effectively scheduling the discharge and charge cycles of fixed and dynamic ES. It integrates the financial, operational, and environmental aspects of energy management to ensure that the energy supply is economically beneficial while aligning with sustainability objectives.

3.2. Fixed ES Constraint

The scheduling of a VESS must adhere to resource constraints to ensure efficient and sustainable energy management. The critical constraint is related to the fixed ES capacity and is denoted as E^F. The sum of the capacities allocated to all participating units must not exceed the total fixed ES capacity. The fixed ES capacity constraint is expressed as [7]

$$\sum_{u \in \mathcal{U}} E_u^F \leq E^F, \tag{8}$$

where E_u^F represents the fixed ES capacity allocated to each unit u.

Moreover, the operation of fixed ES at each participant unit u during any given time t must be within the allocated capacity. The operational conditions for ES, utilized as a battery, include a charging and discharging rate (C-rate) of 0.5. This results in the following power constraints [7]:

$$-0.5\, E_u^F \leq e_{u,t}^F \leq 0.5\, E_u^F, \qquad \forall t \in \mathcal{T}, \forall u \in \mathcal{U}. \tag{9}$$

Additionally, the state-of-charge (SoC) of the battery, which is affected by accumulated actions up to time t, starting from an initial state $s_{u,0}^F$, must not exceed the maximum allowable SoC, represented by E_u^F. This is expressed as [7]

$$0 \leq s_{u,0}^F + \sum_{i=0}^{t} e_{u,i}^F \leq E_u^F, \qquad \forall t \in \mathcal{T}, \forall u \in \mathcal{U}. \tag{10}$$

This constraint ensures that the capacity of the battery is optimally utilized without overcharging or deep discharging, thereby maintaining its longevity and efficiency.

3.3. Dynamic ES Constraint

Dynamic ES, which involves EVs in a V2G service, is limited by the flexibility of the V2G.

Predicting the flexibility of V2G involves analyzing historical data on the charging and discharging behaviors of EVs, such as the duration and frequency at which they are connected to the grid and available for energy services. These data are essential for estimating the potential energy contributions of EVs to the grid.

From existing research [27], the available V2G flexibility at a charging station (CS) equipped with m chargers, V_t, can be quantified as

$$V_t = E^D \Delta E\, m\, \rho_t [1 - P_b(\rho_t, m)], \tag{11}$$

where E^D denotes the charging power of the charger installed at the CS, ΔE indicates the participation time of EVs in the V2G service, ρ_t symbolizes the utilization rate of the CS, and $P_b(\rho_t, m)$ represents the blocking probability, which indicates the likelihood that an EV cannot access a charger due to all chargers being occupied.

The blocking probability is calculated using the formula [28]

$$P_b(\rho_t, m) = \frac{\rho_t{}^m / m!}{\sum_{i=0}^{m} \rho_t{}^i / i!}. \tag{12}$$

To maximize the participation of the EVs in the V2G program, it is necessary to manage the blocking probability of the CS with the following constraint [27]:

$$P_b(\rho_i^+, m) - P_b(\rho_i, m) \leq \gamma_{Th}, \; i \in \{t, t+1, \cdots, T\}, \tag{13}$$

where ρ_i^+ denotes the adjusted utilization rate that reflects the increased EV charging time due to participation in V2G services.

Thus, the dynamic ES capacity for a given time period is defined as

$$E_t^D = E^D \Delta E^*\, m\, \rho_t [1 - P_b(\rho_t, m)], \tag{14}$$

where ΔE^* indicates the maximum V2G service time adjusted by the blocking probability. Equation (14) expresses how the dynamic ES capacity varies with both the historical usage patterns and the current utilization rate of the charging station.

According to the established operational constraints for dynamic ES derived from the V2G capabilities, the available dynamic ES that the participant units can utilize is strictly regulated. This regulation ensures that the dynamic interaction between EVs and the power grid remains within the capacity provided by the V2G service. This constraint is mathematically expressed as follows:

$$0 \leq \sum_{u \in \mathcal{U}} e^{D}_{u,t} \leq E^{D}_{t}, \qquad \forall t \in \mathcal{T}. \tag{15}$$

Unlike fixed ES, which is allocated individually to participant units, Equation (15) expresses that dynamic ES operates on an aggregated basis. This collective management approach is fundamentally different because dynamic ES capitalizes on the variable and mobile nature of EV batteries. In practice, this means that the energy stored in EVs is pooled and managed as a common resource, which participant units draw from or contribute to based on real-time requirements and availability.

The aggregation of dynamic ES allows for greater flexibility and efficiency in managing energy flows. This enables the system to respond more effectively to fluctuations in the energy demand and supply across buildings or grids. For example, during times of peak demand, energy can be dynamically drawn from multiple EVs connected to the system. Conversely, during off-peak times, these vehicles can be collectively recharged, thereby optimizing both the energy usage and cost efficiency.

This method of aggregated usage contrasts with fixed ES, in which each unit has a predetermined storage capacity that is not shared with the other units. Fixed capacity typically contains stationary batteries installed within a building that are used exclusively by the unit to which they are assigned. Although reliable, these systems lack the dynamic scalability and flexibility provided by the pooled resources of dynamic ES.

3.4. VESS Scheduling Method with Fixed and Dynamic ESs

The proposed scheduling method for the VESS, which incorporates both fixed and dynamic ES, was formulated to maximize social welfare, which is defined as the total net benefit from energy savings across all participant units. The optimization problem is expressed as follows:

$$\begin{aligned} &\max_{e^{F}_{u,t}, e^{D}_{u,t}} \mathcal{O}\left(e^{F}_{u,t}, e^{D}_{u,t}\right) = \sum_{u \in \mathcal{U}} N_u\left(e^{F}_{u,t}, e^{D}_{u,t}\right) \\ &\text{subject to} \sum_{u \in \mathcal{U}} E^{F}_{u} \leq E^{F}, \\ &\quad -0.5\, E^{F}_{u} \leq e^{F}_{u,t} \leq 0.5\, E^{F}_{u}, \qquad \forall t \in \mathcal{T}, \forall u \in \mathcal{U}, \\ &\quad 0 \leq s^{F}_{u,0} + \sum_{i=0}^{t} e^{F}_{u,i} \leq E^{F}_{u}, \qquad \forall t \in \mathcal{T}, \forall u \in \mathcal{U}, \\ &\quad 0 \leq \sum_{u \in \mathcal{U}} e^{D}_{u,t} \leq E^{D}_{t}, \qquad \forall t \in \mathcal{T}. \end{aligned} \tag{16}$$

Equation (16) outlines the optimization framework, presenting a convex problem with the objective of maximizing electricity bill savings by strategically allocating fixed and dynamic ES resources during each billing period. The convex nature of the problem suggests that it can be effectively solved using iterative methods, such as the gradient descent technique, which ensures convergence to the global optimum [29]. Additionally, to solve the problem, predicted information for future power demand and EV profiles is utilized. This predictive approach allows for the efficient scheduling of resources based on anticipated demand and availability. However, for real-time operation of the scheduled resources, compensation for uncertainties at each time point is necessary. The model does not currently address these real-time uncertainties, which presents a limitation in its practical application.

The VESS scheduling methods described in Equation (16) comprehensively address the dual aspects of fixed and dynamic ES by leveraging the unique benefits of each type. For fixed ES, using a shared large-capacity battery system enables significant economies of scale. By assigning the specific capacity of this shared battery to individual users for the scheduling duration, the unit cost of energy storage can be substantially reduced. This is a critical advantage, as it not only lowers the financial barriers associated with adopting energy storage technologies but also enhances the overall economic feasibility of the VESS.

In contrast, dynamic ES exploits the benefits of multi-user diversity. By allowing flexible and real-time sharing of energy resources among multiple users, primarily through the use of EVs in a V2G framework, dynamic ES can adaptively respond to fluctuations in energy demand and supply. This dynamic utilization optimizes grid stability and energy efficiency, providing a robust mechanism for harnessing the intermittent and unpredictable nature of renewable energy sources.

Collectively, these strategies highlight a sophisticated approach to managing ES in the context of a commercial building. The integration of fixed and dynamic ES capitalizes on economic efficiency and operational flexibility while contributing significantly to the sustainability goals of ES. By ensuring that energy storage is both cost-effective and dynamically responsive, this VESS scheduling method supports a broader transition toward smarter and more sustainable energy infrastructures.

4. Results and Discussion

4.1. Experimental Environment

To evaluate the proposed VESS scheduling method, a simulation was conducted on a building comprising 24 units. The energy demand data for this simulation were sourced from the Korea Micro Grid Energy Project (K-MEG) and recorded with a 1-h resolution [30]. Figure 2 illustrates the average daily demand distribution across these building units, providing insights into the consumption patterns that are critical for optimizing energy storage and usage.

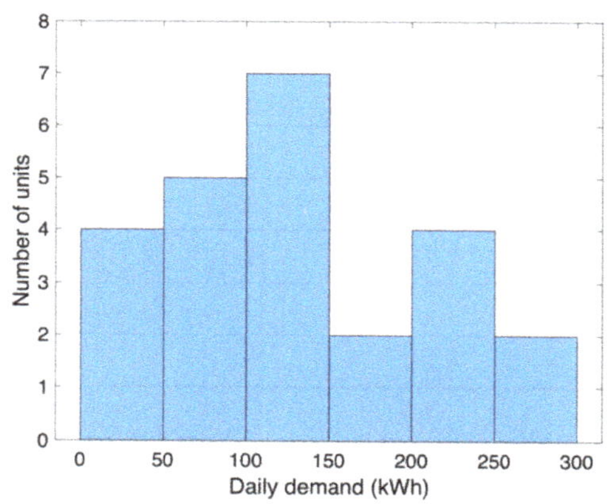

Figure 2. Average daily demand distribution of the units in the building.

To calculate electricity bills in this simulation, the ToU tariff applied was that of the medium general demand-metered service provided by the Pacific Gas and Electric Company (PG&E). Under this tariff, the energy prices for the mid-peak and on-peak periods are considered equal, as detailed in Table 1 [31].

Table 1. ToU tariff of Pacific Gas and Electric Company.

Demand Price (USD/kW)	Energy Price (USD/kWh)		
	Off-Peak	Mid-Peak	On-Peak
24.10	0.27	0.20	0.20

In the simulation, fixed ES was implemented using lithium-ion batteries, reflecting one of the most common choices in current energy storage systems. The cost parameters associated with the fixed ES service, denoted as $\alpha 1$, $\alpha 2$, and $\alpha 3$, were set to 0.17, 0.0004, and 0.043, respectively, to calculate the service price according to the levelized cost of storage (LCOE) formula detailed in [24].

For dynamic ES, the service price was determined based on the average rate of the Business Electric Vehicle tariff proposed by the PG&E, which is 0.15 USD/kWh [32]. This rate effectively reflects the economic considerations associated with utilizing electric vehicles as part of a VESS.

The flexibility offered by dynamic ES was facilitated through a CS with specifics on the EVs used for dynamic ES sourced from the Korea Environment Corporation (KEC) [33], as depicted in Figure 3. The selected charging station was equipped with four DC chargers, each capable of delivering a charging power of 50 kWh. In Figure 3a, the maximum utilization of the CS is approximately 1.6, which suggests that during peak times, an average of 1.6 EVs simultaneously utilize each of the four available CSs. Figure 3b illustrates the associated blocking probability at times of maximum utilization, which was quantified as approximately 0.06. This indicates that out of every 100 EVs that arrive at the CS intending to charge, approximately 6 EVs are unable to do so because all the chargers are occupied. This blocking probability is a critical metric as it quantifies the likelihood of service denial, which is a crucial factor in assessing the adequacy of the charging station infrastructure to satisfy the demand.

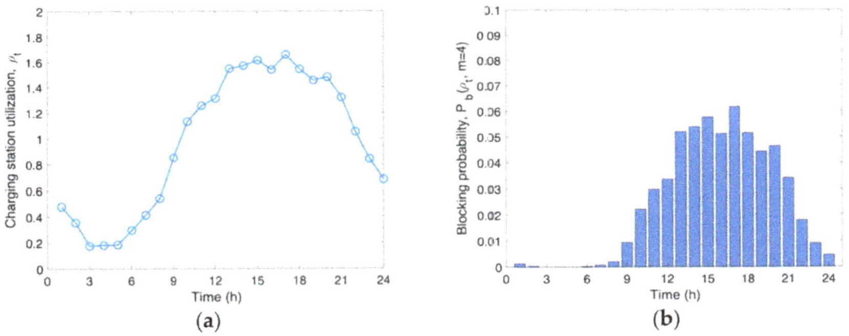

Figure 3. Charging station status. (**a**) Charging station utilization. (**b**) Blocking probability of the charging station.

To clarify the parameters used in our simulations, we have summarized them in Table 2 below.

The total net benefit, defined in Equation (7) and used as the objective function of the scheduling problem in Equation (16), is selected as the system performance metric. In the simulation, the scheduling period is assumed to be one month. Therefore, the total net benefit is measured on a monthly basis. This assumption allows for a detailed analysis of the VESS scheduling method's performance within a practical timeframe, reflecting the typical billing cycle in commercial settings.

To validate the effectiveness of the proposed scheduling method, a comparative analysis with existing methods is necessary. However, methodologies that combine fixed and dynamic ES approaches in the same environment have not yet been developed. Therefore, we compare the proposed method with two existing scenarios: a fixed-ES-only scenario and a dynamic-ES-only scenario. This comparison allows us to evaluate the performance of the proposed method, which integrates both fixed and dynamic ES systems, against established approaches.

By conducting this comparative analysis, we aim to demonstrate the advantages and improvements offered by the combined fixed and dynamic ES approach over traditional methods that utilize either fixed or dynamic ES in isolation. The performance metrics and simulation results will highlight the benefits of integrating these two types of ES, providing a comprehensive evaluation of the proposed scheduling method.

Table 2. Simulation parameters.

Parameter	Value	Description
	Parameters related to BESP	
Scheduling period	One month	Value according to the billing cycle
	Parameters related to participant units	
Demand of unit	Measured value	Recorded value by the K-MEG project [30]
Electricity tariff	A-10 of PG&E	Announced tariff for medium general demand-metered service [31]
	Parameters related to fixed ES	
Battery cost parameters	$\alpha1 = 0.17$, $\alpha2 = 0.0004$, and $\alpha3 = 0.043$	Fitting value of the Li-ion battery's LCOE [24]
Battery capacity	10~500 kWh	Variable to measure performance under different conditions.
	Parameters related to dynamic ES	
EV profile	Measured value	Recorded value by the KEC [33]
EV tariff	BEV of PG & E	Announced tariff for business electricity vehicle [32]
Charging power of chargers	50 kW	Specification of fast charger [27]
Number of CSs	3, 4, 5	Variable to measure performance under different conditions.
Blocking probability threshold	0.02~1.00	Variable to measure performance under different conditions.

4.2. Fixed-ES-Only Scenario

The operation of fixed ES within the proposed VESS is subject to capacity limitations, as defined in Equations (9) and (10), respectively. Figure 4 illustrates the impact of varying fixed ES capacities ranging from 10 to 500 kWh on the objective of maximizing social welfare, measured as the total net benefit to all participating units.

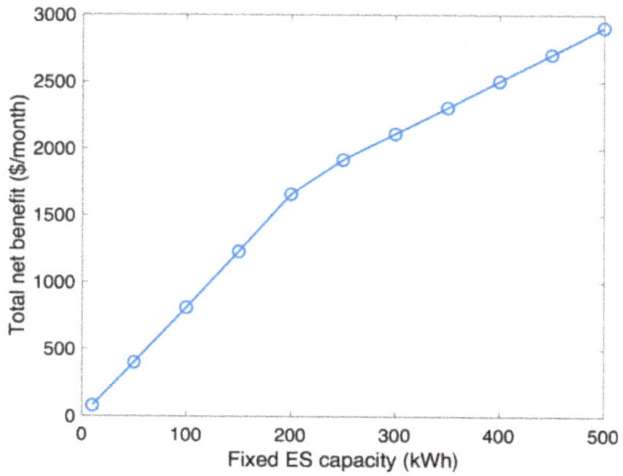

Figure 4. Total net benefit with varying fixed ES capacities in the fixed-ES-only scenario.

As shown in Figure 4, the total net benefit increases linearly as the fixed ES capacity increases to 200 kWh. Beyond this point, the rate of increase in the net benefits begins to decrease. This diminishing return indicates that, while participants gain maximum benefits up to a fixed ES capacity of 200 kWh, additional capacity beyond this threshold results in smaller incremental gains relative to the service fees incurred for using fixed ES. This suggests an optimal capacity threshold at which the cost–benefit ratio of fixed ES utilization stabilizes.

This result also highlights the limitations of using fixed ES. When a fixed ES capacity is allocated, costs are incurred for the entire scheduling period. Participant units need to consider these costs when purchasing capacity. However, there is an inherent limitation in the utilization of fixed ES during periods when it is not being fully used. This under-utilization during off-peak times means that fixed ES may not always be cost-effective, emphasizing the need for careful capacity planning and optimization to ensure that fixed ES provides maximum economic benefit.

To further explore the characteristics driving the gains from fixed ES, Table 3 presents the results of calculating the Pearson's linear correlation coefficient (PLCC) between the participants' demand profiles and their purchased fixed ES capacities, E_u^F [34].

Table 3. Pearson's linear correlation coefficient between the participants' demand profiles and their purchased fixed ES capacities, E_u^F.

200 kWh Fixed ES Capacity Case			500 kWh Fixed ES Capacity Case		
Average Demand	Peak Demand	E_u^F	Average Demand	Peak Demand	E_u^F
0.21	0.77	1.00	0.38	0.88	1.00

The data reveal a strong correlation, with coefficients between the participants' peak demand and the capacity of fixed ES they procure ranging from 0.77 to 0.88. This high correlation underscores the fact that participants primarily invest in fixed ES to mitigate the costs associated with high-demand prices, which are significantly more substantial than variable energy prices. Notably, as the fixed ES capacity increases to 500 kWh, the correlation with the peak demand strengthens compared with that at 200 kWh, indicating that users with higher peak demands tend to utilize larger fixed ES capacities more intensively.

4.3. Dynamic-ES-Only Scenario

Equation (13) shows that the operational scope of dynamic ES is constrained by the blocking probability threshold. The blocking probability, which indicates the likelihood that an EV cannot be charged because all chargers are occupied, significantly affects the system's performance. This probability is also affected by the number of available CSs in Equation (12).

Figure 5 illustrates how changes in the blocking probability threshold and number of CSs affect the total net benefit. The results show that as the blocking probability threshold increases (indicating a relaxation in the blocking constraint) and the number of CSs increases, the total net benefit also increases.

Specifically, with four charging stations, increasing the blocking probability threshold from 0.02 to 0.1 results in a net benefit increase from 882.5 to 2260.3, which is an approximately 156% increase. Adding one more CS under the same conditions boosts the total net benefit to $3092.4, a further increase of approximately 250%.

Additionally, compared to the results shown in Figure 4, Figure 5 does not exhibit discontinuous changes in the total net benefit. This difference is attributed to the nature of dynamic ES, which allows for the purchase of the required capacity at the needed time, unlike fixed ES, which requires upfront allocation for the entire period. This flexibility in dynamic ES usage results in a smoother increase in net benefits.

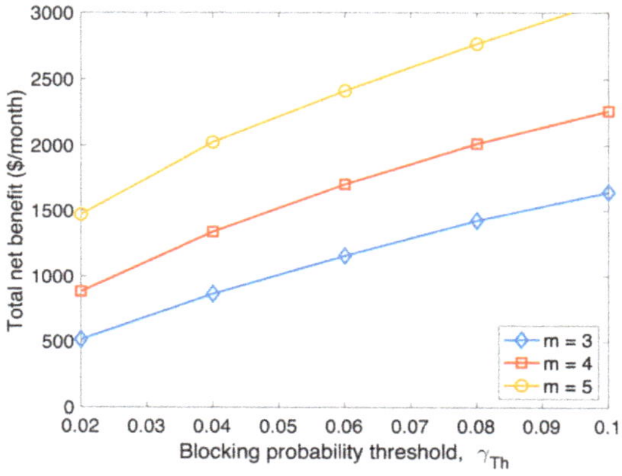

Figure 5. Total net benefit with varying blocking probability threshold and number of charging stations in the dynamic-ES-only scenario.

These findings underscore the critical role of the number of CSs in enhancing the performance of dynamic ES systems. The data demonstrate that both increasing the blocking probability threshold and adding more charging stations significantly boost the total net benefit, highlighting the importance of strategic planning in the deployment of dynamic ES infrastructure.

Table 4 explores the characteristics driving the gains from dynamic ES, similar to the analysis performed for the fixed-ES-only scenario. It presents the results of calculating PLCC between the participants' demand profiles and their total purchased dynamic ES capacity $\sum_{t\in\mathcal{T}} e^D_{u,t}$ with four CSs.

Table 4. Pearson's linear correlation coefficient between the participants' demand profiles and their total purchased dynamic ES capacities, $\sum_{t\in\mathcal{T}} e^D_{u,t}$ with four charging stations.

0.04 Blocking Probability Threshold			0.08 Blocking Probability Threshold		
Average Demand	Peak Demand	$\sum_{t\in\mathcal{T}} e^D_{u,t}$	Average Demand	Peak Demand	$\sum_{t\in\mathcal{T}} e^D_{u,t}$
−0.33	−0.44	1.00	−0.40	−0.54	1.00

Unlike fixed ES, which allows for controlled operation regardless of external conditions, the capacity available from dynamic ES is highly dependent on the state of the EVs. The results indicate a weak negative correlation between the dynamic ES capacity and demand profile characteristics. This suggests that in the operation of dynamic ES, the availability and state of EVs have a more significant impact on performance than the participants' actual energy needs.

4.4. Combining Fixed ES and Dynamic ES Scenario

The proposed VESS scheduling method aims to optimize the combined operation of fixed and dynamic ES to maximize social welfare. This scenario investigates the interplay between these two types of ES and their collective impact on the total net benefit for participants.

Figure 6 illustrates the variation in the total net benefits as the constraints on fixed ES and dynamic ES change, demonstrating how these combined systems can be coordinated to enhance performance. Figure 6a,b presents the results for scenarios with four and five CSs, respectively, illustrating the different VESS configurations.

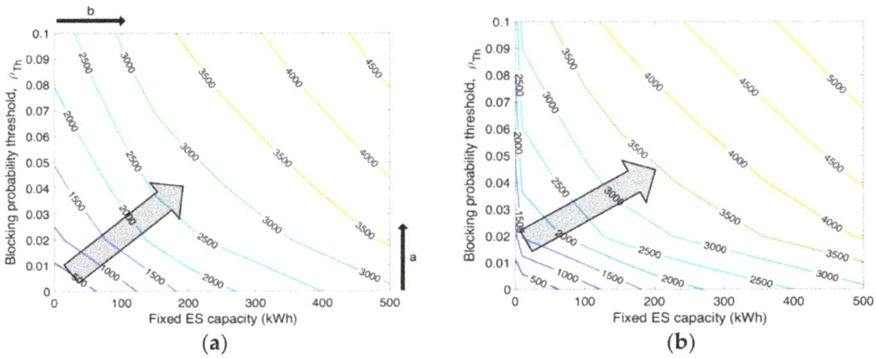

Figure 6. Total net benefit with varying fixed ES capacity and blocking probability threshold in the combining fixed and dynamic ES scenario. (**a**) Four-charging-station case; (**b**) five-charging-station case.

In Figure 6a,b, the axes are defined to reflect the changes in net benefits due to the interaction between fixed and dynamic ES. The y-axis ('a' direction along the y-axis) measures changes in the total net benefits due to the inclusion of dynamic ES, as compared to the scenario where only fixed ES is used. This axis provides insights into the additional benefits of dynamic ES over and above the baseline established by fixed ES alone. Moreover, the x-axis ('b' direction along the x-axis) measures changes in total net benefits due to the inclusion of fixed ES, as compared to the scenario where only dynamic ES is utilized. This axis highlights the incremental benefits that fixed ES brings to a system initially optimized only for dynamic ES.

The results depicted in Figure 6 strongly indicate that the combined VESS operation of fixed and dynamic ES yields greater total net benefits compared to scenarios where only fixed ES or only dynamic ES are used. Specifically, the total net benefit with 200 kWh fixed ES only and dynamic ES only with 0.04 blocking probability scenarios is \$1660.4 and \$1338.3, as shown in Figures 4 and 5, respectively. By combining fixed ES and dynamic ES with 4 CSs, the total net benefit increases to \$2767.1. This enhancement represents a 66.7% increase from the fixed-ES-only scenario and a 106.8% increase from the dynamic-ES-only scenario. This confirms the effectiveness of the proposed VESS scheduling method, which strategically integrates both storage types to optimize overall system performance.

In Figure 6a, the increase in the total net benefits is symmetrically distributed about the origin, as indicated by the gray arrows. This symmetry suggests that fixed and dynamic ES contribute comparably to the gains realized in the combined operational scenario. The balanced enhancement from both storage types underscores the well-coordinated integration within the VESS, leading to a harmonized increase in net benefits.

Conversely, Figure 6b exhibits a more pronounced increase in total net benefits in the 'b' direction. This indicates that in scenarios with a higher number of CSs, dynamic ES plays the predominant role in improving the total net benefits. The difference between Figure 6a,b arises primarily from the number of CSs. As the number of CS increases, as shown in Figure 5, the flexibility provided by dynamic ES also increases. This enhanced flexibility enables dynamic ES to have a more substantial impact on VESS scheduling in the scenarios depicted in Figure 6b, contributing to a steeper increase in benefits.

The results from Figure 6 suggest that the combined approach of using fixed and dynamic ES benefits from resource diversity, which enhances overall performance. The change in the direction of the gray arrow between Figure 6a,b highlights how the dominant resource and the direction of resource utilization can shift based on the available resources. This study presents the relationship between these resources and their combined utilization, but future research should focus on analyzing the impact of unit resource amounts on system performance. Such research could evaluate which resources have a dominant impact and develop methods for optimal resource composition, further improving the system's efficiency and effectiveness.

Overall, the combined use of fixed and dynamic ES demonstrates a scalable and flexible approach, offering higher efficiency and economic gains compared to using either fixed or dynamic ES alone. The results suggest that careful adjustment of the blocking probability threshold and fixed ES capacity is crucial for maximizing total net benefits.

In exploring the synergistic effects of combined fixed and dynamic ES within the VESS, Table 5 plays a crucial role by quantifying the relationship between participants' demand profiles and their capacity purchases from both types of ES.

Table 5. Pearson's linear correlation coefficient between the participants' demand profiles and their total purchased fixed ES capacity E_u^F and dynamic ES capacities, $\sum_{t\in\mathcal{T}} e_{u,t}^D$ with four charging stations.

200 kWh Fixed ES Capacity and 0.04 Blocking Probability Threshold					
Average Demand	**Peak Demand**	E_u^F	**Average Demand**	**Peak Demand**	$\sum_{t\in\mathcal{T}} e_{u,t}^D$
0.38	0.81	1.00	0.27	0.28	1.00

Unlike the dynamic-ES-only scenario presented in Table 4, where a weak negative correlation is observed, Table 5 reveals a positive correlation between the demand profiles and dynamic ES purchase capacity in the combined VESS scheduling scenario. This shift suggests that when dynamic ES is integrated with fixed ES, the purchased capacity is more closely aligned with the actual demand patterns, particularly during peak usage periods.

Furthermore, when compared with the fixed-ES-only scenario results in Table 3, the combined VESS scheduling exhibits a stronger correlation with the demand profiles, particularly with the peak demand profile. This enhanced correlation indicates that the combined scheduling strategy can control and reduce peak demands more effectively than scenarios that utilize fixed or dynamic ES independently. Effective peak demand management is crucial for reducing participant costs and mitigating operational stress on utility grids, which typically face higher costs during peak periods.

The results in Table 5 suggest that the proposed combined VESS scheduling strategy increases the net benefits for participant units by optimizing the energy costs associated with peak demand and provides broader benefits to the utility grid. By smoothing peak loads more efficiently, the combined system helps reduce the necessity for costly peak-time energy production and contributes to overall grid stability and sustainability.

4.5. Discussion Summary

From the results of the scenarios explored in Sections 4.2–4.4, it is clear that integrating both fixed and dynamic ES optimizes the VESS scheduling more effectively than using either system in isolation. Based on these findings, the following key points can be highlighted for the VESS operation:

- Net benefit dynamics: The total net benefits are directly proportional to the combined capacity of the fixed and dynamic ES. As shown in Figure 6, the VESS scheduled under the combined scenario achieves higher benefits owing to the optimized resource utilization and cost efficiency compared to the scenarios where only one type of ES is used. This implies that the synergy between the fixed and dynamic ES significantly enhances the performance of the system.
- Operational efficiency: The operational efficiency of a VESS varies significantly across different setups. In scenarios that utilize only fixed ES, efficiency hinges on matching storage capacity precisely with demand because both oversizing and undersizing can introduce cost inefficiencies. For dynamic ES setups, efficiency is more variable and heavily dependent on the availability and capacity of the charging infrastructure, which can be mitigated by increasing the number of CSs. However, combining fixed and dynamic ES offers the best of both systems, optimizing operational efficiency by

using fixed ES for stability and dynamic ES for flexibility. This combination minimizes energy waste and enhances the ability of the system to respond adaptively to demand fluctuations, thereby showcasing a superior operational model.

- Demand management: Demand management capabilities differ across various VESS scenarios. Fixed ES systems effectively manage base and peak loads within their capacity but lack the flexibility to respond to demands beyond their static limits. Conversely, dynamic ES systems, reliant on EV availability and charging behaviors, tend to struggle with peak demand management when used in isolation because of their dependence on external factors. The integration of both fixed and dynamic ES into a combined system proves to be the most effective, leveraging dynamic ES for its responsiveness to sudden peaks and fixed ES for consistent baseline load management, thereby ensuring a comprehensive demand response strategy.
- Implications for the utility grid and social welfare: The combined VESS operation effectively reduces peak demand, which directly benefits the utility grid by lowering the need for costly peak-time energy production and deferring new infrastructure investments. Moreover, as the combined system enhances the efficiency of both fixed and dynamic ES, it promotes a higher level of social welfare improvement across participating units and broader energy networks.

The results of this study demonstrate the potential benefits of integrating fixed and dynamic energy storage systems (ESSs) into commercial buildings. However, there are several limitations to this research that need to be acknowledged:

- Real-time operation: Our approach focuses on the scheduling of flexible resources and derives solutions as an optimization problem. For real-time operation of the scheduled resources, compensation for uncertainties at each time point is necessary. The model does not currently address these real-time uncertainties.
- Economic impact analysis: The study uses total net benefit to evaluate the economic impact. However, this metric can be influenced by fluctuations in electricity market prices, which may affect the overall benefits. The variability of market conditions needs to be considered for a more comprehensive economic analysis.
- VESS integration into the utility grid: The integration of VESS into the utility grid may introduce additional constraints related to grid stability and operational limitations. These constraints could affect the feasibility and effectiveness of VESS operations, especially during peak demand periods.

According to these limitations, the following future research directions are suggested:

- Integration of renewable energy sources: The potential for integrating renewable energy sources, such as solar and wind, with fixed and dynamic ES within the VESS can be explored. This study focuses on effectively storing the excess energy generated during peak production times and utilizing it during demand peaks or low production periods to enhance sustainability and reduce reliance on non-renewable energy sources. Additionally, real-time uncertainty compensation methods should be developed to address the variability in renewable energy generation.
- Scalability and modular system design: Investigate the scalability of VESS by implementing modular designs that can be easily expanded or adjusted based on the increase in energy demands or changes in the user base. This includes studying the impacts of scaling up on system performance, cost efficiency, and infrastructure requirements. Future research should also consider the integration of uncertainty compensation mechanisms to ensure robust real-time operations as the system scales.
- Economic and regulatory impact analysis: Comprehensive studies can be conducted on the economic impacts of widespread VESS adoption, including cost–benefit analysis, return on investment, and effects on electricity market prices. Sensitivity analyses should be performed to understand how varying market conditions influence the total net benefit. Additionally, frameworks for assessing the regulatory challenges and opportunities should be developed to facilitate the broader deployment of VESS.

- Impact on utility grids: The long-term impacts of large-scale VESS integration on utility grids, including grid stability, load balancing, and the potential to reduce peak load pressures, need to be analyzed. Improvements in these areas will help understand how VESS can be optimized to support the existing grid infrastructure and contribute to grid modernization. Future research should also explore the implications of VESS integration on utility grids, considering constraints such as grid stability and operational limitations. Analyzing how these constraints affect VESS operations will provide insights into optimizing the system for enhanced grid reliability and performance.
- Real-time application of study results: As mentioned in the limitations, our study focuses on scheduling and does not explicitly consider real-time operation. However, the problem addressed can be reformulated as a sequential decision making (SDM) problem, as indicated by the benefit model used in our objective function. Our problem can be tackled using techniques such as dynamic programming or reinforcement learning-based approaches to facilitate real-time operation [35]. Future research should explore this reformulation to enable real-time implementation, addressing the uncertainties at each decision point and optimizing operations dynamically.

5. Conclusions

This study developed an optimized VESS scheduling method for commercial buildings that incorporates both fixed and dynamic ES systems. The VESS model included three main components: fixed ES, dynamic ES, and a utility grid—creating a comprehensive framework for energy management in commercial settings. The VESS scheduling problem, which integrates constraints from both fixed and dynamic systems, was formulated as a mixed-integer linear problem, allowing for optimal solutions through computational techniques, such as gradient methods and dual decomposition. The simulation results, utilizing a real dataset from Korea, demonstrated that the combined VESS scheduling of fixed and dynamic ES provided substantial benefits in terms of total net benefit and operational efficiency. Specifically, the integration of these systems resulted in improved energy cost management and enhanced grid reliability compared with scenarios that utilize only one type of storage system. Furthermore, the combined system showed a significant increase in total net benefits, proving the effectiveness of integrating multiple ES systems.

The basic framework established for the VESS, fixed ES, and dynamic ES can be extended to more complex scenarios. Future research could explore real-time operation scenarios that include active units and their uncertainty, such as prosumers with distributed renewable generators, which would add a layer of complexity and relevance to the VESS model. Additionally, problems related to ownership and the economically optimal sizing of VESS capacities warrant further investigation. Finally, the utility grid model can be expanded to include more detailed power system requirements to provide a more nuanced understanding of grid dynamics and energy management strategies.

Author Contributions: Conceptualization, G.Y.B., E.O. and S.-Y.S.; methodology, G.Y.B., E.O. and S.-Y.S.; software, G.Y.B. and E.O.; validation, E.O. and S.-Y.S.; formal analysis, E.O. and S.-Y.S.; investigation, E.O. and S.-Y.S.; resources, E.O.; data curation, E.O.; writing—original draft preparation, G.Y.B.; writing—review and editing, E.O. and S.-Y.S.; visualization, E.O. and S.-Y.S.; supervision, E.O. and S.-Y.S.; project administration, S.-Y.S.; funding acquisition, S.-Y.S. All authors have read and agreed to the published version of the manuscript.

Funding: This work was supported by the Ministry of Trade, Industry and Energy (Grant No. 20214000000060) supervised by the KETEP (Korea Institute of Energy Technology Evaluation and Planning) and the Ministry of Science and ICT (Grant No. RS-2023-00259004) supervised by the IITP (Institute for Information & Communication Technology Planning & Evaluation).

Data Availability Statement: Publicly available datasets were analyzed in this study. These data can be found at https://www.data.go.kr/index.do (accessed on 14 June 2024).

Conflicts of Interest: The authors declare no conflicts of interest.

References

1. Delmastro, C.; Chen, O. *Buildings*; International Energy Agency (IEA): Paris, France, 2023.
2. Papadakis, N.; Katsaprakakis, D.A. A review of energy efficiency interventions in public buildings. *Energies* **2023**, *16*, 6329. [CrossRef]
3. Hafez, F.S.; Sadi, B.; Safa-Gamal, M.; Taufiq-Yap, Y.H.; Alrifaey, M.; Seyedmahmoudian, M.; Stojcevski, A.; Horan, B.; Mekhilef, S. Energy efficiency in sustainable buildings: A systematic review with taxonomy, challenges, motivations, methodological aspects, recommendations, and pathways for future research. *Energy Strategy Rev.* **2023**, *45*, 101013. [CrossRef]
4. Smdani, G.; Islam, M.R.; Yahaya, A.N.A.; Safie, S.I.B. Performance evaluation of advanced energy storage systems: A review. *Energy Environ.* **2023**, *34*, 1094–1141. [CrossRef]
5. Olajiga, O.K.; Ani, E.C.; Olatunde, T.M.; Sikhakane, Z.Q. Assessing the potential of energy storage solutions for grid efficiency: A review. *Eng. Sci. Technol. J.* **2024**, *5*, 1112–1124. [CrossRef]
6. Oh, E. Fair Virtual Energy Storage System Operation for Smart Energy Communities. *Sustainability* **2022**, *14*, 9413. [CrossRef]
7. Farivar, G.G.; Manalastas, W.; Tafti, H.D.; Ceballos, S.; Sanchez-Ruiz, A.; Lovell, E.C.; Konstantinou, G.; Townsend, C.D.; Srinivasan, M.; Pou, J. Grid-Connected energy Storage Systems: State-of-the-Art and emerging technologies. *Proc. IEEE* **2023**, *111*, 397–420. [CrossRef]
8. Moustafa, A.M.; Abdelghany, M.B.; Younis, A.S.A.; Moness, M.; Al-Durra, A.; Guerrero, J.M. Software-defined control of an emulated hydrogen energy storage for energy Internet ecosystems. *Int. J. Hydrogen Energy* **2024**, *50*, 893–909. [CrossRef]
9. Abdelghany, M.B.; Al-Durra, A.; Zeineldin, H.H.; Gao, F. A Coordinated Multitimescale Model Predictive Control for Output Power Smoothing in Hybrid Microgrid Incorporating Hydrogen Energy Storage. *IEEE Trans. Ind. Inform.* **2024**. [CrossRef]
10. Sati, S.E.; Abdelghany, M.B.; Hamad, B.; Al-Durra, A.; Zeineldin, H.; EL-Fouly, T.H.; El-Saadany, E.F. Economic Power-Sharing and Stability Enhancement for Virtual Synchronous Generators in Islanded MG. *IEEE Trans. Power Syst.* **2024**. [CrossRef]
11. Walker, A.; Kwon, S. Analysis on impact of shared energy storage in residential community: Individual versus shared energy storage. *Appl. Energy* **2021**, *282*, 116172. [CrossRef]
12. Dai, R.; Esmaeilbeigi, R.; Charkhgard, H. The Utilization of Shared Energy Storage in Energy Systems: A Comprehensive Review. *IEEE Trans. Smart Grid* **2021**, *12*, 3163–3174. [CrossRef]
13. Li, L.; Cao, X.; Zhang, S. Shared energy storage system for prosumers in a community: Investment decision, economic operation, and benefits allocation under a cost-effective way. *J. Energy Storage* **2022**, *50*, 104710. [CrossRef]
14. Zhu, H.; Ouahada, K. A distributed real-time control algorithm for energy storage sharing. *Energy Build.* **2021**, *230*, 110478. [CrossRef]
15. Jasim, A.M.; Jasim, B.H.; Flah, A.; Bolshev, V.; Mihet-Popa, L. A new optimized demand management system for smart grid-based residential buildings adopting renewable and storage energies. *Energy Rep.* **2023**, *9*, 4018–4035. [CrossRef]
16. Mediwaththe, C.P.; Shaw, M.; Halgamuge, S.; Smith, D.B.; Scott, P. An Incentive-Compatible Energy Trading Framework for Neighborhood Area Networks with Shared Energy Storage. *IEEE Trans. Sustain. Energy* **2020**, *11*, 467–476. [CrossRef]
17. Gailani, A.; Crosbie, T.; Al-Greer, M.; Short, M.; Dawood, N. On the Role of Regulatory Policy on the Business Case for Energy Storage in Both EU and UK Energy Systems: Barriers and Enablers. *Energies* **2020**, *13*, 1080. [CrossRef]
18. Rimpas, D.; Kaminaris, S.D.; Aldarraji, I.; Piromalis, D.; Vokas, G.; Papageorgas, P.G.; Tsaramirsis, G. Energy management and storage systems on electric vehicles: A comprehensive review. *Mater. Today Proc.* **2022**, *61*, 813–819. [CrossRef]
19. Panchanathan, S.; Vishnuram, P.; Rajamanickam, N.; Bajaj, M.; Blazek, V.; Prokop, L.; Misak, S. A comprehensive review of the bidirectional converter topologies for the vehicle-to-grid system. *Energies* **2023**, *16*, 2503. [CrossRef]
20. Er, G.; Soykan, G.; Canakoglu, E. Designing an optimal multi-energy system with fast charging and hydrogen refueling station under uncertainties. *Sustain. Energy Grids Netw.* **2024**, *39*, 101403. [CrossRef]
21. Jaworski, J.; Zheng, N.; Preindl, M.; Xu, B. Vehicle-to-grid fleet service provision considering nonlinear battery behaviors. *IEEE Trans. Transp. Electrif.* **2024**, *10*, 2945–2955. [CrossRef]
22. Mohammad, A.; Zuhaib, M.; Ashraf, I.; Alsultan, M.; Ahmad, S.; Sarwar, A.; Abdollahian, M. Integration of electric vehicles and energy storage system in home energy management system with home to grid capability. *Energies* **2021**, *14*, 8557. [CrossRef]
23. Güven, A.F. Integrating Electric Vehicles into Hybrid Microgrids: A Stochastic Approach to Future-Ready Renewable Energy Solutions and Management. *Energy* **2024**, *303*, 131968. [CrossRef]
24. Oh, E.; Son, S.Y. Dynamic virtual energy storage system operation strategy for smart energy communities. *Appl. Sci.* **2022**, *12*, 2750. [CrossRef]
25. Lazard's Levelized Cost of Storage. Lazard, Apr. 2023. Available online: https://www.lazard.com/research-insights/2023-levelized-cost-of-energyplus/ (accessed on 30 April 2024).
26. Daneshzand, F.; Coker, P.J.; Potter, B.; Smith, S.T. EV smart charging: How tariff selection influences grid stress and carbon reduction. *Appl. Energy* **2023**, *348*, 121482. [CrossRef]
27. Jang, M.J.; Kim, T.; Oh, E. Data-Driven Modeling of Vehicle-to-Grid Flexibility in Korea. *Sustainability* **2023**, *15*, 7938. [CrossRef]
28. Leon-Garcia, A. *Probability, Statistics, and Random Processes for Electrical Engineering*, 3rd ed.; Pearson: London, UK, 2021.
29. Boyd, S.; Vandenberghe, L. *Convex Optimization*; Cambridge University Press: Cambridge, UK, 2004.
30. Yang, W.; Kang, H.; Kwon, Y.; Yoon, Y. *Energy Moving the World: Micro Energy Grid*; K-MEG RD Center: Seoul, Republic of Korea, 2004; ISBN 978-89-94792-95-815320.

31. A-10 Medium General Demand-Metered Service. Pacific Gas and Electric Company (PG&E). 2021. Available online: https://www.pge.com/tariffs/assets/pdf/tariffbook/ELEC_SCHEDS_A-10.pdf (accessed on 30 April 2024).
32. BEV Business Electric Vehicle. Pacific Gas and Electric Company (PG&E). 2021. Available online: https://www.pge.com/tariffs/assets/pdf/tariffbook/ELEC_SCHEDS_BEV.pdf (accessed on 30 April 2024).
33. Public Data Portal. Ministry of the Interior and Safety, Korea. 2024. Available online: https://www.data.go.kr/index.do (accessed on 30 April 2024).
34. Lee Rodgers, J.; Nicewander, W.A. Thirteen ways to look at the correlation coefficient. *Am. Stat.* **1988**, *42*, 59–66. [CrossRef]
35. Jang, M.J.; Oh, E. Deep-Reinforcement-Learning-Based Vehicle-to-Grid Operation Strategies for Managing Solar Power Generation Forecast Errors. *Sustainability* **2024**, *16*, 3851. [CrossRef]

energies MDPI

Article

Integration of Electric Vehicle Power Supply Systems—Case Study Analysis of the Impact on a Selected Urban Network in Türkiye

Wojciech Lewicki [1,*], Hasan Huseyin Coban [2,*] and Jacek Wróbel [3]

1 Faculty of Economics, West Pomeranian University of Technology Szczecin, 71-210 Szczecin, Poland
2 Department of Electrical and Electronics Engineering, Bartin University, 74110 Bartin, Turkey
3 Department of Bioengineering, West Pomeranian University of Technology Szczecin, 71-210 Szczecin, Poland; jacek.wrobel@zut.edu.pl
* Correspondence: wojciech.lewicki@zut.edu.pl (W.L.); hcoban@bartin.edu.tr (H.H.C.)

Abstract: Undoubtedly, the transition to electromobility with several million new, efficient charging points will have consequences for the energy industry, and in particular for network operators of the distribution infrastructure. At the same time, in the coming years the energy landscape will change into a system in which an increase in decentralized systems based on renewable energy sources will take over the leading function. This transformation process will further increase the complexity and overall pressure for change in energy systems over the next decade. In order to be able to ensure the energy supply and the reliable system operation of the grids in the future as well, communicative networking of generators, storage systems, electrical consumers and grid equipment is indispensable. This study aims to investigate the consequences of including electric vehicles in Istanbul's power system using a unit commitment simulation model. The presented considerations analyze how uncertain and managed charging strategies affect the power system in terms of operating costs and renewable resources. The presented simulations indicate that, in economic terms, the use of a managed charging strategy saves 2.3%, reducing the total cost from USD 66.71 million to USD 65.18 million. The recipients of the presented research are both the demand and supply sides of the future energy transformation based on the idea of synergy of electromobility and renewable energy sources within the framework of the smart city idea.

Keywords: electromobility; optimized charging of electric vehicles; power systems; system integration; smart city

1. Introduction

No researcher doubts that greenhouse gas emissions contribute to climate change, which can have serious consequences for humans and Earth's ecosystems through increased sea and air temperatures and more frequent extreme droughts and floods [1,2]. One way to counteract this change was the Paris Agreement was signed through a UN conference in 2015. The purpose of the Paris Agreement is to work to keep the global temperature increase below 2 °C [3]. One of the countries where this goal has become an important element of energy policy is Türkiye. The country has introduced a series of climate packages, directives and binding targets that aim to minimize greenhouse gas emissions by the end of 2030 and 2050 [4]. It is worth emphasizing the fact that Turkey decided in 2021 that its national net greenhouse gas emissions by 2050 should be zero (Turkey's ambition is to have 100% renewable energy production by 2050, but also to reduce the country's emissions) [5]. As indicated in the literature on the subject, the road transport sector accounts for approximately one-fifth (81 Mt of CO_2e) of Turkey's total national greenhouse gas emissions, calculated as carbon dioxide equivalent [6]. To ensure Turkey's future sustainable development, the road transport sector must be transformed into climate-neutral transport. To make mobility climate neutral, an energy transformation is necessary. The strategic plan to make the transport sector fossil-free is important to achieve climate goals. Especially since the Turkish government is one of the few in this geographical region that invests in the production of its own electric vehicles.

At this stage of considerations, it is important to emphasize that the development of renewable energy sources is another promising way to achieve decarbonization of the fossil fuel-dominated energy system and several approaches for the optimization of renewable energy system deployment exist in the literature [7]. However, the intermittent nature and volatility of renewable electricity generation present significant challenges to the economic functioning and security of the power system [8]. A decrease in renewable resources is inevitable. Therefore, the process of increasing the share of natural renewable energy and the development of energy-sharing paradigms [9] is of utmost importance.

In 2030, it is expected that there will be 2.5 million electric vehicles and 1 million charging stations will be established in Turkey [10]. In other words, 10% of Turkey's passenger car stock will be electric vehicles in 2030, and one of every two vehicles sold will be electric vehicles. Therefore, electrification of the transport sector is a promising way to address climate goals by accelerating toward sustainable energy sources and increasing energy supply security, especially in urban areas.

In the available literature on the subject [11–17], several researchers have pointed out a key aspect regarding the role of integrating energy networks by connecting electromobility with energy management scenarios. The EV charging demand often has to cope with peak demand and operating costs because it operates expensive power plants capable of handling sudden loads. Some studies [18–20] indicate that if EV charging demand is optimally managed, vehicle grid integration can significantly reduce peak demand, provide power system flexibility, and enable frequency regulation [21]. Based on these, the large-scale use of EVs can facilitate the achievement of the country's renewable energy use target and the integration of electricity generation with mutual benefits locally. Entities with energy storage that supply energy to hybrid or fully electric vehicles can offer competitive prices where the energy comes from renewable sources. This mentioned problem seems to be particularly important in urban areas [22] where integration processes in the area of electromobility take place particularly quickly.

As electric vehicles take increasingly larger market shares, more people are connecting to the electricity grid for charging [15]. In the first stage, however, it is necessary to answer the question: what does the charging infrastructure in Turkey look like today, especially in cities, what types of chargers are used, and does electricity come from renewable sources? The Turkish electricity grid is already burdened in parts of Turkey today, especially in larger cities. In response to this situation, Türkiye is expanding renewable energy sources that are highly dependent on the weather, which can result in periods of higher or lower power load on the electricity grid. How does the electricity grid stand against an electric vehicle fleet, will it need to be strengthened in parts of the country or is it possible to route the power loads in a more efficient and smarter way during less active hours to decrease the load? Hence, there are many questions related to the planned energy transformation and the challenges posed by the new technology, but with it there are also proposals for new model solutions in the network–vehicle–energy relationship.

Currently, no one has any doubts that with the rapid growth of the EV population in the next 6–8 years, EVs could be one of the major load demands for the power system. Connectivity options need to be explored today to meet the required energy demands of a fleet of hundreds of vehicles charging simultaneously. However, there are few studies [23–25] which include analyzes of the energy demand for electric vehicles and energy from regional power plants in relation to large metropolises. Consider that the benefits of managed EV charging can help plan large-scale EV adoption and renewable energy generation [23–25].

The aim of this study is to explore and evaluate the impacts on the city of Istanbul's 2030 electricity system, including power plant operating costs, as part of various electric vehicle grid integration strategies. The implementation costs of different electric vehicle grid integration strategies has been investigated and the power system costs resulting from the integration of EVs have been calculated. Hourly simulations for regional electricity generation for the Istanbul power system are used in the electric vehicle charging behavior model. The implementation of the presented tasks will allow the development of a model for implementing the integration of electric vehicle energy systems and assessing the impact on selected energy networks in large metropolises as part of the smart city idea.

Analysis of the literature on the subject showed that this is the first approach to this topic involving a description of a given metropolis of Istanbul and referring to real data. Therefore, the presented research has both theoretical and practical dimensions. In short, the presented article brings a new look at the existing literature in the following areas: (I) integration of power grids with electric vehicles, (II) vehicle charging optimization process, (III) flexible load, (IV) use of renewable energy sources in the process of charging electric vehicles in cites, and (V) smart city concept.

The article is organized as follows. Section 1 provides an introduction to the topic. Section 2 contains a detailed description of the materials and methods—a description of the model. Section 3 describes the EV load estimation and charging strategies. Section 4 presents the experimental results and their interpretations—scenario description. Section 5 contains the final conclusions of the research, indicating its limitations, practical application, and future directions of research in this area.

2. Materials and Methods—Research Model

A unit commitment and economic dispatch simulation model is applied to represent the operating rule of the power system whose purpose is to minimize the overall operating costs of power plants to meet the electrical load. Equation (1) is the cost function of the optimization problem including capital costs of power facilities, fixed annual operation and maintenance costs of all power stations, start-up/down costs, fuel (natural gas, coal) costs, and connecting costs for integrating power plants into transmission lines. Figure 1 depicts the schematic representation of the proposed integrated system with the parameters, uncertainties, and optimal decisions. The main objective is to minimize the overall costs of all unit generators to meet the electricity load with the optimal power generation, and integration of battery electric vehicles. In this study, the unit commitment problem, which is a traditional power system economics problem, also has to deal with the scheduling generation which will be examined. It is important to understand the difference between economic dispatch for units and unit commitment. In the economic dispatch problem, a load and a set of generating units producing synchronously online are given to the system. The question sought to be answered is how much each of these production units must produce to meet this load at minimum cost. However, in power systems, the load does not remain constant. It does not need the same number of generating units when the load is small, and this leads to the unit commitment problem. The unit commitment problem is given a set load profile which is the value of a load for each hour of the day. The model tries to answer the question of when each of these power generators should be started, when should it be stopped, and how much should it generate to meet the load at minimum cost.

The model developed in this study serves as a planning tool for the capacity expansion of the power system, integrating the introduction of EVs. It optimizes the augmentation of capacity for electricity generation facilities, taking into account hourly load demand and renewable energy output profiles while satisfying the projected electricity demand for 2030. The temporal analysis for optimization encompasses both weekdays and weekends. Geographically, the model encompasses the entire power system derived from renewable sources. This model simultaneously optimizes investment decisions and operational allocation to minimize the total costs of electricity generation, adhering to multiple constraints such as power balancing and renewable resource limitations, as illustrated in Figure 1. The model presented in this study is a linear programming framework designed to minimize the aggregate of all investment and operational costs. These costs encompass capital expenditures for generators, fixed Operation and Maintenance (O&M) costs for all generators, transmission line expenses, and fuel costs. The model is governed by two principal sets of constraints: system constraints and unit constraints. Additionally, the model integrates carbon emission constraints and the natural resource limitations of renewable energy sources, while also optimizing for EV charging. The inputs for the model data include detailed information about power plants, fuel costs, capital costs associated with production technologies, and assumptions regarding the O&M costs of these technologies. These inputs are based on renewable electricity production profiles. Through this comprehensive approach, the model aims to provide an optimized strategy for capacity expansion and operational efficiency in the power system.

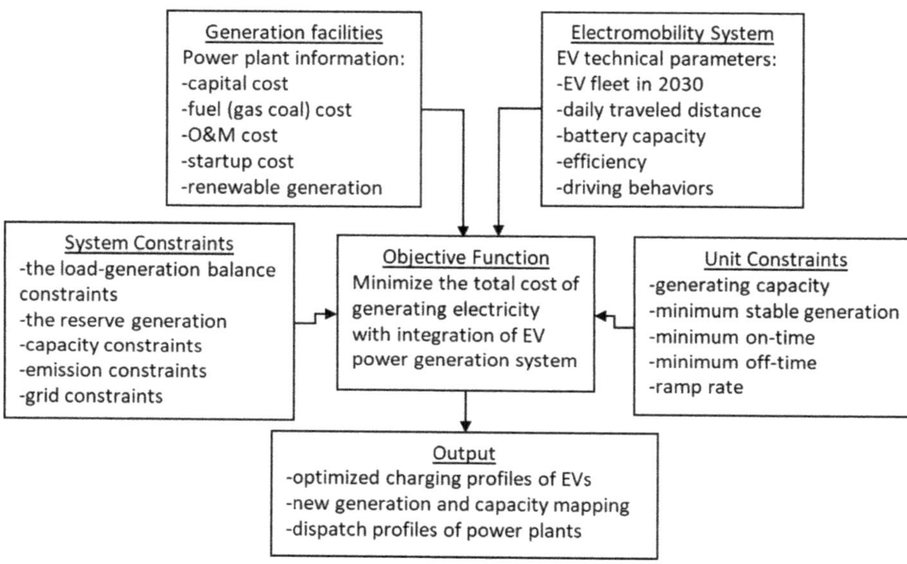

Figure 1. Schematic representations of the proposed system inputs, decisions and uncertainties.

Numerous studies have examined methods to decarbonize the energy sector. For instance, a study [26] proposed the electric road model to decarbonize Turkey's transportation sector, determining that the country could meet its 2045 greenhouse gas emission targets by optimizing its electricity generation mix. They concluded that fostering the development of renewable energy power could significantly reduce national CO_2 emissions. This study proposes a comprehensive evaluation pathway utilizing large-scale electric vehicles to cost effectively support the decarbonization of Turkey's transportation sector. Based on the development trajectory of electric vehicles, this study initially simulates the daily charging profile of EVs and the temporal availability of EVs connected to the grid using the Monte Carlo method. Subsequently, the integration of long-term electric vehicle supply scenarios into the power system under various conditions is examined. Finally, the study assesses the mix of renewable generation capacity, annual costs of the power system, emissions, and the driving distance per EV.

It can make some observations on the basis of power plants. If enough units are not committed to a synchronized system that is ready to generate the power, it will not be able to meet the demand. If there are not enough production units, its demand still can be met, but some of these units may be operating above optimum efficiency points and more money will have to be spent than necessary. On the other hand, if there are too many units committed, some of these units may be operating below optimal economic dispatch. Finally, if there are too many units committed, there will be a problem with minimum generation because it will exceed the demand, and an impossible combination will occur. Another observation has to do with the no-load costs of the generating units. In economic dispatch, the no-load cost does not affect the optimal dispatch, but it affects the choice of optimal combination.

The approach used must have taken all constraints into account in order to fully resolve the unit commit problem. There are two types of restrictions to consider: unit restrictions, and system restrictions. Unit restrictions affect each unit taken individually, while system restrictions affect units taken as a whole. The problem with constraints is that some of these constraints create a link between different periods. That is to say, solving problems separately in each period is not a good solution.

Another thing to consider is startup costs. The startup cost is a cost incurred when we start a production unit, and different production units will have different startup costs [27,28]. Large units will tend to have a high startup cost but will be efficient later on while small power plants will tend to have a lower startup cost but will be less efficient in this case. Therefore, there will be a trade-off to be made between starting and generating power plants or running other higher-output power plants.

$$\min\left[\sum_u C_u^{cap} + \sum_u C_u^{OM}\right] \tag{1}$$

where
C_u^{cap}—the initial capital expenditure,
C_u^{OM}—the operational and management costs.

2.1. Initial Capital Expenditure

This sub-section will detail the initial costs associated with the acquisition and installation of generation technologies, including both traditional and renewable sources. It (Equation (2)) provides a breakdown of costs for equipment, infrastructure, and other capital investments.

$$\sum_u C_u^{cap} = \sum_u (Cap_u^{w} + Cap_u^{pv} + Cap_u^{gas} + Cap_u^{coal}) + \sum_u C_u^{tr} \tag{2}$$

where
Cap_u^{w}—the procurement and installation costs of the wind turbine unit,
Cap_u^{pv}—the procurement and installation costs of the PV units,
Cap_u^{gas}—the capital cost of the gas-fired unit,
Cap_u^{coal}—the capital cost of the coal-fired unit,
C_u^{tr}—the capital cost of the transmission system.

2.2. Operational Startup Costs

This sub-section will focus on the costs incurred during the startup phase of operations. It will cover expenses such as workforce training, initial fuel costs, maintenance, and any other operational expenditures necessary to bring the power generation units online. The annual fixed operation and management cost is calculated as in Equation (3):

$$\sum_u C_u^{OM} = \sum_u (C_u^{w} + C_u^{pv} + C_{u,t}^{gas} + C_{u,t}^{coal}) \tag{3}$$

where
C_u^{w}—the fixed operational cost of the wind turbine unit,
C_u^{pv}—the fixed operational cost of the PV unit,
$C_{u,t}^{gas}$—the startup cost of the gas-fired unit at time t,
$C_{u,t}^{coal}$—the startup cost of the coal-fired unit at time t.

2.3. Unit Constraints

The unit constraints that affect each unit are taken individually. Constraints that affect each power plant individually are maximum generating capacity, minimum stable generation, minimum on time, minimum off time, and ramp rate [29].

- Minimum on and off time

Operators impose some restrictions so that the plant continues to operate for a long time and does not affect the life of the plant. One of these restrictions is what is called the minimum uptime of the plant. The idea is that once a generating unit starts up, it will not be shut down immediately because doing this more or less often will shorten the life of the plant; so mathematically we can formulate it as follows:

1. If $u(i,t) = 1$ and $t_i^{on} < t_i^{on,min}$; $u(i,t+1) = 1$ (then the plant should remain on at the next time period)
   where
   $u(i,t)$—the status of power generator i at the period t;
   $u(i,t) = 1$—the power generator i is on at the period t;

This restriction with minimum downtime has an inverse and essentially, if the generation unit is shut down, it is not allowed to restart it immediately and is expressed mathematically as follows:

2. If $u(i,t) = 0$ and $t_i^{off} < t_i^{off,min}$; $u(i,t+1) = 0$ (then the plant should remain on at the next time period)
   where
   $u(i,t) = 0$—the power generator i is off at the period t;

If the plant is off (must be off for at least the minimum amount of time that must be off), then this plant must remain off for the next time period.

- Ramp rate

Another constraint to consider is the ramp rate which is the rate at which the plant can increase or decrease its generation. This constraint is also imposed to protect the life of the power plant (to avoid damaging the turbine, etc.), and can be expressed mathematically as follows:

$$\Delta P_i^{up,max} \geq x(i,t+1) - x(i,t) \quad (4)$$

Similarly, a maximum ramp-down constraint specifies that which is needed to reduce the output.

$$\Delta P_i^{down,max} \geq x(i,t) - x(i,t+1) \quad (5)$$

where
$x(i,t)$—the generated power by i during the period t.

2.4. System Constraints

The system constraints affect more than one unit. These are mainly the load-generation balance constraints, the reserve generation capacity constraints, emission constraints, and grid constraints.

- The load generation balance constraint: the produced power by the generators must be equal to the load at all times.

$$\sum_{n=1}^{N} u(i,t) \cdot x(i,t) = L(t) \quad (6)$$

where
n, N—a set of available power generators;
u, x—the produced power;
L—a load at times t.

- The reserve generation capacity constraints: It is a restriction imposed to maintain the operational reliability of the system. Every once in a while, the power plant can suddenly go out due to a problem or due to restrictions. This power generation will cause an acceptable frequency drop if not fixed quickly. Therefore, there must be the ability to increase the generation of other units to prevent the frequency from falling beyond acceptable limits. It can be expressed mathematically as follows:

$$\sum_{n=1}^{N} u(i,t)P_i^{max} \geq L(t) + R(t) \tag{7}$$

where
$R(t)$—a set of available power generators.

- Emission Constraints: It defines the CO_2 emission constraints for the power system responsible for generating and transmitting energy to meet the electricity demand over time. CO_2 emissions are produced by coal-fired and gas-fired power plants during power generation and the annual CO_2 emission constraints are formulated as follows:

$$\sum_{t} fuel_{n,\,t} \leq crbn_{lim} \tag{8}$$

$fuel_{n,t}$—the rate of use of fuel (coal, gas) by generator n, during timepoint t,
$crbn_{lim}$—annual CO_2 emissions limits for the power system.

It is the sum of cases where the maximum capacity of all available power generators must be greater than or equal to the load and some reserve requirement. The rapid increase in the generation is only possible if committed power plants are not all operating at their maximum capacity.

Besides all these constraints, the scheduling of production units can be affected by environmental constraints. Environmental constraints were not examined in detail in this study. Also, generation scheduling can be affected by transmission networks. Some switchboards that will not be needed for economic reasons may need to be operated to provide voltage support. On the other hand, the output may be limited as some units that can be very efficient may exceed the transmission capacity of the network. In this study, the grid constraints are neglected.

This unit commitment model enables the investigation of the comprehensive impact of electric vehicle charging modes on the Istanbul electricity system with high penetration levels of renewable resources in 2030.

Electricity consumption, peak demand, and generation capacity are estimated based on historical data [30]. The estimated installed power capacity of the Istanbul power system in 2030 is shown in Table 1. Figure 2 shows the capacity factors of selected power generators.

Table 1. Power capacity of the Istanbul power system.

Power Plant	Hydro	Solar	Wind	Natural Gas	Coal
Capacity	17 GW	1.5 GW	12 GW	25.8 GW	1.5 GW

The hourly variations of hydropower, solar power, electricity demand, and charging demand of EVs are considered. Solar power and hydropower are regarded as zero-cost generation and zero-emission resources. Therefore, operators prefer to use renewable energy sources to reduce operating and management costs and CO_2 emissions. However, the intermittent nature of renewables requires significant reserves for conventional fuel units. The hydroelectric power plant has a ramp facility as it has a dam/reservoir.

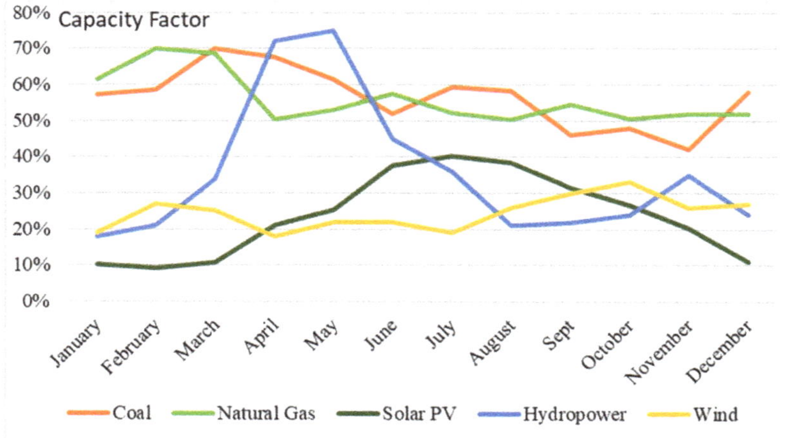

Figure 2. Monthly capacity factors of selected power generators.

3. EV Load Estimation and Charging Strategies

At the beginning of the considerations, a question should be answered. What are the conditions for controlling electric car charging and how is the electricity grid affected by electric car charging in the current situation and by a future growing electric car fleet? This question is crucial from the perspective of an electricity distribution company in Istanbul.

The electricity demand from EVs depends on the size of the car fleet, driving behaviors, wear and tear of the EV, battery health, driving distance, and traffic conditions [31,32]. According to the BloombergNEF Electric Vehicles Outlook 2021 report, there will be more than 169 million EVs on the road by 2030. It is predicted that the number of electric vehicles in Turkey will reach 2.5 million and the number of charging stations will reach 1 million by 2030 [10,33]. The parameters related to the electric vehicles used in this study are summarized in Table 2. Energy consumption for driving purposes only for an EV ranges from 0.16 kWh/km to 0.35 kWh/km [34]. In this study, energy consumption is assumed to be 0.30 kWh/km, since the traffic density in the city of Istanbul is high and the topology of the city is hilly. The annual energy consumption of electric vehicles will be 23.8 TWh, corresponding to 7.2% of electricity consumption. This study focuses on only passenger electric cars rather than electric buses and heavy vehicles.

Table 2. Parameters related to electric cars.

Number of EVs	2.5 million
Mileage per day	Weekday: 25 km Weekend: 35 km
Battery capacity	40 kWh
Energy consumption and emission factor	EVs 0.30 kWh/km Gasoline-driven vehicles 7 L/100 km, (2.3 kgCO_2/L)
Charging rate	Fast charging: 7.2 kW Slow charging: 2.5 kW

Electric cars in Istanbul are mostly used for commuting between workplaces and homes. In this study, it is assumed that electric vehicles are generally used between 7:00 and 9:00 in the morning and between 17:00 and 19:00 in the evening. On weekdays, the charging times for electric vehicles are between arriving home in the evening and leaving home in the morning. On weekends, it is assumed that electric cars are used for non-work purposes such as shopping, entertainment, and sightseeing. It is assumed that the daily driving distance of electric cars is 25 km on weekdays and 35 km on weekends.

The charging behavior of electric vehicles has been produced by the Monte Carlo method based on the probability distributions of departure and arrival times and daily driving distance. The daily driving distance is created depending on whether the day is a weekday or weekend, and the probability is determined according to the charging location.

In this study, two charging strategies are defined, namely the managed charging strategy and the uncertain charging strategy. Uncoordinated charging assumes each EV charges as soon as it is plugged in without any consideration of the electricity supply and it stops charging when the electric car's energy requirement is met. Most research postulates an uncertain charging scenario in which the grid is pressured via many EV drivers coming home from work and plugging in a type at the same time, which occurs to be coincident with the day-by-day peak load. With managed charging which most often occurs unidirectionally, electric car charging is controlled by considering the electricity supply, within the constraints of the user's mobility needs. Charging power limits and cumulative charging energy constraints are applied to represent the energy and power limits of electric cars. This often includes shifting the electric car charging times based on electricity pricing or other incentive signals.

In this section, it will be explained how to solve the unit commitment problem. Decision variables come in two types. The status of each generating unit at each period has to be found.

$$u(i,t) \in [0,\ 1]\ \forall\ i,t \tag{9}$$

In addition, the output of each unit needs to find out what it generates at each period.

$$x(i,t) \in \left[0, P_i^{min}, P_i^{max}\right]\ \forall\ i,t \tag{10}$$

The mathematical formulation of the entire fleet of the electric vehicle's flexibility is shown as follows:

$$E_t^{lower} \leq \sum_{t=1}^{T} \Delta T \cdot x(i,t) \leq E_t^{upper} \tag{11}$$

where

E_t^{lower}—the lower boundaries for the aggregated energy demand by time t;
E_t^{upper}—the upper boundaries for the aggregated energy demand by time t.

The charging demand (EV_{dem}) of EVs is calculated annually by aggregating different vehicle types, accounting for their numbers, average travel distances, and energy consumption rates. This provides a comprehensive view of the overall energy requirements for electric vehicles each year.

$$EV_{dem} = \sum_{p} \frac{NrEV_p \cdot DisEV_p \cdot EffEV_p}{ChEV_p} \tag{12}$$

where,

EV_{dem}—the energy demand of EVs in period p (kWh),
$NrEV$—number of EVs in year p,
$DisEV_p$—the annual travel distance of EVs in period p,
$EffEV_p$—fuel consumption of EVs in period p (km/kWh),
$ChEV_p$—charging efficiency of EVs in period p (%).

At this stage of research, it is important to emphasize that over the years, attempts have been made to develop many techniques to solve the problem of individual involvement. This drive for new techniques has arisen from the need to find solutions that are closer to the optimum, find solutions faster, and improve the way in which the constraint is modeled.

4. Results Analysis—Scenario Description

To solve a research problem, three different scenarios have been created for the unit commitment optimization of the power system with and without electric cars scenario.

1. Managed charging strategy (MCS): In this strategy, where the charging time is not fixed, the electricity supply is taken into account within the constraints of the electric vehicle user's mobility requirements. This type of charging of electric cars can be differentiated by the direction of electricity flow and how charging is controlled.
2. Uncertain charging strategy (UCS): EVs are charged depending on user preferences, vehicle characteristics, and trip requirements; with a fixed charging time, which starts as soon as the EV arrives at home or the workplace.
3. Without electric cars (WEC): No electric cars are connected to the power grid.

The generation profiles of renewables and hourly load profiles are shown in Figure 3.

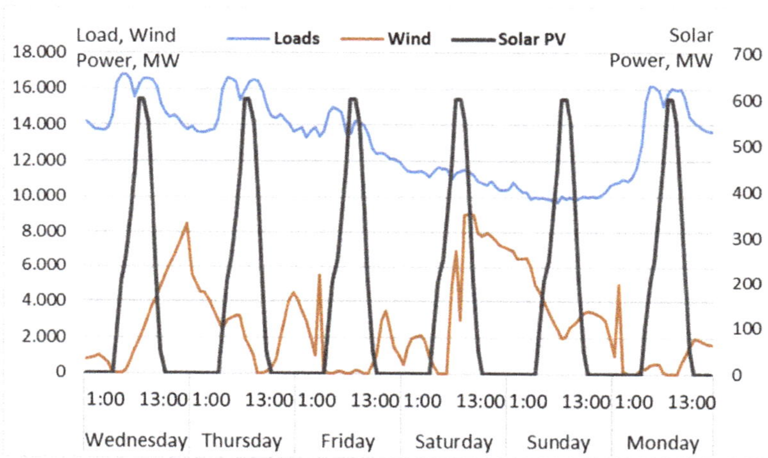

Figure 3. Loads, solar, and wind power profile (26–31 October 2022).

A day is divided into 24 h to find the optimal generation schedule for the entire load profile, assuming the load is constant during these one-hour periods. The question to be asked is which power plants should be committed to generating at minimum cost for each time period. In order to solve the problem, the minimum value of the objective function under the determined constraints is attempted to be found.

Electric vehicles are considered one of the important solutions to balance power fluctuations caused by higher integration of unstable and unpredictable renewable energy sources [35]. However, the integration of electric vehicles acts as additional loads to the electricity grid during battery charging, and these additional loads can create unwanted congestion, frequency ripple, and various negative effects on the distribution grid in Istanbul. All these issues create new challenges for power system operators. Therefore, it is imperative that electric vehicles provide a Managed Charging Strategy (MCS) instead of a simple traditional load.

In this study, the 24-h time frame was examined in 1-h periods. According to the demanded load values, the most suitable unit combinations and the total cost of the system were calculated. The results of all solution methods were compared. Estimated demand values in the system and total operating costs for the three scenarios are summarized in Table 3.

Table 3. Operating costs of scenarios (in $ million).

Scenario	Natural Gas Unit Cost	Coal Unit Cost	Reserve Capacity Cost	Start-Up Cost	Total Cost
MCS	57.1	5.7	1.03	1.35	65.18
UCS	58.1	5.9	1.19	1.52	66.71
WEC	56.4	5.5	1.12	1.31	64.33

Compared to the uncertain charging strategy, managed charging saves 2.3%, reducing the total cost from $66.71 million to $65.18 million. This is because some of these facilities can be considered flexible and their power output can be adjusted within some limits according to the data. When dealing with unit commitment, it is important to realize that not all manufacturing facilities are the same. Thermal generation units such as oil-fired units, coal-fired units, gas-cycle turbines, combined-cycle plants, and storage hydro plants are highly flexible as their condition and power output can be optimized.

On the other hand, there are plants that are inflexible or much less flexible, and these are plants whose output cannot be adjusted for technical or commercial reasons. Renewable energy sources such as wind and sun can only be generated when the wind blows or the sun shines. Combined heat and power plants are often driven by the need for heat rather than their ability to generate power, or their ability to generate power is constrained.

Also, the Uncertain Charge Strategy results in higher evening peak demand than the Managed Charge Strategy and then results in higher load shedding. In the Managed Charging scenario, most of the charging power occurs between 11:00 p.m. and 05:00 a.m. when electric vehicles are charged at home, which decreases the evening peak demand from 17.8 GW to 16.4 GW (7.2% decrease) in the Uncertain Charging scenario. Using the Managed Charging scenario, most of the evening charging demand is optimized to switch to off-peak hours after 01:00 a.m. Thus, a smoother load profile is obtained than using an Uncertain Charging strategy. As seen in Figure 4, Managed Charging also increases the operating efficiency of a fossil fuel-fired power plant.

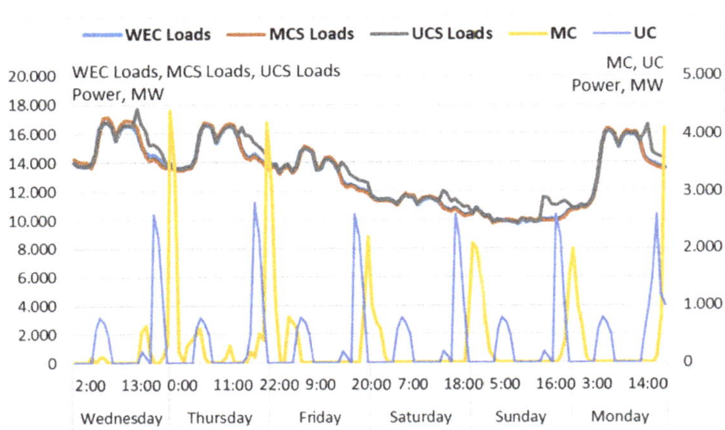

Figure 4. Electric cars' charging profile in the simulated days (26–31 October 2022).

Therefore, proper energy management is a necessary prerequisite for the energy transition. In order to efficiently integrate electric vehicles into it and make their technical flexibilities usable, a systemic approach is required that not only considers the requirements of electric vehicle users, but also the framework conditions of the distribution grids and an energy market oriented towards regenerative generation. Managed charge integration addresses all of these topics and proposes a feasible concept for the intelligent charging of many electric vehicles in private spaces at the low-voltage level, at the same time meeting the requirements of the smart city concept.

5. Conclusions and Discussion

Creating a map for achieving zero-emission road transport in cities is a huge task in the coming years [36–38]. An important postulate in this concept is the development of electromobility processes [39,40]. As many researchers emphasize, one of the determinants is the integration of energy systems with the vehicle and renewable energy sources [15,41].

With the arrival of more electric cars in Istanbul in 2030, the individual commitment model presented in this paper can be used to analyze the interaction between electric car charging and power plants. The uncertain charging strategy coincides with peak electricity demand during the day, while the managed charging strategy is implemented during off-peak hours and during times of low electricity prices to avoid peak demand. This study highlights the importance of charging electric vehicles in a timely manner so that the charging process does not burden the grid. The results of simulations indicate that managed charging is an appropriate strategy for the proper operation of the power system in Istanbul due to its flexibility in reducing peak load and reducing operational costs. In particular, a significant reduction in the operating costs of power plants may result in a decline in electricity prices in the future. The research results presented will provide useful insight into assessing the impact of electric vehicles on the power grid and the use of renewable energy in the process. However, it should be noted that the results depend on the renewable energy generation profiles and from several other factors including the driving behavior of electric vehicles—energy consumption.

In the current unreliable charging method, it has been shown that the temporary charging output overlaps the already peak power output during the day. This worsens the situation in the power grid, which relies on already existing energy peaks, which contributes to greater differences between peaks and valleys in the power curve from the Istanbul grid station. This overlap can cause power quality problems in the future as grid load increases. This situation will certainly result from the need to charge more electric cars at the same time.

However, this does not change the state of affairs that, as the authors have shown, managing the charging of electric vehicles can be a solution and a potential alternative to energy system solutions based on fossil fuels, such as stationary energy storage or peak generators.

It should be emphasized that the presented research focused on assessing the possibility of implementing solutions based on the integration of electric vehicle energy systems by analyzing the impact on selected networks—a case study of a selected city. It is important, both from a scientific and practical point of view, to answer the question of whether the project itself and the research conclusions can be directly implicated or compared with other cities. According to the authors, the implementation of this category of projects should be individual. It is not possible to indicate assumptions and patterns because each of the possible implemented projects is based on different starting assumptions. This does not change the fact that in each case, the implementation of a possible project must include a comprehensive analysis resulting from the impact of the use of electric vehicles on selected energy networks and systems, which are increasingly based on energy from renewable sources. It should therefore be remembered that a possible comprehensive assessment of all consequences resulting from the integration of energy systems will only be possible several years after the implementation and possible launch of individual projects in selected cities.

In the discussion on the integration of energy systems of electric vehicles, economic issues regarding the costs incurred for this process seem to be no less important [42,43]. The current energy transformation is based on the higher use of energy from renewable sources. It is estimated that the introduction of a larger fleet of electric vehicles will certainly lead to an increased demand for energy, i.e., an increase in financial outlays for the construction of appropriate transmission infrastructure, charging stations, and energy storage facilities. However, over time, in the authors' opinion, these costs will be optimized.

The authors support the thesis promoted in the literature that the integration of energy systems in relation to the development of electromobility is an inevitable process and all these activities certainly fit into the currently promoted smart city concept [44,45].

At this stage of the summary, it should be noted that, like every study, the one presented above also has its limitations. The authors focused their research on preconceived data and selected research methods and scenarios described in Sections 2 and 3 of this study. Verification and assessment of the validity of the adopted model assumptions will take place in the coming years. This does not change the fact that the discussed topic certainly requires further in-depth research in the areas of simulation efficiency, identification of limitations, challenges, and the selection of specialized software to analyze changes taking place in this area.

So, the future research in this matter should focus on identifying all barriers in the relationship between renewable energy sources—technical infrastructure—vehicles, taking into account the benefits for all stakeholders of this relationship.

To summarize the presented considerations regarding the integration of energy systems of electric vehicles—the analysis of the impact on selected networks does not fully exhaust the essence of the research problem. They are only an encouragement for research in this matter. This topic certainly requires further in-depth research to fully understand both the essence of the impact of energy transformation on the modern concept of smart city, as well as the role that electromobility and renewable energy sources play in this process. Hence, these types of analyzes will be the subject of future work aimed at determining and identifying key factors for the implementation of plans for the development of the smart city concept based on the development of electromobility and renewable energy sources.

Author Contributions: Conceptualization, W.L., H.H.C. and J.W.; methodology, W.L. and H.H.C.; software, W.L. and H.H.C.; validation, W.L. and H.H.C.; formal analysis, W.L. and H.H.C.; investigation, W.L. and H.H.C.; resources, H.H.C.; data curation, W.L.; writing—original draft preparation, W.L., H.H.C. and J.W.; writing—review and editing, W.L., H.H.C. and J.W.; visualization, W.L. and H.H.C.; supervision, W.L. and H.H.C.; project administration, W.L. and H.H.C.; funding acquisition, W.L., H.H.C. and J.W. All authors have read and agreed to the published version of the manuscript.

Funding: This research received no external funding.

Institutional Review Board Statement: Not applicable.

Informed Consent Statement: Not applicable.

Data Availability Statement: Data are contained within the article.

Conflicts of Interest: The authors declare no conflicts of interest.

References

1. Van Aalst, M.K. The impacts of climate change on the risk of natural disasters. *Disasters* **2006**, *30*, 5–18. [CrossRef] [PubMed]
2. Nema, P.; Nema, S.; Roy, P. An overview of global climate changing in current scenario and mitigation action. *Renew. Sustain. Energy Rev.* **2012**, *16*, 2329–2336. [CrossRef]
3. Rogelj, J.; den Elzen, M.; Höhne, N.; Fransen, T.; Fekete, H.; Winkler, H.; Schaeffer, R.; Sha, F.; Riahi, K.; Meinshausen, M. Paris Agreement climate proposals need a boost to keep warming well below 2 °C. *Nature* **2016**, *534*, 631–639. [CrossRef] [PubMed]
4. Telli, A.; Erat, S.; Demir, B. Comparison of energy transition of Turkey and Germany: Energy policy, strengths/weaknesses and targets. *Clean Technol. Environ. Policy* **2021**, *23*, 413–427. [CrossRef]
5. Şahin, Ü.; Tör, O.B.; Kat, B.; Teimourzadeh, S.; Demirkol, K.; Künar, A.; Yeldan, E. Türkiye'nin Karbonsuzlaşma yol Haritası: 2050'de Net Sıfır. 2021. Available online: https://ipc.sabanciuniv.edu/Content/Images/CKeditorImages/20211026-23105368.pdf (accessed on 1 January 2023).
6. Turkish Statistical Institute. Turkish Greenhouse Gas Inventory 1990–2020. April 2022. Available online: https://unfccc.int/documents/461926 (accessed on 20 January 2023).
7. Minelli, F.; D'Agostino, D.; Migliozzi, M.; Minichiello, F.; D'Agostino, P. PhloVer: A Modular and Integrated Tracking Photovoltaic Shading Device for Sustainable Large Urban Spaces—Preliminary Study and Prototyping. *Energies* **2023**, *16*, 5786. [CrossRef]

8. Li, J.; Liu, F.; Li, Z.; Shao, C.; Liu, X. Grid-side flexibility of power systems in integrating large-scale renewable generations: A critical review on concepts, formulations and solution approaches. *Renew. Sustain. Energy Rev.* **2018**, *93*, 272–284. [CrossRef]
9. Minelli, F.; Ciriello, I.; Minichiello, F.; D'Agostino, D. From Net Zero Energy Buildings to an energy sharing model—The role of NZEBs in Renewable Energy Communities. *Renew. Energy* **2024**, *223*, 120110. [CrossRef]
10. Turkey Projected to Have 2.5 mln Electric Cars by 2030. *Hurriyetdailynews*. 2019. Available online: https://www.hurriyetdailynews.com/turkey-projected-to-have-2-5-mln-electric-cars-by-2030-150242 (accessed on 1 January 2023).
11. Neaimeh, M.; Andersen, P.B. Mind the gap- open communication protocols for vehicle grid integration. *Energy Inform.* **2020**, *3*, 1. [CrossRef]
12. Li, M.; Lenzen, M.; Wang, D.; Nansai, K. GIS-based modelling of electric-vehicle–grid integration in a 100% renewable electricity grid. *Appl. Energy* **2020**, *262*, 114577. [CrossRef]
13. Sovacool, B.K.; Kester, J.; Noel, L.; de Rubens, G.Z. Contested visions and sociotechnical expectations of electric mobility and vehicle-to-grid innovation in five Nordic countries. *Environ. Innov. Soc. Transit.* **2019**, *31*, 170–183. [CrossRef]
14. Li, B.; Ma, Z.; Hidalgo-Gonzalez, P.; Lathem, A.; Fedorova, N.; He, G.; Zhong, H.; Chen, M.; Kammen, D.M. Modeling the impact of EVs in the Chinese power system: Pathways for implementing emissions reduction commitments in the power and transportation sectors. *Energy Policy* **2021**, *149*, 111962. [CrossRef]
15. Das, H.S.; Rahman, M.M.; Li, S.; Tan, C.W. Electric vehicles standards, charging infrastructure, and impact on grid integration: A technological review. *Renew. Sustain. Energy Rev.* **2020**, *120*, 109618. [CrossRef]
16. Brown, M.A.; Soni, A. Expert perceptions of enhancing grid resilience with electric vehicles in the United States. *Energy Res. Soc. Sci.* **2019**, *57*, 101241. [CrossRef]
17. Hildermeier, J.; Kolokathis, C.; Rosenow, J.; Hogan, M.; Wiese, C.; Jahn, A. Smart EV Charging: A Global Review of Promising Practices. *World Electr. Veh. J.* **2019**, *10*, 80. [CrossRef]
18. Nour, M.; Chaves-Ávila, J.P.; Magdy, G.; Sánchez-Miralles, Á. Review of Positive and Negative Impacts of Electric Vehicles Charging on Electric Power Systems. *Energies* **2020**, *13*, 4675. [CrossRef]
19. Coban, H.H.; Lewicki, W.; Sendek-Matysiak, E.; Łosiewicz, Z.; Drożdż, W.; Miśkiewicz, R. Electric Vehicles and Vehicle–Grid Interaction in the Turkish Electricity System. *Energies* **2022**, *15*, 8218. [CrossRef]
20. Fei, L.; Shahzad, M.; Abbas, F.; Muqeet, H.A.; Hussain, M.M.; Bin, L. Optimal Energy Management System of IoT-Enabled Large Building Considering Electric Vehicle Scheduling, Distributed Resources, and Demand Response Schemes. *Sensors* **2022**, *22*, 7448. [CrossRef]
21. Naderi, E.; Asrari, A. Integrated Power and Transportation Systems Targeted by False Data Injection Cyberattacks in a Smart Distribution Network. In *Electric Transportation Systems in Smart Power Grids*; CRC Press: Boca Raton, FL, USA, 2022; pp. 447–472. [CrossRef]
22. An, K.; Song, K.-B.; Hur, K. Incorporating Charging/Discharging Strategy of Electric Vehicles into Security-Constrained Optimal Power Flow to Support High Renewable Penetration. *Energies* **2017**, *10*, 729. [CrossRef]
23. Colmenar-Santos, A.; Muñoz-Gómez, A.-M.; Rosales-Asensio, E.; López-Rey, Á. Electric vehicle charging strategy to support renewable energy sources in Europe 2050 low-carbon scenario. *Energy* **2019**, *183*, 61–74. [CrossRef]
24. Zhu, X.; Xia, M.; Chiang, H.-D. Coordinated sectional droop charging control for EV aggregator enhancing frequency stability of microgrid with high penetration of renewable energy sources. *Appl. Energy* **2018**, *210*, 936–943. [CrossRef]
25. Zhong, J.; He, L.; Li, C.; Cao, Y.; Wang, J.; Fang, B.; Zeng, L.; Xiao, G. Coordinated control for large-scale EV charging facilities and energy storage devices participating in frequency regulation. *Appl. Energy* **2014**, *123*, 253–262. [CrossRef]
26. Coban, H.H.; Lewicki, W. Daily Electricity Demand Assessment on The Example of The Turkish Road Transport System-A Case Study of The Development of Electromobility on Highways. *Transp. Geogr. Pap. PGS* **2022**, *25*, 52–62. [CrossRef]
27. Richter, M.; Oeljeklaus, G.; Görner, K. Improving the load flexibility of coal-fired power plants by the integration of a thermal energy storage. *Appl. Energy* **2019**, *236*, 607–621. [CrossRef]
28. Jha, A.; Leslie, G. Dynamic Costs and Market Power: Rooftop Solar Penetration in Western Australia. *SSRN Electron. J.* **2020**. [CrossRef]
29. Van den Bergh, K.; Delarue, E. Cycling of conventional power plants: Technical limits and actual costs. *Energy Convers. Manag.* **2015**, *97*, 70–77. [CrossRef]
30. YTBS—Load Dispatcher Information System, "Electricity Statistics of Türkiye". Available online: https://ytbsbilgi.teias.gov.tr/ytbsbilgi/frm_hakkinda.jsf (accessed on 1 January 2023).
31. Nandi, A.K.; Chakraborty, D.; Vaz, W. Design of a comfortable optimal driving strategy for electric vehicles using multi-objective optimization. *J. Power Sources* **2015**, *283*, 1–18. [CrossRef]
32. Sanguesa, J.A.; Torres-Sanz, V.; Garrido, P.; Martinez, F.J.; Marquez-Barja, J.M. A Review on Electric Vehicles: Technologies and Challenges. *Smart Cities* **2021**, *4*, 372–404. [CrossRef]
33. BloombergNEF, Electric Vehicle Outlook 2022. 2022. Available online: https://about.bnef.com/electric-vehicle-outlook/ (accessed on 1 January 2023).
34. Cedric, D.C.; Messagie, M.; Sylvia, H.; Thierry, C.; Joeri, V. Electric Vehicle Use and Energy Consumption Based on Realworld Electric Vehicle Fleet Trip and Charge Data and Its Impact on Existing EV Research Models. *World Electr. Veh. J.* **2015**, *7*, 436–446. [CrossRef]

35. Kurucan, M.; Özbaltan, M.; Yetgin, Z.; Alkaya, A. Applications of artificial neural network based battery management systems: A literature review. *Renew. Sustain. Energy Rev.* **2024**, *192*, 114262. [CrossRef]
36. Brdulak, A.; Chaberek, G.; Jagodziński, J. Development Forecasts for the Zero-Emission Bus Fleet in Servicing Public Transport in Chosen EU Member Countries. *Energies* **2020**, *13*, 4239. [CrossRef]
37. Pietrzak, K.; Pietrzak, O. Environmental Effects of Electromobility in a Sustainable Urban Public Transport. *Sustainability* **2020**, *12*, 1052. [CrossRef]
38. Lojano-Riera, B.P.; Flores-Vázquez, C.; Cobos-Torres, J.-C.; Vallejo-Ramírez, D.; Icaza, D. Electromobility with Photovoltaic Generation in an Andean City. *Energies* **2023**, *16*, 5625. [CrossRef]
39. Adenaw, L.; Lienkamp, M. Multi-Criteria, Co-Evolutionary Charging Behavior: An Agent-Based Simulation of Urban Electromobility. *World Electr. Veh. J.* **2021**, *12*, 18. [CrossRef]
40. Tundys, B.; Wiśniewski, T. Smart Mobility for Smart Cities—Electromobility Solution Analysis and Development Directions. *Energies* **2023**, *16*, 1958. [CrossRef]
41. Patil, H.; Nago Kalkhambkar, V. Grid Integration of Electric Vehicles for Economic Benefits: A Review. *J. Mod. Power Syst. Clean Energy* **2021**, *9*, 13–26. [CrossRef]
42. İnci, M.; Savrun, M.M.; Çelik, Ö. Integrating electric vehicles as virtual power plants: A comprehensive review on vehicle-to-grid (V2G) concepts, interface topologies, marketing and future prospects. *J. Energy Storage* **2022**, *55*, 105579. [CrossRef]
43. Lewicki, W.; Niekurzak, M.; Sendek-Matysiak, E. Electromobility Stage in the Energy Transition Policy—Economic Dimension Analysis of Charging Costs of Electric Vehicles. *Energies* **2024**, *17*, 1934. [CrossRef]
44. Hertel, D.; Bräunig, G.; Thürer, M. Towards a green electromobility transition: A systematic review of the state of the art on electric vehicle battery systems disassembly. *J. Manuf. Syst.* **2024**, *74*, 387–396. [CrossRef]
45. Rivera, S.; Goetz, S.M.; Kouro, S.; Lehn, P.W.; Pathmanathan, M.; Bauer, P.; Mastromauro, R.A. Charging Infrastructure and Grid Integration for Electromobility. *Proc. IEEE* **2023**, *111*, 371–396. [CrossRef]

energies

MDPI

Article

Predicting and Forecasting of Vehicle Charging Station Using ECNN with DHFO Algorithm

Rosebell Paul * and Mercy Paul Selvan

Department of Computer Science and Engineering, School of Computing, Sathyabhama Institute of Science and Technology, Chennai 600119, India; mercypaulselvan.cse@sathyabama.ac.in
* Correspondence: rosebellabin@gmail.com

Abstract: The forecast of the optimal placement of a charging station (CS) according to the real-time consumption of electric vehicles is a subject of urgency in this new era. The demand of a charging station in an area based on the trend of consumption can be predicted by means of interpolation and the extrapolation of historical data using a linear function of prediction model. The prediction of the charging station system was performed with distance relevancy methods. An adaptive optimal learning model was proposed to enhance the prediction performance for charging station management and to represent the pattern of vehicles' travelling directions. The proposed model uses Distributional Homogeneity Feature Optimization (DHFO) using artificial intelligence (AI) to categorize and forecast the charging station from the database. The prediction performance of this model is improved more than the conventional classification model by filtering the apt features from all the electric vehicular and charging station attributes in the database. The Enhanced Cladistic Neural Network (ECNN) is used to improve the pattern learning model and increase learning accuracy. By comparing statistical parameters with other state-of-the-art methodologies, the suggested model's overall findings were verified.

Keywords: vehicle charging station prediction; forecasting data; Distributional Homogeneity Feature Optimization (DHFO); Enhanced Cladistic Neural Network (ECNN)

1. Introduction

Faster electric vehicle (EV) adoption all over the world aims to achieve the decarbonation of the entire planet, thereby making Earth a safer and more sustainable place to live. Various research has been conducted all over the world to examine the biggest challenges faced by the automobile industry during this transition phase. The range anxiety is the greatest hindrance for all EV users. The uncertainty of running out of battery power during journeys is the predominant barrier for this quantum change. A myriad of conditional factors were reviewed depending on environmental, economic, social, and technical aspects to resolve this situation. The installation of power supplies at apt locations will reduce the fear of EV users of running out of power. This paved the way for studies and experiments to be carried out on the prediction of EV users' behavior and the optimal placement of charging stations.

Numerous systems are available to forecast the way an electric vehicle will approach a charging station by monitoring the State of the Charge (SOC) of the vehicle [1]. The SOC of the vehicle depends on several environmental and vehicle parameters. It gives an indication of the maximum range that can be covered by a vehicle and helps it to recognize when it is in urgent need of charge. Machine learning algorithms can carry out the forecast process to identify available charging stations using vehicle historical data. The charging stations may not be the most optimal ones as there can be multiple issues while tracking the vehicle's range, traffic conditions, and maximum charging capacity of the charging station, as discussed in [2]. This causes the charging station prediction and other vehicle routing processes to be misclassified. To get around that, the charging station prediction was followed by a historical data reference for relevancy. Apart from that, the lack of sufficient numbers of charging stations is a primary concern hindering the growth of EV adoption. This has triggered the necessity of installing new charging stations. Hence, the identification of the most suitable location for the same is a challenging task.

The Multi-Objective Optimal Placement model is ideally based on the above perceptions to develop an optimal charging station prediction for EV users. The Multi-Objective Optimal Placement model mainly concentrates on three stages—vehicle allocation, vehicle count selection, and end-user vehicle node placement. In this, the vehicle allocation defines the cluster-based coverage area computation of EVs that is in the proximity of the CS. Based on the vehicle distance to the CS and the power requirement, the parameters are validated and assigned to the best CS. The vehicle count selection takes into consideration the traffic conditions that are to be matched against the maximum charging capacity of the CS. This will select the number of vehicles that can be mapped with the available CS by comparing the available power of the CS along with the requirement of the EV to charge. The end-user vehicle node-based optimal placement refers to the parameters of the electric vehicle that will meet the criteria such as State of the Charge (SOC), based on the power demand, distance to travel, availability of terminals, and capacity of CS. Based on these three stages of the optimization model, the vehicle can identify the best place to charge the vehicle with minimum service time.

The proposed system takes into consideration the three stages of the Multi-Objective Optimal Placement model in its design to predict the most optimal charging station from the available ones for an electric vehicle, and forecasts the most suitable location to install a new charging station, taking only the most necessary vehicle parameter values into account. The study in [2] revealed that historical data for an area also play a prominent role in developing an advanced model of feature prediction, and these are adopted in the proposed classification system. To improve the prediction model, an optimization method was used to forecast the data pattern feature analysis [3] from the vehicle data samples and the weight input properties, which led to the selection of the best attributes. To improve the classification performance and increase prediction accuracy, an adequate amount of data samples were taken during the training process for forecasting the power demand as studied in [4].

As in a linear function, a straight line was drawn to predict values outside the dataset—in the proposed system, the route path of the EV from an origin to destination is the standard baseline taken. The prediction of the available charging station system was then performed with the distance relevancy methods across this line. The power demand of the charging station in an area based on the power consumption rate and the vehicle density can be predicted by means of interpolation and the extrapolation of the historical data using the linear function of prediction model. An adaptive optimal learning model was proposed to enhance the performance of prediction of the most apt charging station from the available ones, and the pattern of vehicle movement in the travelling direction was the primary source of data pattern, along with charging station parameters and power demand. This pattern of EV movement and its intersection across the baseline O-D Trip formed the core computation of the prediction. The proposed model used Distributional Homogeneity Feature Optimization (DHFO) and artificial intelligence (AI) to obtain the optimal feature selection for this prediction. The prediction performance of this model was improved more than the conventional classification model by filtering the apt features from all the electric vehicular and charging station attributes in the database. The Enhanced Cladistic Neural Network (ECNN) was then used to improve the pattern learning model and increase learning accuracy to forecast the co-ordinate positions for the installation of a charging station. The pattern analysis for forecasting the co-ordinates of the new charging station mainly concentrated on the power demand of the existing charging stations, along with their availability. By comparing these statistical parameters with other state-of-the-art methodologies, the suggested model's overall findings have been verified.

The following are the main highlights of the proposed framework:

- Distributional Homogeneity Feature Optimization (DHFO) technique, which is a novel optimal feature selection system for vehicle parameters required for the prediction of an apt location for a charging station for an EV, was developed;
- The classification performance was enhanced by organizing the best feature properties that were pertinent to the data;
- The weight and vehicle type of feature attributes were used to anticipate and predict the range of charging station site forecast based on the state of power demand;
- An Enhanced Cladistic Neural Network (ECNN) classifier was implemented for the classification approach;
- The performance of the proposed classification model was validated based on the statistical parameters, and the comparison was conducted with existing approaches.

The overall organization of the paper is as follows. Section 2 examines the various optimization techniques for the Optimal Placement Model to determine the optimal charging station direction selection. The working model of the suggested ECNN with the DHFO optimization algorithm for the optimal charging station direction selection is described in Section 3. Section 4 provides an explanation of the simulation graph charts and validation of the outcomes of the suggested model, along with the comparison results. Section 5 concludes with a discussion of the paper's work and potential improvements.

2. Related Works

This section provides a comprehensive overview of the current techniques for forecasting data and predicting charging stations for electric vehicles The challenges and limitations inferred from these studies, which paved the way to scope out further research, have been summarized in Figure 1. Most of the recent techniques mainly give emphasis to socioeconomic and environmental factors, but vehicle density and state of charge play dominant roles and need to be considered. A study of the influence of the social, economic, and environmental factors on the identification of the charging station location was made in [4–7]. The work [4] proposed a suitable ranking system based on the above influencing factors but lacks the detailing on the technical aspects such as the electric demand and state of charge of the vehicles travelling a particular path that may affect the forecasting of optimal charging station placement. The Charging Station Allocation Data were approximated using the time series and data samples from the Charging Station Time Chart in [8]. In doing so, the production analysis in the basin-scale vehicle level allocation model makes use of the OSM data samples. This also refers to the variation in charging power and the availability forecast based on dataset history. In [9], a gravity-dependent Huff model was described, taking into consideration Sioux-Fall network for illustration. The study's main focus is from the perspective of a private investor of in a charging station and leverages an optimal pricing strategy. A dynamic pricing strategy based on the Markov decision process and multi-agent neural network was proposed in [10]. The study suggested that charging during off-peak times could be more cost effective. In [11], graph-based computations were used for the maximum comfort and ease of EV users. The schedule time of route flow was predicted in this work using data modeling from short-term availability forecasts. This was operated based on historical data samples for station power and demand characteristics, as well as availability.

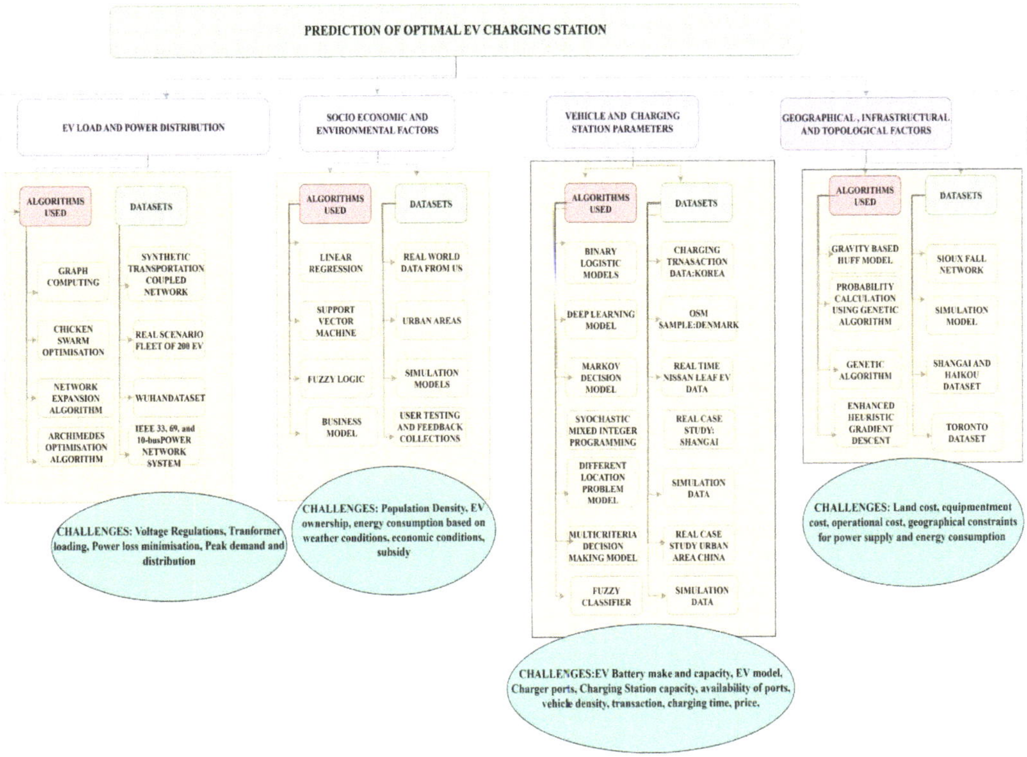

Figure 1. Summary of recent approaches for EV charging station prediction with details of dataset, algorithms, and limitations (Refs. [2–4,8–25]).

Based on the available data and demand forecast, Ref. [12] developed a station availability retrieval model. This analyzed various classification models to compare the efficacy of classification based on statistical variables and to estimate station availability from satellite data. The behavior pattern of EV users for charging was thoroughly analyzed in [13,26–28]. The cost of land invested in by the operators was also taken into consideration for predicting the best charging station location. These are all forecasted based on the power consumption and availability factors for the various station availability circumstances, but lack vehicle parameters and SoC detailing. In a similar vein, Ref. [14] put forth a prediction model for the SOC level in the charging station that considered the uncontrolled power loss and the unequal distribution of the available content load. According to this prediction, after the flooding effect, the status of the station will increase the range of SOC > 6.5 to a maximum value of 7. The construction of the charging station at the proper station is improved by this estimation. With this improvement in mind, Ref. [15] suggested a power logging prediction model for the charging station field to supply route flow-level data. This helps to grow the charging station in the field more by estimating the amount of power supply and other associated features based on the power level. A CNN feature-based path planner for EV was studied in [29]. Based on the traffic flow the waiting time to become charged was predicted. This can be used in further studies to track the most suitable CS when the SOC of a vehicle gets below the minimum threshold.

Several additional techniques have improved the construction range and the other decision plan based on the forecasting data and the station parameter prediction process. Based on this, Ref. [30] suggested using an optimization model to estimate the quality of charging stations as a decision support for corridors. This validates the power availability for the public in highways as special segment of highways are corridors dedicated to EV charging. It classified the CS in corridors as better locations than other CSs. The vehicle level quality was determined by the fuzzy interference system, which also segments each division independently. But this study is suitable only for high density traffic areas. Ref. [17] presented the estimation of geographical variance in the field of land used for charging station installation. During power fills, this forecasts the SOC level of the car. To forecast what kind of car-level charging station can develop there, this also assesses the redox condition of the station. Ref. [18] converted double-layer vehicle levels to ratoon charging station fields to lessen carbon emissions and enhance ecosystem benefits. This will result in a higher annual yield, lower demand, and lower car battery charging costs because of the ratoon vehicle-level (RV) technology. The analytical report of CS construction was used to analyze this together with various data points in the entire article.

Numerous techniques are aimed at estimating the necessity of CSs taking into consideration the OD (Origin Destination) trip and SOC level of the vehicles [17]. Considering this, Ref. [19] suggested a model to allocate optimal CSs for the bus network in Toronto. Depending on the vehicle-level SOC cycle and bus trip route, the fastest charging time with minimum cost was assured. An additional state-of-prediction for determining the allocation of discharge according to station characteristics, vehicle range estimation, and battery SOC conditions was the focus of [20,31]. To make the best choice from the available charging stations, the range of an electric vehicle is forecasted based on a fuzzy classifier. This study provided a report of the maximum distance an EV can travel to reach the most suitable charging station. An intelligent route flow system to increase the CSs based on weight and station availability-level projected from the user preference features was proposed by the author in [32]. Based on the remote sensing data, Ref. [21] suggested a city-level prediction for the deployment of charging stations with a minimum budget. The study gave priority to socioeconomic factors such as customer satisfaction, as well as population density. Using the combination of deep learning with phenological properties, this was estimated and forecasted [33]. An optimal placement of charging station in distribution networks was suggested in [22]. To safeguard the charging station CS, the author of [34] provided research based on several charging station discharge categorization models. The suggested model was a data-based station range prediction based on the station's availability, the cost value of the station's power demand, and other pertinent aspects and parameters. A business approach to a charging station was proposed in [23]. This study led to the motivation of finding the most optimal placement of charging station, but the technical aspects of the prediction algorithm for the same was not elaborated. Although the battery swapping method as in [35,36] is a great trend these days, taking into consideration the time, it may not be practical in all adverse scenarios. A time series-based approach was discussed in [37] regarding charging demand, but the optimal placement of CS is not suggested in that study.

All the studies discussed so far may mainly be categorized into the following domains based on the decision parameters affecting EVs and their charging stations: power distribution, socioeconomic–environmental factors, EV vehicle, CS parameters, and geographical–topological factors. Most of the studies summarized in the existing literature are based on a plethora of features which are specific to regions. The datasets used by most of the studies are real-time or recorded transportation data focused on a particular town or country, taking into consideration the attribute features of that region. A thorough study was conducted based on the literature summarized above to identify the constraints in each of the specified domains as listed in Figure 1. The traffic and vehicle parameters play a prominent role as influencing factors in all situations. Based on these features, a new optimal prediction system was proposed in our study. The most significant features and their relationship to computing the demand of CSs are discussed in detail in the next section. The proposed framework enhances the accuracy of predicting the best CS from existing ones, as well as forecasting the optimal location for installing a CS, taking into consideration the electric vehicle parameter features obtained from the dynamic tracking of vehicle movements, as well as its State of Charge.

The CS characteristics such as power demand, charging piles, as well as frequency of vehicle occurrences and the distance to reach the CS, are also considered. The forecasting model of the CS represents the best location to place a new CS based on the parameters of power demand, frequency of vehicle occurrences, availability of existing CS, and the peak time of vehicle visits. This will predict the best location for placement of CS according to the suggested co-ordinate positions. For this purpose, cutting-edge techniques like the Enhanced Cladistic Neural Network (ECNN) and the Ensemble Modulation Pattern (EMP) system are recommended.

3. Proposed Methodology

This section explains the proposed procedure of the optimal charging station prediction and forecasting model. When an EV travels from an origin to its destination, its initial co-ordinate positions are initialised based on the OSM data samples. As its journey continues, the state of charge (SOC) of the vehicle decreases and the distance that it can travel depends on vehicle parameters such as its battery make—lithium or hybrid—as well as battery power capacity, driving speed, and the traffic density. The existing charging stations near to the EV current position are made available to the EV. The cluster similarity measures form a cluster of CS where EV can place a charging request when the EV battery has a minimum SOC. Vehicles may prefer level1, level2, or DC fast charging, depending on its capability and charger compatibility. Each charging station has its own charging capacity which is divided into the available charging piles. Hence, the demand of a charging station depends on the availability of the number of charging piles, as well as its maximum charging capacity. That is, if a charging station has a maximum of 100 kW and two charging piles, then the charging capacity of that charging station is restricted to a maximum of 50 kW per pile. Hence the vehicle approaching the charging station must ensure this power demand of the CS by checking if the charging piles are available and other EVs are not plugged in at the station.

The demand of CS is computed based on the generated dataset which takes into consideration the number of charging piles, vehicle density, cost, and charging capacity of the charging station.

The power demand 'P_D' in a charging station can be calculated by referring to the below equation.

$$P_D = (V_C \times \beta \times R) - C$$

where, V_C—Vehicle Count (number of vehicles that are connected in the piles in individual CS).

β—Power constant for charging a vehicle.

R—Random value to represent the power consumption of each EV vehicle that are connected to the piles of a CS. Each vehicle will consume different amounts of power based on the battery state and the range of battery, therefore, a random value will refer to the different amount of power consumption for connected EVs, along with the charging constant 'β'.

C—Total capacity of CS that is the ability to supply power to the connected EVs in the piles.

This demand value, if negative, indicates that some piles are empty and vehicles can be connected to the CS. On the other hand, if the demand value is positive, it indicates that the charging stations are fully occupied, and then, based on the historical data and pattern of the feature attributes taken into account in the study, a new charging station location is recommended within the coverage area of that EV. Thus, the forecasting of the CS is based on the vehicle demand and the CS availability, as well as recognizing the area where the vehicle cannot reach the CS with its minimum SOC of battery level.

This prediction is computed using the ECNN algorithm which performs a pattern analysis while tracking the dynamic movement of the EV to available CS with respect to time and the SOC parameters. SOC parameters mainly refer to the percentage of SOC battery level and the distance that can be covered by that EV to a CS. This distance varies from time to time as it depends on the vehicle parameters such as the battery make and capacity, driving speed, and traffic. Hence the proposed system predicts the optimal CS for an EV, and if the CS is already occupied it forecasts an optimal location to install a CS in the future.

Figure 2 displays the main architecture diagram for the proposed charging station prediction model. The methodology is divided into three main stages which are labelled in the Figure 2.

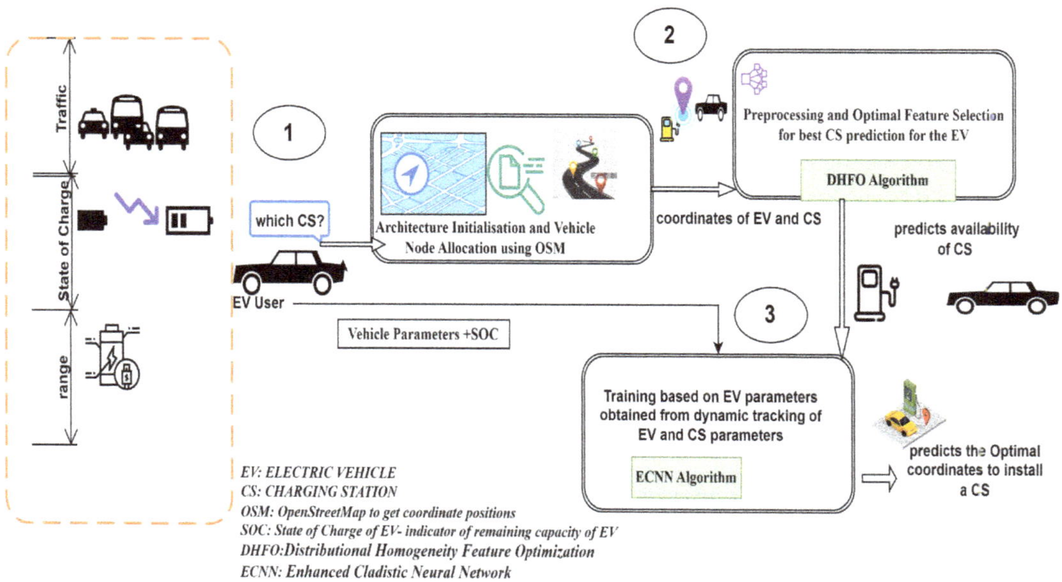

Figure 2. Architecture showcasing the three main phases of the system: initialization, DHFO, and ECNN algorithm.

The first stage is the architecture initialization where the source and destination of the path to be travelled by the user are initialized, along with all the necessary vehicle parameters.

The second stage comprises the DHFO algorithm which mainly concentrates on the optimal feature selection based on the demand computation and vehicle parameters. It recognizes the geographical co-ordinates of the CS that the vehicle can reach with its minimum SOC of battery level. The cluster similarity measures form a cluster of CS where EV can place a charging request when the EV battery has minimal SOC.

In the third stage, the Enhanced Cladistic Neural Network (ECNN) technique optimizes the clustering and feature selection, and makes the final prediction. The ECNN algorithm preprocesses the testing weight data using the concept of normalization. The preprocessed data blocks are then split into groups to extract data patterns. Every block group is then subjected to the Ensemble Modulation Pattern (EMP). This efficiently extracts geometrical characteristics to raise the general level of classification accuracy. Finally, the classifier is utilized to classify whether the location ID is satisfied with the relevance ratio or not. The relevance ratio defines the ratio between the number of known values and the predicted values of the data samples.

Thus, the entire forecasting of the CS is based on the power demand and availability of the CS, and the recognition of the area where the vehicle cannot reach the CS with its minimum SOC of battery level. The cluster similarity measures form a cluster of CS where EV can place a charging request when the EV battery has minimal SOC. This will predict the best location for the placement of a CS by suggesting geographical co-ordinate positions extracted using a map matched tool. Figure 3 is the flowchart of the proposed system.

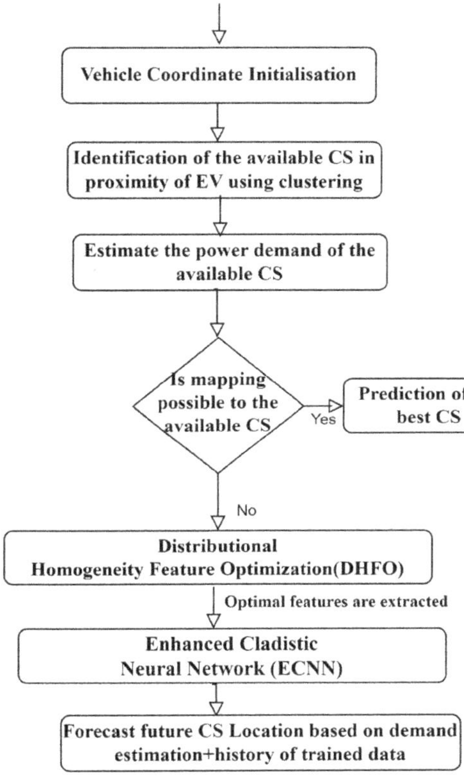

Figure 3. Flowchart of the proposed system.

The proposed algorithms are explained in depth in the ensuing subsections.
Section 3.1: Distributional Homogeneity Feature Optimization (DHFO);
Section 3.2: Enhanced Cladistic Neural Network (ECNN) Algorithm.

3.1. Distributional Homogeneity Feature Optimization (DHFO)

The Distributional Homogeneity Feature Optimization is a multi-objective optimization algorithm which uses the various parameters in different objective conditions to finalize the most suitable combination of parameters that will enhance the classification accuracy. Here, the optimization algorithm is used to reduce the size of feature database (before training) of a classifier to enhance the prediction performance. For that, the input of the DHFO optimization algorithm is the full data of the training feature set. As described in the algorithm below, the input is the geographical co-ordinates of the vehicle obtained from the OSM Data Samples and Demand Cost (M_{NID}) (the entire charging station parameters as in Figure 4) of randomly generated CS. From these inputs, the output of the DHFO algorithm is the optimal selected feature attributes among the overall database. Based on the selected indexes of the training database, the selected attributes are passed as the input for the classifier in both the training and testing processes.

In Algorithm 1, the multi-objective DHFO method's precise steps and equation model are made available. To construct the root direction, the input was selected as the co-ordinate position of the available charging stations, and the starting direction was formed based on the vehicle's geographic latitude–longitude co-ordinates setting of 0.5 if the link difference obtained from the datapoints was less than the defined range as in algorithm below. The coverage area for the randomly allocated CS location in terms of latitude and longitude was ±0.5 from the route path co-ordinate position of EV movement. The number of divisions to allocate the random CS was specified as 10 numbers which split the CS location into different positions. These randomly allocated CS contained the additional parameters along with the latitude and longitude, such as total capacity (kW), number of charging piles, cost (Ps per kWh), vehicle count already connected to the CS piles, and power demand in the CS (kW).

	Latitude(°)	Longitude(°)	Capacity (kW)	# Charging Piles	Cost (Rs per kWh)	Vehicles Count	Demand (kW)
1	12.942887603250835	77.66935414384166	200	20.0	10	6	-52.39533029049156
2	12.992723459216805	77.50163301018188	100	10.0	7	10	47.31281797038656
3	12.926893662422389	77.55497950004221	400	40.0	7	36	870.819695189542
4	13.004989547422753	77.66655346426049	500	50.0	5	38	683.5664814580057
5	13.009454949935652	77.49888319839445	400	40.0	8	20	351.0322820958778
6	12.974351712631394	77.51727852645104	200	20.0	5	20	187.56830227163323
7	12.933626932501177	77.52193767778925	400	40.0	6	10	-262.04632249576997
8	12.97875769463172	77.54550708483049	300	30.0	8	5	-232.41593204638036
9	12.890346678527802	77.62314299300736	400	40.0	8	9	-150.84231007406845
10	12.996239952515312	77.62862668527903	100	10.0	9	10	265.2776603961404
11	12.947948711140501	77.46764008915946	400	40.0	9	9	-81.67697596031968
12	12.866388083137059	77.45860949219548	500	50.0	8	50	918.4446830941838
13	12.864431380910577	77.41394736758173	500	50.0	7	33	44.634081170416266
14	12.961758240516595	77.46165622511381	100	10.0	7	5	-20.819954350083947
15	12.89465535471687	77.58589752948107	400	40.0	7	11	-19.209987676343076

Figure 4. Initialized table of demand data.

Each charging station contained an individual range of all these listed parameters. These formed the training database for the classifier to define the available CS with/without power demand.

The initial value of SOC was set at 20%. The distance covered by the car in terms of SOC was defined as EV-km-per SOC Unit, being 1.5. The critical value of SOC was 5. This was followed by the feature pattern analysis method to extract the most suitable features for prediction of apt CS using the population strength of the vehicle which was generated in this study along with the power capacity of the vehicle that was connected to the electric vehicle model.

The size of the dataset for the N number of attributes in this case is the RN database model. The total dataset size is expressed in terms of M. The key relevancy prediction between feature characteristics is represented by the matrix K_i and K_j. Next, wd_x, wd_y, and wd_{xy} are the particle parameters of K_i and K_j. As indicated by the following equations, the binary representation of these particles is similarly calculated as M_{xy}, N_{xy} and N_{yx}.

The demand cost matrix represents the Euclidean distance between each demand value arranged in the form of a square matrix. In this, Equations (1) and (2) represent the availability matrix and the responsibility matrix, respectively, for the index of '*i*' and '*j*'. This availability matrix is evaluated with the demand cost and with the responsibility matrix to refresh the matrix based on the update of vehicle availability.

$$A_{ij} = \begin{cases} 0.5, & if\ (M_{NID}(i,j) <= 0.5) \\ 0, & Otherwise \end{cases} \tag{1}$$

$$R_{ij} = \begin{cases} M_{NID}(i,j) - A_{ij} & If\ (A_{ij} <= M_{NID}(i,j)) \\ 0 & Otherwise \end{cases} \tag{2}$$

Algorithm 1: DHFO Algorithm

Input: Input Data [co-ordinates and demand Cost (M_{NID})]
Output: Predicted CS and demand, A_{ij}, R_{ij}, C_{id}
Step 1: Construct availability matrix and demand cost matrix
Let S_i and S_j be size of matrix (M_{NID}) and set $K = 2$; where, NID—Node ID
For i = 1 to S_i
  For j = 1 to S_j
    Compute the initial position of CS data, A_{ij} by
    $A_{ij} = \begin{cases} 0.5 & if\ (M_{NID}(i,j) <= 0.5) \\ 0 & Otherwise \end{cases}$
  End for j;
End for i;
Step 2: Construct and update responsibility matrix and availability matrix;
For X_k = 1 to k
  For i = 1 to S_i
    For j = 1 to S_j
      Compute the relevant vector, R_{ij} by
      $R_{ij} = \begin{cases} M_{NID}(i,j) - A_{ij} & If\ \left(A_{ij} <= M_{NID}(i,j)\right) \\ 0 & Otherwise \end{cases}$
    End for j
  End for i
  For i = 1 to S_i
    For j = 1 to S_j
      Let $temp\_R_{ij} = 0$;
      For m = 1 to S_i
        $temp\_R_{ij} = temp\_R_{ij} + R_{im}$
      End for m;
      $temp\_R_{ij} = \begin{cases} temp\_R_{ij} + R_{ij} & if\ \left(temp_{Rij} \leq 0\right) \\ R_{ij} & Otherwise \end{cases}$
      If ($i! = j$), then
      $A_{ij} = \begin{cases} 0 & If\ \left(temp_{Rij} < 0\right) \\ temp\_R_{ij} & Otherwise \end{cases}$
      Else
      $A_{ij} = \begin{cases} temp\_R_{ij} & If\ \left(temp\_R_{ij} > 0\right) \\ 0 & Otherwise \end{cases}$
      End if
    End for S_j
  End for S_i
End for X_k
Step 3: Compute exponential matrix and the average of relevant vector based on the estimated list index and then update in the list by
$Expm_{ij} = \begin{cases} 1 & if\left(A_{ij} + R_{ij}\right) > 0 \\ 0 & Otherwise \end{cases}$
$avg_{idx} = \frac{1}{x}\sum_{x=1}^{size(Idx_{ls})} R_{i(Idx_x)}$
Update $avg_{list} \leftarrow avg_{idx}$;
For y = 1 to S_j
  Compute the related average of overall vector avg_R by
  $avg_R = \frac{\sum_{j=1}^{S_j} R_{ij}}{S_j}$
  Compute the distance list between the average list and the related parameter by
  $dis_{ls} = \sqrt{avg_{List}^2 - (avg_R{}^2)}$
    Update $C_{id} \leftarrow min(dis_{ls})$
End for y;

Next, the ratio of $\frac{1}{\log(\text{Derv}_{xy})}$ and dif_v is used to compute the distance dist_{xy}. As a result, the sum of T_{dist} and dist_{xy} is updated in the T_{dist}.

$$dis_{ls} = \sqrt{avg_{List}^2 - (avg_R{}^2)} \quad (3)$$

$$avg_{idx} = \frac{1}{x}\sum_{x=1}^{size(Idx_{ls})} R_{i(Idx_x)} \quad (4)$$

The values of T_{dist} are then used to update the value of O_{dist}, and lastly, the updated $\frac{O_{\text{dist}}}{M}$ is used to estimate $M_{\text{NID}}(i,j)$.

The charging station architecture optimized the station's orientation and set it up for quicker data transfer based on the evaluation of these criteria. The output of DHFO was the optimal feature database (training database), and the selected attributes of the overall training database were used as the input for the ECNN classifier.

3.2. *Enhanced Cladistic Neural Network (ECNN) Algorithm*

The optimal feature database (training database), which is the output of the above-described DHFO, was the extracted attributes of the overall training database that was used as the input for the ECNN classifier. It was computed like the evolutionary relationship concept in the construction of a cladogram. Cladograms are often used to create a linear representation of hypotheses. They describe the evolutionary relationships between varied features, showing how they are organized depending on co-operative characteristics. The ECNN-based classification algorithm is used to predict the CS position based on the EV and CS features that are optimized from the DHFO algorithm. This will estimate the parameters of the vehicle, and allow the CS module to train itself on a suitable combination of indexes. From that, if a vehicle requests any CS prediction, this will find the best co-ordinate position of CS that is related to the satisfactory level of the vehicle for it to charge its battery. This classification model can also forecast and recommend the location to install a CS later, based on the demand estimation and the history of trained data from the available features. This will update the training database for improving the performance in accuracy and sensitivity levels. The ECNN algorithm takes into consideration the vehicle capacity that can be integrated into the architecture depending on the minimal link coverage distance that is the time it takes to reach the available CS, which may vary from time to time. This algorithm's primary contribution is its ability to forecast the optimal CS, based on the parameters influencing this link coverage. The link coverage varies from time to time as it depends on the vehicle parameters such as its model, the battery make and capacity, driving speed, and traffic.

The model training begins with data preprocessing. The main phases of preprocessing undergone in the proposed framework are (i) data normalization, (ii) feature optimization using DHFO, and (iii) train test split.

3.2.1. Data Normalization

The normalization process aims to scale the numerical values in a dataset to a common range without affecting the range of values. The normalization process applied in the proposed model is min–max normalization applied as in equation,

$$X' = \frac{X - Xmin}{Xmax - Xmin}$$

3.2.2. Feature Optimization

Feature optimization is the process to minimize the number of features to decrease the complexity of computation, as well as to improve the efficiency of the training model. The feature optimization technique used in the proposed model is DHFO which has been described in Section 3.1. Every optimization iteration received the instance's data matrix as input. Here it is used to reduce the size of the dataset before training the classifier to enhance the prediction performance. As mentioned in Algorithm 1, the input is Co-ordinates and Demand Cost (MNID) (the full table parameters are in Figure 4) of randomly generated CS. Thus, the training feature set of the classifier model comprises the collected OSM data co-ordinate positions, the selected attributes from the parameters of CS and EV such as SOC, Km can travel by vehicle, Cost of CS, and Demand in (KW) from the randomly generated CS. The optimal feature database obtained from DHFO and the selected attributes are passed as the input to the ECNN classifier in both the training and testing processes. The data matrix was updated every time the ideal charging station direction was chosen. The training model of the ECNN was used to classify the classes of live moving vehicle parameters.

3.2.3. Train Test Split

The final phase in preprocessing is to split the dataset into train and test sets. The process of splitting the dataset is a typical phase in learning algorithms to address overfitting and for evaluating the performance of the trained model properly on unseen data. In the proposed work, the dataset was split in the ratio of 70:30, where 70% was used for training and remaining 30% was retained for model testing.

3.2.4. Classification Model Description

In this process, the ECNN is a multi-level classifier used for the prediction of the best location for a charging station based on the feature attributes that are collected from the parameters of the CS and EV. The classification result from the ECNN is the prediction of the nearest CS location in the route path based on the parameters of SOC, Km can travel by vehicle, Cost of CS, and Demand in (KW). If there was no CS predicted in the coverage area of the current moving vehicle, then this recommended the location of CS based on the demand level and the distance to the CS in the coverage area. The labels that are considered for the classification are:

(a) CS station available with sufficient power to charge the EV—Class 1;
(b) CS station available with power demand (KW)—Class 2;
(c) CS station not available—Class 3;
    (i) Recommend new CS station;
    (ii) Recommend the longitude and latitude of new CS station.

These classes are defined by the training set for the classifier extracted from the parameters of the CS and EV. These parameters are arranged in the form of feature attributes to construct the training model for the classifier. This training model will be a one-time process while installing the system on the manufacturing side. While driving the EV, the parameters from the live running vehicle will be arranged as the testing feature vector. At this testing stage, the parameters from each time point are used as the input to the classifier for testing. The classifier predicts the class according to the best match with the training database (feature set) and gives out the class label as 1, 2, or 3. If it predicted class 3, then the classifier will again validate the parameters to recommend the best location for a new CS, and this produces the longitude and latitude for the new CS.

The ECNN model's algorithm steps are given in Algorithm 2:

Algorithm 2: ECNN Algorithm

Input: Input vehicle pattern set $T_D(s)$
Output: Classified result $V(k)$
The initial pattern for the prediction is arranged as,
$T_D(s) = \{T_{D1}(s), T_{D2}(s), \ldots, T_{Dm}(s)\}$
In the input layer of the neural network, the data sequence can be formed as the matrix in the equation below.

$$X_D(s) = \begin{bmatrix} T_{D1}(s) \\ T_{D2}(s) \\ \cdots \\ T_{Dm}(s) \end{bmatrix}$$

Algorithm 2: *Cont.*

From the matrix arrangement, the block correlation feature can be estimated by $F(X_D(s).X_D^*(s))$.
This can be represented as
$F(X_D(s).X_D^*(s)) = X_D^*.e^{T-T_m}$
Where, 'T' and 'T_m' represent the attribute values from matrix $X_D(s)$.
Estimate the kernel model of classifier
$K_m = \frac{1}{2^{q-1}}\left(\frac{\sqrt{2q}}{l}\right)^q k_q\left(\frac{\sqrt{2q}}{l}r\right) \quad \forall q = 1, 2, \ldots, N$ //'r' represents the range of feature distance, 'l' represents the length of the feature vector.
Estimate the relevancy using the kernel function with feature points.
$t_n = F^T\omega_n$
Where, 'ω_n' weight value of attributes.
$u_n = F^T\omega_n$
Extract the training features and form the network by
$T_r = \{t_1, t_2, \ldots, t_n\}$
$X_b = \overline{X_b} + \sum_{i=1}^{N} t_i(d)p^i$
Estimate the matching score for the correlated blocks by
$\hat{T}_s = \left(\left(X_b^d - \overline{X_b}\right)^T (P^T)\right)^T$
Where, the relevance factor $X_b^{\overline{d}} \in R^{(T-T_p)M}$ can be written as
$R^{(T-T_p)M} = \hat{T}_s{}^T Q^T + \overline{tt_a}$
Where, 'P' and 'Q^T'—predicted component.
The predicted label can be represented by
$V(k) = \frac{d_{ij}}{R_j - R_i}$
Where, d_{ij}—distance matrix for 'i' and 'j' of the relevance matrix 'R.

4. Results and Discussion

The outcomes of the suggested model for classifying charging stations according to availability and weight characteristics were verified and contrasted with the approaches already in use. The first Section 4.1 was a briefing on the experimental setup and the dataset. The tracking of the map-matched data samples with the output of the experiment was discussed in Section 4.2. The performance indicators used for the study were discussed in the second Section 4.3, followed by the quantitative evaluation based on the metrices used to calculate the performance of the suggested classification model in Section 4.4. The computational complexity of the system was analyzed in the following subsection E. An investigation based on the road map matched with the OSM samples and experimental results obtained was elaborated, followed by the performance comparison with other existing approaches in the last Section 4.5.

4.1. Experimental Setup

Python 3.19 was used for the experimental study and testing. To perform the weight flow analysis, the position and co-ordinate data were chosen based on the coverage area of the electric vehicle.

The data samples for the experimental setup were generated from the OpenStreetMap (OSM). OSM is a map embed API that provides map-matched data for the Origin Destination (O-D) Road Trip as planned for the EV user (https://www.openstreetmap.org/#map=5/21.494/66.665, accessed on 26 January 2024). The geographic data are collected from the traffic co-ordinate positions in the OSM database. Similarly, an archival database of several map-harmonized tours are taken into consideration. According to the direction on the map, the CS co-ordinate positions and the specification of CS are simulated and verified. This includes weight data from the vehicle database, with some data created as placeholders for missing values to forecast and predict the data. In this work, there were five different road trip datasets generated for the locations from Bangalore to Mysore(D1), Bangalore to Pune(D2), Bangalore to Tumkur(D3), Bangalore to Cochin(D4), and Bangalore to Hyderabad(D5) in random directions on the map. These trips interconnecting different

Sates in Southern India were of variable length, as shown in Figure 5a. Figure 5b shows the map direction D1, from the source of Bangalore to the destination of Mysore. Each part of this trip was further fragmented into several road segments. Random position changes and the creation of a situation where data are missing are then generated and simulated in various modules. The Open-Source Routing Machine (OSRM) routing service provided the available charging stations on the routes chosen for the study.

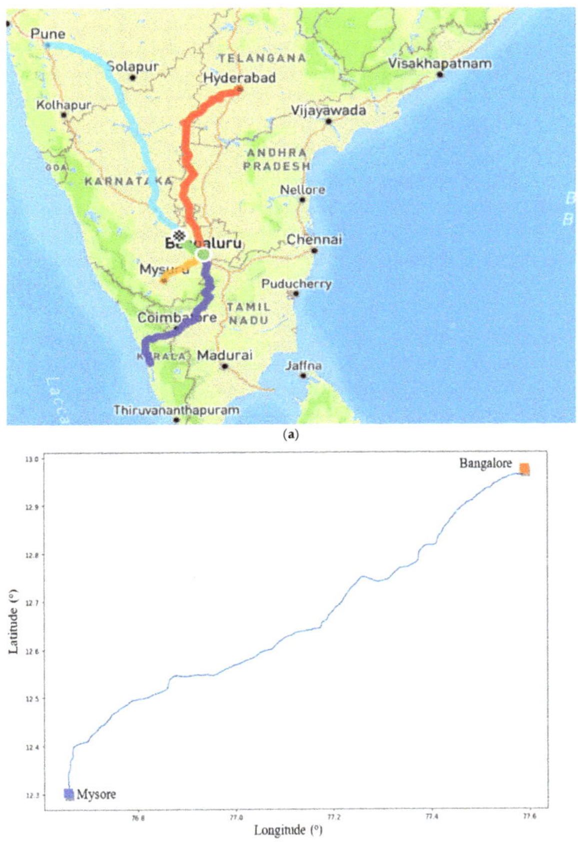

Figure 5. Data Samples taken for the experimental study. (**a**) Five road trips D1 to D5 chosen for study (**b**) path of direction of route Bangalore to Mysore from OSM data.

Figure 4 shows the initialized demand parameters of CS at each co-ordinate position which comprises co-ordinate positions of the CS in terms of latitude and longitude, the maximum output power that a CS provides in kW, number of piles available in the CS, cost per kWh (varying from 5 to 10), and number of vehicles already getting charged. These parameter values are used to compute the power demand of CS. The CS with and without demand are further examined using the DHFO and ECNN algorithm to predict the most optimal charging station for the EV users.

4.2. Track Analysis with Map

The map-matched route trips taken from the OpenStreetMap Database were utilized to perform the proposed study. The data of available charging stations on the trips taken were extracted using the Open-Source Routing Machine (OSRM). All the available charging stations near to the EV were grouped together based on clustering. The demand value of each CS was computed as described in the DHFO algorithm, taking into consideration the vehicle SOC and traffic density. The demand value in the positive range indicates that a greater number of vehicles can be charged there. The negative value of demand suggests unavailability of suitable charging stations, and hence there is an urge to place one there.

Figure 6 shows the simulation update of vehicle movement in all the O-D road trip datasets that were generated for the locations from Bangalore to Mysore (D1), Bangalore to Pune (D2), Bangalore to Tumkur (D3), Bangalore to Cochin (D4), and Bangalore to Hyderabad (D5) in random directions on the map. The dynamic prediction of optimal CS placement is represented by the legend marks. In this, the black color represents the vehicle and the orange color represents the movement tracking of the EV. The green dots represent the co-ordinate positions of all the charging stations adjacent to the route of the electric vehicle, traced using the clustering algorithm. The red dot in the figure indicates the co-ordinate positions of the CS which are of high demand in the path for the specified source to destination. The purple dot indicates the co-ordinate positions of the CS based on the optimization algorithm. It represents the forecasted and recommended location to place the CS. It was obtained based on the demand estimation with vehicle distance. The prediction was done dynamically, based on the classification algorithm taking into consideration the selected feature parameters from the DHFO algorithm and computing the demand MNID (i, j), as described in Algorithms 1 and 2 above. As the EV travels towards to charging station, the SOC reduces. A minimum threshold of 5% was set for the SOC to ensure that the EV would be charged before it was totally drained. The EV, on reaching this minimum threshold of SOC, will identify the predicted CS and become charged. Accordingly, the SOC will be updated and the vehicle travels towards its destination. Each of the data samples of the O-D trip taken (D1–D7 described in the following section) were further divided into more than 50 datapoints, overall. Thus, we considered approximately 350 datapoints. The ensuing subsections address the performance outcomes.

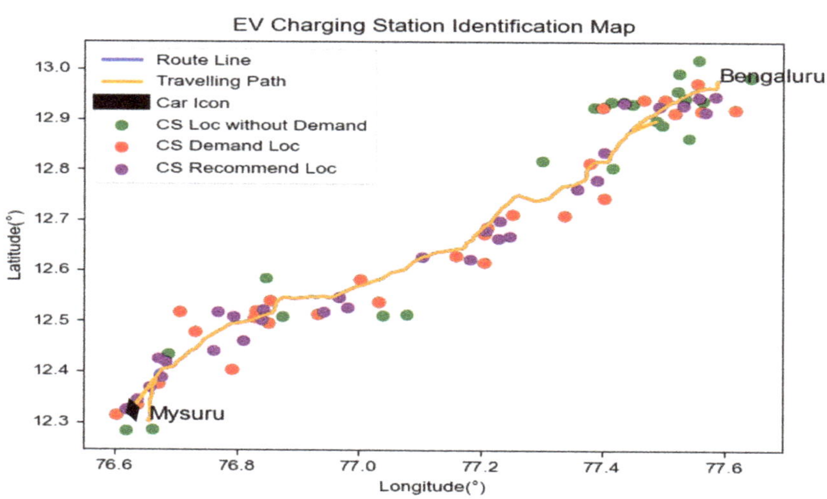

(a)

Figure 6. *Cont.*

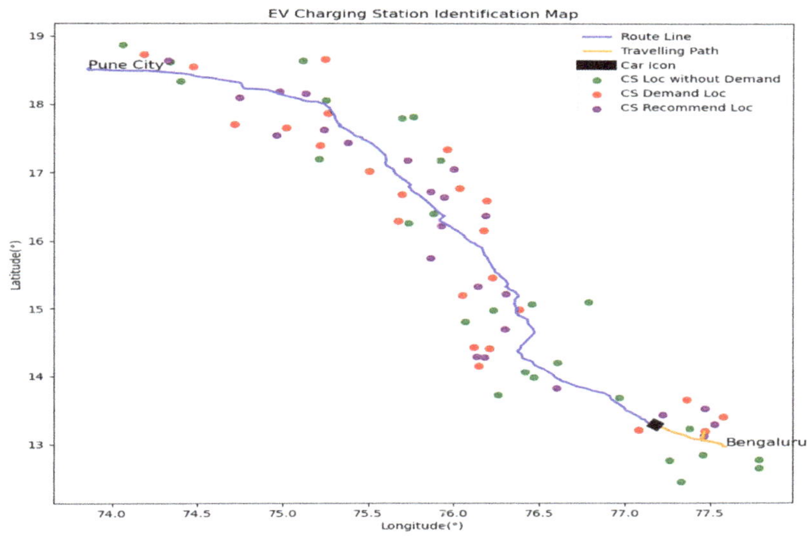

(b)

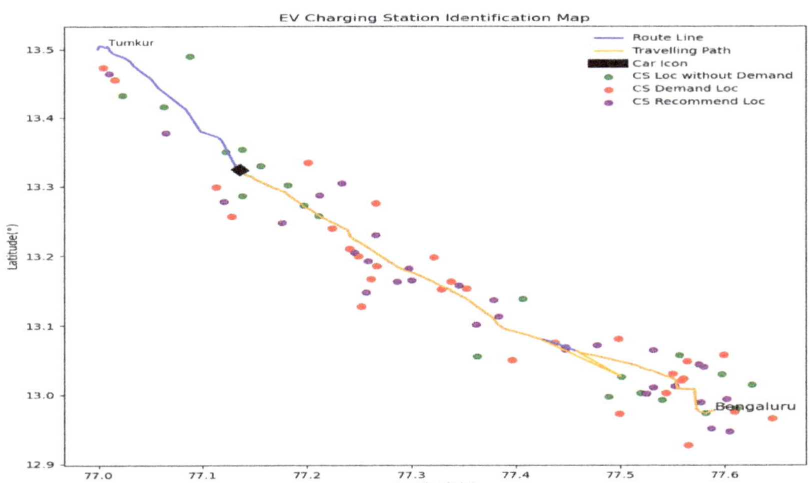

(c)

Figure 6. *Cont.*

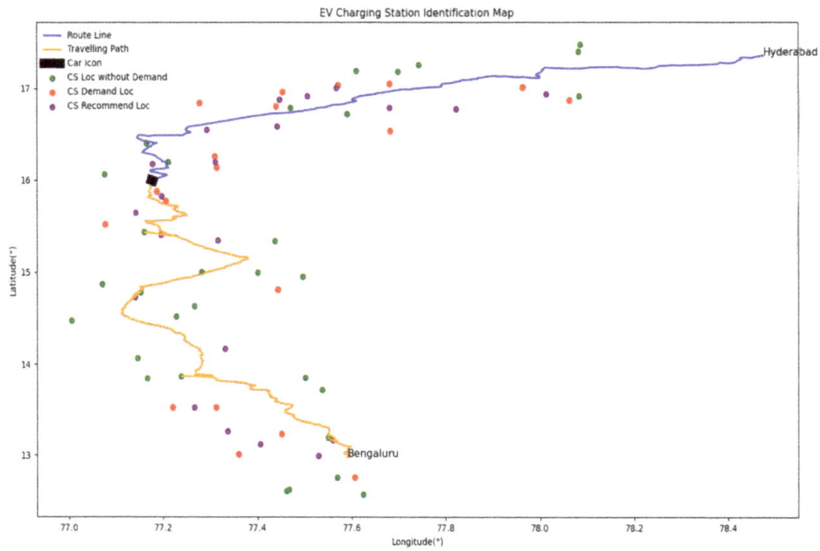

(d)

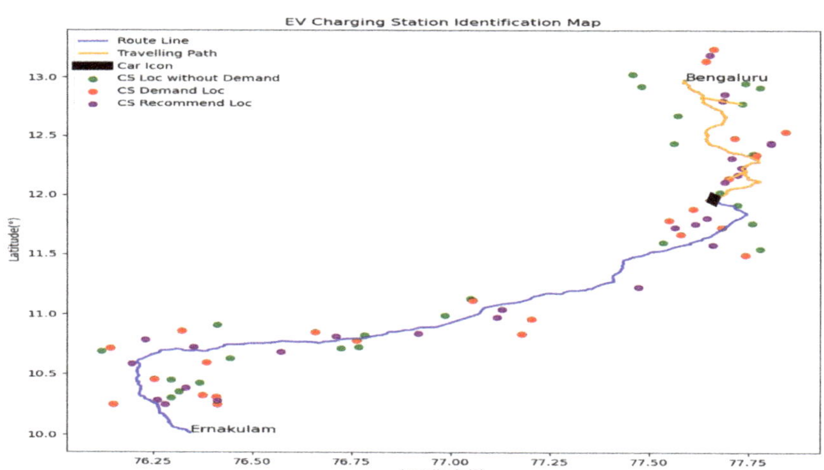

(e)

Figure 6. Simulation of the paths taken for the testing of the prediction. (**a**) Simulation of vehicle tracking and CS prediction for the direction path from Bangalore to Mysore; (**b**) simulation of vehicle tracking and CS prediction for direction path from Pune to Bangalore; (**c**) simulation of vehicle tracking and CS prediction for the direction path from Tumkur to Bangalore; (**d**) simulation of vehicle tracking and CS prediction for direction path from Hyderabad to Bangalore; (**e**) simulation of vehicle tracking and CS prediction for the direction path from Bangalore to Ernakulam.

4.3. Performance Indicators

By calculating the statistical probability between the number of misclassified results and the properly classified data samples, one can determine the parameters utilized in the performance metric-based analytical procedure. The confusion matrix layout is used to generate all of these, which are then assessed by comparing the categorized results to the dataset's ground truth. The true positives and false positives of the predictions made are used to compute the precision, recall, accuracy, and F1 score. Each of these measures is significant, according to the varying situations, to get a broader understanding of the errors and performance of the model. The accuracy is the baseline measure used in the classifier model to get an indication of how efficient the system is. It is computed as in Equation (5).

$$\text{Accuracy}, \quad \text{Acc} = \frac{\text{Total correct labels}}{\text{Total No. of Samples}} \tag{5}$$

To ensure that the system performs well, even when the datasets are not properly balanced, we utilized the precision value as it gives the ratio of true positives across all the positive values. It is beneficial in scenarios when the estimate of false positives is high. The mathematical equation of precision is mentioned in Equation (6) stated below.

$$\text{Precision}, \quad \text{P} = (1 - \text{FPR}) = \frac{\text{TP}}{\text{TP} + \text{FP}} \tag{6}$$

Recall is taken into consideration when the estimate of false negatives is high and the equation to compute it is specified in Equation (7) stated below.

$$\text{Recall, R} = (1 - \text{FNR}) \tag{7}$$

The F1 score is also considered to find an equilibrium between the precision and recall. The F1 score is a single metric that is computed as the harmonic mean of precision and recall, as in Equation (8).

$$\text{F1 Score, F\_S} = \frac{2\text{PR}}{\text{P} + \text{R}} \tag{8}$$

Area under Receiver Operating Characteristic (ROC) curve is another important metric used to evaluate the performance of the model. The ROC curve is plotted as the curve between recall or sensitivity defined in Equation (9) and the false positive rate (FPR) which is (1 − specificity which is mentioned in Equation (10)) or the proportion of incorrect classification made by the model, as in Equation (11).

$$\text{Sensitivity}, \quad \text{TPR} = \frac{\text{True Positive(TP)}}{\text{Total No. of Positive samples}} \tag{9}$$

$$\text{Specificity}, \quad \text{TNR} = \frac{\text{True Negative(TN)}}{\text{Total No. of Negative samples}} \tag{10}$$

$$1 - \text{Specificity} = \text{FPR} = \frac{\text{False Positive(FN)}}{\text{True Negative(TN)} + \text{False Positive(FP)}} \tag{11}$$

4.4. Quantitative Evaluation Based on Performance Measures

The complete outline of the performance of the system was obtained using the above-defined performance metrices. The five defined road O-D trips in Figure 5, as mentioned in the experimental setup, were taken into consideration. The performance measures of precision, recall, F1 score, and accuracy were computed across each of these segments.

The plot showing the tradeoff in these performance measures is visually represented in Figure 7. It was observed that the model obtained a higher recall and accuracy measure across all five defined road trips. The higher recall shows that the system was successful in predicting the maximum number of optimal CSs available for the EV user for a particular road trip. The higher accuracy value across all the road segments ensured that the system's overall performance was high. Hence it can be inferred that the model will ensure that the EV user will be intimated with all possible optimal CS on the path it is travelling on. The ROC curve outlining the performance of the proposed model on the different datasets is plotted in Figure 8. The plot is, again, an indication that the use of optimal feature selection and demand computation of the proposed system results in elevated results (>0.9) in the proposed application. The area under the curves is labelled in the graph.

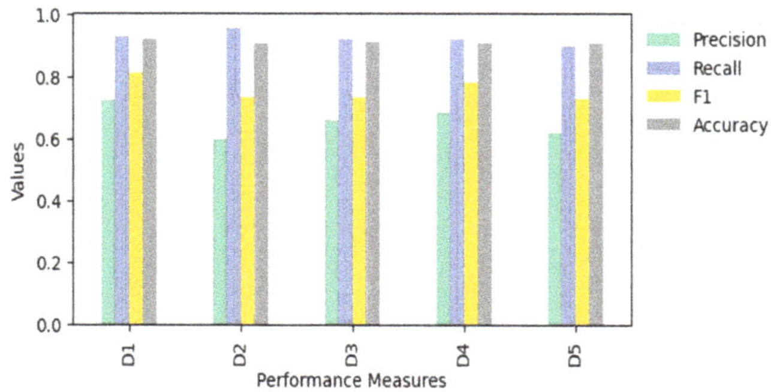

Figure 7. Plot representing the quantitative evaluation based on performance measures.

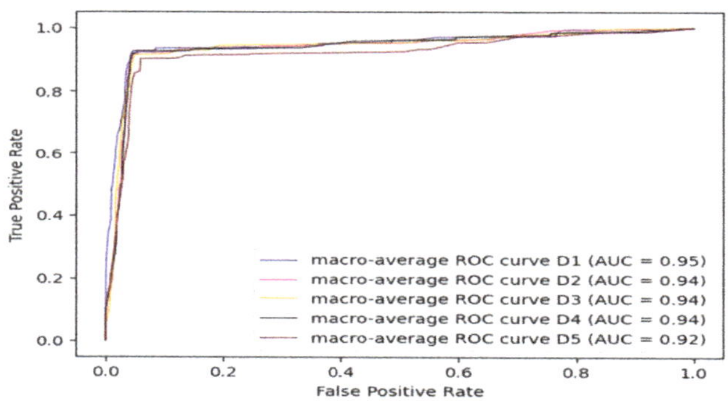

Figure 8. ROC curve of the proposed model on the different datasets D1 to D5.

4.5. Computational Complexity

The proposed study mainly comprises two main algorithms—DHFO and ECNN. The computational complexity of the system depends on the working of the feature optimization and neural network algorithms, as discussed in [38]. The DHFO algorithm takes into consideration the input co-ordinates of the EV user, as well as the charging station. The availability and the demand matrix are updated for each of the nodes taken into consideration. The construction of the relevancy matrix is nested for loop, which is iterated to the maximum of N, where N is the total number of nodes. It is followed by the updating part of the above-constructed matrix based on the relevant vector as well as the vehicle node co-ordinate positions. This modification again includes an iteration within the above-nested for loop, thereby the final computational complexity is O(n^3).

The complexity of the algorithm depends on the training and testing phases. During the training phase, the complexity of the one-time step of the forward pass of each layer involves matrix manipulations resulting in a complexity of O ($n_i \times n_{i+1}$), where i refers to the layers. The updating of the weights in the backward pass also results in O ($n_i \times n_{i+1}$) computational time. The estimation of the matching score for the correlated blocks is O (n^2 R), where R represents the total attributes. Hence, total computational complexity of the ECNN algorithm, taking into consideration k epochs and T time steps, is $O(k.T\sum_{i=1}^{n-1}(n_i * n_{i+1})) + O(n^2.R)$. For the testing phase, the computational complexity is much less linear in nature as it is directly proportional to the input size only.

4.6. Comparison with Existing Approaches

Several studies have been made to resolve the range anxiety problem of EV users. The optimal prediction of charging station is the primary concern of all these studies. The study in [8] examined deep learning methods for optimal prediction. In the experimental study, the OSM road network from Denmark was used, along with weather and digital elevation information, to make the predictions. The speed profile was investigated based on different trip segments. The statistics of the investigated models were tabulated and, being a regression model, the Root Mean Square Error (RMSE) and Mean Absolute Error (MSE) were tabulated. The results showed that the system performed exceptionally well with a RMSE of approximately 0.8 for all kinds of speed profile when compared to other deep learning models.

In our study we proposed an enhanced cladistic neural network classifier algorithm with Distributed Homogenous Feature optimization, thereby chalking out the most significant features for prediction of the CSS. Hence, for analysis, we used the translation function and converted the above regression results setting into a suitable threshold based on the domain area where we could conclude that most of the DNN-based models showed an outstanding performance. Using the threshold of 2.0, the classification performance metrics computed for each regression method indicated that most methods performed similarly, with high precision above 0.9, recall above 0.8, and F1-scores above 0.85, approximately. Figure 9 depicts the route from Denmark to Hannover with the prediction of optimal CS using the proposed system, which is labelled D6.

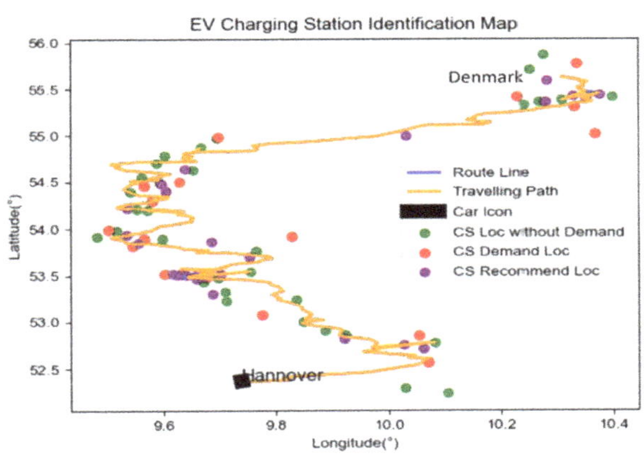

Figure 9. Road trip from Denmark to Hannover with the prediction of optimal CS.

Figure 10 represents the geographical picture of the route trip from the road network in Wuhan which is a case study in [39]. The route chosen was from Wuhan to Wanggang, and labelled D7. In the study in [39], different baseline standards were chosen to analyze the influence of urban area in the prediction. The RMSE and MAE for the different baselines were tabulated. These hypothetical regression results, when converted using the trade-off threshold 2.0, clearly revealed that the effectiveness was high, with an accuracy and precision of approximately 0.9, and a precision and F1 score of almost over 0.85.

The performance measures obtained for these two road trips, D6 and D7, using our system, were also of a high standard in terms of accuracy, precision, and recall. In Table 1, the results are evaluated from the tracking of vehicle movement with respect to time response based on the SOC parameters. This result is simulated for routes D6 and D7. With our proposed approach, the optimal charging stations predicted are highlighted as purple dots in Figures 9 and 10.

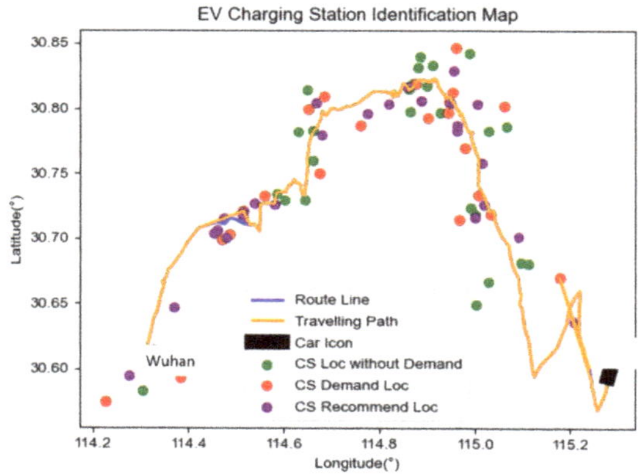

Figure 10. Road trip from Wuhan to Wanggang with the prediction of optimal CS.

The proposed technique could classify the results in a generalized approach by effectively extracting the data patterns of the input image and taking into consideration the most optimized parameters from the OSM sample data. The approaches which have been taken for the analysis in the current study also include additional parameters from the urban region, as well as environmental factors such as weather perspective. This proposed study is not specific, and is independent of whether the samples were taken from urban or rural areas, as well of other environmental factors, but still provides a promising and satisfactory prediction of suitable CS.

In this process, the ECNN classifier worked as the multi-level classifier to predict the best location based on the feature attributes that were collected from the parameters of the CS and EV. While driving the EV, the parameters from the live running vehicle were arranged as the testing feature vector. At this testing stage, the parameters from each time point were used as the input to the classifier for testing. The classifier predicted the class according to the best match with the training database (feature set) and gave out the class label as 1:CS station available with sufficient power to charge the EV (green-colored dots), CS station available with power demand (KW)—Class 2 (red-colored dots), or no CS—Class 3 (purple-colored dots as in Figures 9 and 10). If it predicted class 3, then the classifier would, again, validate the parameters to recommend the best location for a new CS that gives out the longitude and latitude for the new CS.

In this work, the data patterns were efficiently extracted using DHFO, and the moving vehicle was accurately detected by using the ECNN technique. From these CS location predictions and the tracking performance evaluation, the parameters representing the performance of the proposed model were satisfactory when compared with other state-of-art methods. The precision, recall, and accuracy defined the sensitivity of the proposed algorithm in the prediction of CS and the forecasting of the CS location to recommend the position where we can optimally place the CS in the path of direction.

Table 1 and Figure 11 show the accuracy and tracking performance rates of the existing approaches: Data-Driven Context Aware Learning DDCAL [39], Probabilistic Deep learning for EV [8], Machine learning Analysis (MLA) based on Linear Regression and Support Vector Machine [2], Fuzzy-based multi-criteria decision-making NWHFMCGDM [40], and Proposed Techniques. All these existing approaches have been studied using the data samples D6 and D7. These results are evaluated based on the successful classification with less time difference. To track the evaluation, the performance measures' accuracy and tracking performance rates of these approaches were calculated. From the evaluation, it was found that the proposed system provides an efficient result, when compared to the other techniques, and it is stable across all the data routes. To compare the efficiency of feature extraction and classification performance across the systems, the tracking performance rate was used. The tracking performance depends on the computational time taken with respect to the data used, as well as the hyperparameters which are model-specific tunable settings. In this work, the data patterns were efficiently extracted using DHFO, and the moving vehicle could be accurately detected by using the ECNN technique.

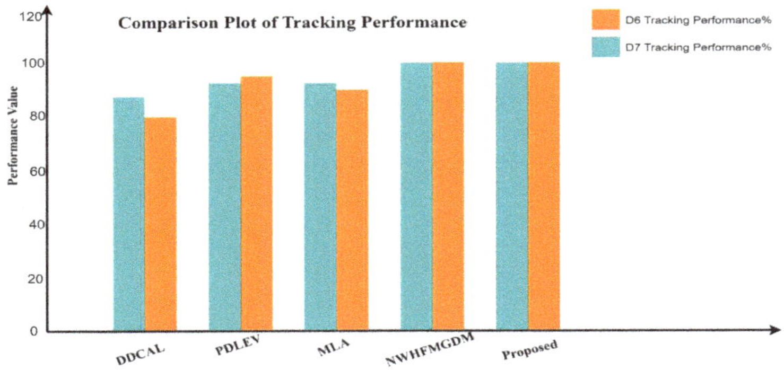

Figure 11. Comparison plot of tracking performance.

From the CS location predictions and the tracking performance evaluation, the parameters representing the performance of the proposed model were better than the other state-of-art methods.

Table 1. Comparative analysis with respect to D6 and D7.

Data Sample Taken	Methods	Accuracy	Tracking Performance %
D6	DDCAL [39]	0.90	80
	PDLEV [8]	0.90	94.45
	MLA [2]	0.90	90.01
	NWHFMGDM [41]	0.89	99.00
	Proposed	0.91	99.89
D7	DDCAL [39]	0.89	89
	PDLEV [8]	0.88	93.29
	MLA [2]	0.89	91.2
	NWHFMGDM [41]	0.88	99.1
	Proposed	0.91	99.95

5. Conclusions

An innovative charging station prediction model was put forth in this era of the Internet of vehicles [42], utilizing an Enhanced Cladistic Neural Network (ECNN) for charging station prediction and an adaptive Distributional Homogeneity Feature Optimization (DHFO) method for optimal feature selection based on pattern analysis. The results were evaluated from the tracking of vehicle movement, in which the EV could reach the CS with an optimal time difference, and it ensured a high score in all standard metrics across all the OSM data samples used, showing the stability of the system's performance. From the evaluation, it was found that the proposed system provided a satisfactory result, when compared to the other existing techniques, by taking into consideration only the most significant features obtained on the pattern of vehicle for prediction of the most optimal existing CS, as well forecasting the geographical co-ordinates apt for the installation of the CS. The additional socioeconomic and geographical factors, as described in other existing approaches, influenced the system, but their impact was considerably small on vehicle and CS parameters.

Limitations and Future Prospects

The study mainly concentrated on the charging station prediction, as well as forecasting of location to install a new CS, assuming the higher adoption rate of EV. The study was predominantly based on dynamic vehicle tracking and was not validated for the standard EV datasets, as most of the available datasets are region-specific and may drastically change with time, depending on the adoption rate of EV in those regions. Although the geographical co-ordinates of a future CS have been forecasted, the power distribution and government policies, as well as guidelines, must also be studied prior to the final installation of the charging stations. Based on feature optimization model advancement, this kind of pattern-based prediction model may be tested and implemented in the future for large data in car patterns. Further studies could also be conducted, based on a time series plot with different vehicle models.

Author Contributions: Conceptualization, R.P.; Methodology, R.P.; Validation, M.P.S.; Writing—original draft, R.P.; Writing—review & editing, R.P. and M.P.S.; Supervision, M.P.S. All authors have read and agreed to the published version of the manuscript.

Funding: This research received no external funding.

Data Availability Statement: The original contributions presented in the study are included in the article, further inquiries can be directed to the corresponding author.

Conflicts of Interest: The authors declare that they have no conflicts of interest.

Nomenclature

CS	Charging Station
EV	Electric Vehicle
AI	Artificial Intelligence
DHFO	Distributional Homogeneity Feature Optimization
ECNN	Enhanced Cladistic Neural Networks
SoC	State of Charge of Electric Vehicle
OSM	OpenStreetMap
CNN	Convolution Neural Network
O-D	Origin Destination
EMP	Ensemble Modulation Pattern
OSRM	Open-Source Routing Machine
P	Precision
R	Recall
FPR	False Positive Rate
FNR	False Negative Rate
ROC	Receiver Operating Characteristic
Acc	Accuracy Measure
RMSE	Root Mean Square Error
MAE	Mean Absolute Error
DNN	Deep Neural Network

References

1. Sasank, T.V.; Akash, K.B.; Siddharth, I.S.; Susila, M. State of Charge Estimation and Analysis For an Electric Vehicle. In Proceedings of the 2023 International Conference on Recent Advances in Electrical, Electronics, Ubiquitous Communication, and Computational Intelligence (RAEEUCCI), Chennai, India, 19–21 April 2023; pp. 1–6. [CrossRef]
2. Alanazi, F.; Alshammari, T.O.; Azam, A. Optimal Charging Station Placement and Scheduling for Electric Vehicles in Smart Cities. *Sustainability* **2023**, *15*, 16030. [CrossRef]
3. Kim, D.; Kwon, D.; Han, J.; Lee, S.M.; Elkosantini, S.; Suh, W. Data-Driven Model for Identifying Factors Influencing Electric Vehicle Charging Demand: A Comparative Analysis of Early- and Maturity-Phases of Electric Vehicle Programs in Korea. *Appl. Sci.* **2023**, *13*, 3760. [CrossRef]

4. Liu, A.; Zhao, Y.; Meng, X.; Zhang, Y. A three-phase fuzzy multi-criteria decision model for charging station location of the sharing electric vehicle. *Int. J. Prod. Econ.* **2020**, *225*, 107572. [CrossRef]
5. Shahriar, S.; Al-Ali, A.R.; Osman, A.H.; Dhou, S.; Nijim, M. Prediction of EV Charging Behavior Using Machine Learning. *IEEE Access* **2021**, *9*, 111576–111586. [CrossRef]
6. Çelik, S.; Ok, Ş. Electric vehicle charging stations: Model, algorithm, simulation, location, and capacity planning. *Heliyon* **2024**, *10*, e29153. [CrossRef]
7. Hecht, C.; Figgener, J.; Sauer, D.U. Predicting Electric Vehicle Charging Station Availability Using Ensemble Machine Learning. *Energies* **2021**, *14*, 7834. [CrossRef]
8. Petkevicius, L.; Saltenis, S.; Civilis, A.; Torp, K. Probabilistic Deep Learning for Electric-Vehicle Energy-Use Prediction. In Proceedings of the 17th International Symposium on Spatial and Temporal Databases (SSTD '21), Virtual, 23–25 August 2021; Association for Computing Machinery: New York, NY, USA, 2021; pp. 85–95. [CrossRef]
9. Zhao, Y.; Guo, Y.; Guo, Q.; Zhang, H.; Sun, H. Deployment of the electric vehicle charging station considering existing competitors. *IEEE Trans. Smart Grid* **2020**, *11*, 4236–4248. [CrossRef]
10. Aljafari, B.; Jeyaraj, P.R.; Kathiresan, A.C.; Thanikanti, S.B. Electric vehicle optimum charging-discharging scheduling with dynamic pricing employing multi agent deep neural network. *Comput. Electr. Eng.* **2023**, *105*, 108555. [CrossRef]
11. Mao, D.; Tan, J.; Wang, J. Location planning of PEV fast charging station: An integrated approach under traffic and power grid requirements. *IEEE Trans. Intell. Transp. Syst.* **2020**, *22*, 483–492. [CrossRef]
12. Li, N.; Jiang, Y.; Zhang, Z.-H. A two-stage ambiguous stochastic program for electric vehicle charging station location problem with valet charging service. *Transp. Res. Part B Methodol.* **2021**, *153*, 149–171. [CrossRef]
13. Cao, W.; Wan, Y.; Wang, L.; Wu, Y. Location and capacity determination of charging station based on electric vehicle charging behavior analysis. *IEEJ Trans. Electr. Electron. Eng.* **2021**, *16*, 827–834. [CrossRef]
14. Li, J.; Liu, Z.; Wang, X. Public charging station location determination for electric ride-hailing vehicles based on an improved genetic algorithm. *Sustain. Cities Soc.* **2021**, *74*, 103181. [CrossRef]
15. Ahmad, F.; Khalid, M.; Panigrahi, B.K. An enhanced approach to optimally place the solar powered electric vehicle charging station in distribution network. *J. Energy Storage* **2021**, *42*, 103090. [CrossRef]
16. Wang, H.; Meng, Q.; Wang, J.; Zhao, D. An electric-vehicle corridor model in a dense city with applications to charging location and traffic management. *Transp. Res. Part B Methodol.* **2021**, *149*, 79–99. [CrossRef]
17. Lee, J.-B. Review on Environmentally-Friendly Vehicle Charging Station Location and Size Problems. *KSCE J. Civ. Eng.* **2022**, *26*, 3741–3751. [CrossRef]
18. Yu, Z.; Wu, Z.; Li, Q.; Bai, Q. A map matching-based method for electric vehicle charging station placement at directional road segment level. *Sustain. Cities Soc.* **2022**, *84*, 103987. [CrossRef]
19. Othman, A.M.; Gabbar, H.A.; Pino, F.; Repetto, M. Optimal electrical fast charging stations by enhanced descent gradient and Voronoi diagram. *Comput. Electr. Eng.* **2020**, *83*, 106574. [CrossRef]
20. Çeven, S.; Albayrak, A.; Bayır, R. Real-time range estimation in electric vehicles using fuzzy logic classifier. *Comput. Electr. Eng.* **2020**, *83*, 106577. [CrossRef]
21. Sun, C.; Li, T.; Tang, X. A Data-Driven Approach for Optimizing Early-Stage Electric Vehicle Charging Station Placement. *IEEE Trans. Ind. Inform.* **2023**. [CrossRef]
22. Nurmuhammed, M.; Akdağ, O.; Karadağ, T. A novel modified Archimedes optimization algorithm for optimal placement of electric vehicle charging stations in distribution networks. *Alex. Eng. J.* **2023**, *84*, 81–92. [CrossRef]
23. Sabyasachi, S.; Singh, A.R.; Godse, R.; Jaiswal, S.; Bajaj, M.; Srivastava, I.; Blazek, V.; Prokop, L.; Misak, S. Reimagining E-mobility: A holistic business model for the electric vehicle charging ecosystem. *Alex. Eng. J.* **2024**, *93*, 236–258. [CrossRef]
24. Zhou, J.; Wu, Y.; Wu, C.; He, F.; Zhang, B.; Liu, F. A geographical information system based multi-criteria decision-making approach for location analysis and evaluation of urban photovoltaic charging station: A case study in Beijing. *Energy Convers. Manag.* **2020**, *205*, 112340. [CrossRef]
25. Chumbi, W.E.; Martínez-Minga, R.; Zambrano-Asanza, S.; Leite, J.B.; Franco, J.F. Suitable Site Selection of Public Charging Stations: A Fuzzy TOPSIS MCDA Framework on Capacity Substation Assessment. *Energies* **2024**, *17*, 3452. [CrossRef]
26. Zhao, T.; Li, N.; Kong, N.; Xie, X. Optimal Location of Electrical Vehicle Charging Stations With Both Self-and Valet-Charging Service. *IEEE Trans. Autom. Sci. Eng.* **2023**. [CrossRef]
27. Shen, X.; Lv, J.; Du, S.; Deng, Y.; Liu, M.; Zhou, Y. Integrated optimization of electric vehicles charging location and allocation for valet charging service. *Flex. Serv. Manuf. J.* **2023**, 1–27. [CrossRef]
28. Almaghrebi, A.; Aljuheshi, F.; Rafaie, M.; James, K.; Alahmad, M. Data-Driven Charging Demand Prediction at Public Charging Stations Using Supervised Machine Learning Regression Methods. *Energies* **2020**, *13*, 4231. [CrossRef]
29. Fang, X.; Xie, Y.; Wang, B.; Xu, R.; Mei, F.; Zheng, J. Research on electric vehicle charging load prediction method based on spectral clustering and deep learning network. *Front. Energy Res.* **2024**, *12*, 1294453. [CrossRef]
30. Erdoğan, S.; Çapar, I.; Çapar, I.; Nejad, M.M. Establishing a statewide electric vehicle charging station network in Maryland: A corridor-based station location problem. *Socio-Econ. Plan. Sci.* **2022**, *79*, 101127. [CrossRef]
31. Kim, S.; Jeong, Y.; Nam, J.-W. Solving Optimal Electric Vehicle Charger Deployment Problem. *Appl. Sci.* **2024**, *14*, 5092. [CrossRef]
32. Zhang, B.; Zhao, M.; Hu, X. Location planning of electric vehicle charging station with users' preferences and waiting time: Multi-objective bi-level programming model and HNSGA-II algorithm. *Int. J. Prod. Res.* **2023**, *61*, 1394–1423. [CrossRef]

33. Zheng, J.; Zhu, J.; Xi, H. Short-term energy consumption prediction of electric vehicle charging station using attentional feature engineering and multi-sequence stacked Gated Recurrent Unit. *Comput. Electr. Eng.* **2023**, *108*, 108694. [CrossRef]
34. Suhail, M.; Akhtar, I.; Kirmani, S. Objective functions and infrastructure for optimal placement of electrical vehicle charging station: A comprehensive survey. *IETE J. Res.* **2023**, *69*, 5250–5260. [CrossRef]
35. Hemmati, R. (Ed.) Chapter 3—Integration of electric vehicles and charging stations. In *Energy Management in Homes and Residential Microgrids*; Elsevier: Amsterdam, The Netherlands, 2024; pp. 79–140, ISBN 9780443237287. [CrossRef]
36. Vani, B.V.; Kishan, D.; Ahmad, W.; Reddy, B.N.K. An efficient battery swapping and charging mechanism for electric vehicles using bat algorithm. *Comput. Electr. Eng.* **2024**, *118*, 109357. [CrossRef]
37. Kim, Y.; Kim, S. Forecasting Charging Demand of Electric Vehicles Using Time-Series Models. *Energies* **2021**, *14*, 1487. [CrossRef]
38. Shah, B.; Bhavsar, H. Time Complexity in Deep Learning Models. *Procedia Comput. Sci.* **2022**, *215*, 202–210. [CrossRef]
39. Yang, Y.; Zhang, Y.; Meng, X. A Data-Driven Approach for Optimizing the EV Charging Stations Network. *IEEE Access* **2020**, *8*, 118572–118592. [CrossRef]
40. Loni, A.; Asadi, S. Data-driven equitable placement for electric vehicle charging stations: Case study San Francisco. *Energy* **2023**, *282*, 128796. [CrossRef]
41. Zhang, P.; Zhang, Z.; Gong, D.; Cui, X. A novel normal wiggly hesitant fuzzy multi-criteria group decision making method and its application to electric vehicle charging station location. *Expert Syst. Appl.* **2023**, *223*, 119876. [CrossRef]
42. Philip, A.O.; U, S.M.; Paul, R.; Saravanaguru, R. Towards intelligent trust-based incident and evidence management models for Internet of Vehicles: A survey. *Comput. Electr. Eng.* **2024**, *117*, 109284. [CrossRef]

energies

MDPI

Review

One Health Ecological Approach to Sustainable Wireless Energy Transfer Aboard Electric Vehicles for Smart Cities

Adel Razek

Group of Electrical Engineering—Paris (GeePs), CNRS, University of Paris-Saclay and Sorbonne University, F91190 Gif sur Yvette, France; adel.razek@centralesupelec.fr

Abstract: This investigation is part of a topical situation where wireless equipment is gradually being used for energy transfer, particularly for autonomous systems and the use of decarbonized energies. A characteristic example of decarbonized autonomous use is linked to the substitution of thermal engine vehicles for electric vehicles (EVs) equipped with energy storage batteries. This response was considered in an ecological context of reducing air pollution and defending planetary biodiversity, which are currently vital. These EVs ultimately operate thanks to the wireless charging of their batteries when stationary or running. By changing long-established means of transport that have become a threat to biodiversity, it is necessary to ensure that innovative replacement solutions protect this biodiversity. In addition, the construction of wireless power transfer (WPT) battery chargers for these EVs must offer an optimal ecology of clean energy saving. In such a context, the two concepts of One Health (OH) and Responsible Attitude (RA) will find their place in the design and control of WPT tools in EVs. This contribution aims to illustrate and analyze the roles of the green and non-wasteful OH and RA approaches in the design and control of WPT embedded in EVs for the smart city (SC) environment. In the paper, WPT tools are first introduced. The design and control of EV battery charging tools are then examined. The biological effects on living tissues due to the electromagnetic field (EMF) radiation of WPT are analyzed. The phenomena and equations governing the design of WPT and the effects of EMF radiation are then exposed. The OH and RA approaches in the SC context are afterward analyzed. The protection against the unsafe effects of WPT tools in the SC environment is consequently explored. The analyses followed in the paper are supported by examples from the literature. The explorations proposed in this contribution have made it possible to highlight certain notions, allowing a more in-depth understanding of the use of EVs with WPT rechargeable batteries for SCs. Thus, the analysis and fusion of these topics are at the heart of this contribution.

Keywords: One Health; Responsible Attitude; electric vehicles; wireless power transfer; smart cities; living tissues; biodiversity; clean energy

1. Introduction

In modern society, various applications are developed for the daily well-being of humans, including comfort, health, safety, etc. These appliances operate using different sources of energy conversion. Such devices, in addition to offering the expected capabilities, produce undesirable side effects. An endless objective has remained at all times to improve the use of these resources. Thus to enhance the expected results and reduce the undesirable side effects. These undesirable effects mainly disturb the health of people as well as that of wildlife, plants, and biodiversity in general [1]. Hence, the notion of OH includes, in the case of the urban context, the health of humans, animals, birds, and plants, all endangered by the disturbances produced by human activities [2,3]. Furthermore, energy and environmental sustainability, which is one of the general issues raised, can ensure a clean energy supply for human well-being. The management of well-being and nuisances related to the use of artificial tools is governed by a concept of RA in the conversion and use of this clean energy. These two methodologies concerning OH and RA would allow an adjusted use of energy for human well-being with reduced dangerous side effects for humans and biodiversity.

Among the devices operating with clean energies associated with human well-being, wireless electromagnetic (EM) energy devices occupy an important place. These are mainly WPT tools and everyday communication devices. These uses, in addition to their authentic functions, similar to those of common devices, have precisely their unexpected effects. Thus, they appear as unexpected suppliers of emitted EMFs, which are correlated with their wireless characteristics generating stray fields. Their effect is focused on the exposed object at an adjacent distance, e.g., a WPT or a mobile phone device [4–6], and homogeneous in case of distant exposure, e.g., a mobile phone tower antenna [7].

Reflecting the WPT tools, they mainly include four classes correlated with the object transfer strategy explicitly magnetic, electric, microwave and laser [8]. The variations of these technologies are mainly related, as well as the transfer methodology used, to various factors. These are the amount of energy transmitted, the transmission distance, as well as the performance, sensitivity, constraints and complexity.

Regarding magnetic WPT, also called inductive power transfer (IPT), two concepts govern its operation: Ampere's law of 1820 and Faraday's rule of magnetic induction of 1831. These fundamentals later helped Nikola Tesla (1856–1943) to propose WPT in the 1890s [9–11]. WPT tools were once originally intended to transport energy over long distances by means of microwave rays [12,13]. Similarly, the idea of WPT proposes the transmission of electrical energy in space from solar energy via satellite broadcasting for use on Earth [14,15]. Only recently, adjacent-range near-field IPT tools have been widely exploited for charging batteries of various everyday devices such as mobile phones, household appliances, drones, and EVs [16–27]. A magnetic WPT or IPT refers to the contactless transfer of power from a source to a load across an air gap. Thus, an IPT scheme comprises a transmitter and a receiver coupled coils.

WPT technology is an important topic of current exploration in general electrical engineering sciences. Possible applications encompass a wide variety of fields, spanning mobility (EV), power generation, device battery charging, biomedical, etc. [28–33]. The explanations for such a wide collection of uses are also numerous and diverse. For example, in a number of medical applications (implants, sensing tools, and pacemakers), WPT develops a safe load stripped of any contact or the need to operate under surgery [34]. Moreover, in many industrial developments, the need for contactless power transmission for moving loads, marking the complications related to sliding connections or moving power cables, has motivated the expansion of contactless energy transfer (CET) links [35].

This contribution aims to examine and explain the behaviors of IPT tools in the contexts of the OH approach containing smart cities biodiversity and the RA concept relating to better clean energy management. The current investigation is positioned in the context of the use of autonomous systems operating with decarbonized energy materialized by the replacement of thermal engine vehicles with EVs [36,37] equipped with electric energy storage batteries. Thus, air pollution is reduced, and the biodiversity of smart cities is protected. The probable use of wireless charging of batteries should ensure that the new solution continues to protect this biodiversity. The two concepts of OH and RA will find their place in this context in the design and management of IPT tools.

In the following sections, we will first present an overview of the IPT tools embedded in EVs and their management for use in SC mobility. Then, the design and control of IPT devices with optimized energy consumption for human well-being and reduced unsafe side effects for humans and biodiversity will be addressed. Next, we will analyze the effects of EMF exposure radiated by IPT tools on living tissues in the SC context. The governing phenomena and their corresponding mathematical equations of the behavior of the IPT tool and its effects will then be informed and analyzed. The roles of OH and RA approaches in the SC context will then be analyzed. This involves the possible charging modes in SCs and their EMF exposure as well as the human in the SC circle of RA and OH approaches.

Then, the procedures for protection against the harmful consequences of EMF exposure will be discussed. The analyses pursued in the article are supported by examples from the literature.

2. Overview of IPT Tools

The construction strategy of an IPT comprises its performances relayed to the input grid, output load, and their related efficiency. Such performances shape the projected consequence relative to human comfort, the EMF drifting radiation involving human and biodiversity unsafety, the power factor, and the diverse losses affecting the grid and environmental sustainability. Regarding the power and frequency of an electric vehicle IPT, they can reach 300 kW (for public transport) and 100 kHz (or even higher for special applications such as bat protection environments). For high-power IPT systems, lower frequencies are often chosen due to the limitations imposed by power electronics, switches, capacitors, and high-frequency electrical wiring. The IPT component accountable for wireless action is its inductive coupler transformer (ICT). Figure 1 exemplifies a succinct schematic illustration of an ICT of an IPT onboard an EV, composed of transmitter and receiver coils possessing inductances of L_1 and L_2, separated by an airgap representing a mutual inductance M_{12}. The two wings of the coupler are compensated by capacities C_1 and C_2. These two roles, power transfer, and capacitive compensation, figure out the process of the IPT tool. The first action linked to the ICT permits wireless transfer through magnetic induction, ensuring a galvanic split of the power source and the battery load. The second action associated with capacitive compensations lets the power electronics connected to ICT (with the source and the load) function at resonance, optimizing the process. Actually, the ICT airgap size is relatively important, and hence, its coupling of two coils is weak. Thus, to transmit the required power, a significant reactive power must be absorbed, hence the essential use of resonant capacitive compensation components [23–26,38–45] on both sides of the ICT to guarantee reliable efficiency. Reactive power compensation can use several topologies (series and parallel) on both sides of ICT, depending on the nature of the load, such as SS, SP, PS, PP, etc. [23,39]. The SS (Series-Series) compensation topology is an economical option [23,45]. In fact, the choice of compensation topology affects the efficiency of power transfer and the ability to provide constant current and voltage at the output. It depends on the application and is related to the operating frequency and the load requirements. Regarding SS compensation, which allows zero-voltage switching and near-zero reactive power without undergoing bifurcation, has more desirable properties than other topologies. For SS topology, only two compensation capacitors are required, resulting in less power losses, smaller size, higher power density, and lower cost. To increase the power transfer efficiency, which is related to a better coupling coefficient, magnetic ferrite sheets are generally used to cover the two coils of the ICT.

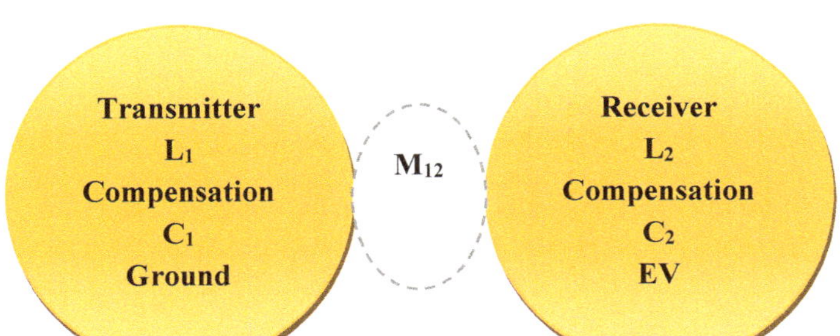

Figure 1. Schematics of a compensated ICT in EV IPT.

The ICT, including its magnetic ferrite sheets, is incorporated in the IPT between the input AC grid via a stage of filtered constant-variable conversion of frequency and voltage and the battery load via a stage of filtered AC-DC conversion. Note that the IPT power regulation is achieved by monitoring the ICT input voltage and frequency. Figure 2 shows a schematic representation of a static IPT charging device for an EV battery. This includes the AC grid, grid frequency AC/DC converter, filter, adjusted voltage and high-frequency DC/AC inverter, ICT, AC/DC load converter, filter, and load.

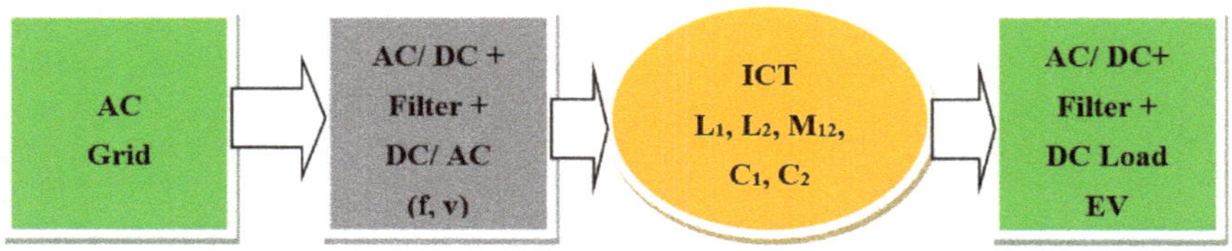

Figure 2. Summarized schematics of IPT between battery load and AC grid.

The RA sustainable design and control of IPT onboard of EVs in SC context could be accomplished within the building of the ICT (coils and ferrites), the compensation features, the involved static convertors and filters. The enhancement could involve losses reduction, superior power factor, bettered coupling and lowered stray field radiation, which impacts OH approach. Such design and control are discussed in the next section.

3. Design and Control of IPT Batteries Charging

The OH and RA approaches are closely related to the IPT design and load (battery) scheduling control. The design involving sustainable optimization of the IPT could be realized for the coil and ferrite structures of the ICT, the static converters and inverters involved, the compensation components needed for resonant operation, and the filters used. Such optimization aims to improve the efficiency and power factor on one side and to better couple the ICT with reduced parasitic radiation on the other side. These two sides of the optimization are mainly related to RA and OH approaches and illustrate their junction. Many works have been published on the topic of optimization of IPT components, for example, for improved performances [38–46] and for reduced radiation effects [47–49]. Load scheduling control enables sustainability linked to the use of clean (carbon-free) energy and economical pricing; see, for example, [50]. In fact, EM energy is considered strictly clean if it comes from a clean energy conversion facility. Since the different energy conversion means are connected to the same grid (for the moment), intelligent load planning allows the use of renewable or other decarbonized energies (hydro or nuclear power plants); see, for example, [51,52]. This problem would disappear when the grid would contain only clean energy conversion deliveries.

4. Biological Effects on Living Tissues Due to EMF Radiation from IPT

The interaction of EMFs with a substance resulting from field radiation causes a dissipation of EM energy in the substance. Various effects are occasioned in the matter due to such dissipation, generally correlated to the frequency range of the EMF.

EMF waves display an extensive frequency range embracing non-ionizing (10^3–10^{14} Hz) and ionizing (10^{15}–10^{22} Hz) extents. Non-ionizing waves are utilized in daily human activities, such as with IPT devices. The mainly common consequence of such exposure, especially in the range (10^5–10^{14} Hz), which comprises utmost IPT uses, is a temperature rise subject to the features of the radiated field and the threatened matter. The exposure characteristics include strength, frequency, nature, and interval of the exposure field. Those of matter relay to its physical belongings, including electric, magnetic, dielectric, mechanic, and heat occurrences. The mentioned customary thermal effect on matter normally arises for lessened radiation due to normalized protection, moderate exposure time, and relative distance from the source. It is worth noting that living tissues exposed to disproportionate field strengths, frequencies, or periods could endure irremediable molecular disorders that can invigorate nerves, muscles, and generally excitable organizations.

As mentioned above, the biological effects (BEs) due to IPT-emitted EMF are commonly thermal due to tissue energy dissipation. These concern the tissues of humans, fauna, and flora. In such circumstances, an instant BE caused by high-frequency radiation would end in a swell in inside tissue temperature. The regular heating persistence of these living tissues is principally fitted to outer heating of the matter, for instance, sun exposure. In this circumstance, heat is slowly spread by conduction and convection within the tissues, which are ordinarily fluid-perfused, permitting correct functioning. Contrariwise, focused heating of the deepest tissues, specifically those weakly irrigated or unwell immersed by blood or sap, could be dangerous contingent on the physiognomies of the radiation field and the tissues. Actually, the various hazardous effects of EMFs radiated by IPTs can be evaluated (as indicated in Section 5) and verified according to the thresholds established by standards. The latter takes into account the nature of the tissues (human, animal, plant, etc., as well as the part of the tissues), the functions (manufacturer, user, etc.), and the exposure conditions; see, for example, for humans and wildlife [53,54] and for plants [55,56]. Regarding the frequency weight in exposure, it should be noted that in near-field applications like IPT for EVs, the used frequencies are relatively low, while high frequencies are normally used for far-field applications like microwave (MW) IPT tools intended to transmit power through free space. Both IPT types implicate EMF radiation, however, unlike BEs. In fact, strengthened fields and frequency produce more severe effects due to deeper internal impacts on tissues. Obviously, reducing near-field IPT frequency could reduce the radiation effects; however, it will worsen the performance of the tool.

5. Phenomena and Equations Governing IPT and Its EMF Radiation Effects

Computation and data are now an integral part of almost every industrial sector. In the automotive industry, mathematical models increasingly compete in terms of functionality, safety, automation, and eco-design [57]. The phenomena governing the design of IPT are related to EM and electric circuit domains. The phenomena involved in the side effects of IPT-radiated EMF are also related to EM and heat transfer (HT) occurrences. In fact, the interaction of EMFs radiated by IPTs with substances produces undesirable thermal BEs, which are governed by the coupled EM and HT phenomena via the EM dissipated power (P_d) in these substances. In the case of EMF exposure to living tissues, a bioheat (BH) phenomenon will govern the temperature rise originated by a heat source P_d. The different mentioned phenomena are ruled by corresponding mathematical equations, which are presented in the next sections.

5.1. Governing Equations

Based on Maxwell's microscopic local behavior, the differential form of the general EMF 4 equations [58] is given by:

$$\nabla \times \mathbf{E} = -\partial_t \mathbf{B} \text{ (Maxwell—Faraday)} \tag{1}$$

$$\nabla \times \mathbf{H} = \sigma \mathbf{E} + \partial_t \mathbf{D} \text{ (Maxwell—Ampère)} \tag{2}$$

$$\nabla \cdot \mathbf{D} = \rho_e \text{ (Maxwell—Gauss)} \tag{3}$$

$$\nabla \cdot \mathbf{B} = 0 \text{ (Maxwell—Thomson)} \tag{4}$$

The HT equation in its differential form is given by:

$$c\, \rho\, \partial T/\partial t = \nabla \cdot (k\, \nabla T) \tag{5}$$

In the case of EMF exposure (radiation) of living tissues, the EMF harmonic fields, the BH, and heat source P_d equations are given as follows:

$$\nabla \times \mathbf{H} = \mathbf{J} \quad (6)$$

$$\mathbf{J} = \mathbf{J}_e + \sigma\, \mathbf{E} + j\, \omega\, \mathbf{D} \quad (7)$$

$$\mathbf{E} = -\nabla V - j\, \omega\, \mathbf{A} \quad (8)$$

$$\mathbf{B} = \nabla \times \mathbf{A} \quad (9)$$

$$c\, \rho\, \partial T/\partial t = \nabla \cdot (k\, \nabla T) + P_d + P_t + c_f\, \rho_f\, p_f\, (T_f - T) \quad (10)$$

$$P_d = \omega \cdot \varepsilon'' \cdot E^2/2 \quad (11)$$

In the above Equations (1)–(11), **H** and **E** are the vectors of the magnetic and electric fields in A/m and V/m, **B** and **D** are the vectors of the magnetic and electric inductions in T and C/m^2, **A** and V are the magnetic vector and electric scalar potentials in W/m and volt. **J** and $\mathbf{J_e}$ are the vectors of the total and source current densities in A/m^2, σ is the electric conductivity in S/m, ρ_e is the volume density of electric charges in C/m^3, and ω is the angular frequency = $2\pi f$, f is the frequency in Hz of the exciting EMF. The symbol ∇ is a vector of partial derivative operators. The symbol ∂_t is the operator of the partial time derivative. The magnetic and electric comportment laws, respectively, between **B/H** and **D/E** are represented by the permeability μ and the permittivity ε in H/m and F/m. The parameters: ε'' is the imaginary part of the complex permittivity of the absorbing material, and ρ is the material density in kg/m^3. E is the absolute peak value of the electric field strength in V/m, c is the specific heat of the substance in J/(kg °C), k is thermal conductivity in W/(m·°C), and T is the substance temperature in °C. The power dissipation in W/m^3 given by Equation (11) relates to the foremost dielectric heating of EMF energy loss. Notice that the imaginary part ε'' of the (frequency-dependent) permittivity ε is a measure of the ability of a dielectric material to convert EMF energy into heat. The volume density of power dissipations given by Equation (11) will be used in the coupling of EMF and BH equations. In the case of living tissues, we consider a self-tissue heat source P_t, convective heat transfer via irrigating fluid of tissue, and external heat source related to the EMF exposure P_d. P_t and P_d are heat sources in W/m^3, T_f, and T are respectively the fluid temperature and the local temperature of tissue in °C, and c_f, ρ_f, and p_f are respectively fluid, specific heat in J/(kg °C), density in kg/m^3, perfusion rate in 1/s.

It should be noted that the source term in the EMF equations is the excitation current density $\mathbf{J_e} = \sigma\, \mathbf{E_e} = j\, \omega\, \mathbf{D_e} = j\, \omega\, \varepsilon\, \mathbf{E_e}$. In addition, the specific absorption rate (SAR) in dielectric materials (biological tissues) is equal to P_d/ρ in W/kg.

Note that Equation (11) relays to bio-heat tissues considering the EMF exposure. This expression is analogous to Penne's bio-heat equation [59–62] related to animal living tissues, including convective heat transfer in blood. The sap fluid in plants plays the role of animal blood. Besides, Phloems and Xylems enclosing sap act the task of veins and arteries, inclosing blood. In Penne's bio-heat equation, the term P_t in (11) is related to animal metabolic heat and corresponds to internal heat in plant tissues. In addition, the last term in (11) acts for convection fluid heat transfer related to animal blood or plant sap.

Considering the design and optimization of ICT discussed in Section 3, based on a better coupling coefficient k according to:

$$k = M_{12} \cdot (L_1 \cdot L_2)^{-1/2} \quad (12)$$

The two resonant circuits, for e.g., SS compensation of the transmitter and receiver, are tuned at the resonant frequency ω_o:

$$\omega_o = (L_1 \cdot C_1)^{-1/2} = (L_2 \cdot C_2)^{-1/2} \quad (13)$$

The structure of the ICT coupler, schematized in Figure 1, is presented in Figure 3. It comprises the transmitter coil (on the ground), the receiver coil (on the EV bottom), and two magnetic ferrite plates that entirely cover the two coils. In addition, the figure includes a steel plate representing the EV chassis. The two ICT coils with their ferrites (pads) are separated by an air gap of distance (d) and coil axes shift (sh).

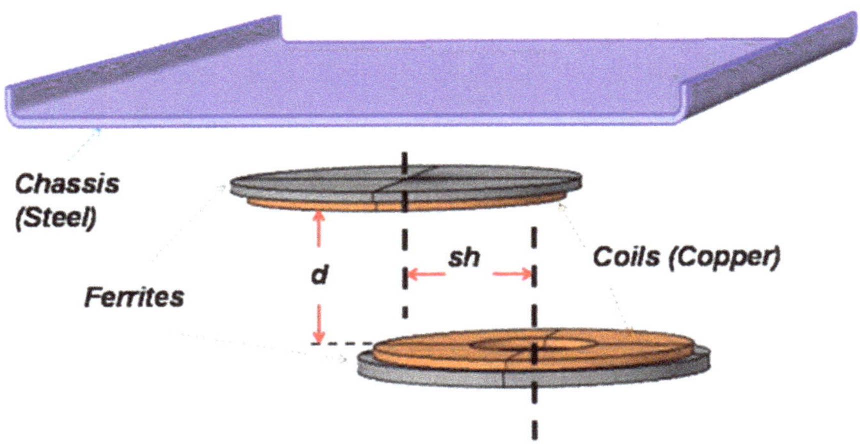

Figure 3. A 3D Structure of an ICT including ground transmitter, EV bottom receiver coils, covered by 2 magnetic ferrites plates, and a steel chassis plate [25].

The EM analysis of the ICT behavior is related to Equations (6)–(9), (12) and (13), where the EM domain (Equations (6)–(9)) will be coupled to the circuit domain via a general circuit equation under the form:

$$v = 1/C. \int i\, dt + r\, i + l.\, di/dt + d\Psi/dt + \mho \quad (14)$$

In (14), v is the source voltage, i is the circuit current, r is the total circuit resistance, l a linear inductance, C a capacitance, $\mho$ a non-linear voltage drop (e.g., a semiconductor component) in the electrical circuit and Ψ the flux linkage. The equations characterizing the EM and circuit domains to be solved are therefore (Equations (6)–(9) and (14)).

5.2. Numerical Solution of EMF, BH, and Circuit Equations

The numerical solution of the different equations relative to the EM, BH, and electric circuit domains, respectively, Equations (6)–(9), (11) and (14) should account for different features of modeled structures. These are geometrical complexity, matter inhomogeneity, variables, nonlinear behaviors, and interdependence of domains. Fulfilling such features imposes a matter local solution indicating the use of discretized 3D methods as finite elements (FEM) or equivalent methods [63–72] associated with suitable equations coupling strategies.

Considering the prediction of tissue BEs, the equations (Equations (6)–(11)) presented in the last section will be solved in a coupled manner. In such circumstances, the EMF and BH discretized equations would be weakly coupled due to the distant values of their time constants [59–61]. Thus, performing an iterative procedure gives the local distributions in the tissue of the fields induced local (i) values $\mathbf{E}_i$, $\mathbf{B}_i$, and $\mathbf{J}_i$, and hence P_{di}, SAR_i, and ΔT_i. The implicated tissue parameters are ε, P_t, c_f, ρ_f, p_f, etc., which could be measured or found in the literature [73–76].

Regarding the IPT design, the governing equations are related to the EM domain (Equations (6)–(9)) and an electric circuit domain (14), which needs a coupled equations solution. Generally, the coupling of EM and electric circuits needs simultaneous strong coupling solution of the equations due to non-linearity of variable behaviors and closeness of the magnetic and electric time constants [77].

5.3. Dedicated Models Adaptations

These latest numerical solutions and equations coupling strategies permit an accurate evaluation of the concerned phenomena. However, the class of needed evaluation depends on the dedicated use and therefore adaptations of the formalism are sometimes necessary to achieve the committed task. This could involve in many cases a recurrent procedure that needs to be executed in a repeated manner. In this case, mathematical models and processing tools are used for simulation, design, optimization and reliability analysis.

In the initial design of the WPT structure, a full coupled model (EM—electrical circuit) (Equations (6)–(9) and (14) associated with Equations (12) and (13)) seems necessary. Similarly, in the prediction of possible adverse effects on living tissues, a full coupled model (EM—BH) (Equations (6)–(11)) seems indispensable. Conversely, in situations where computational time could be a constraint, such as repetitive adaptive optimization approaches or online control procedures, the full-coupled models mentioned above would be penalizing.

Numerical solutions of complex structures could be difficult due to the complex coupled equations involved [77]. Therefore, in the case of structural design, approximate approaches are needed to reduce the model and perform sensitivity analysis. Sensitivity analysis is a phase of the design that allows studying the variation of the calculated quantities according to variations in the inputs and parameters of the model. Reducing a model consists of accelerating its resolution while degrading the precision as little as possible [78]. The difficulty is, therefore, often to find the limit between saving time and deteriorating precision. The numerical experimental plans make it possible to identify a reduced model for optimization purposes. The functions to be identified are generally imposed (radial basis functions, Kriging functions, Bayesian network, etc.), and the data used for identification can come from simulation results [79].

In the context of optimization, a strong constraint is to maintain the dependence between the model outputs and the parameters to be dimensioned. The reduction objective may also come from the need to couple the model with other models for system-level simulation purposes. In this case, numerical processing can be directly applied to the fine model, resulting in the loss of model dependence on physical and geometric parameters. For example, for a dynamic system, reduction often consists of reducing the order of the system, that is, the number of state variables. This approach can be automated by numerical analysis by neglecting the lowest time constants. After applying several algorithms to a model, it is appropriate to approach multi-model optimization and, in particular, multi-level modeling optimization. The main idea comes from the observation that a fine model takes a long time to calculate and much longer to optimize. A reduced model (surrogate model) is then substituted [80,81] for the fine model to carry out a pre-sizing. The solution is then used to reduce the model's fine optimization space.

In the analysis of the compliance of IPT systems with international standards regarding living tissues exposure mentioned in Section 4, one can use stochastic non-intrusive methods (Kriging, Polynomial Chaos, etc.) [79,82] that use 3D FEM computations with a limited set of realizations (learning samples). In this context, Kriging and Polynomial chaos expansions provided efficient meta-models to take into account uncertainties of different physical or geometrical parameters [83].

5.4. Example of Meta-Model Features

In [83], an optimization problem considering material and geometrical features, predictions of the radiated magnetic field have been obtained from two non-intrusive stochastic models for a given WPT system for EV. To illustrate the precision of these meta-models and the gain in computation time, Figures 4 and 5 illustrate the relation of the fields obtained from the 2 meta-models related to the FEM computations, respectively, for 3 and 15 learning samples of 30 all samples computations and Figure 6 shows the error function of the number of these samples. Concerning the computation time required for the meta-models, whatsoever meta-model and learning samples number, the computation time rests insignificant related to all the sample's FEM calculation.

The results given in Figures 4–6 show the efficacy of using meta-models for repetitive possible optimization designs though compliances of IPT with standards. Note that reduced models are also well adapted in control and online monitoring of IPT systems (see Section 8).

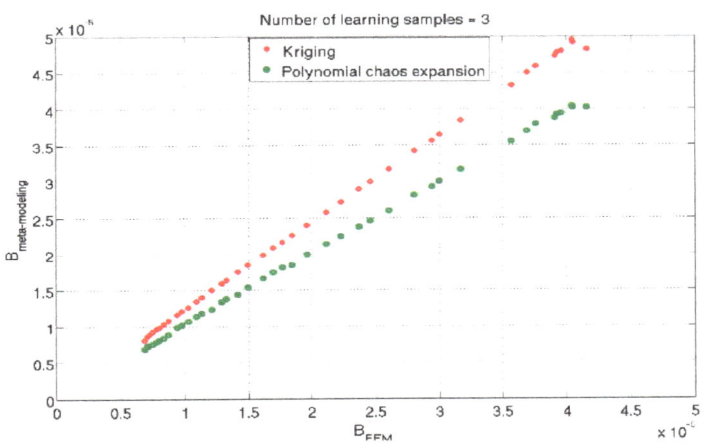

Figure 4. Relation of the B fields obtained from the two meta-models related to the FEM computations for three learning samples case [83].

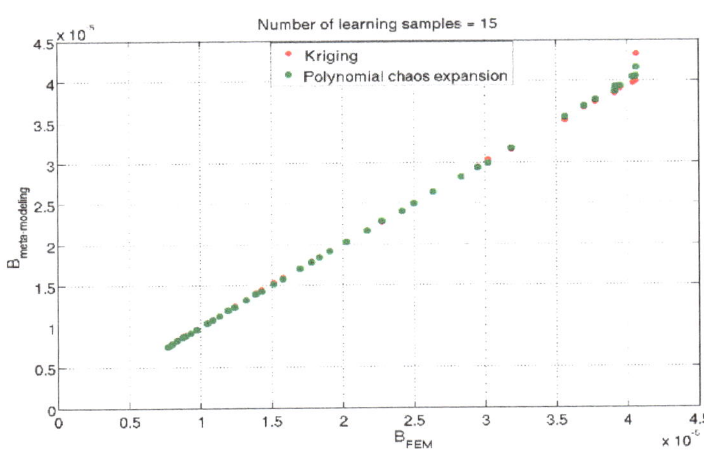

Figure 5. Relation of the B fields obtained from the two meta-models related to the FEM computations for 15 learning samples case [83].

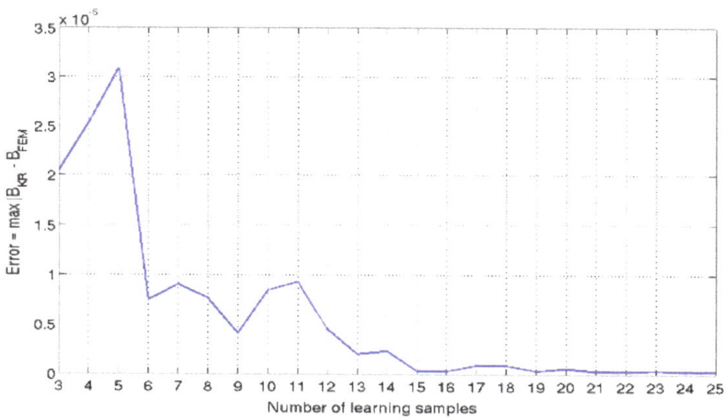

Figure 6. The error of B fields obtained from the Kriging meta-model related to the FEM function of the number of samples [83].

6. OH and RA Approaches in SC Context

IPT wireless charging of electric vehicles is an SC-friendly solution that is relevant for shared and autonomous EVs, public transport (buses, trams, and ships), medium-duty delivery trucks, drones, etc. It can be static charging (in infrastructure stations or at home for personal EVs) or dynamic charging (electric roads). The latter can be wireless via IPT battery charging or connected to the grid via a sliding contact, which is used in particular for public transport. In addition, there is a third intermittent charging mode, which could be discontinuous electric roads or split static charging points.

6.1. Charging Modes Possessions

A full static IPT charging mode corresponds to a limited range over a distance; it requires high battery storage and is well suited for personal EVs and taxis. A mixed dynamic-fixed IPT charging mode is mainly suited for long-distance highways and requires less battery storage than the latter case. A full dynamic mode corresponds, in general, to a fixed trajectory; it does not require battery storage and is well suited for electric buses and trams connected by sliding contact to the grid. A discontinuous dynamic mode corresponds to a partially fixed trajectory, requires low battery storage, and is well suited for buses with particular trajectories or areas, including portions with difficult grid connection infrastructures. The mentioned mode using remote static charging points corresponds to electric buses charging at their stops, for which the battery storage depends on the number of stops, the distance, and the static duration. All the previously mentioned IPT charging modes require safety precautions against EMF exposure (see Section 7).

6.2. RA and OH Approaches of IPT Management in SCs

For all the different charging modes mentioned above, both RA and OH approaches aim at the same goals of eco-design, clean energy usage, minimization of losses, reduction of adverse effects on living tissues, and assessment and protection of biodiversity safety. Moreover, energy management between the grid and the EV could help RA; thus, control algorithms can be used for both grid-to-vehicle (G2V) and vehicle-to-grid (V2G) operating modes [84,85]. Furthermore, it is necessary to ensure interconnection and interoperability between various wireless charging devices [26,86]. Moreover, providing a configurable charging profile enables improved RA [87]. Moreover, connectivity and autonomous driving capability will make EVs (cars and trucks) much safer [88].

6.3. EMF Exposure and Charging Modes

The stray EMFs of ICTs are highly dependent on the relative positions of the ICT coils in 3D geometry and are therefore affected by the charging mode, static or dynamic. The effects of these fields on living tissues are associated with the location of these tissues relative to the position of the ICT, which is different for static and dynamic modes. In dynamic charging mode, the stray fields are variable and take minimum and maximum values depending on the position of the ICT receiver coil at the bottom of the EV relative to the ICT transmitter coil on the ground. In this case, the affected living tissues are mainly those in the passenger compartment inside the EV. In the case of static charging mode, the passenger compartment is normally empty. On the other hand, for such a mode, living tissues, which can be humans, animals, birds, or plants located outside the EV in the vicinity of the ICT under its bottom, could be affected by exposure to stray fields. Open space charging is often practiced for personal EVs at home, and remote static points correspond to electric buses charging at their stops. As mentioned in Section 6.1, exposure safety precautions are necessary with respect to these charging modes. Figure 7 illustrates examples of such threatening situations. The corresponding exposure safety precautions are those mentioned in Sections 4 and 5 relative to the safety threshold [53–56], indicating the SAR involving power and exposure time.

Figure 7. Illustrative examples of threatening situations for animals and plants due to IPT charging tools at the bottom of EVs.

6.4. Human in the SC Circle of RA and OH Approaches

This section summarizes the role of humans in the SC environment in terms of their contribution to the construction of the IPT and the choice of the quality of the energy used. Moreover, how they can benefit from its expected results as well as how they can suffer from its possible side effects. Figure 8 illustrates this causal approach of the design and energy expended in EV-IPT, its expected utility, and its unsolicited effects on humans and the environment.

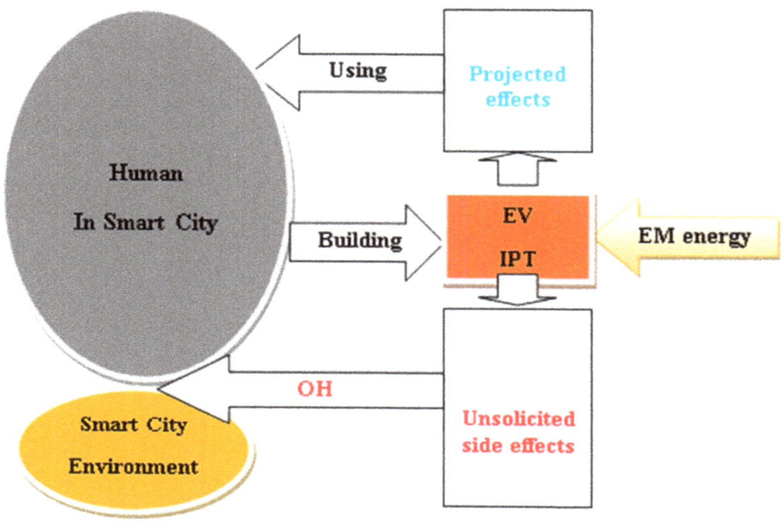

Figure 8. The causal approach of the design and energy expended in EV-IPT, its expected utility, and its unsolicited effects on humans and the environment.

The present work focussed on the approach of RA through the design and use of clean energy in IPT and on the concept of OH by counting whole SC biodiversity involving humans, animal birds and plants in the controlling of adverse EMF exposure influences, which comprise their assessment and monitoring. Figure 9 summarizes such strategy.

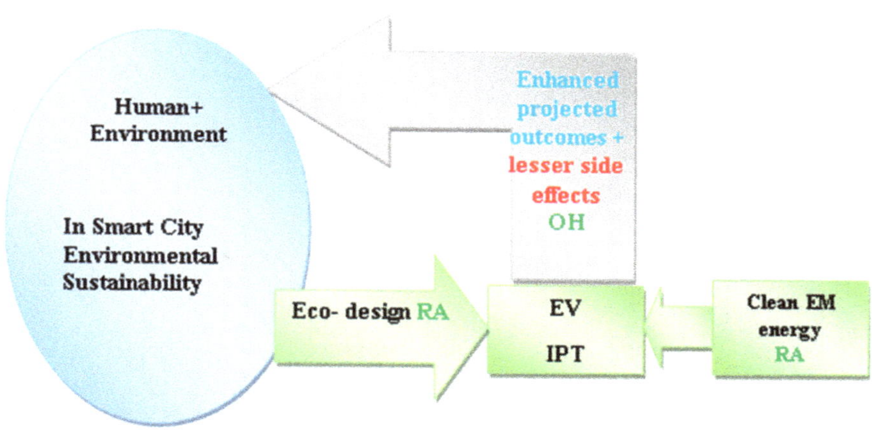

Figure 9. RA and OH approach through the design and use of clean energy in IPT counting whole SC biodiversity by monitoring of adverse EMF effects.

Certainly, the analyses of IPT tools conveyed in this work could be classified as two interconnected actions of RA and OH approaches. The enhanced design of IPT devices using EM clean energy devoted to superior performing and diminished stray fields belong to RA action. On the other hand, the evaluation and control of adverse BE for biodiversity security belong to OH action. Figure 9 shows an illustration of the interaction of these two actions.

7. Safeguard against the Unsafe Effects of SC Environment

As mentioned previously, the enhanced design of an IPT device aims to strengthen the expected power transfer and decrease unwanted EMF radiation that might disturb not only humans but also further associated environmental biodiversity. Such aims could be attained through IPT optimization and monitoring. The reduction of near EMF (target close to ICT) disorders is a difficult mission, particularly in the circumstance of sources associated with wireless devices. Actually, for straight sources emitting EMFs, the target guard could be accomplished via shielding of the source, the target, or both. Note that in the case of EV, ICT (source) shielding (in 3D geometry) is very complicated due to its structure and position, partially in the bottom of the EV and in the ground. This difficulty is stronger for larger or twisted coil air gaps.

7.1. Protection in Dynamic and Static Charging Modes

As mentioned above, reducing the IPT stray field can be difficult. However, shielding the upper part of the ICT, including the passenger compartment, is an easy task using shields of appropriate shape and material [89,90]. Such a technique is mainly used in the case of the presence of people or animals inside the EV, such as in dynamic charging mode or buses with distant static charging points. In the case of static charging mode in an open space, shielding may be necessary only to protect electronic instruments. However, the field radiated by the ICT outside the EV under and around its bottom can affect all living tissues. This effect is more significant for larger air gaps or greater misalignment of the coil axes corresponding respectively to the distances d or sh in Figure 3. Since these parameters are difficult to control, any presence under or near the bottom of the EV should be avoided, especially for a significant duration. In the case of personal EV charging at home, only closed spaces or areas encircled by open space, thus avoiding the presence of humans, animals, birds, plants, etc. near the ICT would be completely safe. Additionally, installing ICT away from the bus entrance and passenger waiting area is necessary for public safety in the case of remote static charging points.

7.2. Control of BEs in Living Tissues

The situations described above could be verified and controlled by routines given in Section 5, checking field values in living tissues compared to thresholds established by universal standards [53–56]. In all charging modes, the above-mentioned routines should include specific living tissue geometry and matter properties. Assessment of living tissue exposure to EMFs requires approaches based on 3D calculations to solve the EM problem (Equations (6)–(9)) containing representations of the ICT system, the vehicle, and the living tissue object (in the vehicle or located nearby).

7.3. Living Tissues Models

The most significant characteristics governing the recognition of such models are the reliability of the physical-biological belongings, the realistic state, and the consistency with the numerical methodology used. Different living tissue models and matter characteristics could be found in literature, e.g., for human tissues [73–76,91,92]. For example, in the human case, Figure 10 shows a structural model of the body and its various organs and tissues.

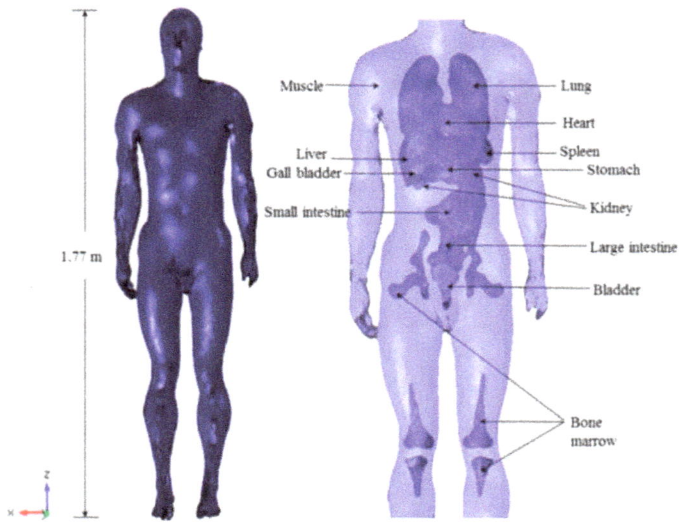

Figure 10. High-resolution anatomical human body model with its different organs and tissues [59].

7.4. Example of Exposure BEs

An illustrative example of ICT radiation exposure relating to the case of a human body positioned horizontally on the ground next to an EV under static charging mode is given here [59]. Figure 11 shows the distributions of induced fields in the body due to exposure to ICT. The used human body for computations corresponds to the high-resolution human anatomical model, compatible with the numerical approach used, presented in Figure 10.

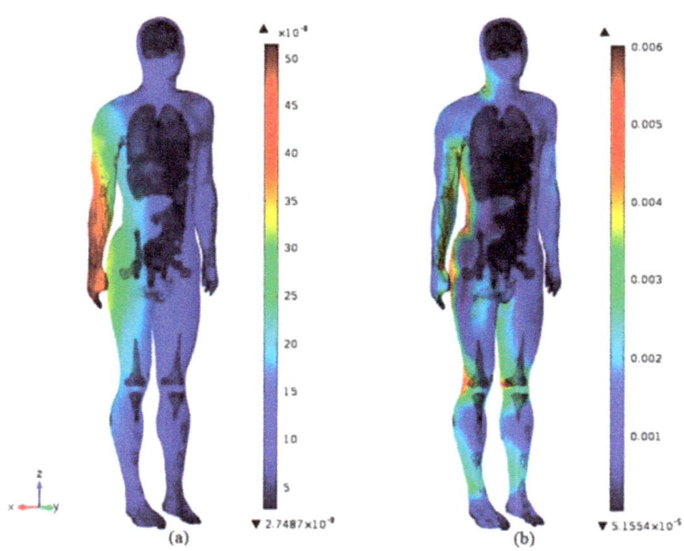

Figure 11. Distribution of induced fields in the body. (**a**) Magnitude of B (T), (**b**) Magnitude of E (V/m), for a 3 kW, 30 kHz IPT [59].

The corresponding results in this case were in agreement with international safety guidelines (27 µT for magnetic induction B and 4.05 V/m for the electric field E).

8. Discussion

In the above analyses, OH and RA approaches have been widely discussed regarding their roles in the design and control of IPTs integrated into EVs for the SC environment. These two approaches are more generally involved or could be extended to more interesting situations.

Interdependence in OH approach and beyond: As defined earlier, the OH approach is a combined, fusing approach that targets to sustainably weigh and enhance the healthiness of man, animals (domestic and wild), plants, and the larger environment (containing ecosystems), which are intimately allied and behave mutually dependent. In addition to the effects comprising human, animal, and plant health, all menaced by disruptions engendered by human activity (EV IPT) analyzed in this work, other situations of interdependences could be remarked. For example, the alliance of some viruses with their hosted organism (virus-host interaction). Thus, the genomes of viruses are composed of genes borrowed from the organisms they infect and can, in certain cases, give their hosts the ability to produce a toxin to destroy their competitors (as in the case of baker's yeast) [93]. Another example that is related to metabolic interactions between bacteria and phytoplankton [94]. Also, the link between certain microorganisms, plants, and nutrient cycles could be found in, e.g., [95].

Exposure to EMFs from multiple IPTs: In the case of static wireless charging stations comprising multiple EVs, spatial EM radiation could reflect a superposition exposure effect [96]. In such a case, the corresponding safety precautions can be predicted by the routines described in Section 7.

Extended RA management in IPT and EVs: As mentioned in Section 6.2, an enhanced RA could involve a planned energy exchange between the grid and the EV (autonomous or shared EV). Thus, G2V is when clean energy is available, and V2G is when energy is expensive. Furthermore, an enhanced RA could result from providing a configurable charging profile. Both strategies could be realized through control algorithms and are closely related to battery storage. This latter could be considered a component of the EV driving system, including its related heating and management systems [97]. The driving range, which is a critical issue in the driving system, is directly related to the capacity of this energy storage. Moreover, the battery condition strongly affects the design and operation of its connected inverter. EV battery condition monitoring, fault detection, and lifetime prediction are important online tasks. Such monitoring could be achieved through digital twin (DT) technology [98]. DT could be described as an easy integration of information between a physical and virtual procedure in both directions. DTs for EV driving systems are commonly used for system condition monitoring, diagnostics, prognosis, optimization, scenarios, and risk assessment [99]. These DTs can be shaped at the system level, subsystem level, specific component level, and many other levels. Thus, the planned energy exchange between the grid and the EV and the configurable charging profile could be monitored by the DT concept as part of battery monitoring for the RA approach.

Moreover, this activity could be corroborated in the context of a DT city. Thus, the amplified number of SCs that are more connected to societies is established; consequently, the need for skills as DTs has increased, which can help the further progress of these SCs [100]. A significant advantage over energy savings in a SC illustrates the way services are disseminated and operated. In a DT city, sensors, cameras, and different digital tools about city infrastructures, resource distribution, movement of people, logistics, and EVs will collect information. Thus, the SC administration would be supervised and managed in a way that makes the city more efficient [101].

To highlight the concept of DT for the supervision of complex procedures such as EV-IPT batteries connected to SC and grid, we detail here its main characteristics [102]. The DT is composed of a real-virtual pair that allows a self-adjusting behavior. The real wing delivers the processed sensed information to the virtual wing while the latter transmits the control orders to the real wing. Such a pairing also decreases uncertainties and reduces unsolicited and threatening operating singularities. The processed data of the real wing delivers detections compared and corrected by external data (IoT) and an acquired history. The compliant product would, once trained, be transmitted in the form of data analysis. These suggestions, along with an appropriate reduced model (see Section 5.3), will be advanced to the virtual tool. Certainly, a fast pairing requires a consistent virtual copy with a low computation time. This can be achieved by reducing the numerical model while keeping the faithful physical representation. Supervision using such a pair allows an adaptive control for a working procedure [103]. Figure 12 summarizes the features of a DT for the supervision of the EV-IPT-Battery operation connected to SC and grid.

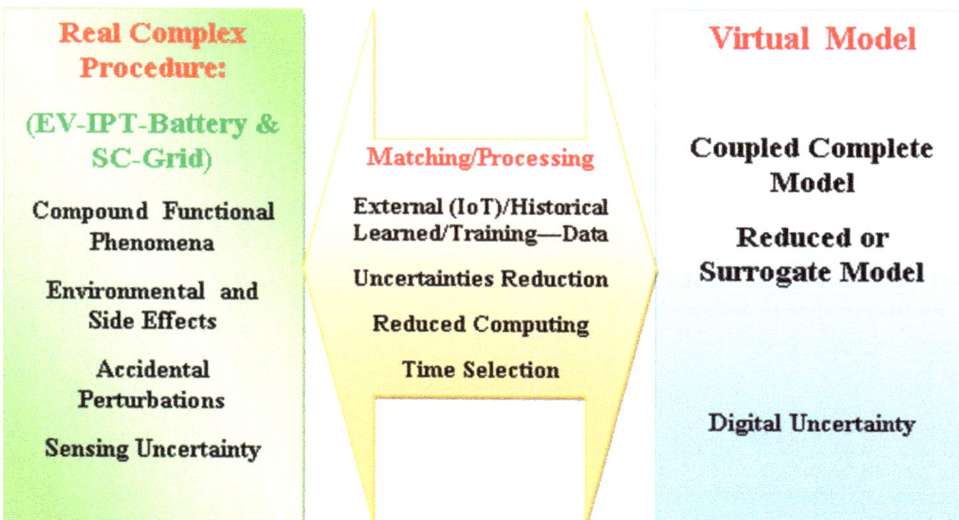

Figure 12. Summarized illustration of a matched monitoring of a complex procedure (EV-IPT-Battery connected to SC-Grid) with its virtual model [103].

The previous analysis illustrated the possibility of using strategies involving a planned energy exchange between the grid and the EV and a configurable charging profile, which could be realized through DT-managed control algorithms as part of the battery monitoring for the RA approach connected to SC DT. It should be noted that the DT concept is also often used in other various general vehicle applications; for example, see [104–111].

Onboard body devices: IPT could disrupt nearby not only living tissues, but also wearable and implanted devices onboard the tissues if they are not protected [112].

9. Conclusions

The analyses discussed in this article have shown the importance of the roles of OH and RA approaches in the design and use of EVs in SC's urban ecology and their associated protection of biodiversity. This particularly concerned the augmenting adaptation of the IPT through its eco-design associated with the planned use of clean energy to allow a healthy ecology and conserved biodiversity. The article has also shown the distinct contributions of the complete coupled mathematical models as well as their reduced versions in the design, optimization, and shielding of the IPT, as well as the prediction of harmful BEs on living tissues in the urban biodiversity of SC. In conclusion, this contribution illustrated the next important points:

The ecological and non-wasteful conducts through OH and RA concepts suggest EVs comprising their IPTs and batteries function in a connected mode. Such a connection would involve SC infrastructure and public grid.

Mathematical modeling in its complete form can help in the initial design and optimization of IPT structures and the prediction of adverse BEs. Contrariwise, in repetitive adaptation tasks and online control and monitoring, reduced models would be more appropriate.

During the operation of IPT, battery charging of cars, buses, trucks, etc., while passengers must be present on the EV board, as during the dynamic mode, charging the shielding under the bottom of the EV is necessary. Also, in the case of the impossibility of using the IPT static mode in a closed space, the battery charging in an open space is conditioned to a distance limitation of position around the vehicle. In these two situations, a check could be practiced by predictive modeling of the BEs in the living tissues of humans, animals, birds, plants, etc.

The investigations intended in this contribution have made it possible to highlight certain notions allowing a more in-depth understanding of the subject related to the use of EVs, with IPT rechargeable batteries, for SCs. Thus, contributing to guide and strengthen studies in the involved topics in the subject.

Funding: This research received no external funding.

Conflicts of Interest: The author declares no conflicts of interest.

References

1. Shivanna, K.R. Climate change and its impact on biodiversity and human welfare. *Proc. Indian Natl. Sci. Acad.* **2022**, *88*, 160–171. [CrossRef]
2. World Health Organization (WHO). One Health—Technical Advisory Group First Meeting. 31 October–1 November 2022, Concept Note. 2023. Available online: https://www.who.int/europe/initiatives/one-health (accessed on 14 March 2024).
3. Deiana, G.; Arghittu, A.; Dettori, M.; Castiglia, P. One World, One Health: Zoonotic Diseases, Parasitic Diseases, and Infectious Diseases. *Healthcare* **2024**, *12*, 922. [CrossRef] [PubMed]
4. Petroulakis, N.; Mattsson, M.-O.; Chatziadam, P.; Simko, M.; Gavrielides, A.; Yiorkas, A.M.; Zeni, O.; Scarfi, M.R.; Soudah, E.; Otin, R.; et al. NextGEM: Next-Generation Integrated Sensing and Analytical System for Monitoring and Assessing Radiofrequency Electromagnetic Field Exposure and Health. *Int. J. Environ. Res. Public Health* **2023**, *20*, 6085. [CrossRef] [PubMed]
5. Cirimele, V.; Freschi, F.; Giaccone, L.; Pichon, L.; Repetto, M. Human Exposure Assessment in Dynamic Inductive Power Transfer for Automotive Applications. *IEEE Trans. Magn.* **2017**, *53*, 5000304. [CrossRef]
6. Tran, N.T.; Jokic, L.; Keller, J.; Geier, J.U.; Kaldenhoff, R. Impacts of Radio-Frequency Electromagnetic Field (RF-EMF) on Lettuce (*Lactuca sativa*)—Evidence for RF-EMF Interference with Plant Stress Responses. *Plants* **2023**, *12*, 1082. [CrossRef] [PubMed]
7. Sivani, S.; Sudarsanam, D. Impacts of radio-frequency electromagnetic field (RF-EMF) from cell phone towers and wireless devices on biosystem and ecosystem—A review. *Biol. Med.* **2012**, *4*, 202–216.
8. Liu, Z.; Li, T.; Li, S.; Mi, C.C. Advancements and challenges in wireless power transfer: A comprehensive review. *Nexus* **2024**, *1*, 100014. [CrossRef]
9. Tesla, N. System of Transmission of Electrical Energy. U.S. Patents 645,576, 20 March 1900.
10. Tesla, N. The transmission of electrical energy without wires. *Electr. World Eng.* **1904**, *1*, 429–431. Available online: https://www.sowp.org/wp-content/uploads/2019/02/LR-1702021903-Tesla-Reprint.pdf (accessed on 1 August 2024).
11. Tesla, N. Apparatus for Transmitting Electrical Energy. U.S. Patents 1,119,732, 1 December 1914.
12. Brown, W. Experiments in the transportation of energy by microwave beam. In Proceedings of the 1958 IRE International Convention Record, New York, NY, USA, 21–25 March 1966; IEEE: New York, NY, USA, 1964; pp. 8–17. [CrossRef]
13. Brown, W.C. The history of power transmission by radio waves. *IEEE Trans. Microw. Theory Tech.* **1984**, *32*, 1230–1242. [CrossRef]
14. Glaser, P.E. Power from the sun: Its future. *Science* **1968**, *162*, 857–861. [CrossRef]
15. Brown, W.C. Satellite power stations: A new source of energy? *IEEE Spectr.* **1973**, *10*, 38–47. [CrossRef]
16. Covic, G.A.; Boys, J.T. Modern Trends in Inductive Power Transfer for Transportation Applications. *IEEE J. Emerg. Sel. Top. Power Electron.* **2013**, *1*, 28–41. [CrossRef]
17. Nutwong, S.; Sangswang, A.; Naetiladdanon, S.; Mujjalinvimut, E. A Novel Output Power Control of Wireless Powering Kitchen Appliance System with Free-Positioning Feature. *Energies* **2018**, *11*, 1671. [CrossRef]
18. Shah, A.; Zada, M.; Shah, S.A.A.; Basir, A.; Yoo, H. Flexible metasurface-coupled efficient wireless power transfer system for implantable devices. *IEEE Trans. Microw. Theory Tech.* **2023**, *72*, 2534–2547. [CrossRef]
19. Patil, D.; Mcdonough, M.K.; Miller, J.M.; Fahimi, B.; Balsara, P.T. Wireless power transfer for vehicular applications: Overview and challenges. *IEEE Trans. Transp. Electrif.* **2018**, *4*, 3–37. [CrossRef]
20. Kod, M.; Zhou, J.; Huang, Y.; Hussein, M.; Sohrab, A.P.; Song, C. An Approach to Improve the Misalignment and Wireless Power Transfer into Biomedical Implants Using Meandered Wearable Loop Antenna. *Wirel. Power Transf.* **2021**, *8*, 6621899. [CrossRef]
21. Zhou, J.; Guo, K.; Chen, Z.; Sun, H.; Hu, S. Design considerations for contact-less underwater power delivery: A systematic review and critical analysis. *Wirel. Power Transf.* **2020**, *7*, 76–85. [CrossRef]

22. Hutchinson, L.; Waterson, B.; Anvari, B.; Naberezhnykh, D. Potential of wireless power transfer for dynamic charging of electric vehicles. *IET Intell. Transp. Syst.* **2019**, *13*, 3–12. [CrossRef]
23. Brahim, M.; Bernard, L.; Pichon, L.; Razek, A.; Houivet, J.; Cayol, O. Advanced modeling of a 2-kw series–series resonating inductive charger for real electric vehicle. *IEEE Trans. Veh. Technol.* **2015**, *64*, 421–430. [CrossRef]
24. Cirimele, V.; Diana, M.; Freschi, F.; Mitolo, M. Inductive Power Transfer for Automotive Applications: State-of-the-Art and Future Trends. *IEEE Trans. Ind. Appl.* **2018**, *54*, 4069–4079. [CrossRef]
25. Razek, A. Review of Contactless Energy Transfer Concept Applied to Inductive Power Transfer Systems in Electric Vehicles. *Appl. Sci.* **2021**, *11*, 3221. [CrossRef]
26. Ibrahim, M.; Bernard, L.; Pichon, L.; Laboure, E.; Razek, A.; Cayol, O.; Ladas, D.; Irving, J. Inductive Charger for Electric Vehicle: Advanced Modeling and Interoperability Analysis. *IEEE Trans. Power Electron.* **2016**, *31*, 8096–8114. [CrossRef]
27. Cirimele, V.; Diana, M.; Bellotti, F.; Berta, R.; El Sayed, N.; Kobeissi, A.; Guglielmi, P.; Ruffo, R.; Khalilian, M.; La Ganga, A.; et al. The Fabric ICT Platform for Managing Wireless Dynamic Charging Road Lanes. *IEEE Trans. Veh. Technol.* **2020**, *69*, 2501–2512. [CrossRef]
28. Trevisan, R.; Costanzo, A. State-of-the-art of contactless energy transfer (CET) systems: Design rules and applications. *Wirel. Power Transf.* **2014**, *1*, 10–20. [CrossRef]
29. Wang, C.S.; Stielau, O.H.; Covic, G.A. Design considerations for a contactless electric vehicle battery charger. *IEEE Trans. Ind. Electron.* **2005**, *52*, 1308–1314. [CrossRef]
30. Russer, J.A.; Dionigi, M.; Mongiardo, M.; Russer, P. A moving field inductive power transfer system for electric vehicles. In Proceedings of the IEEE 43rd European Microwave Conference 2013, Nuremberg, Germany, 7–10 October 2013; pp. 519–522. [CrossRef]
31. Rizzoli, V.; Costanzo, A.; Masotti, D.; Donzelli, F. Integration of numerical and field-theoretical techniques in the design of single- and multi-band rectennas for micro-power generation. *Int. J. Microw. Wirel. Technol.* **2010**, *2*, 293–303. [CrossRef]
32. Costanzo, A.; Romani, A.; Masotti, D.; Arbizzani, N.; Rizzoli, V. RF/baseband co-design of switching receivers for multiband microwave energy harvesting. *Sens. Actuators A Phys.* **2012**, *179*, 158–168. [CrossRef]
33. Ohira, T. Via-wheel power transfer to vehicles in motion. In Proceedings of the 2013 IEEE Wireless Power Transfer Conference 2013, Perugia, Italy, 15–16 May 2013; pp. 242–246. [CrossRef]
34. Pahlavan, S.; Jafarabadi-Ashtiani, S.; Mirbozorgi, S.A. Maze-Based Scalable Wireless Power Transmission Experimental Arena for Freely Moving Small Animals Applications. *IEEE Trans. Biomed. Circuits Syst.* **2024**, *3*, 1–10. [CrossRef]
35. Sample, A.P.; Waters, B.H.; Wisdom, S.T.; Smith, J.R. Enabling seamless wireless power delivery in dynamic environments. *Proc. IEEE* **2013**, *101*, 1343–1358. [CrossRef]
36. Islam, M.; Ross, C.; Lutz, C.W.; Sebastian, T.; Islam, M.S. In-Wheel Motor with Cooling System. U.S. Patent 11,431,228, 30 August 2022. Available online: https://patents.google.com/patent/US11431228B2/en (accessed on 1 August 2024).
37. Boules, N.; Electric Drives for Battery Electric Vehicles. Navigating an Electric Vehicle Future Virtual Workshop, 25–28 October 2021—The US National Academies of Sciences Electric Drives for Vehicles—National Academies of Sciences. Available online: https://www.nationalacademies.org/documents/embed/link/LF2255DA3DD1C41C0A42D3BEF0989ACAECE3053A6A9B/file/D6B5A579A548566915FEC39CBDBB3EAF5E8933C796E8?noSaveAs=1 (accessed on 29 July 2024).
38. Wu, J.; Li, Y.; Dai, X.; Gao, R.; He, M. A Dynamic Power Transfer Route Construction and Optimization Method Considering Random Node Distribution for Wireless Power Transfer Network. *IEEE Trans. Power Electron.* **2024**, *39*, 4858–4869. [CrossRef]
39. Zhang, W.; Mi, C.C. Compensation Topologies of High-Power Wireless Power Transfer Systems. *IEEE Trans. Veh. Technol.* **2016**, *65*, 4768–4778. [CrossRef]
40. Vishnuram, P.; Panchanathan, S.; Rajamanickam, N.; Krishnasamy, V.; Bajaj, M.; Piecha, M.; Blazek, V.; Prokop, L. Review of Wireless Charging System: Magnetic Materials, Coil Configurations, Challenges, and Future Perspectives. *Energies* **2023**, *16*, 4020. [CrossRef]
41. Bi, Z.; Kan, T.; Mi, C.C.; Zhang, Y.; Zhao, Z.; Keoleian, G.A. A review of wireless power transfer for electric vehicles: Prospects to enhance sustainable mobility. *Appl. Energy* **2016**, *179*, 413–425. [CrossRef]
42. Ji, N.; Zhu, R.; Huang, Z.; You, L. An urban-scale spatiotemporal optimization of rooftop photovoltaic charging of electric vehicles. *Urban Inform.* **2024**, *3*, 4. [CrossRef]
43. Afridi, K. The future of electric vehicle charging infrastructure. *Nat. Electron.* **2022**, *5*, 62–64. [CrossRef]
44. Mudgal, Y.; Tiwari, R. Impact analysis of integrated renewable energy sources and EVs charging demand on the distribution system. In Proceedings of the 2023 IEEE IAS Global Conference on Renewable Energy and Hydrogen Technologies, GlobConHT 2023, Male, Maldives, 11–12 March 2023; pp. 1–5. [CrossRef]
45. Zavrel, M.; Kindl, V.; Frivaldsky, M.; Andriukaitis, D.; Navikas, D. Optimization of series-series compensated wireless power transfer system using alternative secondary side rectification. *Sci. Rep.* **2024**, *14*, 1191. [CrossRef]
46. Wu, Y.; Jiang, Y.; Li, Y.; Yuan, H.; Wang, X.; Tang, Y. Precise Parameterized Modeling of Coil Inductance in Wireless Power Transfer Systems. *IEEE Trans. Power Electron.* **2024**, *39*, 11746–11757. [CrossRef]
47. Campi, T.; Cruciani, S.; Maradei, F.; Feliziani, M. Magnetic Field Mitigation by Multicoil Active Shielding in Electric Vehicles Equipped With Wireless Power Charging System. *IEEE Trans. Electromagn. Compat.* **2020**, *62*, 1398–1405. [CrossRef]

48. Liorni, I.; Bottauscio, O.; Guilizzoni, R.; Ankarson, P.; Bruna, J.; Fallahi, A.; Harmon, S.; Zucca, M. Assessment of Exposure to Electric Vehicle Inductive Power Transfer Systems: Experimental Measurements and Numerical Dosimetry. *Sustainability* **2020**, *12*, 4573. [CrossRef]
49. Razek, A. Analysis and control of ornamental plants responses to exposure to electromagnetic fields. *Ornam. Plant Res.* **2024**, *4*, e009. [CrossRef]
50. Etxegarai, G.; Camblong, H.; Ezeiza, A.; Lie, T.T. Design of Three Electric Vehicle Charging Tariff Systems to Improve Photovoltaic Self-Consumption. *Energies* **2024**, *17*, 1806. [CrossRef]
51. Joseph, P.K.; Elangovan, D. A review on renewable energy powered wireless power transmission techniques for light electric vehicle charging applications. *J. Energy Storage* **2018**, *16*, 145–155. [CrossRef]
52. Xie, H.; Huang, R.; Sun, H.; Han, Z.; Jiang, M.; Zhang, D.; Goh, H.H.; Kurniawan, T.A.; Han, F.; Liu, H.; et al. Wireless energy: Paving the way for smart cities and a greener future. *Energy Build.* **2023**, *297*, 113469. [CrossRef]
53. International Commission on Non-Ionizing Radiation Protection. Guide-lines for limiting exposure to time-varying electric and magnetic fields for low frequencies (1 Hz–100 kHz). *Health Phys.* **2010**, *99*, 818–836. [CrossRef]
54. International Commission on Non-Ionizing Radiation Protection. Guidelines for limiting exposure to electromagnetic fields (100 kHz to 300 GHz). *Health Phys.* **2020**, *118*, 483–524. [CrossRef]
55. US Food and Drug Administration. Scientific Evidence for Cell Phone Safety. 2020. Available online: https://www.fda.gov/radiation-emitting-products/cell-phones/scientific-evidence-cell-phone-safety (accessed on 4 January 2024).
56. Council of the European Union. EU Recommendation 1999/519/EC on the Limitation of Exposure of the General Public to Electromagnetic Fields (0 Hz to 300 GHz). 1999. Available online: https://eur-lex.europa.eu/eli/reco/1999/519/oj (accessed on 4 January 2024).
57. Boules, N.; Douglas, K.; Feldman, S.; Fix, L.; Hager, G.; Hailpern, B.; Martial Hebert, M.; Dan Lopresti, D.; Beth Mynatt, B.; Rossbach, C.C.; et al. The future of computing research: Industry-academic collaborations. *arXiv* **2016**, arXiv:1606.09236. [CrossRef]
58. Maxwell, J.C.V.I.I.I. A dynamical theory of the electromagnetic field. *Philos. Trans. R. Soc.* **1865**, *155*, 459–512. [CrossRef]
59. Razek, A. Biological and Medical Disturbances Due to Exposure to Fields Emitted by Electromagnetic Energy Devices—A Review. *Energies* **2022**, *15*, 4455. [CrossRef]
60. Razek, A. Thermal effects of electromagnetic origin from heating processes to biological disturbances due to field exposure—A review. *Therm. Sci. Eng.* **2023**, *6*, 1950. [CrossRef]
61. Razek, A. Assessment of EMF Troubles of Biological and Instrumental Medical Questions and Analysis of Their Compliance with Standards. *Standards* **2023**, *3*, 227–239. [CrossRef]
62. Pennes, H.H. Analysis of tissue and arterial blood temperatures in the resting human forearm. *J. Appl. Physiol.* **1998**, *85*, 5–34. [CrossRef] [PubMed]
63. Nunes, A.S.; Dular, P.; Chadebec, O.; Kuo-Peng, P. Subproblems Applied to a 3-D Magnetostatic Facet FEM Formulation. *IEEE Trans. Magn.* **2018**, *54*, 7402209. [CrossRef]
64. Cicuttin, M.; Geuzaine, C. Numerical investigation of a 3D hybrid high-order method for the indefinite time-harmonic Maxwell problem. *Finite Elem. Anal. Des.* **2024**, *233*, 104124. [CrossRef]
65. Piriou, F.; Razek, A. Numerical simulation of a nonconventional alternator connected to a rectifier. *IEEE Trans. Energy Convers.* **1990**, *5*, 512–518. [CrossRef]
66. Bernard, L. Electrical Characterization of Biological Tissues and Computing of Phenomena Induced in the Human Body by Electromagnetic Fields below 1 GHz. Ph.D. Thesis, Universities of Ecole Centrale de Lyon, Écully, France, Universidade Federal de Minas Gerais, Belo Horizonte, Brazil, 2007. Available online: https://theses.hal.science/tel-00179791v3 (accessed on 1 August 2024).
67. Ren, Z.; Razek, A. Force calculation by Maxwell stress tensor in 3D hybrid finite element-boundary integral formulation. *IEEE Trans. Magn.* **1990**, *26*, 2774–2776. [CrossRef]
68. Freschi, F.; Giaccone, L.; Cirimele, V.; Canova, A. Numerical assessment of low-frequency dosimetry from sampled magnetic fields. *Phys. Med. Biol.* **2018**, *63*, 015029. [CrossRef]
69. Li, C.; Ren, Z.; Razek, A. An approach to adaptive mesh refinement for three-dimensional eddy-current computations. *IEEE Trans. Magn.* **1994**, *30*, 113–117. [CrossRef]
70. Piriou, F.; Razek, A. Calculation of saturated inductances for numerical simulation of synchronous machines. *IEEE Trans. Magn.* **1983**, *19*, 2628–2631. [CrossRef]
71. Madani, S.S.; Schaltz, E.; Kær, S.K. Thermal Analysis of Cold Plate with Different Configurations for Thermal Management of a Lithium-Ion Battery. *Batteries* **2020**, *6*, 17. [CrossRef]
72. Urdaneta-Calzadilla, A.; Galopin, N.; Niyonzima, I.; Chadebec, O.; Meunier, G.; Bannwarth, B. FEM-BEM Modeling of Nonlinear Magnetoelectric Effects in Heterogeneous Composite Structures. *IEEE Trans. Magn.* **2024**, *60*, 7401704. [CrossRef]
73. Gabriel, S.; Lau, R.W.; Gabriel, C. The dielectric properties of biological tissues: II. Measurements in the frequency range 10 Hz to 20 GHz. *Phys. Med. Biol.* **1996**, *41*, 2251–2269. [CrossRef]
74. Barchanski, A.; Steiner, T.; De Gersem, H.; Clemens, M.; Weiland, T. Local grid refinement for low-frequency current computations in 3-D human anatomy models. *IEEE Trans. Magn.* **2006**, *42*, 1371–1374. [CrossRef]

75. Hasgall, P.A.; Di Gennaro, F.; Baumgartner, C.; Neufeld, E.; Lloyd, B.; Gosselin, M.C.; Payne, D.; Klingenböck, A.; Kuster, N. *IT'IS Database for Thermal and Electromagnetic Parameters of Biological Tissues, Version 4.1*; IEEE: New York, NY, USA, 2022. [CrossRef]
76. Makarov, S.N.; Noetscher, G.M.; Yanamadala, J.; Piazza, M.W.; Louie, S.; Prokop, A.; Nazarian, A.; Nummenmaa, A. Virtual Human Models for Electromagnetic Studies and Their Applications. *IEEE Rev. Biomed. Eng.* **2017**, *10*, 95–121. [CrossRef]
77. Razek, A. Coupled Models in Complex Systems Related to Smart Electromagnetic Energy Procedures. *J. Mod. Green Energy* **2024**, *3*, 2. [CrossRef]
78. Razek, A. Strategies for managing models regarding environmental confidence and complexity involved in intelligent control of energy systems—A review. *Adv. Environ. Energ.* **2023**, *2*, 1–17. [CrossRef]
79. Lebensztajn, L.; Marretto, C.A.R.; Costa, M.C.; Coulomb, J.L. Kriging: A Useful Tool for Electromagnetic Device Optimization. *IEEE Trans. Magn.* **2004**, *40*, 1196–1199. [CrossRef]
80. Kudela, J.; Matousek, R. Recent advances and applications of surrogate models for finite element method computations: A review. *Soft Comput.* **2022**, *26*, 13709–13733. [CrossRef]
81. Koziel, S.; Pietrenko-Dabrowska, A. *Fundamentals of Data-Driven Surrogate Modeling*; World Scientific Publishing Co. Pte. Ltd.: Singapore, 2022; pp. 1–37. [CrossRef]
82. Gaignaire, R.; Scorretti, R.; Sabariego, R.V.; Geuzaine, C. Stochastic Uncertainty Quantification of Eddy Currents in the Human Body by Polynomial Chaos Decomposition. *IEEE Trans. Magn.* **2012**, *48*, 451–454. [CrossRef]
83. Razek, A.; Pichon, L.; Kameni, A.; Makong, L.; Rasm, S. Evaluation of Human Exposure owing to Wireless Power Transfer Systems in Electric Vehicles. *Athens J. Technol. Eng.* **2019**, *6*, 239–258. [CrossRef]
84. Aktas, A.; Aydin, E.; Onar, O.C.; Su, G.J.; Ozpineci, B.; Tolbert, L.M. Medium-Duty Delivery Truck Integrated Bidirectional Wireless Power Transfer System with Grid and Stationary Energy Storage System Connectivity. *IEEE J. Emerg. Sel. Top. Power Electron.* 2024, *early access*. [CrossRef]
85. De Lacerda, M.L.S.S.; dos Santos, W.M. Assessment of Time and Frequency Domain Models of Bidirectional LCC-LCC Inductive Power Transfer Systems. *J. Control Autom. Electr. Syst.* **2024**, *35*, 758–768. [CrossRef]
86. Song, K.; Lan, Y.; Zhang, X.; Jiang, J.; Sun, C.; Yang, G.; Yang, F.; Lan, H. A Review on Interoperability of Wireless Charging Systems for Electric Vehicles. *Energies* **2023**, *16*, 1653. [CrossRef]
87. Wang, D.; Qu, X.; Yao, Y.; Yang, P. Hybrid Inductive-Power-Transfer Battery Chargers for Electric Vehicle Onboard Charging With Configurable Charging Profile. *IEEE Trans. Intell. Transp. Syst.* **2021**, *22*, 592–599. [CrossRef]
88. Boules, N. The Automotive Technological Revolution and the Future of Personal Mobility. In Proceedings of the International School on Energetic Efficiency of Connected Vehicles 2018, Bordeaux, France, 17 October 2018. Available online: https://events.vtools.ieee.org/m/324202 (accessed on 25 July 2024).
89. Sagar, A.; Kashyap, A.; Nasab, M.A.; Padmanaban, S.; Bertoluzzo, M.; Kumar, A.; Blaabjerg, F. A Comprehensive Review of the Recent Development of Wireless Power Transfer Technologies for Electric Vehicle Charging Systems. *IEEE Access* **2023**, *11*, 83703–83751. [CrossRef]
90. Quercio, M.; Lozito, G.M.; Corti, F.; Fulginei, F.R.; Laudani, A. Recent Results in Shielding Technologies for Wireless Electric Vehicle Charging Systems. *IEEE Access* **2024**, *12*, 16728–16740. [CrossRef]
91. Harris, L.R.; Zhadobov, M.; Chahat, N.; Sauleau, R. Electromagnetic dosimetry for adult and child models within a car: Multi-exposure scenarios. *Int. J. Microw. Wirel. Technol.* **2011**, *3*, 707–715. [CrossRef]
92. Gjonaj, E.; Bartsch, M.; Clemens, M.; Schupp, S.; Weiland, T. High-resolution human anatomy models for advanced electromagnetic field computations. *IEEE Trans. Magn.* **2002**, *38*, 357–360. [CrossRef]
93. Glingston, R.S.; Yadav, J.; Rajpoot, J.; Joshi, N.; Nagotu, S. Contribution of yeast models to virus research. *Appl. Microbiol. Biotechnol.* **2021**, *105*, 4855–4878. [CrossRef]
94. Bunse, C.; Bertos-Fortis, M.; Sassenhagen, I.; Sildever, S.; Sjöqvist, C.; Godhe, A.; Gross, S.; Kremp, A.; Lips, I.; Lundholm, N.; et al. Spatio-Temporal Interdependence of Bacteria and Phytoplankton during a Baltic Sea Spring Bloom. *Front. Microbiol.* **2016**, *21*, 517. [CrossRef] [PubMed]
95. Maurice, K.; Bourceret, A.; Robin-Soriano, A.; Vincent, B.; Boukcim, H.; Selosse, M.A.; Ducousso, M. Simulated precipitation in a desert ecosystem reveals specific response of rhizosphere to water and a symbiont response in freshly emitted roots. *Appl. Soil Ecol.* **2024**, *199*, 105412. [CrossRef]
96. Mou, W.; Lu, M. Safety evaluation of electromagnetic exposure for compact wireless charging stations of electric vehicles. *J. Chin. Inst. Eng.* **2024**, 1–15. [CrossRef]
97. Salgado, R.M.; Danzi, F.; Oliveira, J.E.; El-Azab, A.; Camanho, P.P.; Braga, M.H. The Latest Trends in Electric Vehicles Batteries. *Molecules* **2021**, *26*, 3188. [CrossRef]
98. Ibrahim, M.; Rassõlkin, A.; Vaimann, T.; Kallaste, A. Overview on Digital Twin for Autonomous Electrical Vehicles Propulsion Drive System. *Sustainability* **2022**, *14*, 601. [CrossRef]
99. Fuller, A.; Fan, Z.; Day, C.; Barlow, C. Digital Twin: Enabling Technologies, Challenges and Open Research. *IEEE Access* **2020**, *8*, 108952–108971. [CrossRef]
100. White, G.; Zink, A.; Codecá, L.; Clarke, S. A digital twin smart city for citizen feedback. *Cities* **2021**, *110*, 103064. [CrossRef]
101. Mohammadi, N.; Taylor, J.E. Smart city digital twins. In Proceedings of the 2017 IEEE Symposium Series on Computational Intelligence (SSCI) 2017, Honolulu, HI, USA, 27 November–1 December 2017; pp. 1–5. [CrossRef]

102. Grieves, M.; Vickers, J. Digital twin: Mitigating unpredictable, undesirable emergent behavior in complex systems. In *Transdisciplinary Perspectives on Complex Systems*; Springer: Cham, Switzerland, 2017; pp. 85–113. [CrossRef]
103. Razek, A. Monitoring Complexity in Clean Energy Systems Applications. *Clean Energy Sustain.* **2024**, *2*, 10007. [CrossRef]
104. Ibrahim, M.; Rjabtšikov, V.; Gilbert, R. Overview of Digital Twin Platforms for EV Applications. *Sensors* **2023**, *23*, 1414. [CrossRef]
105. Polat, A.O.; Erden, B.C.; Kul, S.; Nasiroglu, F. Light Electric Vehicle Performance with Digital Twin Technology: A Comparison of Motor Types. *Arab. J. Sci. Eng.* **2024**, *49*, 7209–7222. [CrossRef]
106. Ali, W.A.; Fanti, M.P.; Roccotelli, M.; Ranieri, L. A Review of Digital Twin Technology for Electric and Autonomous Vehicles. *Appl. Sci.* **2023**, *13*, 5871. [CrossRef]
107. Venturini, S.; Rosso, C.; Velardocchia, M. An automotive steel wheel digital twin for failure identification under accelerated fatigue tests. *Eng. Fail. Anal.* **2024**, *158*, 107979. [CrossRef]
108. Gao, J.; Peng, C.; Yoshinaga, T.; Han, G.; Guleng, S.; Wu, C. Digital Twin-Enabled Internet of Vehicles Applications. *Electronics* **2024**, *13*, 1263. [CrossRef]
109. Li, X.; Niu, W.; Tian, H. Application of Digital Twin in Electric Vehicle Powertrain: A Review. *World Electr. Veh. J.* **2024**, *15*, 208. [CrossRef]
110. Venkatesan, S.; Manickavasagam, K.; Tengenkai, N.; Vijayalakshmi, N. Health monitoring and prognosis of electric vehicle motor using intelligent-digital twin. *IET Electr. Power Appl.* **2019**, *13*, 1328–1335. [CrossRef]
111. Liao, X.; Zhao, X.; Wang, Z.; Zhao, Z.; Han, K.; Gupta, R.; Barth, M.J.; Wu, G. Driver Digital Twin for Online Prediction of Personalized Lane-Change Behavior. *IEEE Internet Things J.* **2023**, *10*, 13235–13246. [CrossRef]
112. Razek, A. Interaction of electromagnetic fields with body-onboard devices. *Explor. Digit. Health Technol.* **2024**, *2*, 124–134. [CrossRef]

energies

MDPI

Article

Method of Determining New Locations for Electric Vehicle Charging Stations Using GIS Tools

Piotr Soczówka [1], Michał Lasota [2,*], Piotr Franke [2] and Renata Żochowska [1,*]

[1] Faculty of Transport and Aviation Engineering, Silesian University of Technology, 40-019 Katowice, Poland; piotr.soczowka@polsl.pl

[2] Faculty of Transport, Warsaw University of Technology, 00-662 Warsaw, Poland; piotr.franke@pw.edu.pl

* Correspondence: michal.lasota@pw.edu.pl (M.L.); renata.zochowska@polsl.pl (R.Ż.)

Abstract: The growing awareness of environmental issues, climate policies, and rapidly developing technologies is contributing to the increasing number of battery electric vehicles (BEVs) around the world. A key requirement for their widespread implementation is providing a charging infrastructure that allows users to operate these vehicles comfortably. Lack of access to charging stations can be a major barrier to the development of electromobility in a given area. Therefore, each additional charging infrastructure can support a change in the structure of the vehicle fleet. One of the key challenges facing this transformation is the selection of suitable locations for charging stations. It is necessary to ensure that they are uniformly distributed so that range anxiety for EV users is reduced and equal access to charging infrastructure is provided to all residents. One of the most important stakeholders in this market is local authorities. Therefore, the objective of this research was to develop a method of determining optimal locations for electric vehicle charging stations (EVCSs) from the perspective of local authorities that also takes into account equal access to the charging infrastructure for all residents, which seems to be a unique approach to this problem. We used commonly available spatial data as input to enable the method to be applied on a larger scale and over an urban area. We carried out our research using a case study: the city of Gliwice in Poland. The city area was divided into hexagonal basic fields, for which potentials for locations of new charging stations were calculated. The analysis was carried out using the geographic information system (GIS) QGIS (ver. 3.34).

Keywords: battery electric vehicles (BEVs); spatial analysis; GIS software; division of the area; hexagonal grid; charging infrastructure; land use

1. Introduction

With the rapid development of automotive technology and growing interest in ecofriendly transportation solutions, electric vehicles (EVs) are becoming increasingly popular worldwide [1]. Reducing greenhouse gas emissions and air pollution, and gradually becoming independent of fossil fuels, are key benefits associated with their deployment [2]. However, one of the main challenges associated with the widespread adoption of electric vehicles is the availability of electric vehicle charging stations (EVCSs) [3–5]. In particular, the proper location of the EV charging infrastructure in urban areas is crucial in ensuring convenience for EV users; therefore, it is crucial in encouraging residents to own such vehicles [6].

Studies in Germany, Brazil, and the United States [7–10] have shown that the problem of unequal access to EVCSs in urban areas is widespread. A correlation has been indicated between some characteristics describing the socioeconomic situation and the availability of EVCSs. For example, the number of existing stations depends on the type of area (urban/rural) [7], the material status and age of inhabitants, and the ownership of dwellings [9,10]. The authors also point out large regional differences, which are often connected to the historical circumstances and developmental lags of the areas concerned. Therefore, locations for EVCSs may not match the demand [11].

The most appropriate actors to address this problem and equalize the availability of EVCSs for all residents are local authorities. Their role may be to fill the gap, indicate the places where public EVCSs should be located, and decide where commercial EVCSs should not be built, for different reasons.

Our review of the literature showed that different criteria and various methods have already been developed to address the problem of the location of EVCSs. However, there is still a research gap in analyzing the problem of determining the locations for EVCSs both from the point of view of local authorities and equal access to charging infrastructure for area residents. Our research is an attempt to solve this problem. The method we propose takes into account the perspective of local authorities and the determinants related to sustainable development that are important to them. Therefore, this is a new approach to determining optimal locations for EVCSs.

The requirement we set for the data needed in our method was that they should be accessible via public and open-access databases. We wanted to use only data that are publicly available so as to not incur any additional costs for local authorities when using our method. Therefore, the tools we use in our approach are based on geographic information system (GIS) and the use of a database that is publicly available in Poland, the Polish National Databases of Spatial Data (the BDOT10k database).

The main aim of this article was to present a GIS-based method that can be a useful tool for local authorities in supporting decisions regarding the locations of EVCSs in urban areas. This article also presents a broad analysis of key factors that have been considered in previously used approaches, which allowed us to highlight the research gap and select factors that are important in supporting different decision-making processes.

The method proposed in the present article places particular emphasis on providing residents with equal access to the EV charging infrastructure. Therefore, it can respond to the challenges of contemporary transport and act as a valuable tool for local authorities in their efforts to increase access to EVCSs. Consequently, it can support the electrification of individual road transport vehicles and the achievement of environmentally friendly policy objectives.

The main hypothesis in this paper is that the potential of a given location for a new and publicly accessible EVCS can be evaluated based on the features of the built environment. We also raised the following research questions:

1. Which factors associated with the features of the built environment will effectively represent the actual potential?
2. How should the stations be spatially distributed to ensure equitable access to them?

This article also presents a case study conducted for the city of Gliwice in Poland. According to 2023 data, there are about 20 EVCSs in the city. As a result of using the proposed method, five additional places were recommended for further EVCSs.

The present article is divided into five parts. The purpose of the developed method and the structure of this article are presented in the Introduction (Section 1) against the general background of the issues analyzed. Section 2 presents a review of the literature on selected areas related to locations for EVCSs, including social, spatial, economic, legal, technical, and environmental aspects. Section 3 presents the general assumptions and a description of the method for determining the locations for EVCSs in urban areas. Section 4 contains the results of the research carried out using the proposed method for the medium-sized city of Gliwice in Poland. The final sections (Sections 5 and 6) include a discussion of the results, directions for further research, and conclusions. Appendix A contains the notation, a formal description of the method, and its algorithm.

2. Literature Review

One of the main challenges of electromobility development in urban areas is the identification of criteria that are required to assess the potential of a given location for an EVCS and to select the best places for building EV charging infrastructure. Figure 1 presents a set of criteria related to the issue of determining the locations of EVCSs, which is the basis of the literature review.

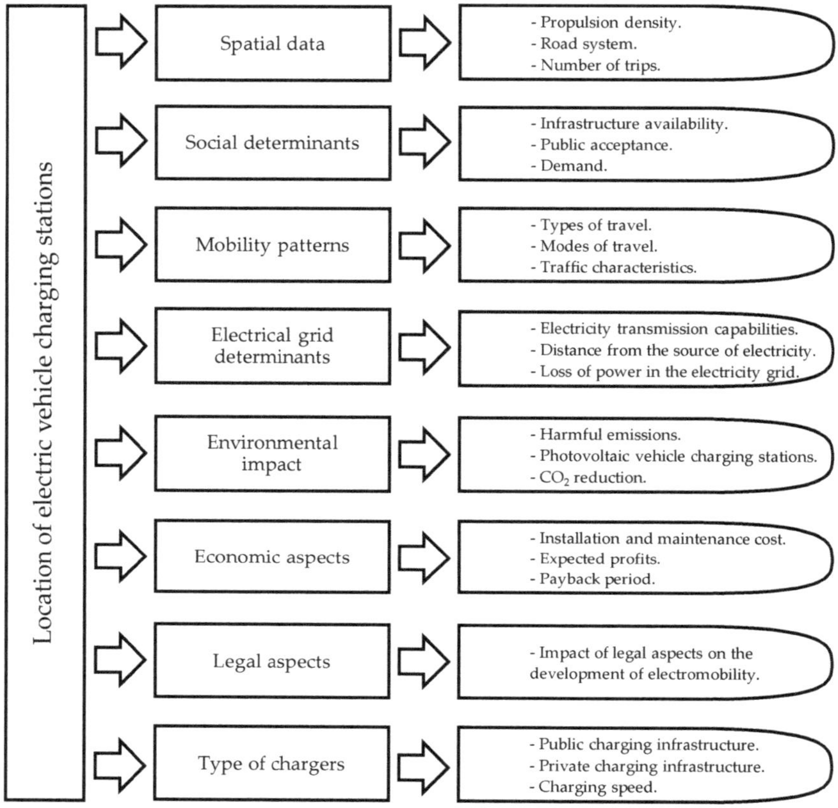

Figure 1. Structure of the literature review in this article (source: own elaboration).

During the literature review, considerable complexity in the issue and a variety of approaches to the problem of determining the locations for EVCSs were observed. The individual aspects presented in Figure 1 are discussed in detail in the following subsections.

2.1. Spatial Data

One of the most commonly used data sources in determining the optimal locations for EVCSs is spatial data. The main advantages of this type of data are their wide availability, reliability, diversity, and ease of use. In the case of the data used in the analysis [12] developed for the example of the city of Łódź, the optimization criteria were as follows:

- Distance from petrol stations;
- Distance from shopping centers and supermarkets;
- Population density;
- Distance from parking lots;
- Location in relation to parks;
- Location in relation to existing EVCSs;
- Distance from the main roads that traverse the city (at least the regional roads).

Point of Interest (POI) data are very often used to assess locations for EVCSs. They are often classified into specific categories, as in article [13], where nine groups are listed based on business categories according to NAICS (North American Industry Classification System). On the other hand, in article [14], a geographic information system (GIS) was used to obtain information on population, jobs, number of cars in high-income areas, and households without cars, as well as the number of students, walking trips, and the total number of trips. The authors of the publication used aggregated data to solve the Set Covering Location Problem (SCLP) and Maximum Location Problem (MCLP). The former is a classic optimization problem consisting of selecting the location for a specified number of charging facilities in such a way as to minimize investment costs and maximize the efficiency of covering a given area. The MCLP problem, in turn, concerns the maximization of benefits from resources that are limited. On the contrary, a case study of Dublin used multi-criteria analysis with GIS [15] and data including population density, the layout of major roads, the number of trips to places of work and study, and the location of parking lots, lamps, housing, and public facilities. The data used came from the National Statistics Office and OSM.

GIS tools were also used to optimize the locations for photovoltaic EVCSs in a study developed for the city of Qingdao in China [16] and in analyses conducted in Salt Lake City [17]. The authors of the publication [18] analyzed the current network of EVCSs in the area of Hahhot, China, based on spatial data on urban land use, charging practices, and travel patterns of electric vehicle users. The second part of their study included a plan for the future development of the vehicle charging system in the area. According to the authors, the most important variables for planning the deployment of EVCSs are the level of charging demand, the coverage of each point in terms of the territory served, and the construction cost [19].

2.2. Social Determinants

Demographic and social factors largely determine the optimal locations for EVCSs. This is evidenced by a study conducted in Hong Kong [20] that assessed the demand for vehicle charging based on surveys, taking the propensity to purchase an electric vehicle as a reference point. The influence of socio-demographic factors such as age, gender, education level, marital status, employment status, household size, income, and place of residence was examined. The results showed that groups of older respondents ($\geq$45 years) were less likely to purchase an electric vehicle compared with the younger age group (18–34 years). Respondents who were self-employed, married, and living in owner-occupied houses and apartments were also more likely to purchase an electric vehicle. Consequently, the owners of electric cars will also become a group of EVCS users. This is also reflected in a study from South Korea [21], which analyzed the factors within a radius of about 914 m from existing stations, including the number of registered EVs and the value of the land, which can approximate the income level.

In turn, the authors of [22] proposed selecting locations for EVCSs based on the application of the graph convolutional network (GCN) method, which is the basis of deep learning technology. This technology was used to analyze the distribution of EVCSs based on the distribution of social classes in China. The study showed that the business area should have the infrastructure with the fastest charging capability to meet the requirements of the most demanding customers. The combination of fast and slow chargers is beneficial in residential areas.

2.3. Mobility Patterns

The analysis of traffic patterns and driver behavior is another important aspect. Models that take these factors into account often use data collected from GPS systems, traffic surveys, and big data analysis to identify the locations with the greatest potential. A study conducted in Canada [23] using data from the operation of over 7000 EVCSs also showed that user behavior patterns were very important factors influencing EVCS usage, including usage at particular times of the day. This also affects the operation and stability of the electric power system. The research confirmed that charging patterns and times reinforce the "duck curve" problem of energy production, which presents a growing challenge to ensure the stability of the network.

Logistic patterns were also analyzed in a simulation model developed for Gothenburg [24]. The study presented a methodology for the deployment of fast EVCSs for the city's logistics operators. The main objective of the study was to minimize total costs by optimizing the locations and number of chargers at each station. The authors showed that the most beneficial locations for the deployment of EVCSs were private EVCSs in vehicle parking areas (because of long overnight stops) and shared or public EVCSs in important and frequently visited locations along delivery routes.

The study [25] focused on optimizing the locations for EVCSs to facilitate long-distance travel and meet urban demands. Based on the research, traffic volume, distance from the nearest EVCS, and the range of services available on the market were indicated as the main variables influencing the use of EVCSs on expressways in Budapest. Furthermore, studies in Austria [26] found that 88% of vehicle charging took place when the user was at home. Almost ten times less charging took place in workplaces, and only 1.7% used public EV charging stations.

2.4. Electrical Grid Determinants

Another important factor considered in decision-making models is the location of an EVCS in the electricity network. Since the charging infrastructure has to ensure that the parameters of the distribution network are adequate, cooperation and coordination with power suppliers are essential. For this reason, many studies also present an optimization approach in terms of local energy supply capacity or the costs associated with providing such capacity. For example, the paper [27] proposed the optimal placement of fast chargers at different locations in the electricity distribution system using Multiple Objective Particle Swarm Optimization (MOPSO), so that power losses and voltage deviations are kept to a minimum. Attempts were also made to combine transportation and network-related criteria in optimization models. A large-scale study on the example of Thailand [28] also considered energy efficiency, measured by network power loss. A broad review of approaches to optimizing the locations for EVCSs is presented in the article [29]. The authors paid particular attention to the need for a comprehensive approach to this problem and proposed an authoritative, multi-stage model for optimizing the locations and sizes of EVCSs.

The optimal deployment of EVCSs requires an analysis of potential charging demand and vehicle traffic patterns to estimate energy consumption. In addition, the projection and implementation of the planned locations for EVCSs must be based on an analysis of the local electricity distribution network to ensure the optimal efficiency of the designed solution and an acceptable level of power. The article [30] provided an extensive review of the CSLP (Charging Station Location Problem), focusing on modeling and solving problems from different perspectives. According to the authors, the following aspects are important when designing a network of EVCSs:

- Estimation of the charging demand;
- Range of the electric vehicle;
- Efficiency of the currently used vehicle charging system;
- Distance of the designed charging point from the power source;
- Strategic decisions and their impacts on the development of the charging network.

2.5. Environmental Impact

The need to minimize environmental impact is also reflected in the current research. Considerations include the use of renewable energy sources to power EVCSs, as well as reducing air and noise pollution [31]. For example, in the model [32] developed for the city of Tunis, the main criteria considered were maximizing the use of energy from photovoltaic panels and minimizing the use of the power network. Another example can be found in the article [33], which included a study on the implementation of batteries in electric buses used for public transport. The researcher developed a model to optimize the cost of implementing the above solution and its impact on greenhouse gas emissions in the city of Guelph in southern Canada.

The authors of the publication [34] addressed the integration of EVCSs with randomly distributed rooftop photovoltaic systems. The aim was to optimize the placement of EVCSs in a way that consumed the excess energy transferred to the network by local photovoltaic systems.

2.6. Economic Aspects

Economic factors such as the cost of installing and maintaining a station, as well as the expected profits, are integral parts of the decision-making process. Cost–benefit analyses often also consider the time costs of station users, such as the need to travel to an EVCS. The Multi-Objective Simulated Annealing Particle Swarm Optimization (MOSAPSO) algorithm [35] was used to solve the problem with a multi-criteria function consisting of criteria that described the investment cost, the operation and maintenance cost, the payback period, and the time cost of the user of a fast EVCS. The analysis also considered the difficulty (time) of reaching the EVCS and the classification of the development function based on the location of POIs. The extensive multi-criteria analysis proposed in the article [31] on the example of the city of Zagazig in Egypt accounted for 19 decision criteria, including the number of utilities, operation and maintenance costs, consumption level, construction costs, and land and equipment costs. However, in the article [36], which considered the multi-criteria Analytic Hierarchy Process (AHP) method in combination with the objective programming method, the conversion goal programming model (CAGM) was used to allocate EVs to the most appropriate EVCSs, minimize the level of energy consumption, minimize operation time, and minimize charging costs and battery degradation. Researchers in Germany [37], using a simulation model, investigated the deployment of an EVCS under the assumption that 100% of the vehicles used in the Berlin area would be electric. The main criteria were capital costs and average overage of users.

2.7. Legal Aspects

The authors of an article that proposed the deployment of EVCS along the TEN-T road network in Poland [38] also paid attention to the need to account for legal aspects when choosing locations for EVCSs. The first aspect is the regulation on alternative fuel infrastructure—AFIR [39]—which assumes, among others, the construction of fast electric vehicle compartments (EVCSs) for passenger cars and trucks, as well as hydrogen refueling stations on the entire TEN-T network. The method adopted by the authors consisted of several steps. The first was the potential existing sites along the road network where EVCSs could be built. Then, the operators of Rest and Service Areas (RSAs) were asked to indicate the locations they assessed as the most needed. In the next step, data collected in a study commissioned by the European Automobile Manufacturers Association (ACEA) and conducted by Fraunhofer ISI [40] were used. Data on the GPS coordinates of around 400,000 trucks in Europe allowed the identification of regular truck stops as optimal locations for EVCSs. In addition, traffic volumes on routes within the TEN-T network in Poland were taken into account. Through consultations with electricity network operators, the necessary investment costs in the distribution networks were estimated to be able to connect the required connection capacity. The time required to realize the above infrastructure investments was also assessed for individual locations.

2.8. Type of Chargers

A study conducted in Switzerland [41] also indicated the significant role of public policies in shaping and developing electromobility. The authors divided chargers into the following types: private (home), public in residential areas, public slow, and public fast. They analyzed the development of the car fleet structure until 2050 depending on the development and changes in the availability of the different types of EVCSs. The authors indicated the important role of private and public chargers in residential areas in changing the structure of the car fleet to one that is less carbon-intensive. The authors of the model developed for Brussels [42] emphasized a similar problem. In the first step, they estimated the demand for charging electric vehicles based on aggregated mobile data. Then, they analyzed what the segmentation of demand for charging EVs for different charging technologies should look like. Based on OSM data, the classification of urban development types, and POIs, they presented a model for optimizing the occupancy rate of stations according to the type of charger, i.e., normal, semi-fast, or fast-charging, thus maximizing their economic efficiencies.

2.9. Summary of the Literature Review

The choice of optimal locations for EVCSs is a key issue in the development of EV infrastructure. Deployment decisions affect the efficiency of the charging system and power network, the convenience for users, the general acceptance of electric vehicles, and, in the future, the stability of the energy system. There are several factors considered when selecting the locations for EVCSs, which are discussed in more detail below. A summary of the decision criteria found in the literature review is shown in Table 1.

Table 1. Decision criteria used in selected models.

No.	Category	Criteria Applied	Literature
1.	Economical	The value of property	[21,22]
2.		Operation and maintenance costs of the station	[24,31,33,35]
3.		Rate of consumption	[31]
4.		Construction costs	[18,31–33,35,37,38]
5.		Land and equipment costs	[31]
6.		Payback period	[35]
7.		Cost of charging station user's time	[26,35]
8.	Environmental	Noise generation	[31]
9.		Air pollution	[31,33]
10.		Petrol and transport stations	[16,31,38]
11.		Maximize the use of energy from photovoltaic panels	[32,34]
12.	Spatial	Number of public facilities	[15,31]
13.		Number and type of POIs	[11,13,16,17,19,22,25,28,32,35,42,43]
14.		Residential apartments and houses	[15]
15.		Type of urban development areas	[17,19,21,42]
16.	Social	Adverse impact of electromagnetic fields	[31]
17.		Population/population density	[11,14,15,21,25,31,43]
18.		Number of jobs	[14]
19.		Factors influencing the purchase of an electric car	[20,43]
20.		Number of cars owned in high-income areas	[14]
21.		Number of households not owning a car	[14]
22.		Number of students	[14]
23.		The use of electric cars	[21,23,28,43]
24.	Technical	EVCS (number/capacity)	[15–18,20,24,31]
25.		Capacity and use of the energy network	[17,23,27,28,31,32,34]
26.		Lamp posts	[15]
27.		Installation permits and spatial coordination	[31,38]
28.	Political	Legal framework for the implementation of tenders	[31]
29.		Incentive strategies and subsidies for EVCS	[31]
30.		Political strategies and legal requirements	[38]
31.	Traffic	Number of roads (road intersections), density of road network	[15,16,19,21,22,28,31,43]
32.		Traffic density on the road	[21,22,31,32,42]
33.		Walkability level	[14]
34.		Number of trips	[14,15,18]
35.		Parking areas	[11,15,20,31,38]
36.		Difficulties (time) in reaching the EVCS	[25,28,35]

Source: own elaboration.

Taking into account the above literature review, a great variety of approaches can be observed, indicating the considerable complexity of the problem of locating charging stations for electric vehicles in urban areas. However, among the analyzed publications, we did not find methods that approach the problem from the perspective of local authorities or consider the uniform distribution of charging stations, concerning the criteria related to the locations of residential buildings, access to public transport, and public facilities. Therefore, the contribution of our research is as follows:

- Our method takes into account the perspective of local authorities, striving to ensure the availability of electric vehicle charging infrastructure for residents and guests using public facilities;
- Our method considers the locations of public transport infrastructure, which may encourage owners of electric vehicles to use the services of this system in their urban travel, thus contributing to the sustainable development of transport in cities;
- Our method is based on GIS tools and a publicly available spatial database, which significantly reduces the costs of its use;
- Our method can be used by local authorities as a decision support tool in selecting the best locations for electric vehicle charging stations, taking into account existing facilities of this type.

3. Materials and Methods

3.1. General Method Assumptions

The study area was divided into identical basic fields. There are many ways to divide an area into regular basic fields [43]. In this approach, the hexagon shape was chosen. For each hexagon, the neighborhoods of the 1st and 2nd levels were determined according to the principle presented in Figure 2. This shape of the basic fields was selected because of the several positive features of hexagons. Hexagonal fields are more similar to circles than square and triangular fields. Therefore, each point inside a hexagon is located closer to the center than in the case of the basic field of another shape. Moreover, the use of a hexagonal grid allows for determining a set of neighborhood fields in a quite precise way, and the center of each neighboring hexagon is within the same distance.

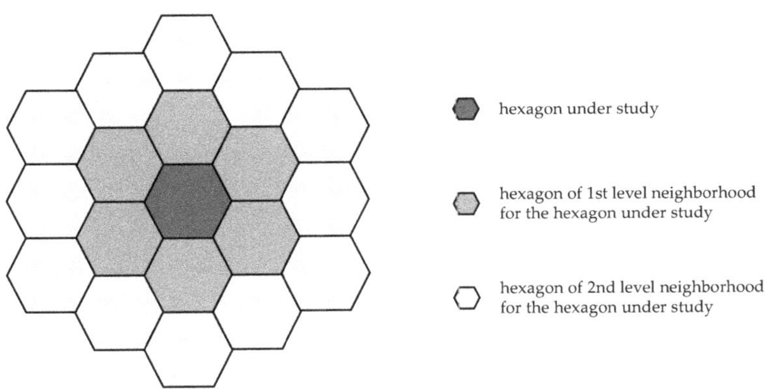

Figure 2. The hexagon under study and hexagons of its 1st and 2nd neighborhood levels.

The purpose of the method is to support decisions on the selection of new locations for EVCSs, taking into account the characteristics of the spatial development of the area under study. Publicly available data collected from the BDOT10k database were used. This object-oriented database contains the spatial location of topographic objects along with their characteristics. The content and details of the BDOT10k database correspond to a topographic map at a scale of 1:10,000. The method used GIS tools, which are increasingly used in transport decision-making problems, for example, for forecasting transport demand [44], assessing the accessibility of public transport [45–47], assessing the spatial integration of bike-sharing stations [48], and even for conducting analyses of innovative solutions introduced in the future [49] and risk analyses [50].

Table 2 provides information on the layers used for the analysis.

Table 2. Information on the layers used for the analysis.

Layer	Type	Source
Map background	Raster	openstreetmap.org
Existing charging stations in Gliwice	Vector	https://www.plugshare.com/pl (accessed on 18 June 2024)
Hexagonal grid	Vector	Own research
Built-environment objects	Vector	Polish National Databases of Spatial Data (BDOT10k)
Public transport infrastructure	Vector	Polish National Databases of Spatial Data (BDOT10k)

Source: own research.

The assumptions adopted in the proposed approach are presented in three main groups as follows:

- Assumptions regarding the division of space, including the following:
  - The area is divided into basic hexagonal fields;
  - Only the basic fields that are entirely within the administrative boundaries of the area under study are considered in the analysis;
  - The first level of the neighborhood is considered when determining the potential of the hexagon for new locations for EVCSs.
- Assumptions regarding available data, including the following:
  - The potential of each hexagon for the new location for EVCS is determined based on selected data available in the BDOT10k database;
  - When determining new locations for charging stations, the locations of existing stations are taken into account, which requires information on these locations;
  - The number of new locations for EVCSs is known, and it is one type of input data;
  - The method does not take into account the number of charging sites in individual locations.
- Assumptions regarding the criteria for selecting recommended locations for EVCSs, including the following:
  - The locations for EVCSs should be evenly distributed across the studied area;
  - The potential of a single hexagon for location for EVCS includes three types of development as follows: residential area (taking into account the number of floors), connection to public transport (e.g., transfer centers, bus stations, bus stops, and railway stations), and public facilities (e.g., hospitals, sports facilities, museums, cinemas, offices, post offices, schools, etc.).

The recommended locations for EVCSs are determined based on the potentials of the basic fields for new locations for EVCSs. As a consequence of the assumptions adopted, the locations for EVCSs are determined with hexagonal accuracy, which means that the method indicates a set of hexagons in which new locations for EVCSs are recommended. In our method, we assume that ensuring equal access to the electric vehicle charging infrastructure will contribute to the greatest possible coverage of this area, with EVCSs and increase the accessibility of such facilities, which may have a positive impact on the decision-making process regarding the ownership of EVs by residents of a given area and people using its facilities [51].

The method is intended to act as a tool that supports local authorities in making decisions on the locations for EVCSs. Therefore, special emphasis is placed on ensuring the accessibility of this type of infrastructure for residents while maintaining the principles of sustainable development, including promoting the use of public transport services when traveling.

3.2. Description of the Method

The general scheme of the method is shown in Figure 3. The method consists of five stages. The final stage is iterative.

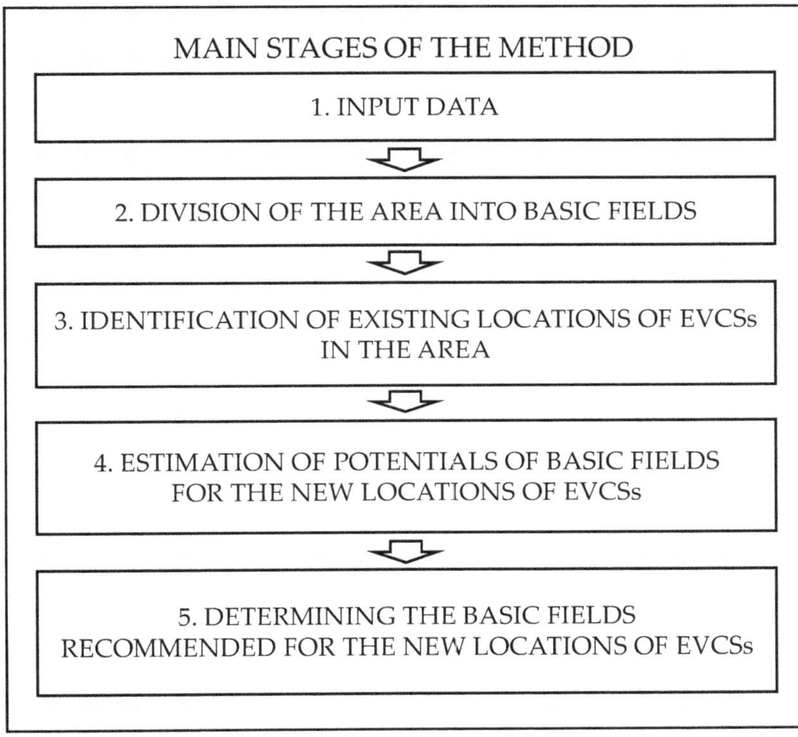

Figure 3. The main stages of the method.

This section only discusses the general operation principle of the method. Appendix A contains the notation and a detailed flowchart of the algorithm as well as a description of the procedures considered in the proposed method in each of the stages.

In the stage when input data are entered (stage 1), it is required to determine the spatial boundaries of the area, select the method of dividing the space (i.e., the shape and size of the basic fields), and specify the number of new locations for EVCSs. The size of the basic field depends primarily on the availability of spatial data that enable the identification of development facilities in a single field and acceptable access (mainly walking time) to an EVCS by potential users of these facilities. Hexagonal shapes of the basic fields were adopted for the analysis.

In the second stage, the space of the analyzed area is divided into basic fields based on the assumptions adopted. As a result of this process, a set of basic hexagonal fields is created along with sets of hexagons on the first and second levels of the neighborhood according to the principle presented in Figure 2.

Stage 3 of the method consists of identifying the locations of existing EVCSs and assigning them appropriate hexagons based on the previously performed space division.

When determining the potential of a single hexagon for the new location for an EVCS (stage 4), the characteristics of various types of objects are considered. They are classified into three main groups as follows:

- Group 1—objects related to residential buildings;

- Group 2—objects related to public transport;
- Group 3—objects related to public utility facilities.

Then, the range of variability in each parameter is divided into classes, which allows for assigning each hexagon to the appropriate class. Each hexagon under study is rated according to the Formulas (A21)–(A23) (see details in Appendix A). On this basis, the potential of each basic field is first determined without taking into account the potential of its neighborhood (i.e., $\boldsymbol{P}_0(h)$), and then, this value is corrected (according to Formula (A26)) by adding the potentials of the hexagons of the first level of the neighborhood, which gives the value of the potential $\boldsymbol{P}(h)$ for each basic hexagonal field under study.

The process of determining the basic fields recommended for new locations for EVCSs (stage 5) is iterative and includes a number of steps corresponding to the number of new locations to recommend. In each step, the hexagon with the greatest potential is identified and recommended for the new location for EVCS.

4. Results

An example of the application of the proposed method was carried out for the city of Gliwice. All five stages of the methods were included. Gliwice is one of the biggest cities of Metropolis GZM (GZM)—a large metropolitan area in southern Poland. The city is inhabited by about 170,000 residents and constitutes the center of the western part of the Metropolis.

In the first stage of the method, input data and the boundaries of the area under study were established. The spatial boundaries of the area, as well as the location of Gliwice in the background of GZM, are presented in Figure 4.

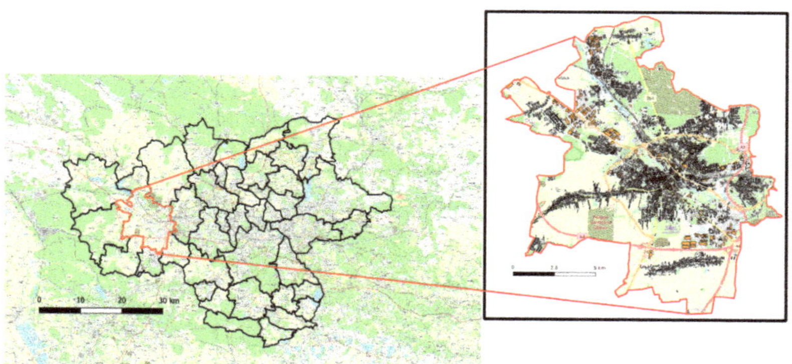

Figure 4. The boundaries of the area of analysis.

The city of Gliwice is also characterized by a large area. In addition to the densely built-up central part, there are also suburban districts with single-family housing. Numerous important facilities are located in Gliwice, such as cultural and recreational objects and the biggest university of technology in the Silesian region.

The next step in the first stage of the analysis was to establish the number N of new locations for EVCSs that should be determined. The number N was assumed to be equal to five.

The second stage of the analysis was to divide the area under study into basic fields. Hexagonal basic fields were selected. The distance between the parallel sides of each hexagon was 300 m. This value was chosen based on the assumption that EV users leave their vehicles at charging stations and then walk to their destinations. Therefore, the size of the hexagon accounted for the distance that people are willing to walk. The hexagonal grid was overlayed on the Gliwice area, and the resulting division is presented in Figure 5.

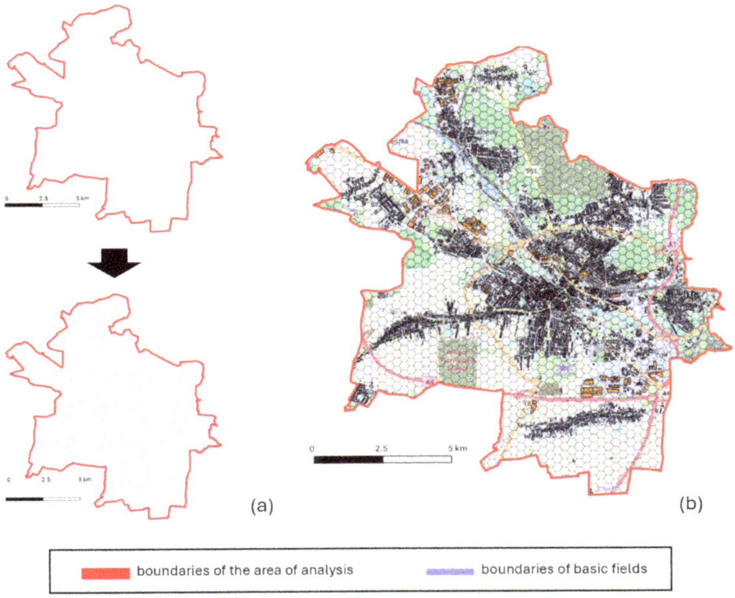

Figure 5. The spatial division of the area: (**a**) the division of the area into hexagonal basic fields, (**b**) the hexagonal grid on the background of land use in Gliwice.

Subsequently, set $\boldsymbol{H}$, which contains the numbers of hexagons in the area under study, and sets $\boldsymbol{H}^1(h)$ and $\boldsymbol{H}^2(h)$ for each hexagon $h \in \boldsymbol{H}$ were created. The area of Gliwice was divided into 1890 hexagonal basic fields.

The third stage was the identification of existing locations for EVCSs in the area. These locations are presented in Figure 6.

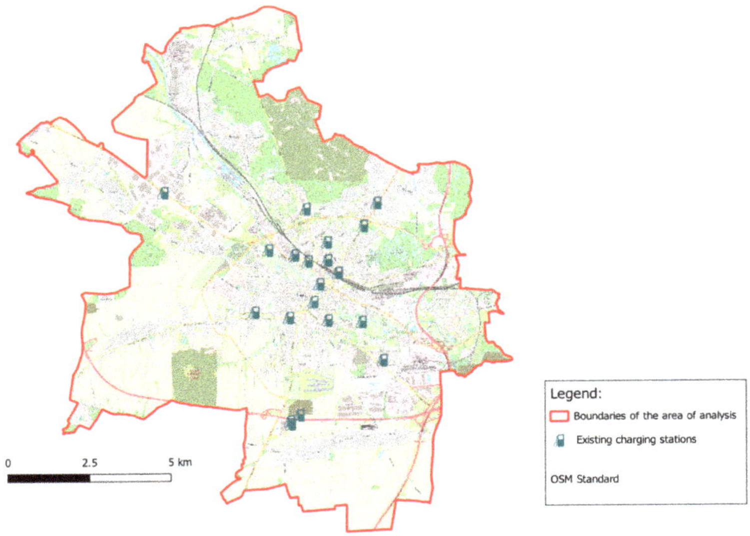

Figure 6. Identification of the existing locations of EVCSs in the area.

In total, 20 existing EVCSs were identified in Gliwice, according to [52]. They are located mostly in the central part of the city, and a few operate in the suburban areas, mainly in the vicinity of large shopping centers. The existing EVCSs differ in terms of the types of chargers in use; however, the selection of the types of these devices is outside the scope of this study. Therefore, detailed features of the chargers were not analyzed. Subsequently, a set of numbers of hexagons in which the existing EVCSs are located $\boldsymbol{H}_0$ as well as sets $\boldsymbol{H}_0^1$ and $\boldsymbol{H}_0^2$, containing numbers of hexagons in the first and second levels of the neighborhood for all hexagons in which the existing EVCSs are located were obtained. The basic fields from sets $\boldsymbol{H}_0$, $\boldsymbol{H}_0^1$, and $\boldsymbol{H}_0^2$ are presented in Figure 7.

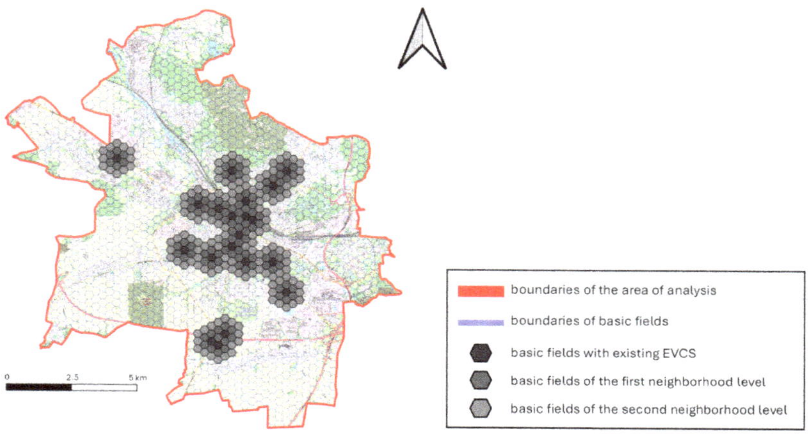

Figure 7. Basic fields with existing EVCSs and basic fields of the 1st and 2nd levels of the neighborhood.

According to the assumptions of the proposed method, all basic fields from sets $\boldsymbol{H}_0$, $\boldsymbol{H}_0^1$, and $\boldsymbol{H}_0^2$ were excluded from the consideration as fields where new EVCS could be located.

The fourth stage of the method was to estimate the potential of the basic fields for the new locations for EVCSs. This potential was estimated for all hexagons except for the basic fields from sets $\boldsymbol{H}_0$ and $\boldsymbol{H}_0^1$.

The potential of each hexagon was determined based on the characteristics of the selected types of built-environment objects. The specific elements of the sets of objects for each group are presented in Table 3.

In the case of group $S1$, i.e., objects related to residential buildings, the total area of these objects was taken into account to calculate the potential for each field. However, in the case of group $S2$, i.e., objects related to public transport, and group $S3$, i.e., objects related to public utility facilities, only the presence of these types of objects was considered (binary variable taking the value 0 or 1).

After performing the necessary calculations, the potential for all basic fields of set $\boldsymbol{HP}$ was obtained. In total, set $\boldsymbol{HP}$ included the numbers of 1625 basic fields.

In the case of 565 basic fields, there were no built-environment objects from groups $S1$–$S3$; therefore, the potential for the location of the new ECVS in these fields was equal to 0. The number of basic fields in the selected ranges of values of potential is presented in Table 4. The maximum value of the potential $\boldsymbol{P}(h)$ was 2.93, and it was the only basic field with a value of potential higher than 2.5.

Table 3. Elements of the sets considered in the analysis.

Group	Number of a Type of Objects	Description
S1—objects related to residential buildings	s1 = 1	single-family housing
	s1 = 2	double-family housing
	s1 = 3	multi-family housing
S2—objects related to public transport	s2 = 1	railway stations or stops
	s2 = 2	bus stations
	s2 = 3	interchange centers
	s2 = 4	public transport stops
S3—objects related to public utility facilities.	s3 = 1	kindergartens
	s3 = 2	elementary schools
	s3 = 3	high schools
	s3 = 4	universities
	s3 = 5	hospitals
	s3 = 6	cinemas
	s3 = 7	theatres
	s3 = 8	museums
	s3 = 9	zoological gardens
	s3 = 10	post offices
	s3 = 11	sport or recreational facilities
	s3 = 12	city halls
	s3 = 13	local government buildings

Source: own research.

Table 4. Distribution of basic fields in selected value ranges of the potential for a location of the new ECVS.

Range of Values of Potential $P(h)$ [-]	Number of Basic Fields
0	565
<0–0.5)	695
<0.5–1.0)	247
<1.0–1.5)	75
<1.5–2.0)	33
<2.0–2.5)	9
<2.5–3.0)	1

Source: own research.

In the fifth stage of the proposed method, the basic fields recommended for new locations for EVCSs were determined using an iterative procedure. The number of steps in the procedure was equal to N = five. The results of the procedure and the selection of the basic fields recommended for the locations of the new EVCSs in Gliwice are presented in Figure 8.

In each step of the procedure, one recommended basic field was selected, based on the highest value of potential $\boldsymbol{P}(h)$, and sets of basic fields of the first and second neighborhood levels for these basic fields were created. All these basic fields were then excluded from consideration in the following steps of the iterative procedure.

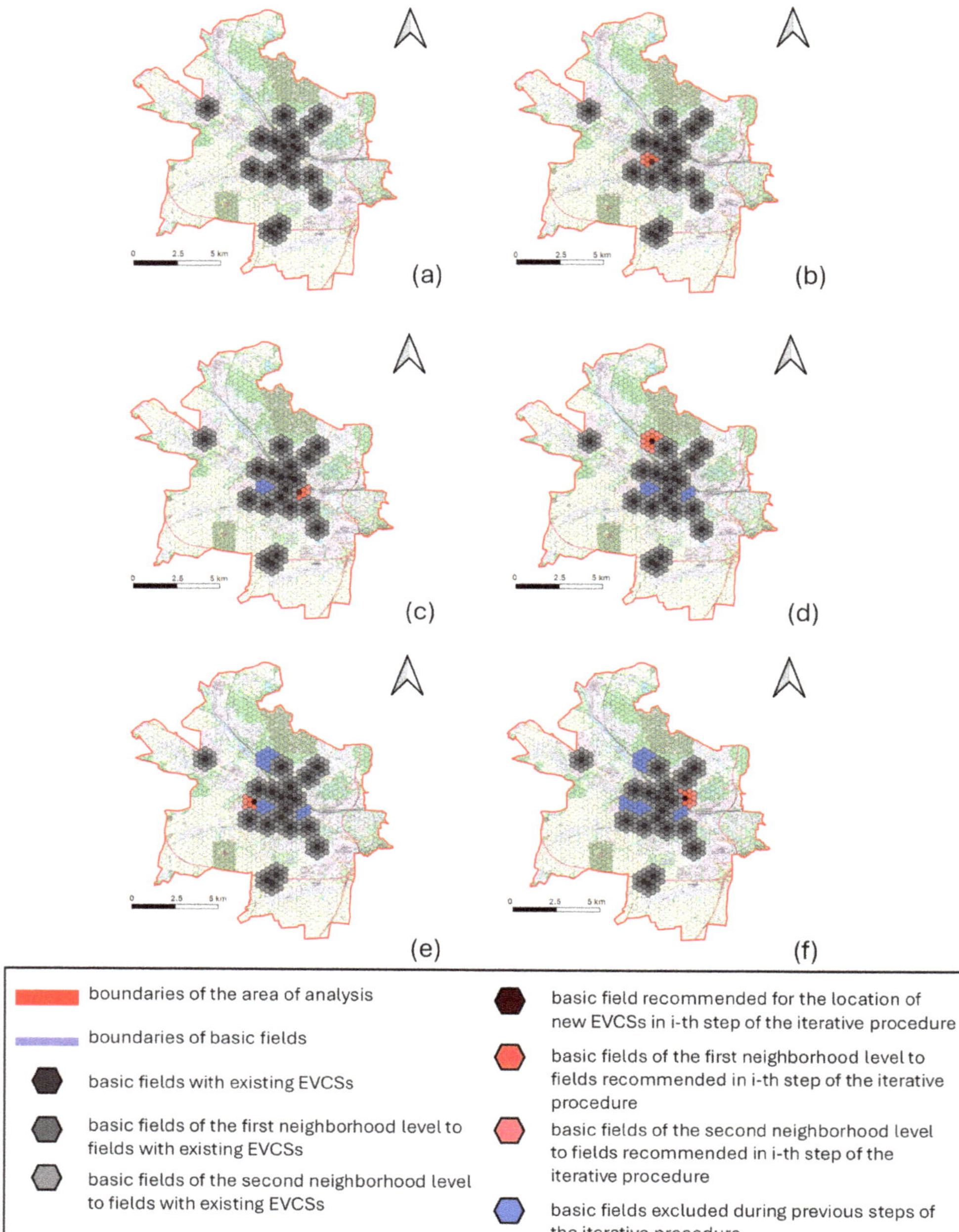

Figure 8. Results of the iterative procedure for selecting basic fields recommended for the locations of new EVCSs: (**a**) state before the application of the iterative procedure, (**b**) 1st step of the iterative procedure, (**c**) 2nd step of the iterative procedure, (**d**) 3rd step of the iterative procedure, (**e**) 4th step of the iterative procedure, and (**f**) 5th step of the iterative procedure.

5. Discussion of the Results

An overview of the recommended fields for the locations of new EVCSs in the area under study is presented in Figure 9.

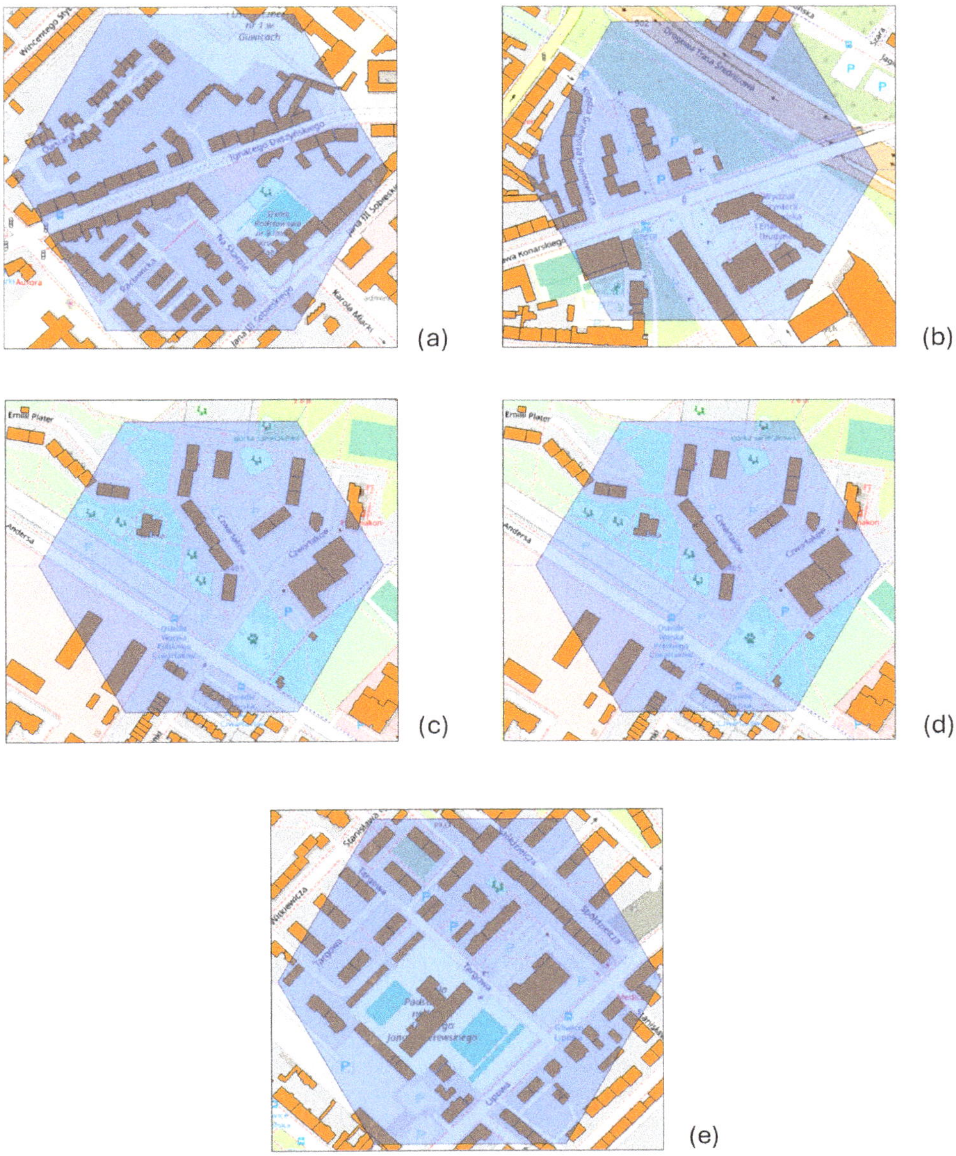

Figure 9. Overview of recommended basic fields: (**a**) basic field chosen during the 1st step of the iterative procedure, (**b**) basic field chosen during the 2nd step of the iterative procedure, (**c**) basic field chosen during the 3rd step of the iterative procedure, (**d**) basic field chosen during the 4th step of the iterative procedure, and (**e**) basic field chosen during the 5th step of the iterative procedure.

The hexagonal basic field chosen in the first step of the iterative procedure is located in the central part of the city. It is characterized by dense development with the following functions: residential buildings, educational buildings, and public facilities. The basic field selected in the second step of the iterative procedure contains the buildings of the Silesian University of Technology, the largest technical university in the Silesian region. The basic fields chosen in the third and fourth steps of the procedure are characterized by large areas of residential buildings, mostly multi-floor houses. Finally, the basic field selected in the fifth step of the procedure is located in a densely populated area, and there are educational buildings located within its boundaries. The detailed results for these final locations for new EVCSs are provided in Table 5.

Table 5. Main characteristics of the recommended locations for EVCSs.

Characteristic	Step Number of the Iterative Procedure				
	First	Second	Third	Fourth	Fifth
$h_i^{\max}$	2197	2596	2187	2024	2707
$rate\_f1(h_i^{\max})$	0.6	0.4	1.0	0.6	0.6
$rate\_f2(h_i^{\max})$	0.3	0.0	0.0	0.3	0.3
$rate\_f3(h_i^{\max})$	0.8	0.4	0.2	0.0	0.4
$P_0(h_i^{\max})$	0.57	0.27	0.40	0.30	0.43
$P(h_i^{\max})$	2.93	2.07	2.07	2.00	2.00

Source: own research.

The values of potential $P(h_i^{\max})$ for the recommended basic fields range from 2.0 to 2.93. These fields are located mainly in the outer parts of the central district of Gliwice. In all the indicated locations, there is a relatively high level of residential development. The highest density of residents is in the location selected in the third step of iteration (the $rate\_f1(h_i^{\max})$ value is equal to 1.0). It is also worth noting that not all selected locations have public transport infrastructure (the $rate\_f2(h_i^{\max})$ value is equal to 0.0 for the locations selected in the second and third step of the iterative procedure). Similarly, based on the available data, no public utility facilities are found in the hexagon selected in the fourth iteration (the $rate\_f3(h_i^{\max})$ value is equal to 0.0).

The results presented allowed us to confirm the hypothesis that the potential of the location for a new and publicly accessible EVCS can be assessed based on the features of the built environment. The potential was calculated for each basic field in the area under study and—on this basis—conclusions regarding the proposed locations for the EVCSs were formulated.

The performed analysis also allowed us to answer the first research question and to create a set of built-environment object types that should be considered in the process of estimating potential. That set contains objects from the following three main groups: objects related to residential buildings, objects related to public transport, and objects related to public utility facilities.

The introduction of two neighborhood levels of basic fields allowed us to determine new locations for EVCSs that are evenly distributed in space in such a way that ensured their equal accessibility for users and answered the second research question. The presented approach also differs from other methods, as it leads to the exclusion of basic fields that are characterized by high potential but do not allow for a uniform distribution of EVCSs in the studied area. Nevertheless, it is important to point out that such a basic field can still be included when all stations are equally distributed to increase the density of EVCSs in a given area and provide even better conditions for users.

Determining appropriate locations for EV charging stations is particularly difficult because of the multifaceted nature of this issue, as complex legal, financial, technological, and organizational aspects must be considered [38]. This article presents a five-step approach to determining the locations for electric vehicle charging stations in an urban area, which takes into account the following three factors related to spatial development: the presence of residential buildings, connection to the public transport system, and access to selected public facilities. These factors are related to the directions of sustainable transport development [53–61]. The proposed method may assist local authorities in making decisions on new locations for charging stations in the case of limited financial resources.

The developed model also has some limitations in its application. The accuracy of the location is determined by the basic field. Therefore, it depends on the size of this field. In addition, the proposed algorithm gives very high priority to the even distribution of charging stations, thus ensuring equal access to the charging infrastructure for all residents and users of EVs. At the same time, it does not verify the technical possibilities of building infrastructure, obtaining appropriate connection capacities, or administrative permits in a given area. Therefore, further analysis in the context of the practical application of the method is necessary, and these issues should be included in the constraints of the model.

Moreover, in a situation where the number of new locations for EVCSs to be determined is greater than the maximum number of possible (and justified) hexagons to be used, we can assume that the condition of s uniform distribution is met. It is then worth including the hexagons located on the second level of the neighborhood in the analysis. In the next step, hexagons on the first neighborhood level can also be included, gradually densifying the network of EVCSs. It is also necessary to define the constraints on the reasonable designation of new proposed locations for EVCSs. One idea could be to determine the required minimum level of potential for a given hexagon, below which new locations will no longer be proposed (e.g., hexagons located in the middle of green areas).

There are also potential limitations when using the method in countries other than Poland. The proposed method requires detailed data on the built environment. These data are easily accessible for Polish cities with the use of national databases. In different countries, spatial data could be inconsistent with Polish standards. Therefore, it is necessary to conduct research to facilitate the use of the method in different countries.

Previous studies also included dividing the area into basic fields to evaluate their potential for locating new EVCSs. In [25], the authors assessed the potential of hexagonal basic fields; however, they used different variables related to the population in each hexagon and traffic characteristics. They also considered the influence of existing EVCSs in the area. The case study presented in this article was prepared for a different city and a different country, but we investigated how the new EVCSs were spatially distributed as one of the results of our analysis. The results obtained by the authors of the other study suggested that new EVCSs should be equally distributed, also in the basic fields close to the edges of the area under study. Our results, on the other hand, suggested locating new EVCSs closer to the city center. However, it is important to keep in mind that this could be caused by differences in the characteristics of the built environment in each city.

In [62], the authors also used hexagonal-shaped basic fields. However, their approach was quite different because the authors divided the analysis into two stages as follows: macroscopic, to assess potentials in larger administrative units, and microscopic, to choose basic fields in which EVCSs should be located. The potential was calculated based on the number of registered electric vehicles, average income, and selected features associated with the built environment.

The method presented in this study is universal, which means that it can also be applied in other geographical locations and urban areas with different residential developments. In the case of other public utility facilities, set S3 containing numbers of these types of objects can be modified to take into account the specificity of the given area. The resolution of the hexagonal grid can also be adjusted for the access to EVCSs acceptable to their users, which is mainly related to their travel behavior.

6. Conclusions

The application of the proposed method allowed us to determine a set of five basic fields in Gliwice, which are characterized by the highest values of potential for new locations for EVCSs. These basic fields contain built-environment objects that have a positive impact on the potential, and they also fulfill the requirement of the uniform spatial distribution of basic fields to ensure equal access to EVCSs.

An important result of the performed analysis was answering two research questions. The first research question pertained to the factors associated with the features of the built environment, which can effectively represent the actual potential of the new locations for EVCSs. During the analysis, it was determined that built-environment objects of three main groups should be considered in the method, i.e., objects related to residential buildings, objects related to public transport, and objects related to public utility facilities. The first group contains single-family housing, double-family housing, and multi-family housing. In the case of the second group, these objects include railway stations or stops, bus stations, interchange centers, and public transport stops. The last group contains the widest set of objects that concern generally understood public utility facilities such as kindergartens, primary schools, secondary schools, and universities, as well as hospitals, cinemas, theatres, museums, zoos, post offices, sports and recreational facilities, city halls, and local government buildings. The second research question referred to the spatial distribution of new EVCSs to provide equitable access to them. During the analysis, two levels of the neighborhood of basic fields were created, which allowed us to fulfill that requirement.

The proposed model requires further work and development. The next steps need to include the option of densifying EVCSs in the basic fields on the first and second levels of the neighborhood, as well as in the fields with existing charging stations. The extended model could also be used to determine the specific number of chargers within each station. The use of GIS data in this model also provides the possibility of a preliminary assessment of the conditions on the side of the power network.

Another direction of future studies includes the development of dedicated software that facilitates the calculations and the entire process of applying the proposed method. Such software could be especially useful in the case of large areas and when a large number of charging stations are to be located.

It is also necessary to consider different methods of dividing the area under study in future research. Although the hexagonal grid offers significant positive features, it could also be beneficial to consider, for example, the Voronoi diagram.

Future studies should also be conducted on the behavior and trends among users of EVCSs. In particular, this would involve taking into account data such as charging frequencies, charging times, and length of charging sessions, which could be specified for different types of facilities or types of urban areas. The research could show when and how often users charge their vehicles as well as which charging patterns they prefer—charging at night, during working hours, or while shopping. In addition, the length of the charging session could illustrate whether users prefer fast charging to cover a given distance or whether full charging is more popular. Analyzing such aspects could improve the efficient planning and optimization of new locations for EVCSs in future studies.

Author Contributions: Conceptualization, P.S., M.L., P.F. and R.Ż.; methodology, P.S., M.L., P.F. and R.Ż.; software, P.S.; validation, P.S. and R.Ż.; investigation, P.S., M.L., P.F. and R.Ż.; resources, P.S., M.L. and P.F.; data curation, P.S.; writing—original draft preparation, P.S., M.L., P.F. and R.Ż.; writing—review and editing, P.S., M.L., P.F. and R.Ż.; visualization, P.S., M.L., P.F. and R.Ż.; supervision, R.Ż.; project administration, P.S., M.L., P.F. and R.Ż.; funding acquisition, M.L. All authors have read and agreed to the published version of the manuscript.

Funding: This article is the result of work carried out under the Dean's grant entitled: "Decision model for route planning in road transport of oversize loads, taking into account risk minimization aspects" awarded in November 2023. The grant was financed by the Dean of the Faculty of Transport of the Warsaw University of Technology. Number of the awarded grant: 1/WT/2023.

Data Availability Statement: The raw data supporting the conclusions of this article will be made available by the authors on request.

Conflicts of Interest: The authors declare no conflicts of interest.

Appendix A

Notation:

$cl\_f1(h), cl\_f2(h), cl\_f3(h)$	—Class of the h-th hexagon concerning the characteristics, respectively, $f1(h)$, $f2(h)$, and $f3(h)$;
F1	—Matrix containing the values of the function $f1_{s1}(h)$ with the interpretation related to the residential buildings of the $s1$-type in the h-th hexagon;
$f1(h)$	—The aggregated value of characteristics $f1_{s1}(h)$ for all types of objects in the **S1** set for the h-th hexagon;
F2	—Matrix containing the values of the function $f2_{s2}(h)$ with the interpretation related to the presence of facilities of public transport of the $s2$-type in the h-th hexagon;
$f2(h)$	—The aggregated value of characteristics $f2_{s2}(h)$ for all types of objects in the **S2** set for the h-th hexagon;
F3	—Matrix containing the values of the function $f3_{s3}(h)$ with the interpretation related to the presence of public utility facilities of the $s3$-type in the h-th hexagon;
$f3(h)$	—The aggregated value of characteristics $f3_{s3}(h)$ for all types of objects in the **S3** set for the h-th hexagon;
$\boldsymbol{H}$	—Set of hexagon numbers in the area under study;
$\boldsymbol{H}^1(h)$	—Set of hexagon numbers of the first neighborhood level for the h-th hexagon;
$\boldsymbol{H}^2(h)$	—Set of hexagon numbers of the second neighborhood level for the h-th hexagon;
$\boldsymbol{H}_0$	—Set of hexagon numbers in which the existing EVCSs are located;
$\boldsymbol{H}_0^1$	—Set of hexagon numbers of the first neighborhood level for all hexagons in which the existing EVCSs are located;
$\boldsymbol{H}_0^2$	—Set of hexagon numbers of the second neighborhood level for all hexagons in which the existing EVCSs are located;
$h_i^{\max}$	—Number of the hexagon with the highest value of potential for the new location of an EVCS in the i-th iteration;
$\boldsymbol{HL}_i$	—Set containing the numbers of hexagons considered for the new location of an EVCS in the i-th iteration;
$\boldsymbol{HP}$	—Set of hexagon numbers for which the potential for the new location of an EVCS is determined;
$\boldsymbol{HR}_i$	—Set of hexagon numbers recommended for the new location of an EVCS in the i-th iteration;
i	—Iteration number;
N	—Number of new locations of EVCSs to be determined;
num_cl	—The number of classes into which the ranges of variability in spatial development characteristics are divided;
$\boldsymbol{P}_0(h)$	—The potential of a single h-th hexagon for the new location of an EVCS without taking into account the potential of neighboring hexagons;
$\boldsymbol{P}(h)$	—The potential of a single h-th hexagon for the new location of an EVCS taking into account the potential of hexagons located at the first level of the neighborhood;
$range\_f1, range\_f2, range\_f3$	—Ranges of the characteristics included in the matrices, respectively, **F1**, **F2**, and **F3**;
$rate\_f1(h)$, $rate\_f2(h)$, $rate\_f3(h)$	—Estimation of the potential of the h-th hexagon for the new location of an EVCS in terms of characteristics, respectively, $f1(h)$, $f2(h)$, and $f3(h)$;
$S1$	—Set containing the number of object types related to residential buildings taken into account when determining the potential of the hexagon for the new location of an EVCS;
$S2$	—Set containing the number of object types related to public transport taken into account when determining the potential of the hexagon for the new location of an EVCS;
$S3$	—Set containing the number of object types related to public utility facilities taken into account when determining the potential of the hexagon for the new location of an EVCS.

The algorithm of the proposed method is presented in Figure A1.

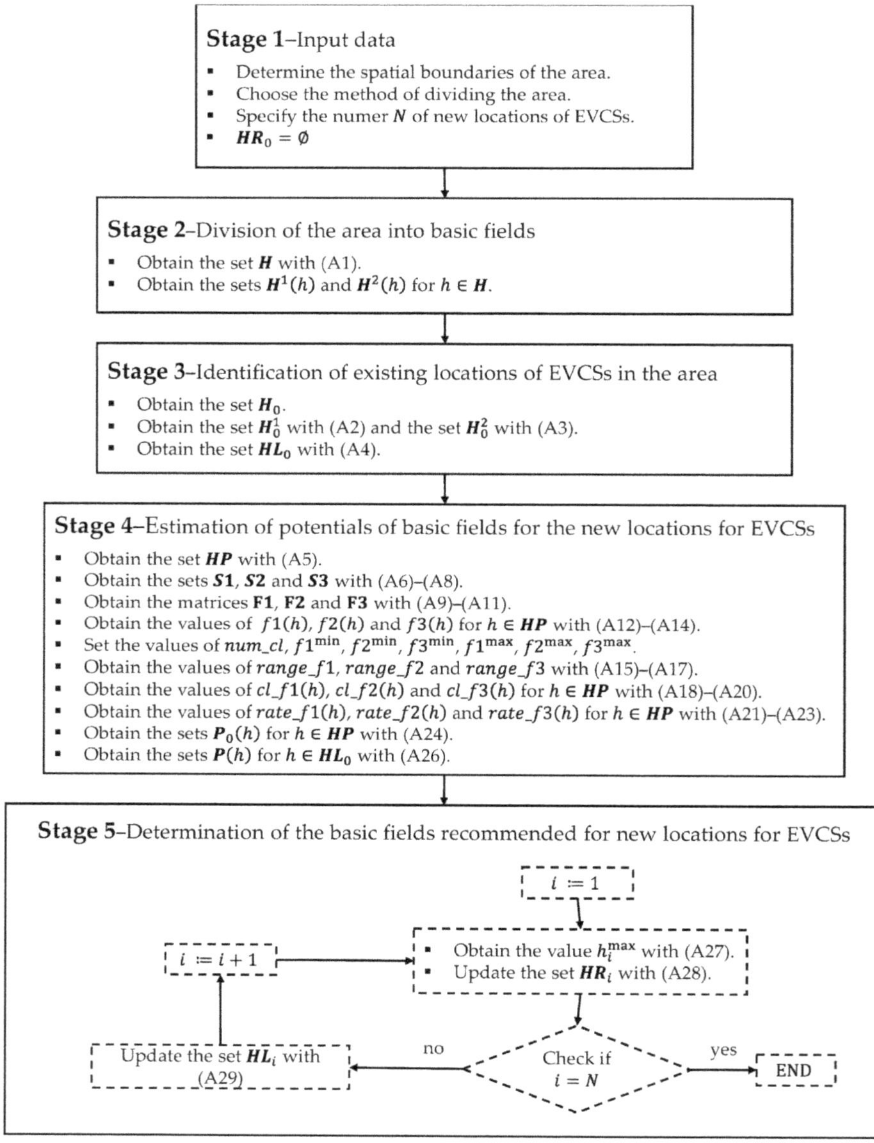

Figure A1. The flowchart of the proposed method.

STAGE 1—INPUT DATA

Among the input data, the most important are the following:

- The number of new locations for EVCSs to be established (N);
- The spatial scope of the area that requires precise delineation of its boundaries.

Moreover, in this stage, the size and shape of the basic fields should also be determined.

In this stage, the set of numbers of basic fields recommended for new locations of EVCSs is empty, i.e., $HR_0 = \emptyset$.

STAGE 2—DIVISION OF THE AREA INTO BASIC FIELDS

By dividing the space of the area under study into basic fields, we obtain a set of hexagons with specific numbers, which is written as follows:

$$\boldsymbol{H} = \{1, 2, \ldots, h, \ldots, \overline{H}\} \tag{A1}$$

where $\overline{H}$ denotes the size of the set $\boldsymbol{H}$, i.e., the number of all hexagons in the area.

Then, for each hexagon $h \in \boldsymbol{H}$ the sets of hexagon numbers of the first and second levels of the neighborhood are determined according to the principle presented in Figure 2. These sets are marked as $\boldsymbol{H}^1(h)$ and $\boldsymbol{H}^2(h)$, respectively.

STAGE 3—IDENTIFICATION OF THE EXISTING LOCATIONS OF EVCS IN THE AREA

In this stage, the locations of the existing EVCSs in the area under study are identified and assigned to the hexagons in which they are located. In this way, the set $\boldsymbol{H}_0 \subset \boldsymbol{H}$, which contains hexagons with the existing locations of EVCSs, is built.

Then, sets $\boldsymbol{H}_0^1$ and $\boldsymbol{H}_0^2$, containing, respectively, the hexagon numbers of the first and second neighborhood levels for all $h \in \boldsymbol{H}_0$, are determined according to the formulas:

$$\boldsymbol{H}_0^1 = \bigcup_{h \in \boldsymbol{H}_0} \boldsymbol{H}^1(h) \tag{A2}$$

$$\boldsymbol{H}_0^2 = \bigcup_{h \in \boldsymbol{H}_0} \boldsymbol{H}^2(h) \tag{A3}$$

Basic fields that are elements of the sets $\boldsymbol{H}_0$, $\boldsymbol{H}_0^1$, and $\boldsymbol{H}_0^2$ are not considered for new locations of EVCSs. Therefore, the initial set, $\boldsymbol{HL}_0$, which contains hexagon numbers in which new locations for EVCSs are considered, is defined as follows:

$$\boldsymbol{HL}_0 = \left\{h : \quad h \in \boldsymbol{H} \wedge h \notin \boldsymbol{H}_0 \wedge h \notin \boldsymbol{H}_0^1 \wedge h \notin \boldsymbol{H}_0^2\right\} \tag{A4}$$

STAGE 4—ESTIMATION OF POTENTIALS OF BASIC FIELDS FOR THE NEW LOCATIONS OF EVCSs

When estimating the potential of basic fields for new locations of EVCSs, hexagons creating the first level of the neighborhood are also taken into account. Therefore, set $\boldsymbol{HP}$, which contains the hexagon numbers for which the characteristics are used to estimate the potential of hexagon $h \in \boldsymbol{HL}_0$, is defined as follows:

$$\boldsymbol{HP} = \left\{h : \quad h \in \boldsymbol{H} \wedge h \notin \boldsymbol{H}_0 \wedge h \notin \boldsymbol{H}_0^1\right\} \tag{A5}$$

The potential of a single hexagon is determined based on the characteristics of various types of objects classified into three main groups as follows:

- Group 1—objects related to residential buildings;
- Group 2—objects related to public transport;
- Group 3—objects related to public utility facilities.

Therefore, three sets containing the numbers of object types taken into account when determining the potential of basic fields for the new locations of EVCSs are defined as follows:

$$S1 = \{1, 2, \ldots, s1, \ldots, \overline{S1}\} \tag{A6}$$

$$S2 = \{1, 2, \ldots, s2, \ldots, \overline{S2}\} \tag{A7}$$

$$S3 = \{1, 2, \ldots, s3, \ldots, \overline{S3}\} \tag{A8}$$

where $\overline{S1}$, $\overline{S2}$, and $\overline{S3}$ denote the number of object types in a given group (i.e., the sizes of the sets $S1$, $S2$, and $S3$), respectively.

For each group of object types, one characteristic is selected as an additive variable, which can be determined for each basic field quantitatively for all types defined within each group. For the group of object types included in set $S1$, the characteristic $f1_{s1}(h)$ with the interpretation of the residential development area, taking into account the number of floors of objects of type $s1$ in the hexagon $h \in \boldsymbol{HP}$, is determined. For set $S2$, the characteristic $f2_{s2}(h)$ related to the presence of public transport objects of type $s2$ in the hexagon $h \in \boldsymbol{HP}$ is chosen, and for the group $S3$, the characteristic $f3_{s3}(h)$ identifying the presence or absence of an object of type $s3$ in the hexagon $h \in \boldsymbol{HP}$ is chosen. The values of these functions are arranged in matrix form as follows:

$$\mathbf{F1} = [f1_{s1}(h): \quad s1 \in S1, h \in \boldsymbol{HP}] \tag{A9}$$

$$\mathbf{F2} = [f2_{s2}(h): \quad s2 \in S2, h \in \boldsymbol{HP}] \tag{A10}$$

$$\mathbf{F3} = [f3_{s3}(h): \quad s3 \in S3, h \in \boldsymbol{HP}] \tag{A11}$$

To determine the potential of the basic field for the new location of an EVCS, the aggregated values of the characteristics defined by Formulas (A9)–(A11) are determined for each hexagon as follows:

$$f1(h) = \sum_{s1=1}^{\overline{S1}} f1_{s1}(h), \quad h \in \boldsymbol{HP} \tag{A12}$$

$$f2(h) = \sum_{s2=1}^{\overline{S2}} f2_{s2}(h), \quad h \in \boldsymbol{HP} \tag{A13}$$

$$f3(h) = \sum_{s3=1}^{\overline{S3}} f3_{s3}(h), \quad h \in \boldsymbol{HP} \tag{A14}$$

The parameter num_cl means the number of classes into which the range of variability in each variable determined by Formulas (A12)–(A14) is divided. This number is the same for all characteristics. The class ranges for individual characteristics are as follows:

$$range\_f1 = \frac{f1^{\max} - f1^{\min}}{num\_cl} \tag{A15}$$

$$range\_f2 = \frac{f2^{\max} - f2^{\min}}{num\_cl} \tag{A16}$$

$$range\_f3 = \frac{f3^{\max} - f3^{\min}}{num\_cl} \tag{A17}$$

where the minimum values $f1^{\min}$, $f2^{\min}$, and $f3^{\min}$ and the maximum values $f1^{\max}$, $f2^{\max}$, and $f3^{\max}$ are determined based on the assumed variability of each characteristic.

Determining the range of classes for individual characteristics allows for the allocation of a hexagon to classes $cl\_f1(h)$, $cl\_f2(h)$ and $cl\_f3(h)$, respectively, according to the following rule:

$$\begin{gathered} \forall \ f1(h) > f1^{\min} \wedge \ f1(h) \leq f1^{\max} \\ cl\_f1(h) = \left\lceil \frac{f1(h) - f1^{\min}}{range\_f1} \right\rceil, \quad h \in \boldsymbol{HP} \end{gathered} \tag{A18}$$

$$\begin{aligned} &\forall \ f2(h) > f2^{\min} \wedge f2(h) \le f2^{\max} \\ &cl\_f2(h) = \left\lceil \frac{f2(h) - f2^{\min}}{range\_f2} \right\rceil, \ h \in \boldsymbol{HP} \end{aligned} \tag{A19}$$

$$\begin{aligned} &\forall \ f3(h) > f3^{\min} \wedge f3(h) \le f3^{\max} \\ &cl\_f3(h) = \left\lceil \frac{f3(h) - f3^{\min}}{range\_f3} \right\rceil, \ h \in \boldsymbol{HP} \end{aligned} \tag{A20}$$

For the values $f1(h)$, $f2(h)$, and $f3(h)$ that do not meet the conditions indicated in Formulas (A18)–(A20), no classification is carried out. Then, each hexagon is rated by $rate\_f1(h)$, $rate_{f2(h)}$, and $rate_{f3(h)}$, respectively, depending on the class assigned for a given characteristic according to the following rule:

$$rate\_f1(h) = \begin{cases} 0, & f1(h) \le f1^{\min} \\ cl\_f1(h)/num\_cl, & f1(h) > f1^{\min} \wedge f1(h) \le f1^{\max}, \\ 1, & f1(h) > f1^{\max} \end{cases} \quad h \in \boldsymbol{HP} \tag{A21}$$

$$rate\_f2(h) = \begin{cases} 0, & f2(h) \le f2^{\min} \\ cl\_f2(h)/num\_cl, & f2(h) > f2^{\min} \wedge f2(h) \le f2^{\max}, \\ 1, & f2(h) > f2^{\max} \end{cases} \quad h \in \boldsymbol{HP} \tag{A22}$$

$$rate\_f3(h) = \begin{cases} 0, & f3(h) \le f3^{\min} \\ cl\_f3(h)/num\_cl, & f3(h) > f3^{\min} \wedge f3(h) \le f3^{\max}, \\ 1, & f3(h) > f3^{\max} \end{cases} \quad h \in \boldsymbol{HP} \tag{A23}$$

The potential of a single basic field $h \in \boldsymbol{HP}$ without taking into account the potential of neighboring fields is marked as $\boldsymbol{P}_0(h)$ and determined according to the formula:

$$\boldsymbol{P}_0(h) = w_1 \cdot rate\_f1(h) + w_2 \cdot rate\_f2(h) + w_3 \cdot rate\_f3(h), \qquad h \in \boldsymbol{HP} \tag{A24}$$

where w_1, w_2, and w_3 denote the weights adopted for individual groups of object types, with:

$$w_1 + w_2 + w_3 = 1 \tag{A25}$$

The potential of a single basic field, taking into account the potential of hexagons located at its first level of the neighborhood, which is the basis for determining the set of hexagon numbers recommended for new locations for EVCSs, is determined as follows:

$$\boldsymbol{P}(h) = \boldsymbol{P}_0(h) + \sum_{h \in \boldsymbol{H}^1(h)} \boldsymbol{P}_0(h), \qquad h \in \boldsymbol{HL}_0 \tag{A26}$$

STAGE 5—DETERMINATION OF THE BASIC FIELDS RECOMMENDED FOR NEW LOCATIONS OF EVCSs

Step 5 contains an iterative procedure with the number of steps equal to N, corresponding to the number of new locations of EVCSs to be determined. In each i-th iteration, the following actions are performed:

- Selecting the hexagon number $h_i^{\max}$ with the highest potential value from the set of hexagon numbers $\boldsymbol{HL}_{i-1}$ according to the rule:

$$h_i^{\max} : \boldsymbol{P}(h_i^{\max}) = \max_{h \in \boldsymbol{HL}_{i-1}} \{\boldsymbol{P}(h)\}, \tag{A27}$$

- Updating the set of $\boldsymbol{HR}_i$ that contains the hexagon numbers recommended for the new locations of EVCSs with the basic field number $h_i^{\max}$, i.e.,:

$$\boldsymbol{HR}_i = \boldsymbol{HR}_{i-1} \cup \{h_i^{\max}\}, \tag{A28}$$

- Updating the set $\boldsymbol{HL}_i$, which contains the numbers of basic fields in which new locations of EVCSs are considered in the next iteration; removing the hexagon number $h_i^{\max}$ along with the hexagon numbers of its first and second levels of the neighborhood from the set $\boldsymbol{HL}_{i-1}$, which can be expressed as follows:

$$\boldsymbol{HL}_i = \Big((\boldsymbol{HL}_{i-1} \setminus \{h_i^{\max}\}) \setminus \boldsymbol{H}^1(h_i^{\max}) \Big) \setminus \boldsymbol{H}^2(h_i^{\max}) \tag{A29}$$

The algorithm is completed after N iterations, giving the output set $\boldsymbol{HR}_N$ containing the hexagon numbers in which new locations for EVCSs are recommended.

References

1. IEA. *Global EV Outlook 2024*; IEA: Paris, France, 2024; Available online: https://www.iea.org/reports/global-ev-outlook-2024 (accessed on 7 June 2024).
2. Altenburg, T.; Bhasin, S.; Fischer, D. Sustainability-Oriented Innovation in the Automobile Industry: Advancing Electromobility in China, France, Germany and India. *Innov. Dev.* **2012**, *2*, 67–85. [CrossRef]
3. Calvo-Jurado, C.; Ceballos-Martínez, J.M.; García-Merino, J.C.; Muñoz-Solano, M.; Sánchez-Herrera, F.J. Optimal location of electric vehicle charging stations using proximity diagrams. *Sustain. Cities Soc.* **2024**, *113*, 105719. [CrossRef]
4. Ademulegun, O.O.; MacArtain, P.; Oni, B.; Hewitt, N.J. Multi-Stage Multi-Criteria Decision Analysis for Siting Electric Vehicle Charging Stations within and across Border Regions. *Energies* **2022**, *15*, 9396. [CrossRef]
5. Skaloumpakas, P.; Spiliotis, E.; Sarmas, E.; Lekidis, A.; Stravodimos, G.; Sarigiannis, D.; Makarouni, I.; Marinakis, V.; Psarras, J. A Multi-Criteria Approach for Optimizing the Placement of Electric Vehicle Charging Stations in Highways. *Energies* **2022**, *15*, 9445. [CrossRef]
6. Morocho-Chicaiza, W.; Barragán-Escandón, A.; Zalamea-León, E.; Ochoa-Correa, D.; Terrados-Cepeda, J.; Serrano-Guerrero, X. Identifying locations for electric vehicle charging stations in urban areas through the application of multicriteria techniques. *Energy Rep.* **2024**, *12*, 1794–1809. [CrossRef]
7. Neumeier, S.; Osigus, T. Germany's Public EV Charging Points: Analysing the 2023 State of Expansion. *Geogr. J.* **2024**, *190*, e12570. [CrossRef]
8. Machado, C.A.S.; Takiya, H.; Yamamura, C.L.K.; Quintanilha, J.A.; Berssaneti, F.T. Placement of Infrastructure for Urban Electromobility: A Sustainable Approach. *Sustainability* **2020**, *12*, 6324. [CrossRef]
9. Ermagun, A.; Tian, J. Charging into Inequality: A National Study of Social, Economic, and Environment Correlates of Electric Vehicle Charging Stations. *Energy Res. Soc. Sci.* **2024**, *115*, 103622. [CrossRef]
10. Min, Y.; Lee, H.W. *Social Equity of Clean Energy Policies in Electric-Vehicle Charging Infrastructure Systems*; ASCE: Reston, VA, USA, 2020; pp. 221–229.
11. Çelik, S.; Ok, Ş. Electric vehicle charging stations: Model, algorithm, simulation, location, and capacity planning. *Heliyon* **2024**, *10*, e29153. [CrossRef]
12. Szterlik-Grzybek, P.; Kucharski, A. Assessment of the Possibility of Locating Electric Car Charging Stations Using Fuzzy AHP and GIS—The Case of Łódź, Poland. *Econ. Environ.* **2023**, *84*, 134–148. [CrossRef]
13. Gazmeh, H.; Guo, Y.; Qian, X. Understanding the Opportunity-Centric Accessibility for Public Charging Infrastructure. *Transp. Res. Part D Transp. Environ.* **2024**, *131*, 104222. [CrossRef]
14. Iravani, H. A Multicriteria GIS-Based Decision-Making Approach for Locating Electric Vehicle Charging Stations. *Transp. Eng.* **2022**, *9*, 100135. [CrossRef]
15. Charly, A.; Thomas, N.J.; Foley, A.; Caulfield, B. Identifying Optimal Locations for Community Electric Vehicle Charging. *Sustain. Cities Soc.* **2023**, *94*, 104573. [CrossRef]
16. Zhao, H.; Gao, J.; Cheng, X. Electric Vehicle Solar Charging Station Siting Study Based on GIS and Multi-Criteria Decision-Making: A Case Study of China. *Sustainability* **2023**, *15*, 10967. [CrossRef]
17. Yi, Z.; Liu, X.C.; Wei, R. Electric Vehicle Demand Estimation and Charging Station Allocation Using Urban Informatics. *Transp. Res. Part D Transp. Environ.* **2022**, *106*, 103264. [CrossRef]
18. Rabl, R.; Reuter-Oppermann, M.; Jochem, P.E.P. Charging Infrastructure for Electric Vehicles in New Zealand. *Transp. Policy* **2024**, *148*, 124–144. [CrossRef]
19. Zhang, H.; Shi, F. A Multi-Objective Site Selection of Electric Vehicle Charging Station Based on NSGA-II. *Int. J. Ind. Eng. Comput.* **2024**, *15*, 293–306. [CrossRef]
20. He, S.Y.; Kuo, Y.-H.; Sun, K.K. The Spatial Planning of Public Electric Vehicle Charging Infrastructure in a High-Density City Using a Contextualised Location-Allocation Model. *Transp. Res. Part A Policy Pract.* **2022**, *160*, 21–44. [CrossRef]

21. Gong, Y.; Kim, I. Optimization and Observation of EV Charging Station Deployment in the Republic of Korea: An Analysis of the Charging History and Correlation With Socioeconomic Factors. *IEEE Access* **2024**, *12*, 68285–68302. [CrossRef]
22. Qin, J.; Qiu, J.; Chen, Y.; Wu, T.; Xiang, L. Charging Stations Selection Using a Graph Convolutional Network from Geographic Grid. *Sustainability* **2022**, *14*, 16797. [CrossRef]
23. Jonas, T.; Daniels, N.; Macht, G. Electric Vehicle User Behavior: An Analysis of Charging Station Utilization in Canada. *Energies* **2023**, *16*, 1592. [CrossRef]
24. Fu, J.; Nåbo, A.; Bhatti, H.J. Locating Charging Infrastructure for Freight Transport Using Multiday Travel Data. *Transp. Policy* **2024**, *152*, 21–28. [CrossRef]
25. Csonka, B.; Csiszár, C. Determination of Charging Infrastructure Location for Electric Vehicles. *Transp. Res. Procedia* **2017**, *27*, 768–775. [CrossRef]
26. Baresch, M.; Moser, S. Allocation of E-Car Charging: Assessing the Utilization of Charging Infrastructures by Location. *Transp. Res. Part A Policy Pract.* **2019**, *124*, 388–395. [CrossRef]
27. K, V.S.M.B.; Chakraborty, P.; Pal, M. Planning of Fast Charging Infrastructure for Electric Vehicles in a Distribution System and Prediction of Dynamic Price. *Int. J. Electr. Power Energy Syst.* **2024**, *155*, 109502. [CrossRef]
28. Prakobkaew, P.; Sirisumrannukul, S. Optimal Locating and Sizing of Charging Stations for Large-Scale Areas Based on GIS Data and Grid Partitioning. *IET Gener. Transm. Distrib.* **2024**, *18*, 1235–1254. [CrossRef]
29. Kandil, S.M.; Abdelfatah, A.; Azzouz, M.A. Optimization Approaches for Fast Charging Stations Allocation and Sizing: A Review. *IEEE Access* **2024**, *12*, 46741–46763. [CrossRef]
30. Kchaou-Boujelben, M. Charging Station Location Problem: A Comprehensive Review on Models and Solution Approaches. *Transp. Res. Part C Emerg. Technol.* **2021**, *132*, 103376. [CrossRef]
31. Abdel-Basset, M.; Gamal, A.; Hezam, I.M.; Sallam, K.M. Sustainability Assessment of Optimal Location of Electric Vehicle Charge Stations: A Conceptual Framework for Green Energy into Smart Cities. *Environ. Dev. Sustain.* **2024**, *26*, 11475–11513. [CrossRef]
32. Mehouachi, I.; Trojette, M.; Grayaa, K. MONNA: Multi-Objective Neural Network Algorithm for the Optimal Location of Electric Vehicle Charging Infrastructure in Tunis City. *J. Clean. Prod.* **2023**, *431*, 139837. [CrossRef]
33. Foda, A.; Abdelaty, H.; Mohamed, M.; El-Saadany, E. A Generic Cost-Utility-Emission Optimization for Electric Bus Transit Infrastructure Planning and Charging Scheduling. *Energy* **2023**, *277*, 127592. [CrossRef]
34. Tounsi Fokui, W.S.; Saulo, M.J.; Ngoo, L. Optimal Placement of Electric Vehicle Charging Stations in a Distribution Network With Randomly Distributed Rooftop Photovoltaic Systems. *IEEE Access* **2021**, *9*, 132397–132411. [CrossRef]
35. Bian, H.; Zhou, C.; Guo, Z.; Wang, X.; He, Y.; Peng, S. Planning of Electric Vehicle Fast-Charging Station Based on POI Interest Point Division, Functional Area, and Multiple Temporal and Spatial Characteristics. *Energy Rep.* **2022**, *8*, 831–840. [CrossRef]
36. Algafri, M.; Alghazi, A.; Almoghathawi, Y.; Saleh, H.; Al-Shareef, K. Smart City Charging Station Allocation for Electric Vehicles Using Analytic Hierarchy Process and Multiobjective Goal-Programming. *Appl. Energy* **2024**, *372*, 123775. [CrossRef]
37. Fadranski, D.; Syré, A.M.; Grahle, A.; Göhlich, D. Analysis of Charging Infrastructure for Private, Battery Electric Passenger Cars: Optimizing Spatial Distribution Using a Genetic Algorithm. *World Electr. Veh. J.* **2023**, *14*, 26. [CrossRef]
38. Mazur, M.; Dybała, J.; Kluczek, A. Suitable Law-Based Location Selection of High-Power Electric Vehicles Charging Stations on the TEN-T Core Network for Sustainability: A Case of Poland. *Arch. Transp.* **2024**, *69*, 75–90. [CrossRef]
39. Regulation (EU) 2023/1804 of the European Parliament and of the Council of 13 September 2023 on the Deployment of Alternative Fuels Infrastructure, and Repealing Directive 2014/94/EU; Official Journal of the European Union L 234/1. Available online: https://eur-lex.europa.eu/legal-content/EN/TXT/PDF/?uri=CELEX:32023R1804 (accessed on 18 June 2024).
40. Speth, D.; Sauter, V.; Plötz, P. Where to Charge Electric Trucks in Europe—Modelling a Charging Infrastructure Network. *World Electr. Veh. J.* **2022**, *13*, 162. [CrossRef]
41. Luh, S.; Kannan, R.; McKenna, R.; Schmidt, T.J.; Kober, T. How, Where, and When to Charge Electric Vehicles—Net-Zero Energy System Implications and Policy Recommendations. *Environ. Res. Commun.* **2023**, *5*, 095004. [CrossRef]
42. Radermecker, V.; Vanhaverbeke, L. Estimation of Public Charging Demand Using Cellphone Data and Points of Interest-Based Segmentation. *World Electr. Veh. J.* **2023**, *14*, 35. [CrossRef]
43. Wu, R.; Wang, T.; Wang, Z.; Fang, Z.; Zhou, X.; Xu, N. Spatial Adaptability Evaluation and Optimal Location of Electric Vehicle Charging Stations: A Win-Win View from Urban Travel Dynamics. *Energy Strategy Rev.* **2023**, *49*, 101122. [CrossRef]
44. Lopes, S.B.; Brondino, N.C.M.; Rodrigues da Silva, A.N. GIS-Based Analytical Tools for Transport Planning: Spatial Regression Models for Transportation Demand Forecast. *ISPRS Int. J. Geo-Inf.* **2014**, *3*, 565–583. [CrossRef]
45. O'Sullivan, D.; Morrison, A.; Shearer, J. Using desktop GIS for the investigation of accessibility by public transport: An isochrone approach. *Int. J. Geogr. Inf. Sci.* **2000**, *14*, 85–104. [CrossRef]
46. Żochowska, R.; Kłos, M.J.; Soczówka, P.; Pilch, M. Assessment of Accessibility of Public Transport by Using Temporal and Spatial Analysis. *Sustainability* **2022**, *14*, 16127. [CrossRef]
47. Soldatke, N.; Sydorów, M.; Żukowska, S. Assessment of the accessibility of public transport in the Tricity (Poland): Analytical use of geographical information systems (GIS) in the context of selected public transport measures. *Int. J. Digit. Earth* **2024**, *17*, 2344586. [CrossRef]
48. Żochowska, R.; Jacyna, M.; Kłos, M.J.; Soczówka, P. A GIS-Based Method of the Assessment of Spatial Integration of Bike-Sharing Stations. *Sustainability* **2021**, *13*, 3894. [CrossRef]

49. Settou, B.; Settou, N.; Gouareh, A.; Negrou, B.; Mokhtara, C.; Messaoudi, D. GIS-based method for future prospect of energy supply in Algerian road transport sector using solar roads technology. *Energy Procedia* **2019**, *162*, 221–230. [CrossRef]
50. Younes, A.; Mohamadi, B.; AbuGhazala, M.O. Terrorism in Egypt: A Comprehensive Spatial, Spatio-Temporal, and Statistical Analysis. *GeoJournal* **2023**, *88*, 6339–6364. [CrossRef]
51. Soczówka, P.; Żochowska, R.; Karoń, G. Method of the Analysis of the Connectivity of Road and Street Network in Terms of Division of the City Area. *Computation* **2020**, *8*, 54. [CrossRef]
52. Available online: https://www.plugshare.com/pl (accessed on 28 June 2024).
53. Izdebski, M.; Jacyna, M.; Bogdański, J. Minimisation of the Energy Expenditure of Electric Vehicles in Municipal Service Companies, Taking into Account the Uncertainty of Charging Point Operation. *Energies* **2024**, *17*, 2179. [CrossRef]
54. Jacyna, M.; Żochowska, R.; Sobota, A.; Wasiak, M. Decision support for choosing a scenario for organizing urban transport system with share of electric vehicles. Scientific Journal of Silesian University of Technology. *Ser. Transp.* **2022**, *117*, 69–89. [CrossRef]
55. Mrozik, M.; Merkisz-Guranowska, A. Modeling of material and energy inputs in the life cycle of a vehicle. *Arch. Transp.* **2024**, *70*, 117–136. [CrossRef]
56. Chamier Gliszczyński, N. Sustainable Operation of a Transport System in Cities. *Key Eng. Mater.* **2011**, *486*, 175–178. [CrossRef]
57. Nosal, K.; Starowicz, W. Evaluation of influence of mobility management instruments implemented in separated areas of the city on the changes in modal split. *Arch. Transp.* **2015**, *35*, 41–52. [CrossRef]
58. Chamier-Gliszczyński, N.; Bohdal, T. Mobility in Urban Areas in Environment Protection. *Rocz. Ochr. Sr.* **2016**, *18*, 387–399.
59. Jacyna-Gołda, I.; Żak, J.; Gołębiowski, P. Models of traffic flow distribution for various scenarios of the development of proecological transport system. *Arch. Transp.* **2014**, *4*, 17–28. [CrossRef]
60. Kierzkowski, A.; Tubis, A.A. Transportation Systems Modeling, Simulation and Analysis with Reference to Energy Supplying. *Energies* **2023**, *16*, 3586. [CrossRef]
61. Kamińska, M.; Szymlet, N.; Fuć, P.; Rymaniak, Ł.; Lijewski, P.; Grzeszczyk, R. Analysis of harmful compounds concentrations in the exhaust behind a vehicle with compression ignition engine. *Combust. Engines* **2024**, *197*, 50–55. [CrossRef]
62. Csiszar, C.; Csonka, B.; Foldes, D.; Wirth, E.; Lovas, T. Urban public charging station locating method for electric vehicles based on land use approach. *J. Transp. Geogr.* **2019**, *74*, 173–180. [CrossRef]

energies

MDPI

Article

Challenges for Implementing Vehicle-to-Grid Services in Parking Lots: A State of the Art

Antonio Comi * and Elsiddig Elnour

Department of Enterprise Engineering, University of Rome Tor Vergata, 00133 Rome, Italy; elnour@ing.uniroma2.it
* Correspondence: comi@ing.uniroma2.it; Tel.: +39-06-7259-7061

Abstract: Electric vehicles (EVs) play a vital role in the transition to renewable energy and decarbonisation, and there is increasing global interest in expanding their use. However, the growing number of EVs poses challenges to the electricity grid due to increasing energy demand. Vehicle-to-grid (V2G) technology can address these issues by allowing for EVs to charge and discharge energy, thus helping to balance the grid when needed. Aggregating vehicles in designated parking areas optimises energy transfer, making it crucial to identify suitable parking locations and forecast the energy available from parked vehicles. A spatial–temporal framework ensures that V2G services operate efficiently considering both the location and the timing of vehicle parking. This paper reviews studies on temporal–spatial V2G parking demand, identifying high-demand areas such as shopping centres and workplaces, where vehicles park for extended periods. Strategic locations of V2G hubs in these areas ensures seamless integration into existing mobility patterns without disrupting users' routines. In addition, this review examines user acceptance, technical feasibility, and V2G's role in reducing grid demand peaks. The findings indicate the potential of effectively implemented V2G services to enhance electricity grid stability and efficiency while minimising disruptions to EV users.

Keywords: electric vehicle; V2G; vehicle to grid; energy transfer; temporal–spatial parking demand; controlled charging and discharging; parking lot

1. Introduction

The rapid acceleration of decarbonisation and the reduction of greenhouse gas emissions, especially in the transport sector, have generated significant interest in electric vehicles (EVs). EVs are now considered one of the most efficient tools for achieving sustainability in urban mobility [1,2]. Recent studies on passenger transport in large European cities indicate that a substantial part of the population uses a private vehicle to commute [3,4]. Consequently, the high rate of passenger transport by road contributes to traffic congestion and environmental pollution, with the highest emissions of CO, CO_2, and NH_3 originating from traffic emission [5–7]. To address these challenges, the European Union (EU) is implementing policies aimed at achieving full decarbonisation by 2050, including the phasing out of internal combustion engine (ICE) vehicles, or traditional vehicles (TVs), by 2035 [1,8,9]. This transition is expected to significantly increase the demand for EVs as alternatives, leading to higher electric energy consumption; combined with the rapid pace of urbanisation, where more than 68% of the global population is projected to live in urban areas by 2050 [10], this underscores the urgent need for expanded energy production and more efficient solutions to meet the growing demand. In this context, researchers are increasingly exploring the potential of EVs not only as energy consumers but also as energy providers, depending on the needs of the power grid [11,12]. Studies show that vehicles used for systematic travel (e.g., home-to-work or home-to-school trips) within urban and metropolitan areas are parked for approximately 6–8 continuous hours at workplace parking lots or 95% of their average life cycle, with an average trip distance of less than 5 km [13,14]. This presents an opportunity to utilise EVs as distributed energy stores (DESs), enabling their batteries to give back to the power grid.

Vehicle-to-grid (V2G) technology emerges as a critical innovation in this context, first proposed by Kempton and Steven [15]. V2G facilitates bidirectional energy transfer between the grid and EVs. In this system, EVs fulfil their owners' mobility needs while supplying surplus energy back to the grid during peak demand periods. This process provides auxiliary services and offsets the variability of renewable energy sources. Technically, this is facilitated through a bidirectional power inverter connected to the vehicle battery and the grid [16]. However, for V2G to be effective, especially given that individual vehicles cannot participate alone, it is essential to identify suitable parking locations where EVs can be aggregated for energy transfer to the grid. This necessitates a temporal–spatial framework that considers both the location and the timing of vehicle parking. By incorporating a temporal dimension to the vehicle location, it becomes possible to determine when and where parked vehicles can participate in V2G services effectively. This paper focuses on temporal–spatial parking demand analysis, which is crucial for optimising controlled charging and discharging processes in V2G services. Understanding parking demand patterns helps identify high-demand zones where vehicles are likely to park for extended periods, such as near shopping centres, workplaces, or other specific land use types [17]. V2G services can be integrated into existing systematic mobility patterns without requiring users to significantly alter their daily agendas, enhancing the efficiency and feasibility of V2G implementation.

Due to the relevance of such a topic and the studies implemented for supporting the real implementation of V2G, this paper proposes a new literature review aiming to systematise the works developed and to identify the road ahead for supporting city planners in its real implementation. In fact, the novelty of this literature review lies in its emphasis on temporal–spatial parking demand studies to strategically implement V2G services in areas with the highest potential for energy transfer. The existing literature on V2G is extensive, covering aspects such as grid stability, user behaviour, and the environmental benefits of integrating EVs into the power grid. For instance, Comi et al. [18] conducted a systematic literature review highlighting the ecological transition and behavioural barriers affecting V2G adoption. Their work underscored critical issues like battery degradation and user privacy concerns, which hinder widespread adoption, but did not specifically address the spatial–temporal aspects of EV aggregation. Similarly, Yuvaraj et al. [19] provided a comprehensive analysis of optimal charging station placements and their impacts on distribution networks; however, their emphasis was primarily on infrastructure allocation rather than on temporal patterns of EV usage.

This review builds upon these foundational works by incorporating a temporal–spatial dimension into the analysis, which is crucial for optimising controlled charging and discharging processes in V2G services. By leveraging origin–destination data, stop locations, and battery levels, the framework addresses key gaps in the literature by ensuring that EVs participating in V2G retain sufficient energy for future trips while maximising grid contributions. This approach goes beyond existing methodologies by integrating V2G into established mobility patterns without requiring significant behavioural changes from EV users. In doing so, it not only enhances the feasibility and efficiency of V2G services but also mitigates user concerns about accessibility and convenience to this new and advanced service.

The rest of this paper is structured as follows: Section 2 presents the proposed analysis methodology. Section 3 discusses V2G activity in the context of technology and important challenges. Section 4 deals with V2G implementation challenges through spatial–temporal dynamics and the three main influencing categories on parking demand, i.e., land use and physical environment, parking supply characteristics, and parking operations policies, and management. The user acceptance for V2G and the impacts of V2G adoption are also discussed. Finally, Section 5 concludes and summarises this paper.

2. The Proposed Methodology of Analysis

This section describes the methodology used to search, compile, and categorise the literature on the main challenges in implementing V2G services. The review points out V2G demand from both temporal and spatial perspectives, with the goal of identifying optimal locations that can be leveraged for V2G services. The analysis focusses on three key categories, i.e., (1) the interaction between land use and trip characteristics; (2) parking supply characteristics, and (3) parking operation policies, regulations, and management. Additionally, this review examines user acceptance and V2G impacts.

The methodology uses a multi-step approach, as shown in Figure 1, consisting of the following steps:

1. Search Step
   It focuses on identifying and retrieving studies from established academic databases; this yielded a total of 62 papers for initial consideration; specifically, it consists of three main tasks:
   - identification of *databases*; the task involves selecting the appropriate databases to provide the literature on the topic; for this study, databases such as IEEE, Scopus, and JSTOR were chosen due to their extensive coverage of engineering, technology, and urban planning research;
   - identification of *keywords*; this involves defining a set of keywords that align with the research topic; the used keywords are ("V2G services" or "V2G infrastructure") AND ("parking demand" or "land use" or "trip characteristics" or "parking supply" or "user parking choices" or "parking operations policies" or "parking regulations" or "parking management"), and, in addition, "V2G impact", "user acceptance";
   - *search criteria for inclusion*; this task filters the search results according to specific inclusion criteria, outlined in Table 1, to ensure that the literature is relevant and of high quality; criteria include factors like the type of document (peer-reviewed journals, conference proceedings, or book chapters), with the time interval from 2013 to 2024;
2. Selection Step
   In this step, the 62 *retrieved papers* from the search step undergo a screening process to identify studies most relevant to this review; it includes the following:
   - *elimination of duplicates*; initially, the retrieved papers are screened to remove any duplicates given that more than one database has been used;
   - *screening and relevance assessment*; after duplicates are eliminated, the remaining papers are evaluated based on their titles, abstracts, and complete list of keywords; this step assesses each paper's relevance to the core topic of challenges in implementing V2G services; only studies deemed relevant proceed to the final selection.

Table 1. Inclusion criteria for literature review.

Criterion	Inclusion
Database	IEEE, Scopus, JSTOR
Analysed field	Titles, abstracts, keywords
Document type	Journal papers, conferences proceedings papers, others [1]
Time interval	2013–2024

[1] *book chapters and reports.*

Following this multi-step process, a total of 42 papers were selected for an in-depth review. Figure 2 provides the details on the frequency and distribution of studies over the specified time interval.

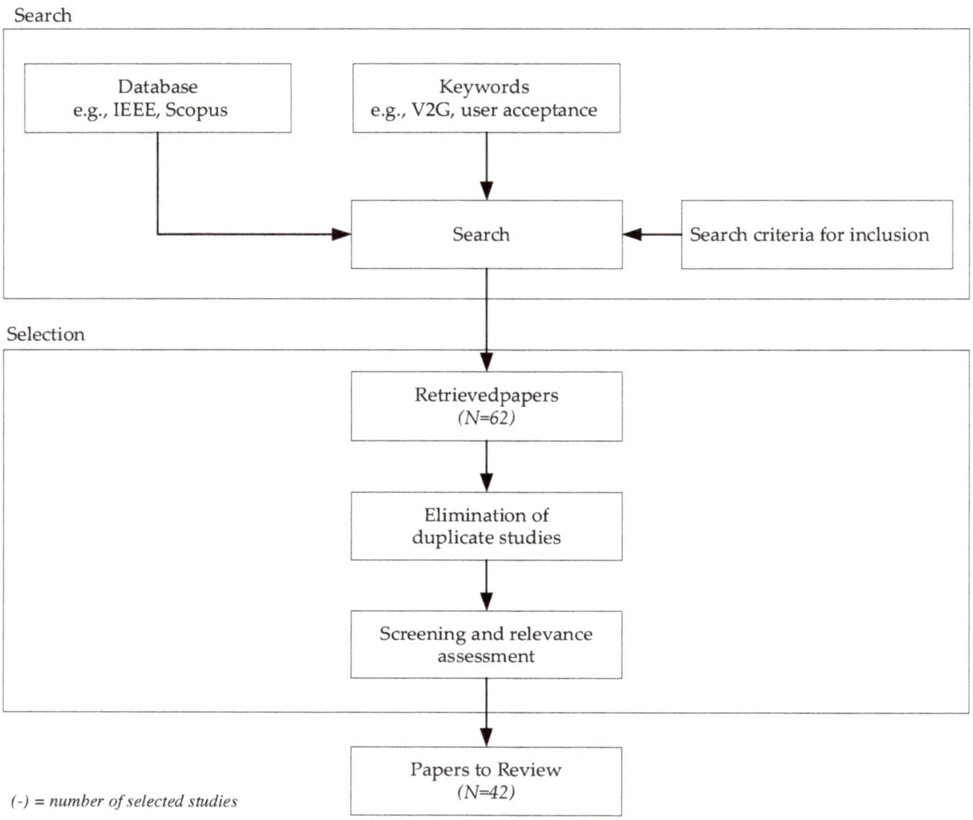

Figure 1. The proposed methodology.

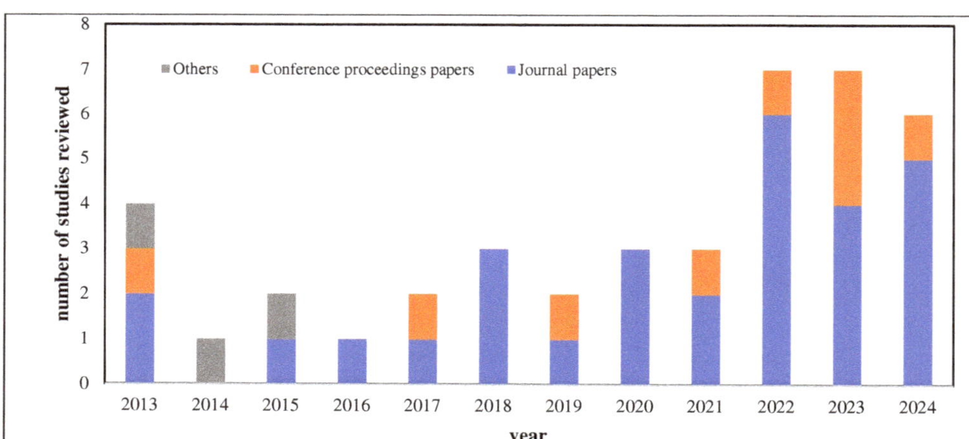

Figure 2. Frequency distribution of the selected studies in the last twenty years.

3. Vehicle-to-Grid Working Scheme: Background and Long-Term Perspective

V2G technology enables EVs to interact dynamically with the power grid. It leverages the energy storage capabilities of EV batteries, allowing for a bidirectional flow of electricity. Vehicles can charge from the grid (grid-to-vehicle, G2V) or discharge stored energy back into the grid when demand is high V2G [20,21]. This concept transforms parked EVs into mobile energy assets, offering a decentralised, flexible approach to energy management, which is especially crucial as the adoption of renewable energy sources rises.

The primary function of V2G is to provide ancillary services to the power grid [22]. This includes frequency regulation, voltage stabilisation, and peak load shaving [23,24]. V2G helps maintain grid stability by acting as a supplementary energy source, especially during times of high electricity demand [25]. Integration of renewable energy sources, such as solar and wind, with V2G systems further enhances the reliability and efficiency of the energy grid [26]. PHEVs (plug-in hybrid electric vehicles) [25,27] or full EVs connected to V2G systems can discharge energy to nearby buildings or feed power directly into the utility grid, assisting in voltage regulation and reducing the necessity for additional power plants [28].

Vertical parking structures are particularly suitable for V2G deployment, as they increase vehicle density and, consequently, the amount of available energy storage within a limited ground footprint [24]. These structures, when integrated with V2G capabilities, can be part of a community energy management system (CEMS) to optimise local energy use and promote green energy solutions.

V2G services involve a series of steps, ensuring efficient energy transfer while maintaining grid stability, as well as the maximisation of the total energy transferred to the grid from the EVs parked at given parking lots, subject to some constraints such as the vehicle availability timeframe, state of the charge (SOC), vehicle charging/discharging features, V2G infrastructure features, and grid demand. Vehicle availability timeframe depends on the arrival and departure times of each EV, enabling energy transfers only when the EV is parked and remains parked for a timeframe compatible for V2G. The SOC ensures that EVs maintain adequate charge for their mobility requirements after participation, accounting for energy consumed before parking, energy transferred to the grid, and the reserve energy set by users. Vehicle and infrastructure features refer to the capacity of transferring or taking energy to/from the grid. Additionally, grid demand ensures that the energy transferred to the grid does not exceed the grid demand.

Figure 3 illustrates the operational flow of a V2G-enabled system in a scheduled and controlled energy transfer environment:

- *vehicle energy input* (EV in charging–discharging infrastructure); an EV enters a designated area equipped with V2G infrastructure, enabling both charging and discharging, and the EV is plugged into V2G infrastructure;
- *battery status check*; the system checks the current battery status of the EV to determine if there is enough charge to support energy transfer;
- sufficient *charge*,
- Yes—if the battery has sufficient charge, the system moves to the next step;
- No—if the battery does not have sufficient charge, the system prioritises charging the vehicle or waits until there is adequate charge before energy transfer;
- *grid demand signal* for scheduled and controlled energy transfer: if the battery has sufficient charge, the grid sends a demand signal to initiate energy transfer based on current energy needs;
- *grid request*; the system receives the grid request and prepares to transfer energy from the EV to the grid;
- *energy transfer to the grid* (V2G); controlled energy transfer occurs, allowing for the EV to discharge energy back into the grid;
- *grid stabilisation*; the energy provided by the EVs helps stabilise the electricity grid, particularly during peak demand periods.

This scheme allows for V2G systems to dynamically interact with the power grid; however, it has to point out the time when starting the transfer from the EV, the transfer rate allowed for by the technological feature of the EV, as well as the time when the EV leaves the parking lot. Furthermore, the starting check can be integrated with the evaluation of energy needed by users for not modifying their daily routines after participation in V2G [29]. For example, if an EV is parked in a parking lot while the user is at work, the V2G system, in order to forecast the energy that can be transferred from the vehicle assuring to maintain the minimum level of charge, also has to take into account the time when the user will come back to reuse his/her vehicle. By leveraging the energy storage capabilities of EV batteries, vehicles can charge from the grid (grid-to-vehicle, G2V) or discharge stored energy back into the grid during peak demand periods (vehicle-to-grid, V2G). This approach transforms parked EVs into decentralised, mobile energy assets, offering a flexible solution for energy management.

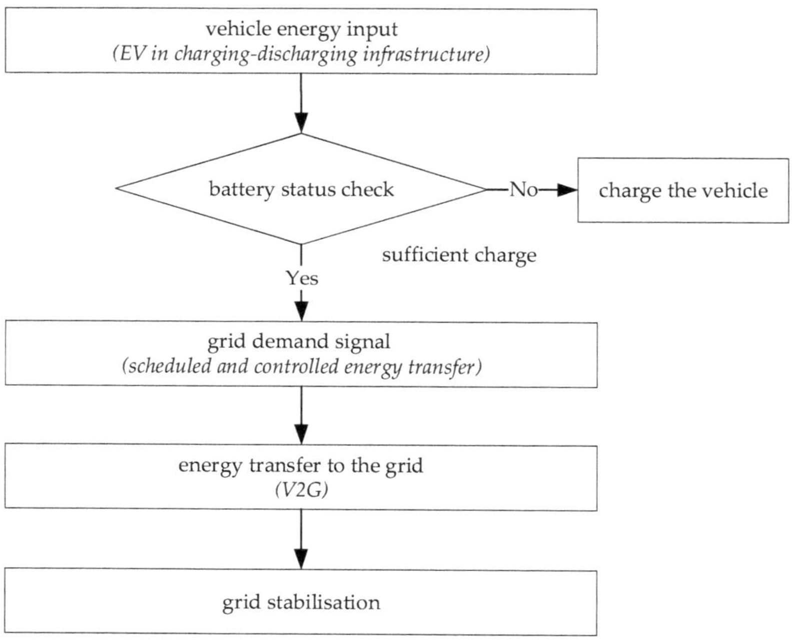

Figure 3. Scheme of V2G working.

Modern EVs, such as the Nissan Leaf, Hyundai Ioniq 5, and several Tesla models, are equipped with V2G capabilities [25,30]. The Nissan Leaf, in particular, has been actively utilised in V2G testing programmes in the United Kingdom due to its compatibility with bidirectional chargers, with Nissan commercial V2G charger technology anticipated for release in 2026. These chargers are essential for V2G operation, enabling energy exchange between EVs and the grid. However, the deployment of bidirectional charging infrastructure remains limited, as the majority of current EV charging stations lack this functionality.

Despite existing limitations, numerous pilot projects have demonstrated the feasibility and advantages of V2G technology. For example, the collaboration between OVO Energy and Kaluza in the United Kingdom highlights the potential of V2G to optimise energy flows by leveraging artificial intelligence and market signals. The project demonstrated that EVs can charge during periods of low energy demand or high renewable energy generation and discharge during peak demand, therefore stabilising the grid while offering financial incentives to EV owners [31].

By 2030, it is projected that the United Kingdom will have approximately 11 million EVs on the road, with around half being V2G-enabled. This level of adoption could unlock over 22 TWh of flexible discharging capacity annually, equivalent to approximately 16 GW of daily grid support [31].

In the context of 2050, V2G technology is expected to play a transformative role in global energy systems. Nations worldwide are striving to meet ambitious climate goals, such as the "Fit for 55" initiative by the European Union, which aims to cut greenhouse gas emissions by 55% by 2030. The integration of EVs into energy grids will become indispensable in achieving these objectives. V2G is uniquely positioned to address critical challenges, including the integration of renewable energy sources, the reduction of fossil fuel dependency, and the mitigation of infrastructure expansion costs.

While V2G presents transformative potential, several challenges have to be addressed to facilitate its widespread adoption:

- *battery degradation*; one of the most critical concerns related to V2G technology is the impact of frequent charging and discharging on the lifespan of EV batteries; the cyclic stress placed on batteries during energy exchanges can reduce their overall capacity and efficiency, prompting the need for compensation mechanisms to encourage EV owners to participate [28];
- *state of charge (SOC) management*; V2G systems should carefully manage the SOC of connected EVs to ensure they retain sufficient energy for the vehicle primary function transportation; real-time scheduling and dynamic energy management algorithms are essential to maintain optimal battery levels without impacting the driver's needs; advanced building energy management systems (BEMSs) can forecast building load demand and schedule EV charging or discharging accordingly to maintain grid balance [11,23];
- *energy infrastructure and grid compatibility*; successful V2G deployment requires significant upgrades to existing grid infrastructure; the grid has to be capable of handling bidirectional energy flows and accommodating large-scale integration of distributed energy resources, as pointed out by Ebrahimi et al. [32]; the complexity of synchronising energy flows between vehicles and the grid is a logistical challenge; additionally, ensuring compatibility between different EV models, charging standards, and grid systems is necessary for seamless V2G operation;
- *spatial–temporal dynamics*; identifying optimal locations and times for V2G implementation is crucial; high-density urban areas, where EVs are parked for long durations (e.g., near office complexes or shopping centres), are ideal for V2G hubs; the methodology proposed in Aydinoglu et al. [33] emphasises the use of geographic information systems (GIS) to analyse spatial data, identifying key zones for energy storage and transfer based on parking patterns and trip characteristics; the interaction between land use and parking demand plays a pivotal role in V2G planning [34];
- *user acceptance and behavioural factors*; convincing EV owners to participate in V2G programmes hinges on several factors, including financial incentives, assurance of vehicle availability, and minimal inconvenience; concerns such as range anxiety, a fear of running out of charge, are exacerbated when participating in V2G services, especially if vehicles are needed for unexpected trips; research by Esmaili et al. [35] emphasises the need for systems that prioritise a minimum SOC to alleviate these concerns and encourage greater participation;
- *regulatory and policy frameworks*; implementing V2G technology on a large scale requires supportive regulatory frameworks and standardised protocols; policies have to address energy fares, compensation for V2G participation, and data security measures to protect user privacy [36].

4. Analysis of V2G Implementation Challenges Through Spatial–Temporal Dynamics

Successful integration of V2G services into urban infrastructure requires a multifaceted approach that considers various spatial and temporal factors. This analysis delves into five critical categories:

- the interaction between land use and trip characteristics;
- parking supply characteristics;
- parking operation policies, regulations, and management;
- users' acceptance;
- V2G impact.

Additionally, an analysis of the significant literature highlights that user acceptance and the assessment of V2G impacts are essential for identifying optimal locations and strategies for effective V2G deployment. These factors play a key role in implementing successful V2G services.

4.1. Land Use and Trip Characteristics

Understanding land use patterns is pivotal in determining the feasibility and optimisation of V2G services. High-accessibility areas, characterised by proximity to employment centres and essential services, often exhibit increased parking facilities [37]. This suggests that such areas present significant opportunities for V2G integration, as parked EVs can serve as distributed energy resources for the grid.

The interaction between land use, the physical environment, and parking demand varies across zones. For instance, residential areas typically experience peak parking demand during evening hours, while commercial and office zones peak during business hours [38]. This temporal variation aligns with energy demand patterns, indicating that V2G services can be optimised by targeting specific land use types at appropriate times. Tailored parking demand models that consider local land use patterns and transportation options are essential for accurate forecasting [39]. Variables such as dwelling units, vehicle ownership, and floor space have strong correlations with spatial–temporal parking demand. By leveraging these models, planners can identify optimal locations with predictable long-term parking, ideal for establishing V2G hubs.

Recently, GIS has played a crucial role in identifying optimal parking locations based on spatial data and transportation networks [33], with new trends to integrate GIS with temporal profiles of trips and parking demand data. Such an integration enables the strategic placement of V2G infrastructure, ensuring that energy transfer capabilities are aligned with areas of high demand.

Urban density also impacts EV parking and V2G potential, as dense areas typically require fewer parking spaces due to better access to public transportation [40]. This characteristic reduces the need for extensive parking infrastructure, allowing for a more efficient use of existing spaces for V2G services without contributing to urban sprawl or congestion.

4.2. Parking Supply Characteristics

Parking supply characteristics, including accumulation, volume, capacity, turnover, duration, and saturation, directly influence the potential for V2G services [17,40]. High-turnover parking areas, particularly off-street facilities, present unique opportunities for energy transfer between EVs and the grid [41]. These areas facilitate efficient V2G operations by minimising congestion and allowing for seamless vehicle movement.

Addressing parking supply shortages is critical, as imbalances between supply and demand can exacerbate urban congestion and hinder participation in V2G [42]. Implementing shared parking systems and dynamic allocation strategies can optimise parking utilisation, enhancing the viability of V2G services in densely populated areas [43].

The impact of parking turnover on traffic delays underscores the importance of strategically selecting V2G locations [44]. High-turnover zones enable frequent energy exchanges, making them suitable for V2G integration. Furthermore, understanding user preferences on parking supply, through advanced gamification models and real-time data, can inform the placement of V2G-enabled hubs, aligning the infrastructure with demand patterns [45].

User preferences are a critical factor in the successful implementation of V2G technology; parking choices are influenced by socioeconomic factors, security concerns, convenience, and access to real-time information [46,47]. Higher-income individuals often prefer off-street, reserved parking, which can be equipped with V2G capabilities to attract users willing to participate in energy exchange programmes.

Real-time parking information and reservation systems significantly impact user choices by reducing uncertainty and enhancing the appeal of V2G enabled facilities [48]. Providing transparent and accessible information increases the likelihood of users opting for parking options that are compatible with V2G services.

Demographic factors, such as age and education level, also play a role in parking behaviour and EV adoption [49]. Tailoring V2G initiatives to specific user groups can improve participation rates and maximise the benefits of this service.

4.3. Parking Operation Policies, Regulations, and Management

Effective parking management policies and regulations are instrumental in shaping user behaviour and optimising parking resources for V2G deployment. Differentiated pricing strategies can balance parking demand across various zones, incentivising EV owners to park in locations conducive to V2G participation [50]. By adjusting parking fees based on demand, cities can encourage EV owners to use V2G-equipped facilities in high-demand areas.

The implementation of intelligent parking systems (IPSs) and parking guidance systems (PGSs) leverages technology to reduce cruising times and congestion by directing drivers to available spaces equipped with V2G infrastructure [51,52]. These systems enhance the efficiency of parking operations, ensuring that EVs are readily available for energy exchange without contributing to traffic bottlenecks.

Urban planning initiatives that limit parking supply and enforce maximum parking caps can support V2G integration by reducing TV dependency and promoting sustainable mobility [53]. By reallocating space from traditional parking to V2G-enabled facilities, cities can foster a more resilient and flexible energy grid. These measures not only alleviate congestion, but also encourage the use of alternative transportation modes.

Dynamic pricing and flexible operating hours for points of interest can optimise parking turnover and availability aligning parking patterns with the temporal aspects of energy demand [41].

4.4. User Acceptance of V2G

The commercialisation of V2G technology depends heavily on user acceptance, as the willingness of EV owners to participate is essential for system success. Financial compensation emerges as a primary motivator, as users are more likely to participate in V2G services if adequately remunerated for the energy provided to the grid [54]. Ensuring that the financial benefits outweigh the perceived inconveniences that can significantly increase participation rates.

Battery degradation concerns also significantly affect user acceptance. Frequent charging and discharging cycles associated with V2G operations can lead to increased wear on EV batteries [55]. Addressing these concerns through technological advancements that mitigate degradation, or compensation for battery wear, can alleviate users' concerns and encourage participation.

Range anxiety, defined as the fear of depleting the EV battery before reaching a charging station or destination, is another critical factor influencing acceptance [56]. Implementing smart energy management systems that ensure a minimum state of charge can reduce this anxiety, making users more comfortable with their vehicles participating in V2G services without compromising their mobility needs.

Other factors influencing acceptance such as user interface design, control over charging and charging processes, privacy, trust, and perceived societal contribution of V2G systems play key roles in user acceptance [54].

Clear communication, transparency, and user-friendly platforms can improve trust and encourage broader adoption of V2G technology. Furthermore, empowering users with control over their data and charging preferences promotes a positive user experience, thereby increasing participation.

4.5. V2G Impact

V2G technology offers significant benefits for power grid stability and renewable energy integration. As the number of EVs increases, the potential strain on the power grid due to higher electricity demand also rises. V2G addresses this challenge by enabling EVs not only to consume, but also to supply electricity back to the grid, effectively functioning as distributed energy storage units [57].

By facilitating bidirectional energy flow, V2G systems help balance supply and demand, smoothing out fluctuations in grid demand [58]. During peak demand periods, EVs can discharge stored energy to support the grid, reducing the need for additional generation from Peaker plants. Conversely, during periods of low demand or high renewable energy production, EVs can absorb excess energy, preventing waste and enhancing renewable resource utilisation [25].

V2G technology contributes to voltage and frequency stability within the power grid. Studies have shown that integrating V2G systems can maintain voltage levels and reduce frequency deviations, even under varying load conditions and high EV penetration rates [58]. This stabilisation prevents blackouts and ensures consistent power quality.

In addition to technical benefits, V2G supports the global mission to combat climate change by optimising the use of renewable energy and reducing fossil fuel dependency. By enabling more EVs to connect to the grid without causing congestion, V2G systems facilitate the integration of clean energy sources, thereby contributing to a reduction in greenhouse gas emissions [57].

5. Conclusions

An in-depth review of the key challenges and opportunities associated with the implementation of V2G services is provided, with particular emphasis on their integration into urban network planning. Given that EVs cannot participate alone in such a service, emphasis is placed on the planning of parking lots as a means of aggregation. The dual role of EVs as both transportation assets and distributed energy resources creates a unique opportunity to address energy grid challenges, particularly in managing peak electricity demand and enhancing renewable energy integration. V2G systems facilitate the bidirectional energy transfer between EVs and the grid, where parked vehicles can provide energy during peak hours and store energy during periods of excess electricity demand and of high numbers of parked EVs. This can sustainably contribute to the grid's stability and support renewable energy integration without compromising the mobility needs of vehicle owners.

The effective deployment of V2G services relies on the spatial–temporal dynamic, therefore, the review underscores the importance of identifying strategic locations for V2G compatible parking lots, especially in high-demand zones such as workplaces, shopping centres, and residential areas. By leveraging areas with consistent parking patterns and longer durations, V2G infrastructure can be optimised to provide substantial energy support to the grid. Findings indicate that urban areas with extended EV parking durations, such as residential neighbourhoods or business districts, are particularly suitable for implementing V2G, as these locations align well with grid energy demand cycles.

Summarising, V2G helps build a greener, more efficient, and resilient energy system that is aligned with global sustainability goals, with significant benefits, i.e., support to the electricity grid (i.e., it balances energy supply and demand, reducing the impacts due to consumption peaks, which could require extra effort by the energy production), economic benefits (i.e., it allows for electricity providers to reduce/optimise the production costs as well as EV owners to sell excess energy to the grid, generating revenue and reducing operational costs), integration of renewables (i.e., it uses electric vehicles as "mobile batteries" to store and return renewable energy), city sustainability (i.e., it contributes to the decarbonisation of the energy system by reducing the use of fossil fuels for obtaining energy in particular time periods), innovation (i.e., it allows for EVs to return energy to the grid, helping to make smart grids more efficient and supporting a more sustainable decentralised energy system).

However, several challenges remain to be addressed for widespread V2G adoption. User acceptance is a critical barrier, often influenced by concerns about battery degradation, range anxiety, and the need for sufficient financial compensation. Users fear that frequent battery cycling could decrease battery lifespan, and there is hesitation about whether participating in V2G services might leave vehicles without enough charge for unexpected trips. To encourage user participation, well-designed compensation business models, improved charging management systems, and advances in battery technology are essential. Additionally, regulatory frameworks must evolve to support the integration of V2G services within existing energy markets, ensuring that both grid operators and EV owners derive mutual benefits.

Technological advancements are also necessary to support V2G implementation. Parking facilities equipped with bidirectional charging capabilities are a fundamental requirement, yet such infrastructure is not yet widely available. The integration of IPSs can play a pivotal role in dynamically managing parking spaces and optimising energy transfer. By incorporating real-time data for parking availability, pricing, and energy needs, an IPS can help ensure the most effective use of V2G enabled parking lots.

The potential benefits of V2G are vast. These systems can significantly improve energy efficiency, reduce dependence on traditional power plants, and support the integration of renewable energy by balancing fluctuations in electricity supply. In densely populated urban centres, V2G-equipped parking hubs can provide dual benefits: supporting mobility needs while acting as energy storage units. Such parking hubs can provide ancillary services to the grid, such as frequency regulation and peak shaving, contributing to a more stable and resilient energy system.

Looking ahead, future research should explore detailed spatial–temporal mapping to identify zones with high V2G potential; for instance, GIS and real-time data analytics could pinpoint optimal parking hubs in urban centres, taking into account the EV parking patterns and peak energy demand periods. To fully harness the potential of V2G, future research should also focus on refining dynamic pricing models to incentivise user participation, improving EV battery longevity to address degradation concerns, and developing robust infrastructure capable of managing the complex demand for bidirectional energy flows. Integrating advanced technologies such as real-time parking information systems, dynamic pricing, and autonomous vehicles will be instrumental in shaping the future of V2G hubs, improving both flexibility and sustainability. Additionally, shared and vertical parking facilities, especially in high-density urban areas, should be considered to meet parking demand and facilitate V2G services. By aligning research on parking infrastructure with the development of V2G systems, urban planners and policymakers can create efficient, resilient, and sustainable energy systems while addressing the growing mobility needs of urban environments.

Author Contributions: Conceptualisation, A.C.; methodology, A.C. and E.E.; software, E.E.; validation, A.C. and E.E.; formal analysis, A.C. and E.E.; investigation, A.C. and E.E.; resources, A.C.; data curation, A.C. and E.E.; writing—original draft preparation, E.E.; writing—review and editing, A.C. and E.E.; visualisation, A.C. and E.E.; supervision, A.C.; project administration, A.C.; funding acquisition, A.C. All authors have read and agreed to the published version of the manuscript.

Funding: This research was supported by "Progetto ACCUMULO—1.2 Progetto Integrato Tecnologie di accumulo elettrochimico e termico—LA2.12-Analisi dell'offerta territoriale per la realizzazione di modelli di predizione della capacità aggregata fornita da veicoli elettrici a supporto delle esigenze della rete elettrica", Ministero dell'Ambiente e della Sicurezza Energetica MASE (ex MiTE), Consiglio Nazionale delle Ricerche, Italy, CUP E87H23001620005.

Data Availability Statement: Dataset available on request from the authors.

Conflicts of Interest: The authors declare no conflicts of interest.

References

1. Thiel, C.; Nijs, W.; Simoes, S.; Schmidt, J.; van Zyl, A.; Schmid, E. The impact of the EU car CO_2 regulation on the energy system and the role of electro-mobility to achieve transport decarbonisation. *Energy Policy* **2016**, *96*, 153–166. [CrossRef]
2. Polimeni, A.; Donato, A.; Belcore, O.M. Urban freight distribution with electric vehicles: Comparing some solution procedures. *Front. Future Transp.* **2024**, *5*, 1491799. [CrossRef]
3. Cirianni, F.M.M.; Leonardi, G. Analysis of transport modes in the urban environment: An application for a sustainable mobility system. *WIT Trans. Ecol. Environ.* **2006**, *93*, 637–645. [CrossRef]
4. Corselli-Nordblad, L.; Jere, N.; Ford-Alexandraki, E.; Xenellis, G. *Key Figures on European Transport: 2022 Edition*; Publications Office of the European Union: Luxembourg, 2022.
5. MElser, M.; El-Haddad, I.; Maasikmets, M.; Bozzetti, C.; Wolf, R.; Ciarelli, G.; Slowik, J.G.; Richter, R.; Teinemaa, E.; Hüglin, C.; et al. High contributions of vehicular emissions to ammonia in three European cities derived from mobile measurements. *Atmos. Environ.* **2018**, *175*, 210–220. [CrossRef]
6. CO2 Emissions from Cars: The Facts A Report by Transport & Environment Acknowledgements. 2018. Available online: www.transportenvironment.org (accessed on 20 September 2024).
7. Comi, A.; Cirianni, F.M.M.; Cabras, L. Sustainable Mobility as a Service: A Scientometric Review in the Context of Agenda 2030. *Information* **2024**, *15*, 637. [CrossRef]
8. Franke, T.; Neumann, I.; Bühler, F.; Cocron, P.; Krems, J.F. Experiencing Range in an Electric Vehicle: Understanding Psychological Barriers. *Appl. Psychol.* **2012**, *61*, 368–391. [CrossRef]
9. Erbach, G.; Jensen, L.; Chahri, S.; Claros, E. BRIEFING Towards Climate Neutrality. Fit for 55 Package. 2021. Available online: https://www.europarl.europa.eu/thinktank/en/document/EPRS_BRI(2022)733513 (accessed on 20 September 2024).
10. Nations, U.; Economic, D.O.; Affairs, S.; Division, P. *World Urbanization Prospects The 2018 Revision*; United Nations Publications: New York, NY, USA, 2018.
11. Patanè, L.; Sapuppo, F.; Xibilia, M.G.; Napoli, G. A comprehensive data analysis for aggregate capacity forecasting in Vehicle-to-Grid applications. In Proceedings of the 2024 International Conference on Control, Automation and Diagnosis (ICCAD), Paris, France, 15–17 May 2024. [CrossRef]
12. Comi, A.; Hriekova, O.; Crisalli, U.; Napoli, G. A methodology based on floating car data for forecasting the available capacity for vehicle-to-grid services. *Transp. Res. Procedia* **2024**, *78*, 47–54. [CrossRef]
13. Comi, A.; Polimeni, A.; Nuzzolo, A. An Innovative Methodology for Micro-Mobility Network Planning. *Transp. Res. Procedia* **2022**, *60*, 20–27. [CrossRef]
14. Nigro, M.; Comi, A.; De Vincentis, R.; Castiglione, M. A mixed behavioural and data-driven method for assessing the shift potential to electric micromobility: Evidence from Rome. *Front. Futur. Transp.* **2024**, *5*, 1391100. [CrossRef]
15. Kempton, W.; Letendre, S.E. Electric vehicles as a new power source for electric utilities. *Transp. Res. Part D Transp. Environ.* **1997**, *2*, 157–175. [CrossRef]
16. Shipman, R.; Roberts, R.; Waldron, J.; Naylor, S.; Pinchin, J.; Rodrigues, L.; Gillott, M. We got the power: Predicting available capacity for vehicle-to-grid services using a deep recurrent neural network. *Energy* **2021**, *221*, 119813. [CrossRef]
17. Parmar, J.; Das, P.; Dave, S.M. Study on demand and characteristics of parking system in urban areas: A review. *J. Traffic Transp. Eng. (Engl. Ed.)* **2020**, *7*, 111–124. [CrossRef]
18. Comi, A.; Idone, I. The Use of Electric Vehicles to Support the Needs of the Electricity Grid: A Systematic Literature Review. *Appl. Sci.* **2024**, *14*, 8197. [CrossRef]
19. Yuvaraj, T.; Devabalaji, K.R.; Kumar, J.A.; Thanikanti, S.B.; Nwulu, N.I. A Comprehensive Review and Analysis of the Allocation of Electric Vehicle Charging Stations in Distribution Networks. *IEEE Access* **2024**, *12*, 5404–5461. [CrossRef]
20. Singh, S.; Saket, R.K.; Khan, B. A comprehensive state-of-the-art review on reliability assessment and charging methodologies of grid-integrated electric vehicles. *IET Electr. Syst. Transp.* **2023**, *13*, e12073. [CrossRef]
21. Alam, M.S.; Singh, M.; Agarwal, P. Advancing V2G Technology: Design of an Adaptive Bi-directional Three Phase AFE Converter for High-Power Applications. *J. Energy Storage* **2024**, *92*, 112138. [CrossRef]
22. Mehdizadeh, M.; Nordfjaern, T.; Klöckner, C.A. Estimating financial compensation and minimum guaranteed charge for vehicle-to-grid technology. *Energy Policy* **2023**, *180*, 113649. [CrossRef]
23. Kumar, K.N.; Sivaneasan, B.; Cheah, P.H.; So, P.L.; Wang, D.Z.W. V2G capacity estimation using dynamic EV scheduling. *IEEE Trans. Smart Grid* **2014**, *5*, 1051–1060. [CrossRef]

24. Atmaja, T.D.; Susanti, V.; Mirdanies, M.; Muharam, A. V2G Development on Public Vertical Parking Lot to Support Community Energy Management System. *MATEC Web Conf.* **2018**, *164*, 01046. [CrossRef]
25. Kumar, M.; Vyas, S.; Datta, A. A Review on Integration of Electric Vehicles into a Smart Power Grid and Vehicle-to-Grid Impacts. In Proceedings of the 2019 8th International Conference on Power Systems (ICPS), Jaipur, India, 20–22 December 2019.
26. Onishi, V.C.; Antunes, C.H.; Trovão, J.P.F. Optimal energy and reserve market management in renewable microgrid-PEVs parking lot systems: V2G, demand response and sustainability costs. *Energies* **2020**, *13*, 1884. [CrossRef]
27. Van den Bossche, P. Electric Vehicle Charging Infrastructure. In *Electric and Hybrid Vehicles*; Elsevier: Amsterdam, The Netherlands, 2010; pp. 517–543. [CrossRef]
28. Drude, L.; Junior, L.C.P.; Rüther, R. Photovoltaics (PV) and electric vehicle-to-grid (V2G) strategies for peak demand reduction in urban regions in Brazil in a smart grid environment. *Renew. Energy* **2014**, *68*, 443–451. [CrossRef]
29. Comi, A.; Crisalli, U.; Sportiello, S. Forecasting the vehicle energy potential to support the needs of electricity grid: A floating car data-based methodology. *Front. Futur. Transp.* **2024**, *5*, 1500224. [CrossRef]
30. Pereirinha, P.G. Electric vehicles. In *Encyclopedia of Electrical and Electronic Power Engineering*; Elsevier: Amsterdam, The Netherlands, 2002; Volume 1–3, pp. 350–387. [CrossRef]
31. Maher-McWilliams, C. Case Study (UK). Electric Vehicle-to-Grid (V2G) Charging_Ofgem. 2021. Available online: https://www.ofgem.gov.uk/publications/case-study-uk-electric-vehicle-grid-v2g-charging (accessed on 20 November 2024).
32. Ebrahimi, M.; Ebrahimi, M.; Shafie-Khah, M.; Laaksonen, H. The Role of EV Parking Lots for Supporting the Distribution System Operation Considering EV Uncertainties. *Transp. Res. Procedia* **2023**, *70*, 263–270. [CrossRef]
33. Aydinoglu, A.C.; Senbil, M.; Saglam, D.; Demir, S. Planning of Parking Places on Transportation Infrastructure by Geographic Information Techniques. In Proceedings of the 2015 3rd International Istanbul Smart Grid Congress and Fair (ICSG), Istanbul, Turkey, 29–30 April 2015. [CrossRef]
34. Hooper, K.G. *Parking Generation Manual*, 5th ed.; Institute of Transportation Engineers: Washington, DC, USA, 2019.
35. Esmaili, M.; Shafiee, H.; Aghaei, J. Range anxiety of electric vehicles in energy management of microgrids with controllable loads. *J. Energy Storage* **2018**, *20*, 57–66. [CrossRef]
36. Inci, E. A review of the economics of parking. *Econ. Transp.* **2015**, *4*, 50–63. [CrossRef]
37. Hansen, W.G. How Accessibility Shapes Land Use. *J. Am. Plan. Assoc. J. Am. Inst. Plan.* **1959**, *25*, 73–76. [CrossRef]
38. Habib, K.M.N.; Morency, C.; Martin, M.T. Integrating parking behaviour in activity-based travel demand modelling: Investigation of the relationship between parking type choice and activity scheduling process. *Transp. Res. Part A Policy Pr.* **2012**, *46*, 154–166. [CrossRef]
39. Al-Sahili, K.; Hamadneh, J. Establishing parking generation rates/models of selected land uses for Palestinian cities. *Transp. Res. Part A Policy Pract.* **2016**, *91*, 213–222. [CrossRef]
40. Ketut, N.; Sukawati, S.A.; Agung, G.; Suryadarmawan, G. Parking land evaluation and planning at the Kereneng market, Denpasar city. *Int. J. Appl. Sci. Sustain. Dev.* **2022**, *4*, 39–48.
41. Kuo, P.F.; Hsu, W.T.; Putra, I.G.B.; Sulistyah, U.D. The proposed model for analyzing off-street parking Dynamics: A case study of Taipei City. *Transp. Res. Part A Policy Pr.* **2024**, *180*, 103965. [CrossRef]
42. Peng, T.; Chen, J.; Liu, K.; Qiu, Z.; Fu, Z.; Huang, Y. Examining the relationship between built environment and urban parking demand from the perspective of travelers. *J. Clean Prod.* **2023**, *385*, 135766. [CrossRef]
43. Xiong, J.; Shi, H.; Guo, G.; Liu, X.; Xiao, C.; Zhu, S. Research on intelligent transportation and shared parking spaces inurban areas based on multi-source data integration. *Comput. Electr. Eng.* **2023**, *110*, 108835. [CrossRef]
44. Borovskoy, A.; Yakovleva, E. Simulation Model of Parking Spaces Through the Example of the Belgorod Agglomeration. *Transp. Res. Procedia* **2017**, *20*, 80–86. [CrossRef]
45. Wang, B.; Nourinejad, M.; Bascuñán, O.; Roorda, M.J. Parking Survey Design Using Gamification. *Transp. Res. Procedia* **2024**, *76*, 596–607. [CrossRef]
46. Macea, L.F.; Serrano, I.; Carcache-Guas, C. A reservation-based parking behavioral model for parking demand management in urban areas. *Socio-Econ. Plan. Sci.* **2023**, *86*, 101477. [CrossRef]
47. Ibeas, A.; Dell'Olio, L.; Bordagaray, M.; Ortúzar, J.D.D. Ortúzar. Modelling parking choices considering user heterogeneity. *Transp. Res. Part A Policy Pract.* **2014**, *70*, 41–49. [CrossRef]
48. Alogdianakis, F.; Gkartzonikas, C.; Dimitriou, L. Development of a credit scheme for managing mobility in university communities: Results from a feasibility study. *Res. Transp. Bus. Manag.* **2024**, *53*, 101106. [CrossRef]
49. Zhang, Y.; Luo, X.; Wang, Z.; Mai, Q.; Deng, R. Estimating charging demand from the perspective of choice behavior: A framework combining rule-based algorithm and hybrid choice model. *J. Clean Prod.* **2022**, *376*, 134262. [CrossRef]
50. Rodríguez, A.; Alonso, B.; Moura, J.L.; Dell'olio, L. Analysis of user behavior in urban parking under different level of information scenarios provided by smart devices or connected cars. *Travel Behav. Soc.* **2024**, *37*, 100847. [CrossRef]
51. Aditya, A.; Anwarul, S.; Tanwar, R.; Koneru, S.K.V. An IoT assisted Intelligent Parking System (IPS) for Smart Cities. *Procedia Comput. Sci.* **2022**, *218*, 1045–1054. [CrossRef]
52. Reddy, D.S.; Babu, K.V.G.; Murthy, D.L.N. Transportation Planning Aspects of a Smart City-Case Study of GIFT City, Gujarat. *Transp. Res. Procedia* **2016**, *17*, 134–144. [CrossRef]
53. Campisi, T.; Caselli, B.; Rossetti, S.; Torrisi, V. The Evolution of Sustainable Mobility and Urban Space Planning: Exploring the factors contributing to the Regeneration of Car Parking in Living Spaces. *Transp. Res. Procedia* **2022**, *60*, 76–83. [CrossRef]

54. van Heuveln, K.; Ghotge, R.; Annema, J.A.; van Bergen, E.; van Wee, B.; Pesch, U. Factors influencing consumer acceptance of vehicle-to-grid by electric vehicle drivers in the Netherlands. *Travel Behav. Soc.* **2021**, *24*, 34–45. [CrossRef]
55. Kempton, W.; Pearre, N.S.; Guensler, R.; Elango, V.V. Influence of Battery Energy, Charging Power, and Charging Locations upon EVs' Ability to Meet Trip Needs. *Energies* **2023**, *16*, 2104. [CrossRef]
56. Kim OT, T.; Tran, N.H.; Nguyen, V.; Kang, S.M.; Hong, C.S. Cooperative between V2C and V2V Charging: Less Range Anxiety and More Charged EVs. In Proceedings of the 2018 International Conference on Information Networking (ICOIN), Chiang Mai, Thailand, 10–12 January 2018; p. 965.
57. Guille, C.; Gross, G. A conceptual framework for the vehicle-to-grid (V2G) implementation. *Energy Policy* **2009**, *37*, 4379–4390. [CrossRef]
58. Harb, A.; Hamdan, M.; Hamdan, M. Power Quality and Stability Impacts of Vehicle to Grid (V2G) Connection. In Proceedings of the 2017 8th International Renewable Energy Congress (IREC), Amman, Jordan, 21–23 March 2017.

energies

MDPI

Article

Optimizing Electric Cold-Chain Vehicle Scheduling for Sustainable Urban Logistics: A Novel Framework Balancing Freshness and Vehicle Charging

Zhenkun Gan [1], Peiwu Dong [1], Zhengtang Fu [2,*], Yanbing Ju [3] and Yajun Shen [1]

[1] School of Economics, Beijing Institute of Technology, Beijing 100081, China; ganzhenkun@bit.edu.cn (Z.G.); dongpeiwu@bit.edu.cn (P.D.); 18801032250@163.com (Y.S.)
[2] School of Environment, Tsinghua University, Beijing 100084, China
[3] School of Management, Beijing Institute of Technology, Beijing 100081, China; juyb@bit.edu.cn
* Correspondence: zhengtangbit@163.com

Abstract: Cold-chain logistics, characterized by high energy consumption and significant emissions, pose a critical challenge for the green transformation of global transportation. Electric cold-chain vehicles have emerged as a promising solution to reduce carbon emissions in urban logistics. However, their scheduling is highly complex due to the need to balance freshness and charging requirements, presenting operational challenges for cold-chain companies. To address this issue, this paper proposes an optimization model and algorithm for the efficient scheduling of these innovative electric cold-chain vehicles. First, we define the unique features of these vehicles and establishes an operational framework tailored to cold-chain logistics. Subsequently, we develop a mixed-integer programming model to optimize freshness preservation. Additionally, we design a state-of-the-art algorithm based on an improved genetic algorithm to solve the scheduling model effectively. Numerical experiments conducted using operational data from Shanghai, China, validate the proposed method and algorithm. This study provides valuable insights and tools to support the green transformation of urban cold-chain logistics and contributes to the reduction of urban carbon emissions.

Keywords: electric cold-chain vehicles; carbon reduction; charging constraints; freshness loss; mixed-integer programming; green transformation

1. Introduction

With the rise in residents' living standards, customers now have higher expectations for the quantity and quality of cold-chain products [1,2]. However, traditional fuel-powered cold-chain vehicles face restrictions in many cities due to their high pollution emissions, excessive noise, and other drawbacks, which limit their operating time and range [3]. Notably, fuel-powered refrigerated vehicles are substantial energy consumers, which not only results in considerable carbon emissions but also poses significant challenges for pollution control. As a result, the management of these high-emission vehicles has become a priority for governments worldwide [4,5]. Hence, these limitations generate significant challenges for delivering time-sensitive cold-chain products. Fortunately, the emergence of electric cold-chain vehicles has accelerated the integration and development of cleaner and more efficient solutions in this field.

Electric vehicles offer numerous advantages, such as low noise, zero emissions, and environmental friendliness [6,7], enabling them to conduct cold-chain distribution at any time and in any area of the city. These benefits are significant for building green cities and reducing urban air pollution [8]. At the same time, as customer expectations for cold-chain services grow, there is a demand for both refrigerated and frozen goods to be delivered in a single shipment to minimize unloading time and enhance operational efficiency. However, traditional single-temperature electric cold-chain vehicles cannot meet these evolving needs. Therefore, optimizing the distribution and charging scheduling of electric cold-chain vehicles to achieve multi-temperature co-distribution has become a critical focus for cold-chain company managers.

Compared with traditional cold-chain logistics research, the main contributions of this study are as follows. Firstly, this paper addresses the business needs of an intercity cold-chain delivery company in Shanghai by proposing the use of two-compartment electric cold-chain vehicles for cold-chain distribution. Secondly, considering the constraints of compartment capacity and vehicle battery capacity, a mixed-integer programming (MIP) model is constructed to minimize freshness loss. To solve this model, a state-of-the-art genetic algorithm is employed, taking into account the complexity of integrating vehicle delivery routes and charging schedules in a multi-temperature co-distribution environment. During the chromosome decoding process, a rollback strategy is introduced to enhance the feasibility of infeasible solutions, and a comparative study is conducted with mainstream penalty strategies.

This study holds both practical significance and theoretical contributions, offering efficient decision-making support for cold-chain company managers and providing a new research framework for scholars in the field of sustainable cold-chain logistics. The remainder of this paper is organized as follows: Section 2 reviews the related works. Section 3 defines the optimization problem and proposes the model. Section 4 develops the novel algorithm to solve the model. Section 5 conducts the case study in Shanghai. Section 6 proposes the managerial insights derived from the above experiment. Section 7 summarizes the paper and proposes future research directions.

2. Literature Review

2.1. Electric Vehicle Scheduling

Nowadays, electric vehicles (EVs) have become very popular for eco-friendly transportation [9,10]. In this field, most scholars mainly focus on vehicle path optimization and optimal scheduling strategies for vehicle charging and swapping. In terms of vehicle path optimization, Seyfi et al. [11] handled the problem of electric vehicle path optimization in multi-objections. They constructed a MIP model and used a large-scale search algorithm to solve the model. Hermann et al. [12] studied the routing optimization problem of EVs and made improvements to the genetic algorithm, which improved its performance in local search and global optimization. Wang et al. [13] researched the routing optimization issue of electric vehicles in multi-yard collaboration, incorporating time window constraints and shared electric pile features into the model construction, and developed effective algorithms to solve the model.

EV scheduling strategies for charging and swapping are also hot topics for researchers [14,15]. Hof et al. [16] introduced the battery-swapping process of electric vehicles into the scheduling model and used an adaptive variable neighborhood search algorithm for model solving. Basso et al. [17] considered the energy prediction problem in the path optimization process and applied machine learning methods to the path optimization modeling process of electric vehicles. Bezzi et al. [18] researched the routing optimization issue of EVs under multiple technology integration modes and proposed an improved branch pricing algorithm. The experimental results verified the progressiveness of the proposed model. Montoy et al. [19] considered the nonlinear relationship between the quantity and time during charging and constructed a heuristic algorithm to handle the model. Jie et al. [20] considered battery replacement, taking into account the electric vehicle load, battery capacity, and replacement cost. They constructed an integer programming mathematical model and used a hybrid-heuristic algorithm to get the solutions. Schiffer et al. [21] analyzed the charging behavior of pure electric logistics fleets and constructed a multi-objective MIP model to minimize fleets' total distance and total charging cost. At the same time, the influence of charging station location on vehicle path optimization was also considered.

2.2. Cold-Chain Logistics

Cold-chain logistics is a significant topic in academia, focusing on the transportation of vaccines, food, and other products closely tied to people's daily lives [22,23]. Scholars primarily concentrate on supply chain coordination for cold-chain products and the environmental impact of cold-chain transportation. In recent years, the multi-temperature coordination problem has emerged as a key research hotspot. Regarding supply chain coordination, Awad et al. [24] explored collaborative distribution within cold-chain transportation, systematically analyzing existing models, processes, and methods for supply chain transportation and distribution management, thereby laying a foundation for subsequent research. Shi et al. [25] investigated vehicle collaborative optimization in cold-chain supply chains, developing a multi-objective optimization model validated with real-world data. Zhang et al. [26] addressed the green transformation of cold-chain transportation by constructing a site selection optimization model for urban cold-chain logistics under carbon emission constraints. Their case study analysis provided policy recommendations and scientific support for government departments to optimize urban cold-chain transportation systems.

2.3. Optimization Model and Methodology

High-quality optimization models and methods are essential tools for enhancing the operational efficiency of enterprises [27,28]. Wei et al. [29] proposed a cold-chain network collaborative delivery model based on consumer demand and solved it using a genetic algorithm, verifying the applicability of the model. Considering the environmental impact of cold-chain transportation, Saif et al. [30] considered the impact of greenhouse gases generated during cold-chain transportation on the environment, constructed a mathematical model to minimize greenhouse gas emissions, and proposed a new hybrid simulation optimization algorithm for problem-solving. Hariga et al. [31] considered the carbon emissions on the social environment during the whole lifecycle of transportation and storage and constructed a cold-chain transportation model that takes into account carbon tax regulatory policies. Zheng et al. [32] studied the truck scheduling issue in cold-chain transportation scenarios and solved the cold-chain truck scheduling model with the heuristic algorithms.

Supply chain optimization management is a critical application scenario of optimization methodology, and many scholars have made significant contributions [33,34]. Ren et al. [35] studied the collaborative optimization issue of a large-scale cold-chain transportation supply chain, analyzed and optimized the e-commerce cold-chain operations using LNS methods, and verified the algorithm through practical cases. While studying the optimization of the cold-chain distribution aspect, Chen et al. [36] used a large-scale neighborhood search algorithm based on the actual problem background to solve the scheduling optimization problem of multi-compartment cold-chain vehicles. Wu et al. [37] studied the dynamic optimization problem of cold-chain vehicles, solved the transportation scheduling with a mixed ant colony algorithm, and obtained an effective and reliable method in polynomial time.

2.4. Research Gap

From the literature reviewed, it is evident that existing studies have primarily focused on the characteristics of cold-chain logistics and electric vehicles (EVs) separately. Only a limited number of studies have integrated the analysis of multi-temperature cold-chain logistics with the unique features of electric vehicles. However, as the industry evolves, the multi-temperature requirements in cold-chain logistics have become an urgent issue that demands resolution. Additionally, the complexity of route optimization for multi-temperature cold-chain vehicles presents a significant opportunity for advancing the theoretical development of vehicle routing optimization. Therefore, this study will investigate the optimization of multi-temperature EV cold-chain vehicle routing to address the existing research gap.

3. Problem Description and Modeling

3.1. Model Specification

In this paper, a distribution path optimization problem for pure electric cold-chain vehicles is proposed, taking into account the capacity constraints, charging costs, refrigeration costs, and transportation costs of vehicles. Cold-chain logistics is a door-to-door logistics service that involves the coordinated operation of vehicles and service points, thus presenting a certain level of complexity [38]. As shown in Figure 1, N = {1, 2, 3..., n} is the set of customer points, T is the set of charging stations, and O is the starting point. Then, the problem can be defined as the existence of a directed graph G = (V,S), where $V = N \cup T \cup O$ represents the set of all points, and $(i,j) \in S$ represents the set of all arcs in the graph, where $i \neq j$. Cold-chain vehicles start from the distribution center, travel between different points while meeting capacity constraints, and complete the distribution works of all customer points. Each customer point has at least one type of goods demand (refrigerated or frozen goods) and is only served once. During the service process, the vehicle can be charged multiple times [39]. The two compartments of the vehicle have their maximum capacity limits, and each vehicle does not restock midway, with no returns occurring. The number of vehicles at the delivery station is fixed.

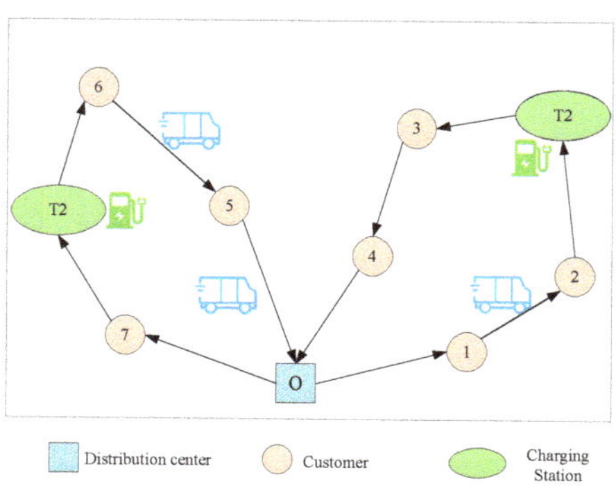

Figure 1. Cold-chain EV delivery network diagram (the icon is supported by https://www.iconfont.cn/, accessed on 15 January 2025).

Meanwhile, due to the high requirement for freshness in cold-chain transportation goods, we also considered the variation of goods' freshness over time. Figure 2 shows the example decrease curve in the freshness of cold-chain goods in short-term cold-chain logistics and long-term cold-chain logistics ($a_1 < a_2 < 0$, $a_3 > a_4 > 1$, b_1, b_2, b_3, b_4 is the constant variable). Short-term cold-chain delivery (e.g., urban daily delivery) could be set as linear, but in long-term cold-chain delivery (e.g., shipping cold-chain logistics) the decrease could be set as nonlinear [40,41]. This paper investigated the daily cold-chain delivery, which falls under the category of short-term cold-chain delivery. Hence, a linear degradation (Figure 2a) was adopted.

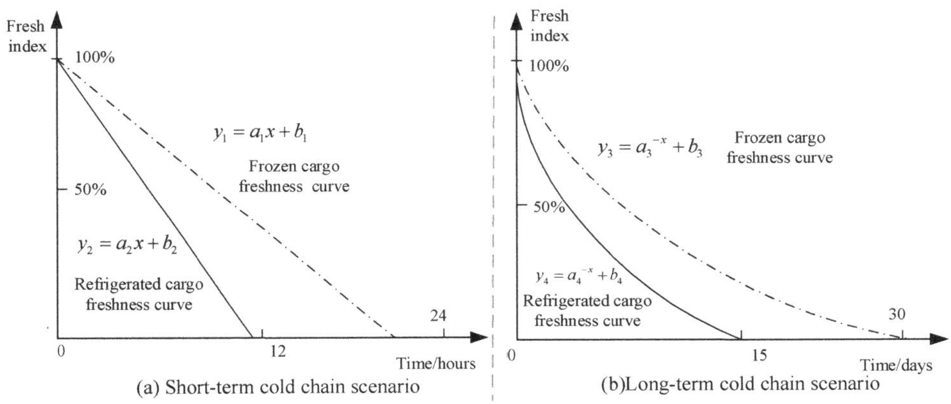

Figure 2. Freshness curve of cold-chain goods in one day.

It is noteworthy that the determination of the freshness degradation curve is inherently complex, involving multiple factors, such as cold-chain delivery distance, product type, and seasonal temperatures, etc. Hence, the linear degradation employed in this study serves merely as an example. In practical applications, it is essential to design the degradation curve function according to specific scenarios.

3.2. Mathematic Model Proposed

In the process of model construction, this paper follows the basic assumptions:

(1) The electricity consumption of vehicles is constant per kilometer.
(2) Cold-chain vehicles have the same unloading time at different customer points.
(3) When a cold-chain vehicle is charging at a charging station, the refrigeration system needs to work properly.
(4) The driver strictly follows the established delivery plan route.
(5) The demand for goods of customer nodes could be known in advance.
(6) No returns occur during the delivery process.

3.2.1. Notations

(1) Sets

N denotes the set of customer points,
T denotes the set of charging stations,
O denotes the starting point,
K denotes the set of cold-chain vehicles,
V denotes the set of all points in the directed graph, $V = N \cup T \cup O$

(2) Variables and parameters

d_{ij} is the distance between nodes i and j,
q_i^1 is the demand for refrigerated goods at site i,
q_i^2 is the demand for frozen goods at site i,
g_{jk} is the cumulative electricity consumption of the k-th vehicle at the station,
e_1 is the electricity consumption per unit distance,
e_2 is the electricity consumption per unit of time in the refrigerated compartment,
e_3 is the electricity consumption per unit of time in the frozen compartment,
C is the total battery capacity of cold-chain vehicles,
v is the average speed of cold-chain vehicles,
Q_1 is the capacity of the vehicle's refrigerated compartment,
Q_2 is the capacity of the vehicle's frozen compartment,
α_1 is the freshness sensitivity factor of refrigerated goods,
α_2 is the freshness sensitivity factor of frozen goods,
T_1 is the freshness date of refrigerated goods,
T_2 is the freshness date of frozen goods,
f_{kj}^1 is the freshness of refrigerated goods delivered to customer j on the k-th EV,
f_{kj}^2 is the freshness of frozen goods delivered to customer j on the k-th EV,
w_i is the dwell time of the vehicle at point i.

(3) Decision variables

x_{ijk} is the 0, 1 variable. If EVs k pass through arc (i, j), it is 1; otherwise, it is 0.
t_{ki} is the time when vehicle k reaches node i.

3.2.2. Mathematical Model

Due to this paper primarily focusing on the impact of introducing electric vehicles in cold-chain logistics distribution, the objective function was established to minimize the loss of freshness in cold-chain products.

Equation (1) shows the optimization objection, which represents minimizing the freshness loss when customers receive goods at their points:

$$Min \sum_{\in V} \sum_{j \in N} \sum_{k \in K} x_{ijk} q_j^1 (1 - f_{kj}^1) + \sum_{i \in V} \sum_{j \in N} \sum_{k \in K} x_{ijk} q_j^2 (1 - f_{kj}^2) \quad (1)$$

Equation (2) shows the relationship between the freshness of refrigerated goods and time:

$$f_{kj}^1 = 1 - \frac{\alpha_1 t_{ki}}{T} \; \forall i, j \in N \quad (2)$$

Equation (3) is the function relationship between the freshness of frozen products and time:

$$f_{kj}^2 = 1 - \frac{\alpha_2 t_{ki}}{T} \quad (3)$$

Equation (4) is the maximum capacity constraint for compartment 1 (refrigerated) of the vehicle:

$$\sum_{i \in V} \sum_{j \in V} q1_i x_{ijk} \leq Q_1 \quad \forall k \in K \quad (4)$$

Equation (5) represents the maximum capacity constraint for compartment 2 (frozen) of the vehicle:

$$\sum_{i \in V} \sum_{j \in V} q2_i x_{ijk} \leq Q_2 \quad \forall k \in K \quad (5)$$

Equation (6) indicates that the customers are only served by one vehicle and served once:

$$\sum_{i \in V} \sum_{k \in K} x_{ijk} = 1 \quad \forall j \in N \tag{6}$$

Equation (7) is a constraint on the number of vehicles, indicating that there are m EVs participating in the delivery operation:

$$\sum_{i \in \{O\}} \sum_{j \in V} \sum_{k \in K} x_{ijk} = \sum_{i \in V} \sum_{j \in \{O\}} \sum_{k \in K} x_{ijk} \tag{7}$$

Equation (8) represents the cumulative quantity of electricity consumption of the k-th cold-chain vehicle, and specifies that when the vehicle is at the starting point or after leaving the charging station, it is fully charged:

$$g_{jk} = \begin{cases} g_{ik} + \sum\limits_{i \in N} \sum\limits_{j \in N} \sum\limits_{k \in K} e_0 d_{ij} x_{ijk} + \sum\limits_{i \in N} \sum\limits_{j \in N} \sum\limits_{k \in K} (\frac{x_{ijk}}{v} + w_{kj}) \cdot (e_1 + e_2), \forall i, j \in N, k \in K, i \neq j \\ 0, \forall j \in T \cup \{O\}, k \in K \end{cases} \tag{8}$$

Equation (9) represents the quantity constraint of electricity consumption during driving:

$$g_{jk} + e_0 d_{ij} x_{ijk} + \frac{x_{ijk}}{v}(e_1 + e_2) \leq C \qquad \forall i \in N, j \in T \cup \{O\}, k \in K \tag{9}$$

Equations (10) and (11) represent the range of values for decision variables:

$$x_{ijk} = \begin{cases} 1, \text{ vehicle k passes through } arc(i, j) \\ 0, \text{ otherwise} \end{cases} \tag{10}$$

$$t_{ki} \geq 0 \tag{11}$$

Equation (12) represents the range of values for other variables:

$$0 \leq i, j \leq n, 1 \leq t \leq T, 1 \leq k \leq m \tag{12}$$

4. Algorithm Framework

Considering that this problem is an MIP problem, the selection of charging stations also needs to be considered. Therefore, the optimization scope of the problem was quite big, and the complexity was high, making it an NP-hard problem. Therefore, constructing an effective algorithm is crucial to get the perfect solutions. The genetic algorithm is widely used to address NP-hard problems for its great parallelism and clear coding logic. Therefore, we also adopted a genetic algorithm for model solving. To improve the computational performance and problem adaptability of the algorithm, a rollback strategy was adopted for the processing of some infeasible solutions in the chromosome decoding process, thereby transforming infeasible solutions into feasible solutions and improving the searchability of the algorithm, as shown in Figure 3.

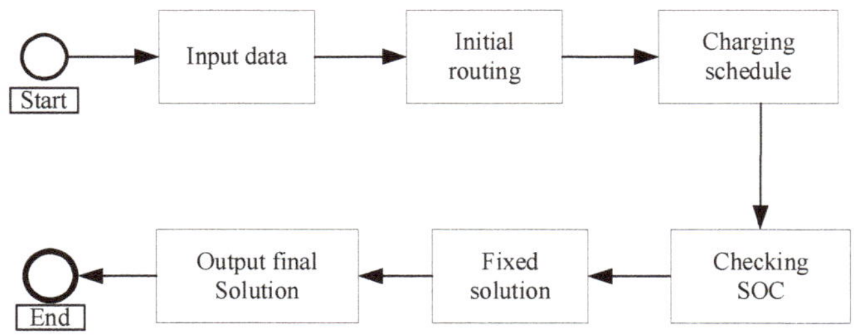

Figure 3. Diagram of the main process of the algorithm.

4.1. Algorithm Step Process

In practical applications, it is essential to encode the solution to the problem in a manner that enables computer recognition and facilitates efficient computational processing. The initial solution to this problem is represented as a sequence of numbers. During the encoding process, relevant constraints must be taken into account, such as the capacity limits of the two compartments and the vehicle's energy consumption constraints. The details are shown in Figure 4.

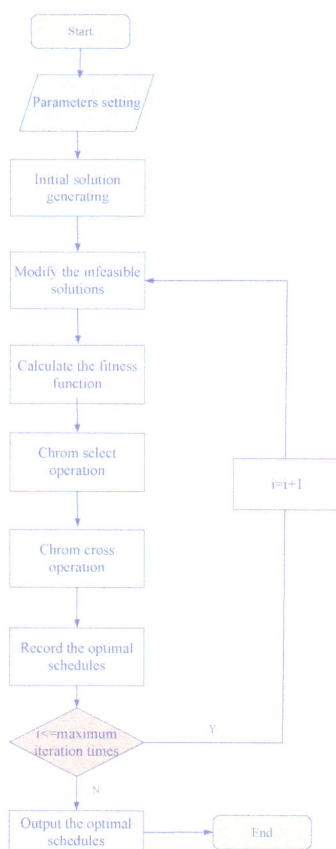

Figure 4. Genetic algorithm flowchart with a modification operator.

Step 1. Generate the initial solution. According to the capacity constraint of each compartment, divide the specific customer points served by different vehicles. If any storage space exceeds the capacity limit, the current customer point will be allocated to the next vehicle.

Step 2. Calculate the charging location. Calculate the location $\xi_i, \{i = 1,2,3\ldots n\}$ of charging demand generation based on the vehicle quantity constraint of electricity consumption.

Step 3. Modify the infeasible solutions. Based on the location ξ_i of charging demand generation, determine whether the vehicle can travel from this point to the nearest charging station. If the remaining battery cannot support reaching the nearest charging station, execute a rollback strategy, adjust the charging position, and generate a feasible solution.

Step 4. Calculate the fitness function. Based on the path results of Step 3, calculate the freshness loss values of the current refrigerated and frozen goods, respectively, and sum them to obtain the total freshness loss.

Step 5. Perform selected operation. Adopt a roulette wheel strategy for selection of operations, select a scheduling strategy with high adaptability, and enter the next round of operations.

Step 6. Perform cross-operation. Adopt a two-point crossover mechanism to exchange genes in the crossover region, followed by conflict detection operations.

Step 7. Solution mutation. Perform mutation operation with a certain probability and randomly select two positions in the chromosome for exchange operation.

Step 8. Record the optimal schedules. Select and record the vehicle delivery plan, charging location, and total freshness loss value for the best fitness individual.

Step 9. Determine termination conditions. Determine whether the current genetic algebra meets the set maximum genetic algebra. If it does not meet the stopping condition, return to Step 5.

4.2. Algorithm Decoding Process

The decoding process of chromosomes is a crucial step in enabling genetic algorithm solutions. This process can be broken down into three key components. First, customer points are grouped based on their demands and the compartment's capacity constraints, with each vehicle assigned to one group. Next, following the delivery sequence derived from the grouping results, the algorithm identifies the necessary charging points for the electric vehicles (EVs) and inserts them into the route. Finally, using the established delivery and charging routes, the fitness values of the chromosomes are calculated, completing the decoding process.

(1) Demand nodes classification

As shown in Figure 5, the algorithm utilizes the demand point number as chromosome encoding information, where the numbers represent the demand points. By using this encoding method, it can effectively ensure that each customer point is only served once. Meanwhile, we should ensure that the refrigerated and frozen compartments of each vehicle do not exceed their respective maximum capacity constraint (Constraints 4 and 5), to sequentially allocate different demand points to different vehicles and complete the initial correspondence between vehicles and customer points.

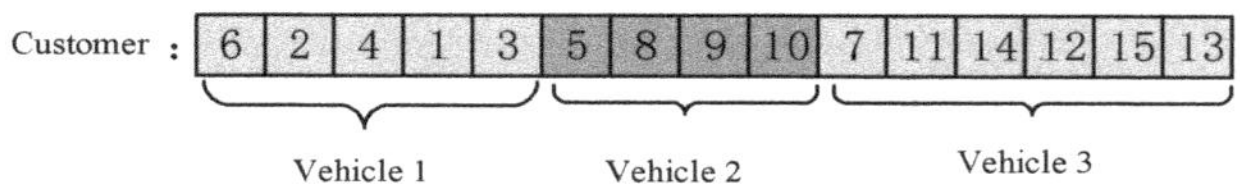

Figure 5. Diagram of vehicle grouping.

(2) Insert charging station

Calculate the charging positions $i \in N$ of different vehicles based on their battery capacity constraints. After completing the service at node i, the remaining battery of the vehicle is no longer able to support its journey to the next destination (customer point or distribution center). The vehicle needs to drive to the nearest charging station, wait for the charging to be completed, and then continue with the operation, as shown in Figure 6. Unlike electric passenger vehicles, which only need to consider the electricity consumption of driving, we comprehensively considered the consumption of the driving and cooling systems during the vehicle's driving process when calculating the cumulative electricity consumption g_{ik} of the k cold-chain vehicle.

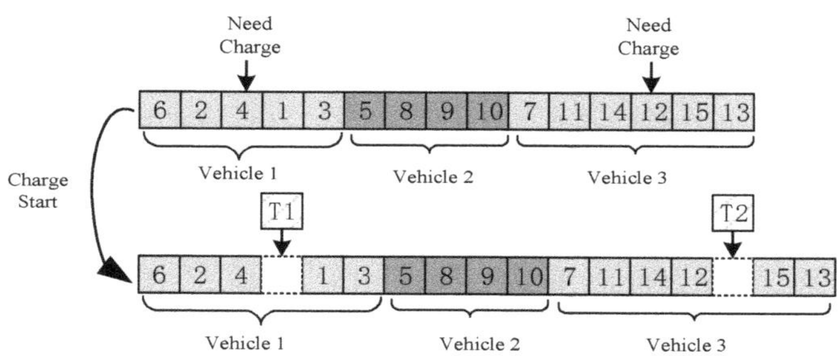

Figure 6. Charging station insertion diagram.

The detailed processes for selecting the EVs' charging station are as follows:

Step 1. Initialize the cumulative electricity consumption of each vehicle, $g_{ik} = 0, \forall k \in K$.

Step 2. Calculate the charging position. Accumulate the driving distance and cooling time of the vehicle, calculate the cumulative electricity consumption $g_{(i+1)k}$ from the position i to the next target point $i + 1$, and determine whether the cumulative electricity consumption exceeds the battery capacity $g_{(i+1)k} > C$. If it is met, execute Step 3. If it is not met, continue to accumulate electricity consumption.

Step 3. Select the most suitable charging station. According to the charging location provided in Step 2, calculate the closest charging station to that point. Determine whether the current remaining battery can support the vehicle to reach the charging station. If it cannot be reached, move the charging position back by one point until a charging station that meets the conditions is found.

Step 4. After the charging process, the cumulative electricity consumption is reset to zero, $g_{(i+1)k} = 0$, indicating that it is fully charged.

Step 5. Return to Step 2 and select the charging station for the later stage of charging.

(3) Calculate fitness value

Based on the results of vehicle distribution grouping and the charging schedule, calculate the time t_{ki} for different vehicles to go to the customer node. With the time sequence, we could calculate the freshness loss for the current refrigerated and frozen goods. As it is a loss minimization problem, we assigned the reciprocal of the loss value

$\frac{1}{loss}$ to the fitness of the corresponding chromosome. The smaller the loss value, the greater the fitness, and the better the individual can enter the next iteration process.

5. Case Study

5.1. Case Background

To validate the correctness of the proposed MIP model and the developed algorithm, we conducted a case study using actual delivery data from a cold-chain company in Shanghai. The case represents the real-world distribution scenario of the cold-chain logistics company in Shanghai. The distribution starts from a location in Jiading District, Shanghai, and covers the entire city, including a total of 117 customer points. The goods requested by customers are categorized as refrigerated and frozen, with demand quantities known in advance. The demand distribution across customer points is illustrated in Figure 7.

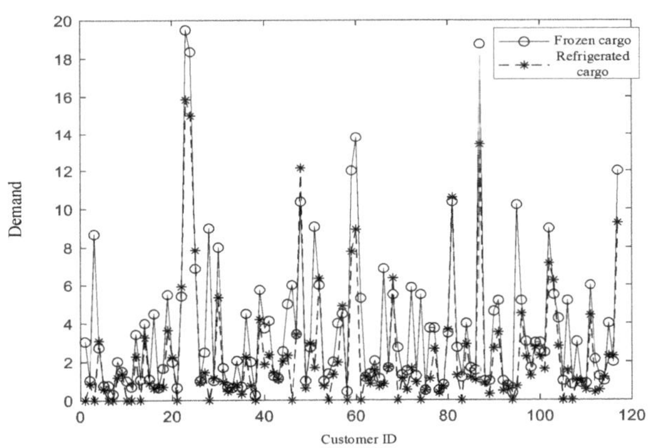

Figure 7. Customers' demand for goods in line chart.

To better contrast the differences in the distribution of refrigerated cargo versus frozen cargo, we constructed box plots for analysis, as Figure 8 shows. Compared to refrigerated cargo, the demand for frozen cargo exhibited more dispersed data points, indicating an uneven demand pattern that necessitates careful distribution planning.

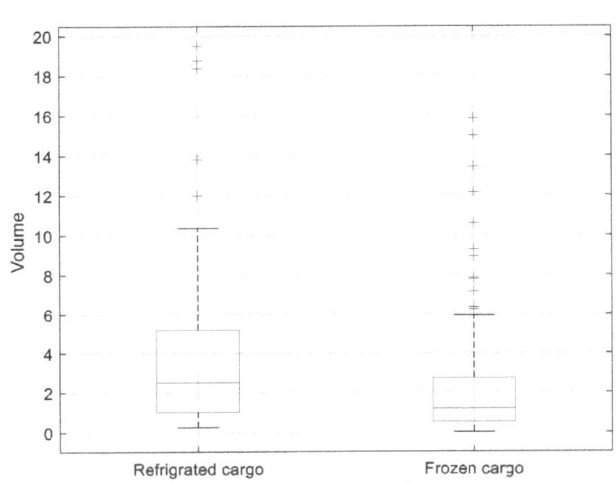

Figure 8. Customers' demand for goods in box diagram.

The locations of charging stations in Shanghai were obtained from the Baidu Map API. The simulation was performed on a computer with a Win10 64-bit operating system, 16 GB of RAM, and a CPU frequency of 2.6 GHz.

5.2. Parameter Settings

This experiment was conducted using Matlab 2017b, with the genetic algorithm parameters set as follows: population size of 50, 500 generations, crossover probability of 0.9, and mutation probability of 0.05. To validate the effectiveness of the proposed method, real business data from a major cold-chain company in Shanghai were utilized. The distribution data were categorized by range, as detailed in Table 1, and three cold-chain distribution scenarios were constructed: small scale (R1), medium scale (R2), and large scale (R3). Additional experimental parameters are provided in Table 2.

Table 1. Distribution case.

Case Name	Distribution Area	Number of Customers	Number of Charging Stations
R_1	Jiading District, Baoshan District, Hongkou District, Jingan District	28	25
R_2	Jiading District, Baoshan District, Hongkou District, Jingan District, Huangpu District, Xuhui District	49	33
R_3	All regions of Shanghai	117	78

Table 2. Parameter settings.

Parameter	Value	Unit
Capacity of refrigerated compartment	30	box
Capacity of frozen compartment	20	box
Electricity capacity	40	kWh
Driving electricity consumption	0.3	kWh/km
Refrigerated system power	2.5	kW
Unloading time	20	min
Charging time	60	min

5.3. Case Study Results

The experiment in this paper was divided into two parts. (1) We compared the distribution of two-compartment cold-chain electric vehicles with that of single-compartment cold-chain electric vehicles. By analyzing relevant distribution indicators, the advantages and disadvantages were compared. (2) We compared the genetic algorithm with the roll-back strategy and the genetic algorithm based on the traditional penalty strategy, using the distribution data of the entire city of Shanghai (R_3 case) as input, to verify the merits and drawbacks of the improved algorithm.

A comparative analysis of the indicators across various vehicle models is detailed in Table 3, revealing the presented observations.

Table 3. Comparison of experimental results.

Case Name	Single-Temperature Cold-Chain EV			Multi-Temperature Cold-Chain EV		
	Freshness Loss Value	Loss Rate/%	Charging Cycles	Freshness Loss Value	Loss Rate/%	Charge Cycles
R_1	19.36	10.86%	3	7.57	4.25%	1
R_2	38.24	13.26%	9	14.84	5.15%	2
R_3	186.35	26.55%	46	53.03	7.86%	22

In terms of algorithm performance analysis, we evaluated the decoding process of genetic algorithms using both the rollback strategy and the traditional penalty operations. As shown in Figure 9, the implementation of the rollback strategy effectively identified better computational results, enhanced the efficiency of genetic algorithm operations, and improved the overall algorithm performance. The traditional genetic algorithm does not incorporate a rollback operation [42]. The primary reason for this improvement is that the rollback strategy enabled the transformation of a portion of chromosomes from infeasible solutions into feasible solutions, allowing them to be retained and proceed to the next iteration. In contrast, under the traditional penalty strategy, infeasible solutions are often unable to advance to the next iteration, leading to the loss of potentially valuable chromosomes. Consequently, compared to traditional penalty strategies, the rollback strategy offered greater flexibility in chromosome decoding and was well suited for solving NP-hard problems under complex constraints.

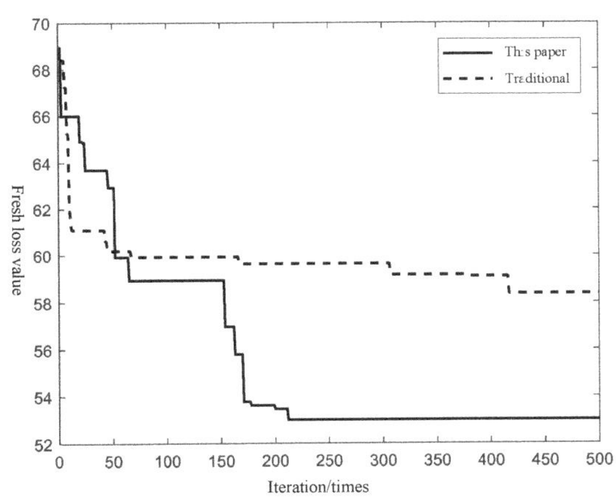

Figure 9. Computational simulation process for minimizing fresh loss.

To test the robustness of the algorithm, we conducted sensitivity analysis experiments, focusing on the vehicle's range and population size for the sensitivity tests, as Figure 10 shows. Figure 10a illustrates the impact of the vehicle's range on the outcomes. As can be seen from the graph, the algorithm consistently yielded satisfactory charging and distribution methods across different ranges, optimizing the reduction in delivery freshness loss. Figure 10b demonstrates the influence of the algorithm's population size on the results. The graph reveals that a larger population size can significantly enhance the algorithm's convergence rate. Therefore, it is advisable to consider moderately increasing the population size during the algorithm parameter configuration process.

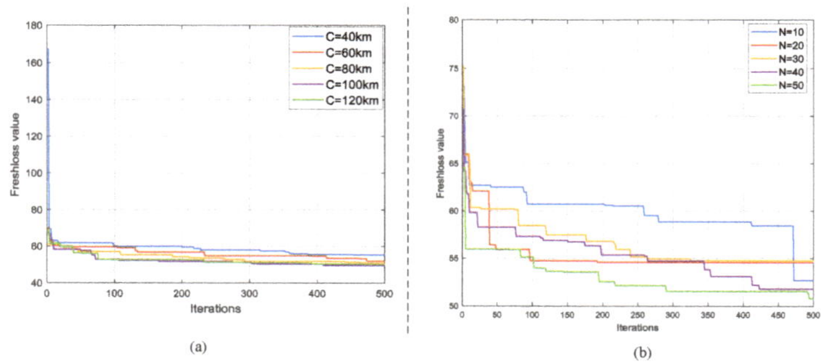

Figure 10. Sensitivity analysis experiments. (**a**) vehicle's range parameter analysis (**b**) algorithm's population size parameter analysis.

6. Managerial Insights

(1) Adopting multi-compartment cold-chain EVs could reduce charging frequency and minimize cargo loss.

Under the same delivery scenario, the multi-compartment electric cold-chain vehicle demonstrated significant advantages in reducing the freshness loss of goods and lowering the loss rate. In terms of freshness loss, compared to single-compartment cold-chain vehicles, the two-compartment model achieved a notable reduction in freshness loss, which could save significant costs, thereby enhancing user satisfaction. Additionally, regarding charging performance, the two-compartment cold-chain vehicle not only saved a significant amount of total freshness loss but also reduced the charging cycles compared to its single-compartment counterpart.

(2) The theoretical method proposed in this study applies to cold-chain logistics scenarios of various scales and could be adopted by multiple types of customers to support sustainable development.

Through analysis of different delivery scenarios, it can be seen that compared to delivery cases R_1 and R_2, there was a significant increase in both the loss of goods and the number of charges, especially in the case R_3. This is due to the delivery range of cases R_1 and R_2 being relatively close to the delivery starting point (Jiading District, Shanghai), and because the battery capacity of the vast majority of vehicles could run back to the starting point after a single delivery. The case R_3 involved distributions throughout the city, with a significant increase in the distribution radius. Some distribution routes span a considerable distance, extending from the northern part of Jiading District in Shanghai to the southern part of Pudong District, crossing the entire Shanghai urban area. Consequently, when formulating large-scale, long-distance local distribution plans, it is essential to incorporate reasonable optimization of vehicle charging strategies.

(3) Logistics enterprises should actively embrace new-energy vehicles and emerging optimization technologies to reduce carbon emissions for the world.

Cold-chain logistics enterprises are characterized by high energy consumption and high carbon emissions. The introduction of multi-compartment new-energy cold-chain vehicles can help reduce corporate carbon emissions and enhance corporate social responsibility. Although the scheduling optimization of electric cold-chain vehicles is complex, requiring consideration of both charging and operational aspects, this case study demonstrated that by adopting reasonable and effective distribution optimization theories and models, it is possible to balance the charging and operational processes during the generation of scheduling plans. This approach not only reduces operational costs and improves customer satisfaction but also contributes to the global green transition.

7. Conclusions

Research on routing optimization for cold-chain vehicles has been a critical issue in the logistics field, but EV cold-chain logistics research is emerging in interdisciplinary research. The multi-temperature co-distribution of electric vehicle (EV) cold-chain systems not only addresses practical needs but also enriches the classical vehicle routing problem (VRP) theoretical model. This paper introduced the novel concept of "freshness loss" to quantify the value degradation of cold-chain products over the transit time and proposed the use of two-compartment cold-chain EVs for delivery operations, thereby optimizing the business processes of cold-chain logistics. In terms of model construction, this paper aimed to minimize the total freshness loss of all goods while comprehensively considering constraints, such as vehicle range and compartment capacity, ultimately formulating a mixed-integer programming model. For algorithm design, an improved genetic algorithm was employed to solve the model. During chromosome decoding, a rollback strategy was introduced, which significantly enhanced the algorithm's performance compared to the traditional penalty strategy, yielding a more effective delivery decision scheme. Finally, the feasibility of the proposed method and algorithm was validated using real-world distribution data from cold-chain logistics companies.

However, there are still certain challenges in the model construction process. Among these, the availability of data is a significant challenge. The construction of the model requires detailed and reliable data for support. Additionally, it is necessary to consider whether the construction of charging infrastructure can meet the future charging needs of large-scale electric cold-chain vehicles. At the same time, corresponding legal regulations also need to be updated accordingly.

For future research, we recommend that researchers could integrate other realistic constraints, which may influence the EV charging process. These constraints may include congestion queues at charging stations during peak hours, nonlinear battery power consumption patterns, and the refrigeration performance on goods' freshness. The introduction of these constraints will further enrich the mathematical model, facilitating better solutions to practical problems. Meanwhile, further studies could also explore innovative operational patterns and business models, particularly in the context of cold-chain logistics companies' operation innovations in the business ecosystem aspect. Besides, the further development of artificial intelligence technologies, such as large language models like Deepseek and GPT, can also provide support for cold-chain customer transportation services, enhancing the intelligence and sustainability of cold-chain logistics services.

In summary, we hope that this study can expand the research content on new-energy vehicle management and inspire more scholars to contemplate how algorithms and soft technologies can serve the sustainable development of humanity.

Author Contributions: Z.G., methodology; P.D., conceptualization and supervision; Z.F., writing—original draft; Y.J., conceptualization and validation; Y.S., validation. All authors have read and agreed to the published version of the manuscript.

Funding: This work is supported by the Postdoctoral Fellowship Program of CPSF under grant number GZB20240381 and the Natural Science Foundation of China under grant number 71873015.

Data Availability Statement: All relevant data are presented in the manuscript.

Conflicts of Interest: The authors declare no conflict of interest.

References

1. Iyer, P.; Robb, D. Cold chain optimisation models: A systematic literature review. *Comput. Ind. Eng.* **2025**, 110972. [CrossRef]
2. Wang, Y.-Z.; Fan, Y.-W.; Li, X.-L.; Yang, J.-G.; Zhang, X.-R. Experimental Validation of a Novel CO_2 Refrigeration System for Cold Storage: Achieving Energy Efficiency and Carbon Emission Reductions. *Energies* **2025**, *18*, 1129. [CrossRef]
3. Lin, H.J.; Chen, P.C.; Lin, H.P.; Hsieh, I.Y.L. Quantifying carbon emissions in cold chain transport: A real-world data-driven approach. *Transp. Res. Part D Transp. Environ.* **2025**, *142*, 104679. [CrossRef]
4. Jiang, H.; Wang, J.; Tian, M.; Zhao, C.; Zhang, Y.; Wang, X.; Liu, J.; Fu, M.; Yin, H.; Ding, Y. Assessment of identification performance for high emission heavy-duty diesel vehicles by means of remote sensing. *Sci. Total Environ.* **2024**, *912*, 168851. [CrossRef]
5. Zhang, S.; Fu, M.; Zhang, H.; Yin, H.; Ding, Y. Emission control status and future perspectives of diesel trucks in China. *J. Environ. Sci.* **2025**, *148*, 702–713. [CrossRef]
6. Nyamayoka, L.T.-E.; Masisi, L.; Dorrell, D.; Wang, S. Techno-Economic Feasibility and Optimal Design Approach of Grid-Connected Hybrid Power Generation Systems for Electric Vehicle Battery Swapping Station. *Energies* **2025**, *18*, 1208. [CrossRef]
7. Cavus, M.; Dissanayake, D.; Bell, M. Next Generation of Electric Vehicles: AI-Driven Approaches for Predictive Maintenance and Battery Management. *Energies* **2025**, *18*, 1041. [CrossRef]
8. Lopes da Paixão, J.; da Rosa Abaide, A.; Danielsson, G.H.; Sausen, J.P.; da Silva, L.N.F.; Neto, N.K. Optimized Strategy for Energy Management in an EV Fast Charging Microgrid Considering Storage Degradation. *Energies* **2025**, *18*, 1060. [CrossRef]
9. Ameer, H.; Wang, Y.; Chen, Z. A density-based spatial clustering and linear programming method for electric vehicle charging station location and price optimization. *Energy* **2025**, *317*, 134581. [CrossRef]
10. Zhang, X.; Martin, E.; Shaheen, S. The Dynamics of the California Electric Grid Mix and Electric Vehicle Emission Factors. *Energies* **2025**, *18*, 895. [CrossRef]
11. Seyfi, M.; Alinaghian, M.; Ghorbani, E.; Çatay, B.; Sabbagh, M.S. Multi-mode hybrid electric vehicle routing problem. *Transp. Res. Part E Logist. Transp. Rev.* **2022**, *166*, 102882. [CrossRef]
12. Hiermann, G.; Richard, F.H. Routing a mix of conventional, plug-in hybrid, and electric vehicles. *Eur. J. Oper. Res.* **2019**, *272*, 235–248.
13. Wang, Y.; Zhou, J.; Sun, Y.; Fan, J.; Wang, Z.; Wang, H. Collaborative multidepot electric vehicle routing problem with time windows and shared charging stations. *Expert Syst. Appl.* **2023**, *219*, 119654. [CrossRef]
14. Marchesano, M.G.; Popolo, V.; Rozhok, A.; Cavalaglio, G. Performance Evaluation of Battery Swapping Stations for EVs: A Multi-Method Simulation Approach. *Energies* **2024**, *17*, 5969. [CrossRef]
15. Ren, L.; Liao, W.; Chen, J. Systematic Design and Implementation Method of Battery-Energy Comprehensive Management Platform in Charging and Swapping Scenarios. *Energies* **2024**, *17*, 1237. [CrossRef]
16. Hof, J.; Schneider, M. Solving the battery swap station location-routing problem with capacitated electric vehicles using an AVNS algorithm for vehicle-routing problems with intermediate stops. *Transp. Res. Part B* **2017**, *97*, 102–112.
17. Basso, R.; Kulcsár, B.; Sanchez-Diaz, I. Electric vehicle routing problem with machine learning for energy prediction. *Transp. Res. Part B Methodol.* **2021**, *145*, 24–55.
18. Bezzi, D.; Ceselli, A.; Righini, G. A route-based algorithm for the electric vehicle routing problem with multiple technologies. *Transp. Res. Part C Emerg. Technol.* **2023**, *157*, 104374.
19. Montoya, A.; Guéret, C. The electric vehicle routing problem with nonlinear charging function. *Transp. Res. Part B* **2017**, *103*, 87–110. [CrossRef]
20. Jie, W.C.; Yang, J. The two-echelon capacitated electric vehicle routing problem with battery swapping stations: Formulation and efficient methodology. *Eur. J. Oper. Res.* **2019**, *272*, 879–904.
21. Schiffer, M.; Walther, G. The electric location routing problem with time windows and partial recharging. *Eur. J. Oper. Res.* **2017**, *206*, 995–1013. [CrossRef]
22. Huang, W.; Zou, J.; Yin, M.; Liu, F.; Wang, X.; Zhang, X. Multi-scale energy-carbon assessment for oyster cold chain: An empirical study. *J. Clean. Prod.* **2025**, *486*, 144538.
23. Luo, H.; Ma, T.; Li, Z. An exact branch-price-and-cut algorithm for the time-dependent cold chain pickup and delivery problem with incompatibility constraints. *Comput. Oper. Res.* **2025**, *178*, 107007. [CrossRef]
24. Awad, M.; Ndiaye, M.; Osman, A. Vehicle routing in cold food supply chain logistics: A literature review. *Int. J. Logist. Manag.* **2021**, *32*, 592–617.
25. Shi, Y.; Lin, Y.; Lim, M.K.; Tseng, M.L.; Tan, C.; Li, Y. An intelligent green scheduling system for sustainable cold-chain logistics. *Expert Syst. Appl.* **2022**, *209*, 118378.

26. Zhang, S.; Chen, N.; She, N.; Li, K. Location optimization of a competitive distribution center for urban cold-chain logistics in terms of low-carbon emissions. *Comput. Ind. Eng.* **2021**, *154*, 107120. [CrossRef]
27. Ryu, Y.M.; Lee, E.H. Development of dam inflow prediction technique based on explainable artificial intelligence (XAI) and combined optimizer for efficient use of water resources. *Environ. Model. Softw.* **2025**, *187*, 106380.
28. Sun, D.; Dong, J.; Gu, X.; Chen, Z. Scenario simulation and regulation policy optimization of industrial enterprise production water considering scale heterogeneity: A case study in the chemical industry. *Comput. Ind. Eng.* **2025**, *202*, 110961.
29. Wei, C.; Gao, W.W. Assigning customer dependent travel time limits to routes in a cold-chain inventory routing problem. *Comput. Ind. Eng.* **2019**, *133*, 275–291.
30. Saif, A.; Elhedhli, S. Cold supply chain design with environmental considerations: A simulation-optimization approach. *Eur. J. Oper. Res.* **2016**, *251*, 274–287.
31. Hariga, M.; As'ad, R.; Shamayleh, A. Integrated economic and environmental models for a multi stage cold supply chain under carbon tax regulation. *J. Clean. Prod.* **2017**, *166*, 1357–1371.
32. Zheng, F.; Pang, Y.; Xu, Y.; Liu, M. Heuristic algorithms for truck scheduling of cross-docking operations in cold-chain logistics. *Int. J. Prod. Res.* **2021**, *59*, 6579–6600.
33. Wang, Y.; Ji, X.; Lang, Y.; Zhang, Z. Supply chain network optimization of power battery based on the new EU Battery Regulation under uncertainty. *Comput. Ind. Eng.* **2025**, *201*, 110930.
34. Hussain, J.; Lee, C.C.; Lev, B. Optimizing AI-based emission reduction efficiency and subsidies in supply chain management: A sensitivity-based approach with duopoly game dynamics. *J. Clean. Prod.* **2025**, *494*, 144991.
35. Ren, J.; Li, H.; Zhang, M.; Wu, C. Massive-scale graph mining for e-commerce cold-chain analysis and optimization. *Future Gener. Comput. Syst.* **2021**, *125*, 526–531.
36. Chen, L.; Liu, Y.; Langevin, A. A multi-compartment vehicle routing problem in cold-chain distribution. *Comput. Oper. Res.* **2019**, *111*, 58–66.
37. Wu, L.J.; Shi, L.; Zhan, Z.H.; Lai, K.K.; Zhang, J. A buffer-based ant colony system approach for dynamic cold-chain logistics scheduling. *IEEE Trans. Emerg. Top. Comput. Intell.* **2022**, *6*, 1438–1452.
38. Lu, F.; Du, Z.; Wang, Z.; Wang, L.; Wang, S. Towards enhancing the crowdsourcing door-to-door delivery: An effective model in Beijing. *J. Ind. Manag. Optim.* **2025**, *21*, 2371–2395. [CrossRef]
39. Yang, X.F.; Yao, Y.; Fu, Q. Multi objective optimization model of cold chain logistics distribution based on freshness. *Appl. Res. Comput.* **2016**, *33*, 1050–1074.
40. Wang, S.; Jiang, Y.; Mou, J. Inventory and pricing decision of an integrated cold chain based on freshness. *Chin. J. Manag. Sci.* **2018**, *26*, 132–144.
41. Fang, X.; Yuan, F.J.; Cai, J.R. An optimized coordination model of freshness-keeping efforts and shelf service for fresh produce supply chain. *Chin. J. Manag. Sci.* **2023**, *31*, 142–152.
42. Alhijawi, B.; Awajan, A. Genetic algorithms: Theory, genetic operators, solutions, and applications. *Evol. Intell.* **2024**, *17*, 1245–1256.

energies

MDPI

Review

Electric Vehicles Empowering the Construction of Green Sustainable Transportation Networks in Chinese Cities: Dynamic Evolution, Frontier Trends, and Construction Pathways

Dacan Li [1], Albert D. Lau [2] and Yuanyuan Gong [3,*]

[1] Institute of Social Governance, Hebei University of Economics and Business, Shijiazhuang 050061, China; dacan@hueb.edu.cn
[2] Urban Development Research Center, Yangtze Delta Region Institute of Tsinghua University, Jiaxing 314006, China; e0973901@u.nus.edu
[3] School of Public Administration, Hebei University of Economics and Business, Shijiazhuang 050061, China
* Correspondence: yuanyuan@hueb.edu.cn

Abstract: As the global ecological environment faces serious challenges and extreme climate change threatens the survival of humankind, the promotion of green development has become the focus for all countries in the world. As one of the world's major greenhouse gas emitters, China has put forward the "twin goals" of achieving carbon peaking and carbon neutrality and is committed to promoting the green and low-carbon transformation of its cities. As the core of economic and social development, cities are the main source of carbon emissions. In response to the dual challenges of carbon emission control and traffic growth, it is particularly important to promote the development of green transportation. With the acceleration of urbanization, urban traffic pollution is becoming more and more serious. As a zero-emission transportation mode, electric vehicles have become a key way to achieve the carbon peak and carbon neutrality targets. In order to deeply analyze the research status of electric vehicles in the field of the green and low-carbon transformation of urban transportation in China and to explore the research hot spots, evolution trends, and their roles and strategies in the construction of green transportation networks, this paper uses the CiteSpace, VOSviewer, and Tableau analysis tools to review and analyze the 2460 articles and reviews in the Web of Science Core Collection (WOS) and 2650 articles and reviews in the China National Knowledge Infrastructure (CNKI), including the "publication volume and publication trend", "subject citation path", "countries cooperation and geographical distribution", "author cooperation and institution cooperation", "keyword co-occurrence and keywords clusters", and the "evolution trend of research hot spots in timeline". The results show that: (1) Since 2010, the research focus on electric vehicles has gradually increased, and especially in the past three years, the number of such publications has increased significantly. (2) China holds the lead in research output regarding electric vehicles and related fields, but its international cooperation needs to be strengthened. (3) In recent years, the research has focused on "energy transformation", "energy-saving technology", "carbon emissions", "battery recycling", and other relevant topics. The promotion and development of electric vehicles will continue to usher in new opportunities concerning technological innovation, policy support, and market expansion. Finally, based on the research hot spots and evolution trends of electric vehicles in the field of urban green transportation and low-carbon transportation in China, this paper discusses the key paths and strategies for electric vehicles to promote the transformation of urban transportation in China to green and low-carbon types and looks forward to future research directions. The research in this paper can provide theoretical support and practical guidance for China to promote electric vehicles, build low-carbon cities, and realize green transportation. It is expected to act as a useful reference for relevant policy formulation and academic research.

Keywords: electric vehicle; green transportation; low-carbon city; carbon emission; carbon peak

1. Introduction

As global ecological damage is becoming increasingly serious, extreme climate change poses a great threat to the survival of mankind, and the contradiction between world economic development and the deterioration of the natural environment is becoming more and more prominent; thus, green development has become a focus for all countries in the world [1].

As one of the world's major emitters of greenhouse gases, China has actively assumed international responsibility and proposed the "twin goals" of achieving carbon peaking and carbon neutrality [2,3]. The goals are not only a concrete action by China to deal with climate change, but are also important measures to promote urban green and low-carbon transformation. As the center of national and regional economic and social development, cities are one of the main sources of carbon emissions, which makes cities the important carrier for promoting low-carbon transformation and achieving high-quality economic and social development. Urban transportation is a significant source of urban carbon emissions; thus, under the dual challenges of achieving strict control of carbon emissions and meeting the increasing traffic demand, promoting green transportation is of great significance for facilitating the green and low-carbon transformation of transportation [4].

Green transportation plays an important role as one of the important ways to achieve the "twin goals". There is no unified definition in the academic community regarding the concept and connotation of "green transportation". As early as 1994, Canadian scholar Chris Bradshaw put forward the concept of the "green transportation hierarchy" [5]. Since then, green transportation has become a hot topic for scholars, both domestically and internationally. The scholars generally believe that green transportation represent a completely new concept of transportation development and city growth [6], which is the embodiment of people-oriented sustainable development in the transportation field. It aims to construct a diversified and comprehensive transportation system that coordinates the environment, resources, and society, with the goal of reducing traffic congestion, lowering environment pollution, decreasing greenhouse gas emissions, and promoting social equity [7–11]. Green transportation emphasizes the realization of low-carbon, environmentally friendly, and efficient development in transportation and reduces greenhouse gas emissions by optimizing the transportation structure, promoting clean energy, and improving transportation efficiency, which will not only help to achieve the goals, but also promote the sustainable development of cities [12].

With the rapid advancement of urbanization in China, urban traffic pollution has become more and more serious, which has been one of the major obstacles restricting the sustainable development of cities. Particularly in the field of transportation, especially urban transportation, the realization of the "twin goals" (carbon peaking and carbon neutralization) is facing severe challenges. Vehicle exhaust is one of the main sources of urban carbon emissions. According to relevant data, carbon dioxide emissions from the transportation industry account for about 10% of the total carbon emissions in the country [13]. Therefore, reducing the use of traditionally fueled vehicles is the key to reducing carbon emissions [14]. As a zero-emission transportation mode [15], electric vehicles provide many benefits, such as energy conservation, energy storage, and emission reduction. Promoting electric vehicles is an effective way to simultaneously ensure energy security and achieve the "twin goals" [16]. In view of this, the promotion of electric vehicles, the construction of low-carbon cities, and the realization of green transportation have become urgent issues.

In order to cope with the above challenges, the Chinese government has issued a series of policy measures. For instance, the Chinese government released a document titled, "Working Guidance for Carbon Dioxide Peaking and Carbon Neutrality in Full and Faithful Implementation of the New Development Philosophy" in September 2021 [17]. This document advocated for the vigorous development of new energy vehicles and promoted their deep integration with next-generation information technology and high-end equipment to form a green manufacturing system. Additionally, this document emphasized accelerating the construction of a "convenient, efficient and moderately advanced" charging and swapping network to provide favorable conditions for the usage of new energy vehicles. In October of the same year, the Chinese government also released the "Action Plan for Carbon Dioxide Peaking Before 2030" [18], which proposed vigorously supporting the key technology breakthrough of new energy vehicles (such as power batteries and energy storage technology) and promoting the demonstration and application of key technologies. Furthermore, it called for pilot projects in carbon peaking cities and encouraged innovation in electric vehicle application scenarios. These policies introduced by the Chinese government require that carbon peaking be integrated into the entire process and all aspects of economic and social development, with a focus on implementing key tasks such as green and low-carbon energy transformation and green and low-carbon transportation. This provides vast space and significant opportunities for China to build an "urban green transportation network".

In recent years, an increasing number of scholars have begun paying attention to the research on "green transportation". How to develop urban green transportation has become a hot topic in academia. Countries and international organizations like the United States, the United Kingdom, and the European Union boast the theoretical research results and practical experience that can be used for reference in the construction of green transportation networks [19]. At present, scholars at home and abroad mainly focus on research topics such as "decision making", "lithium-ion battery", "green and low-carbon", "indicator system", "carbon neutrality", "carbon peak", and "emission reduction" in the field of green transportation [20–32]. However, there are few comprehensive analyses achieved through bibliometrics, especially in the context of China's vigorous promotion of electric vehicles; it is extremely rare to use bibliometrics to explore the strategies and paths for building urban green transportation networks.

In order to fill the above research gaps, this paper takes the urban green transportation vehicle, "electric vehicle", as the main research object and systematically reviews and analyzes the relevant literature on electric vehicles in the field of urban green transportation in the WOS and CNKI through the CiteSpace, VOSviewer, and Tableau analysis tools to try to answer the following core questions:

Q1: What is the number of studies regarding electric vehicles in the field of urban green transportation, their publication trend, and the discipline citation path?

Q2: What is the status of cooperation among the main research authors and the research institutions currently involved? Which countries have participated in the studies? What is the geographical distribution of the literature, and what is the mode of cooperation?

Q3: Which topics are the focuses and hot spots in the research of electric vehicles in the field of urban green transportation, and what are the future development and research trends?

Q4: What are the main roles, strategies, and key paths regarding electric vehicles in the construction of an urban green transportation network?

The main research achievements of this paper are as follows: firstly, taking 2650 articles and reviews in the CNKI and 2460 articles and reviews in the WOS as the research objects, the authors use CiteSpace, VOSviewer, and Tableau to analyze the publication trend, author and institution cooperation, keyword co-occurrence and keyword clusters, the evolution trend of burst terms in the timeline, and the content of the literature on electric vehicles in the field of urban green transportation and low-carbon transportation, aiming to review the development process of this field and explore the research hot spots and frontier fields with the help of a visual knowledge map, hoping to achieve a more profound and comprehensive understanding of this field and to provide accurate guidance and support for China to promote electric vehicles, build low-carbon cities, and realize green transportation. In addition, the research in this paper also provides a Chinese solution for building a green transportation network for electric vehicles globally, which not only helps to enhance China's influence in the international academic community, but also provides practical experience for other countries and regions. In summary, on the one hand, this study enriches the academic research in the field of electric vehicles and green transportation. On the other hand, it provides important theoretical support for the formulation and practice of relevant policies, promoting the progress and development of the relevant disciplines.

The follow-up study in this paper is organized as follows: Section 2 elaborates the data sources and analysis methods; Sections 3–6 systematically review and deeply analyze the research hot spots, trends, and progress concerning the involved literature; Sections 7 and 8 comprise the discussion and conclusion portions, analyzing the key role of electric vehicles in building urban green transportation networks.

2. Materials and Methods

2.1. Data Sources and Retrieval Strategies

The data in this paper are mainly obtained from the China National Knowledge Infrastructure (CNKI) and the Web of Science Core Collection (WOS). On 31 August 2024, with (Subject: electric vehicle) AND (Subject: low carbon + low-carbon travel + low-carbon transportation + low-carbon cities + green travel + green transportation + green cities + environmental protection + energy conservation + sustainable development) as search terms, a total of 3526 Chinese studies were retrieved from the CNKI. On this basis, excluding duplicated data, news information, and invalid literature with low correlation, a total of 2650 studies highly relevant to the research theme of this paper were finally screened out.

In WOS, our search string was as follows: (TS = ("electric vehicle*" OR "EV" OR "EVs")) AND TS = ("low carbon" OR "low-carbon travel" OR "low-carbon transportation" OR "low-carbon cities" OR "green travel" OR "green transportation" OR "green cities" OR "environmental protection" OR "energy conservation" OR "sustainable development"). The search scope is the literature published in all years. The language of the literature is English, and the types are articles and reviews. A total of 2460 articles and reviews highly related to the theme of this study were retrieved from WOS.

2.2. Research Methods and Tools

This paper combines bibliometrics and systematic review as the research methodology. Bibliometrics can effectively mitigate potential research bias that may arise in traditional literature reviews [33]. During the research process, we used CiteSpace (6.2.R3), VOSviewer (1.6.20), and Tableau (2024.1.0) software to conduct a visual analysis of the literature [34–39], as these tools complement each other in bibliometric analysis. Specifically, CiteSpace is used for publication trends analysis, dual-map overlay analysis of disciplinary citation paths, author and institutional collaboration analysis, keyword co-occurrence analysis, keyword clustering analysis, and timeline evolution trends, thereby effectively identifying frontier areas and emerging trends regarding the research topics. Meanwhile, VOSviewer is employed to analyze the co-citation patterns in the literature and journals, as well as to study the collaboration relationships among publishing countries or regions. Additionally, we leveraged Tableau to perform a geographic distribution analysis of the international collaborations in the literature and to generate visual maps of the publication volumes and geographic distributions at the country level. The research process of the paper is shown in Figure 1.

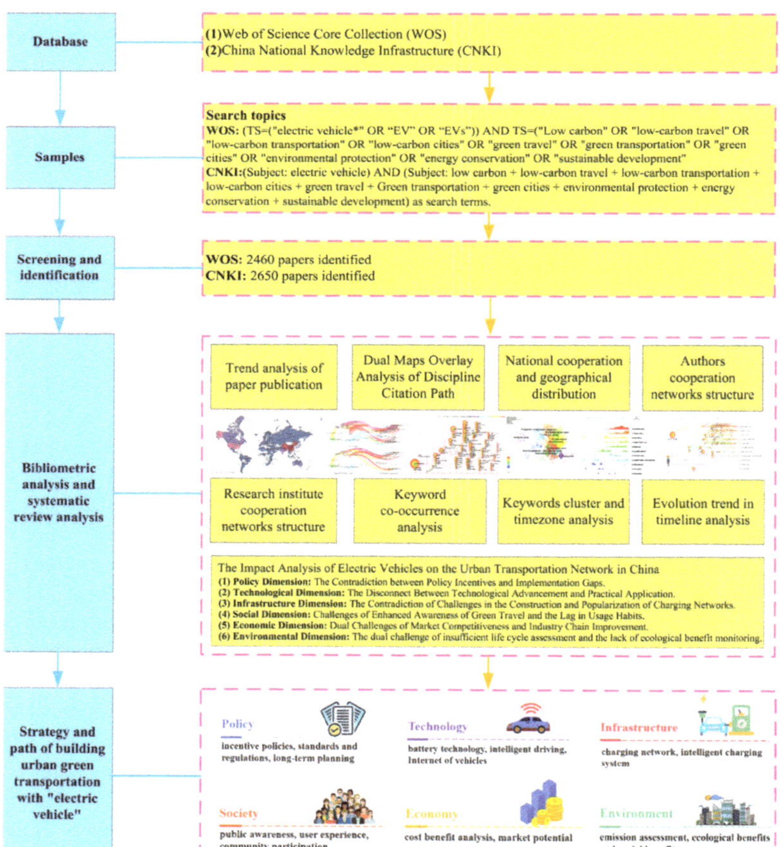

Figure 1. Research process of the paper.

3. Publication Trend, Country Cooperation, and Geographical Distribution

3.1. The Number and Trend of Publications

The number of annual publications is an important indicator for analyzing the development process and predicting the future development trends of the research field, which can directly reflect the dynamic changes in the input of the research strength in the field. Via the analysis of the annual publications trend, it can also help researchers to identify the development stage and the future development trend of the field.

The annual number of publications and the publication trend in the research field in the CNKI and WOS are shown in Figure 2 and Figure 3, respectively.

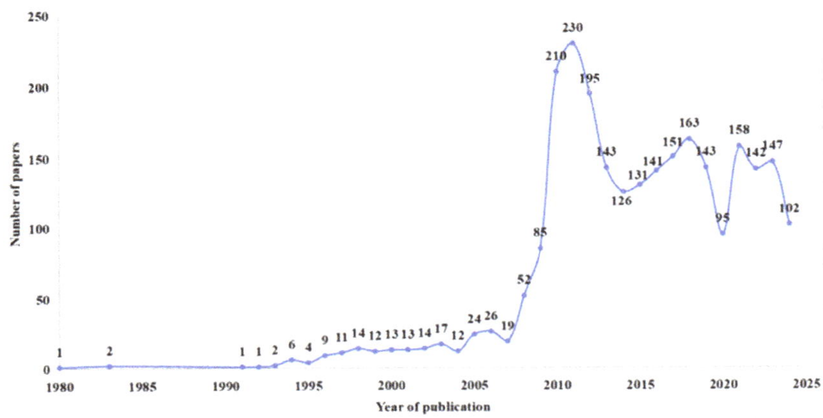

Figure 2. Annual number of publications and publication trend in CNKI.

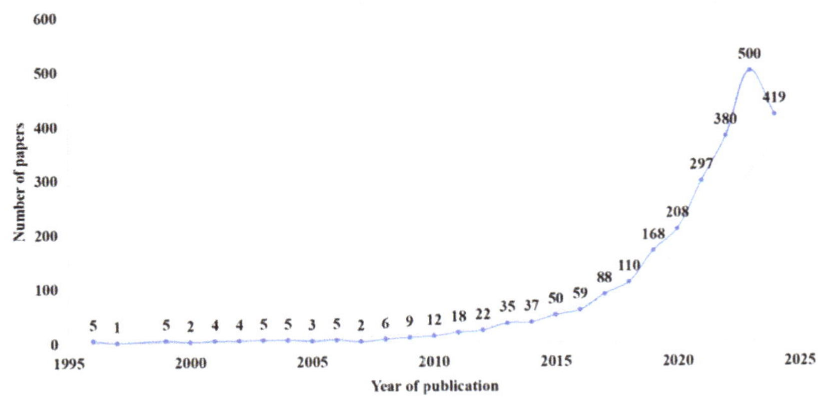

Figure 3. Annual number of publications and publication trend in WOS.

After completing the "Remove Duplicates" step in CiteSpace, statistical analysis of the number of annual publications in the CNKI literature was conducted. In the CNKI, the keyword "electric vehicle" first appeared in 1980. The number of annual publications was less than 10 from 1980 to 1996, and increased year by year in 1997–2007, with approximately 20 examples appearing in 2007. From 2008 to 2011, the number of annual publications in the field of electric vehicle in China increased rapidly from 19 to 230, while in 2012–2014, the number decreased year by year, with only 126 in 2014. Then, from 2015 to 2018, the annual number of publications again increased year by year, with 163 in 2018. Currently, from 2019 to 2024, the number appears to be in a fluctuating state, in which it decreased to 95 in 2020, increased to 102 in 2024 (January to August), and then the annual number of publications in 2019, 2021, 2022, and 2023 stabilized at about 150.

In the WOS, literature in the research field of this paper first appeared in 1996, and the annual publication number was below 10 from 1996 to 2009. In 2010–2018, the annual number of publications increased year by year, with 110 in 2018. Since 2019, the number has increased sharply, reaching 500 in 2023, and in 2024 (January to August), the number was 419.

From the data above, it can be seen that the number of publications in the research field of this paper increased rapidly around 2010, both in the CNKI and WOS, indicating that "electric vehicle" has increasingly become the hot spot for building green, low-carbon cities and developing green, low-carbon transportation. Therefore, this paper chooses 2010 as the starting time, focusing on analyzing literature from 2010 to 2024. This choice is based on the considerations that since 2010, the Chinese government has significantly increased its support for the electric vehicle industry and introduced a series of encouraging policies [40]. Notably, the "Ten Cities, Thousands of Vehicles" program (launched in 2009) entered its large-scale promotion phase in 2010. These initiatives marked the government's systematic strategic layout, injecting strong momentum into the research and application of electric vehicles [41]. Additionally, from the perspective of data availability and completeness, literature published after 2010 is more abundant and comprehensive in the relevant databases, enabling a more accurate reflection of research trends and developments in this field.

3.2. Dual-Map Overlay Analysis of Disciplinary Citation Paths

In this article, the dual-map overlay function reveals the distribution of main journals in the Web of Science database regarding "research on electric vehicles supporting urban green transportation", as well as the journals that cite this research, thereby illustrating the flow of knowledge among relevant journals and disciplines in the database. The dual-map overlay consists of two parts: the citing region on the left and the cited region on the right, with curves representing citation connections. The dual-map overlay is shown in Figure 4, and the details of the citation pathways are presented in Table 1.

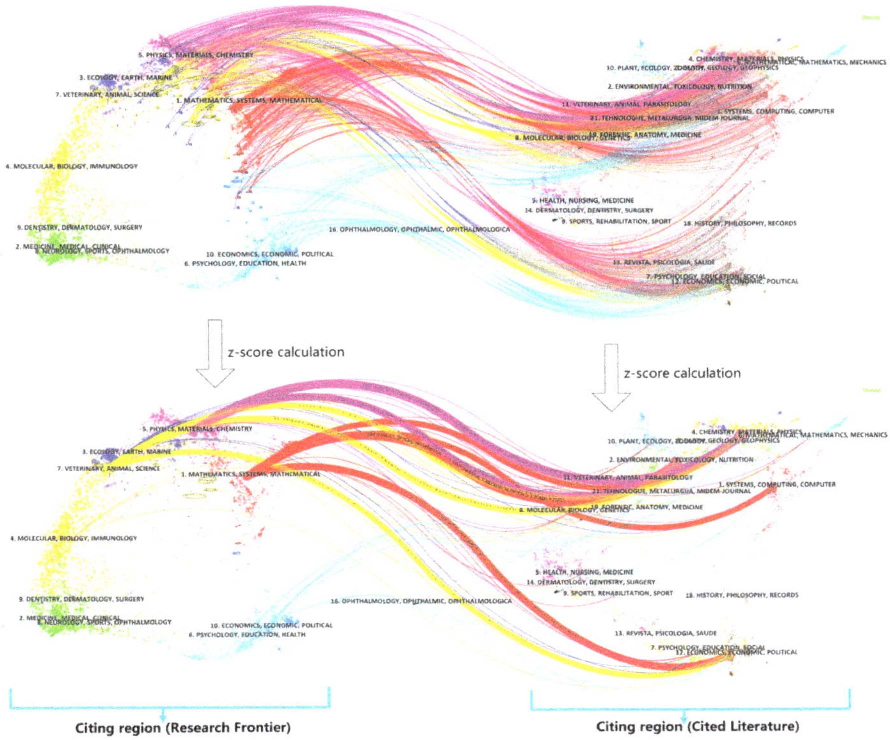

Figure 4. Dual-map overlay diagram.

Table 1. Domain-level citation patterns in this paper.

No.	Citing Region	Cited Region	Z-Score
1	physics, materials, chemistry	chemistry, materials, physics	6.058974
2	physics, materials, chemistry	environmental, toxicology, nutrition	2.334004
3	veterinary, animal, science	chemistry, materials, physics	2.0093758
4	veterinary, animal, science	environmental, toxicology, nutrition	2.493567
5	veterinary, animal, science	economics, economic, political	1.8910793
6	mathematics, systems, mathematical	chemistry, materials, physics	2.1854453
7	mathematics, systems, mathematical	systems, computing, computer	1.9405988
8	mathematics, systems, mathematical	environmental, toxicology, nutrition	3.233609
9	mathematics, systems, mathematical	economics, economic, political	2.4495487

According to Figure 4 and Table 1, research on "electric vehicles supporting urban green transportation" in the WOS database primarily focuses on three disciplinary clusters: "physics, materials, chemistry", "veterinary, animal, science", and "mathematics, systems, mathematical". There are nine outward citation paths, among which the pathway from "physics, materials, chemistry" to "chemistry, materials, physics" has the highest citation count and the highest z-score (z-score = 6.058974). This indicates that research in the field of "electric vehicles supporting urban green transportation" is predominantly concentrated in disciplines such as "physics, materials, chemistry".

Furthermore, by analyzing the knowledge flow between these disciplinary clusters, we can identify key research areas and potential research gaps. For instance, although the "veterinary, animal, science" cluster has fewer citation pathways compared to those in the "physics, materials, chemistry", cluster, it may offer unique research perspectives and innovations within specific subfields. Meanwhile, the "mathematics, systems, mathematical" cluster could provide theoretical foundations and methodological support for the optimization and system integration of electric vehicle technologies. By delving into the intersections and collaborations among these disciplines, researchers can gain a better understanding of the multidisciplinary integration trends in the development of electric vehicle technology, thus providing a scientific basis for future research directions and policy formulation in the field of urban green transportation.

3.3. Co-Citation Analysis of References and Journals

In this section, VOSviewer software was used to analyze the co-citation networks of references and journals. Figure 5 displays the co-citation network map of the literature. Using VOSviewer, we analyzed a total of 120,226 cited references and set the minimum citation threshold of 20, resulting in 107 references meeting this threshold for co-citation analysis, as shown in Figure 5a. Subsequently, we plotted the citation density map, where the higher the citation frequency of the literature, the brighter the map appears, as demonstrated in Figure 5b.

The aforementioned literature is mainly divided into three clusters, each represented by a distinct color: red represents Cluster 1, which includes 39 publications that primarily analyze and summarize the barriers and driving factors for the promotion of electric vehicles from the consumer perspective; green represents Cluster 2, which consists of 33 publications that focus on the latest technologies in electric vehicles and their impacts and prospects; blue represents Cluster 3, which also includes 33 publications that primarily investigate the battery technology for electric vehicles and their environmental impacts.

Using VOSviewer software, we filtered out journals with more than 200 citations, totaling 100. Subsequently, we constructed a high-citation journal map and a high-citation density map that each include three clusters. These maps illustrate the research directions and focuses of "electric vehicles" in the fields of "building green and low-carbon cities" and "developing green and low-carbon transportation," as shown in Figure 6.

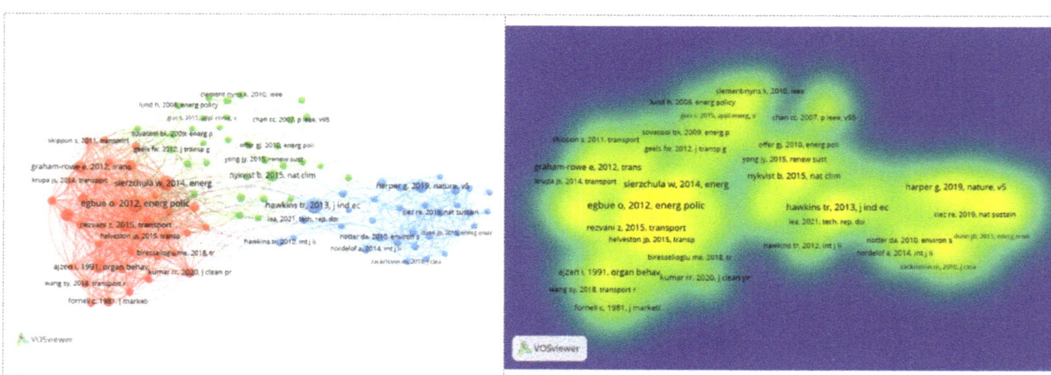

(a) (b)

Figure 5. (**a**) References co-citation network map (citation weights); (**b**) references co-citation density visualization map (citation weights) [42–70].

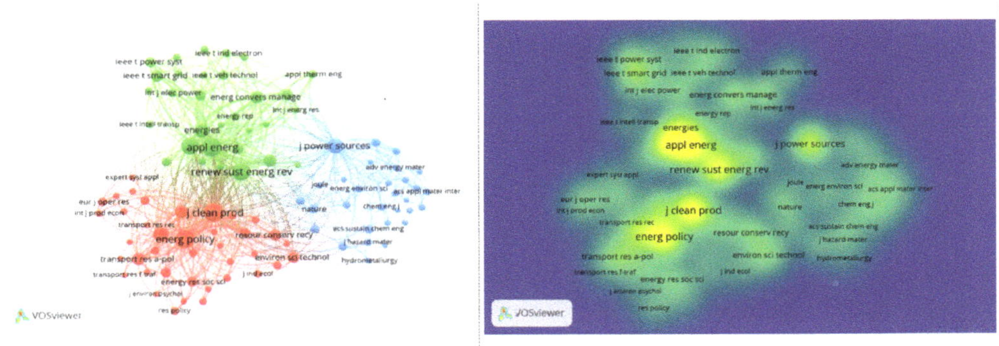

(a) (b)

Figure 6. (**a**) Journal co-citation network map (citation weights); (**b**) journal co-citation density visualization map (citation weights).

As shown in Figure 6a, *Applied Energy* and the *Journal of Cleaner Production* have the highest citation counts, reaching 4202 and 4201 citations, respectively. They are followed by journals such as *Energy Policy*, *Energy*, and *Renewable and Sustainable Energy Reviews*. Figure 6b comprises the journal co-citation density visualization map, where the greater the number of citations for a journal, the higher its density. The top 15 journals with the highest citations related to "electric vehicles" in the fields of "building green low-carbon cities" and "developing green low-carbon transportation" are detailed in Table 2.

Table 2. Citation counts of journals (top 15).

No.	Source	Citations	Total Link Strength
1	*Applied Energy*	4202	187,096
2	*Journal of Cleaner Production*	4201	180,334
3	*Energy Policy*	3852	141,272
4	*Energy*	3475	165,398
5	*Renewable and Sustainable Energy Reviews*	3466	186,540
6	*Journal of Power Sources*	2660	160,651
7	*Transportation Research Part D: Transport and Environment*	2090	75,219
8	*Energies*	1932	88,780
9	*Sustainability*	1592	60,449
10	*International Journal of Hydrogen Energy*	1409	61,645
11	*Transportation Research Part A: Policy and Practice*	1325	49,658
12	*Energy Conversion and Management*	1275	78,757
13	*IEEE Transactions on Smart Grid*	1213	43,028
14	*Environmental Science & Technology*	1026	36,463
15	*Journal of Energy Storage*	1023	67,205

3.4. Country Cooperation and Geographical Distribution

In today's wave of research regarding electric vehicles in urban green transportation, the scientific research strength and contributions of every country can be vividly demonstrated according to the countries cooperation view of the WOS literature. It not only reveals the number of publications of every country in this field, but also reflects their scientific research influence and international cooperation situation. An analysis of national collaborations is conducted using VOSviewer software, and the collaboration map of the countries and regions is shown in Figure 7.

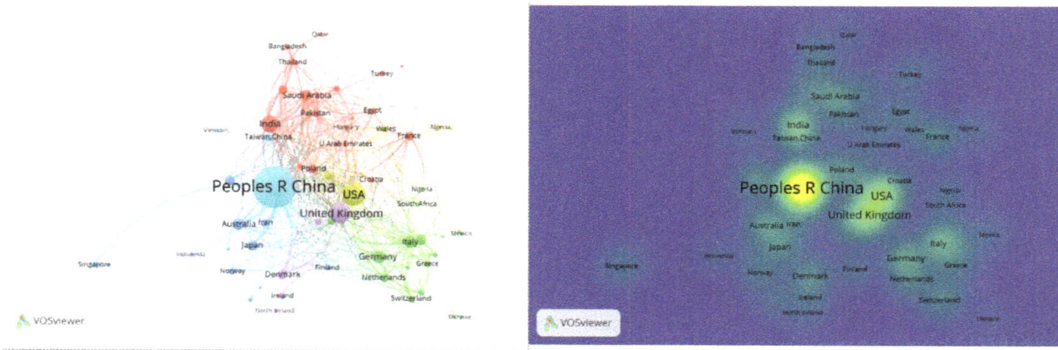

(a) (b)

Figure 7. (**a**) National and regional network collaboration map (document weights); (**b**) collaboration density visualization map (document weights).

In Figure 7a, the size of the nodes represents the number of publications from each country, with larger nodes indicating a higher volume of publications. According to VOSviewer, 61 countries and regions have published more than five papers, divided into six clusters, indicating collaborative relationships among them. Figure 7b illustrates the cooperation density among countries, showing that countries such as China, the United States, and the United Kingdom display high densities, which indicates a substantial number of publications.

An analysis of the publication volume and geographical distribution of the countries, based on WOS literature, is conducted by using CiteSpace and Tableau software. It is noteworthy that China stands out, with an impressive publication volume of 1100 papers, far surpassing that of other countries. Specifically, the United States follows with 313 publications, showing its strong strength and continuous investment in scientific and technological innovation. The United Kingdom ranks third, with 250 of publications, and its profound academic background and scientific research environment provides strong support for the research of electric vehicles in green transportation. India and Canada rank among the top five, with 170 and 103 publications, respectively, showing their positive efforts and remarkable results in promoting the green transformation of transportation. It is worth noting that there are 5 countries with more than 100 of publications, and the number of countries with more than 50 reaches 15, which shows that the research on electric vehicles in urban green transportation is widely employed and highly valued around the world. The geographic distribution of the number of published papers is shown in Figure 8. The number of publications and citation status by country are shown in Table 3.

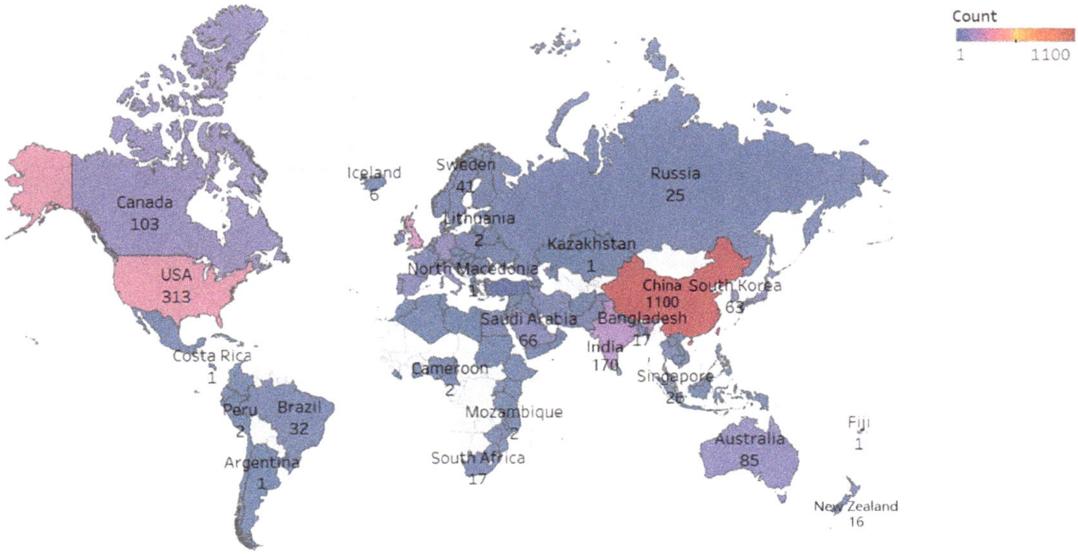

Figure 8. Geographical distribution chart showing the number of published papers.

Table 3. The number of publications in and citation status of the WOS literature.

No.	Countries	Count	Centrality	First Published Year	Times Cited (Average per Item)	H-Index	Citation Curve (Figure 9)
1	Peoples R China	1100	0.03	1999	29.08	84	Figure 9a
2	USA	313	0.24	1999	41.45	60	Figure 9b
3	UK	250	0.25	1999	47.49	60	Figure 9c
4	India	170	0.07	1999	18.74	30	Figure 9d
5	Canada	103	0.34	1999	46.92	35	Figure 9e
6	Germany	98	0.43	2006	46.89	35	Figure 9f
7	Italy	89	0.13	1996	31.06	31	Figure 9g
8	Australia	85	0.18	2008	50.72	34	Figure 9h
9	Spain	66	0.22	2013	35.76	26	Figure 9i
10	Saudi Arabia	66	0.1	2014	25.93	18	Figure 9j
11	Japan	66	0.05	1996	33.28	26	Figure 9k
12	South Korea	63	0.04	1997	37.98	25	Figure 9l
13	Denmark	61	0	2011	55.26	30	Figure 9m
14	France	51	0.64	1996	31.47	22	Figure 9n
15	Netherlands	50	0	2008	41.44	27	Figure 9o

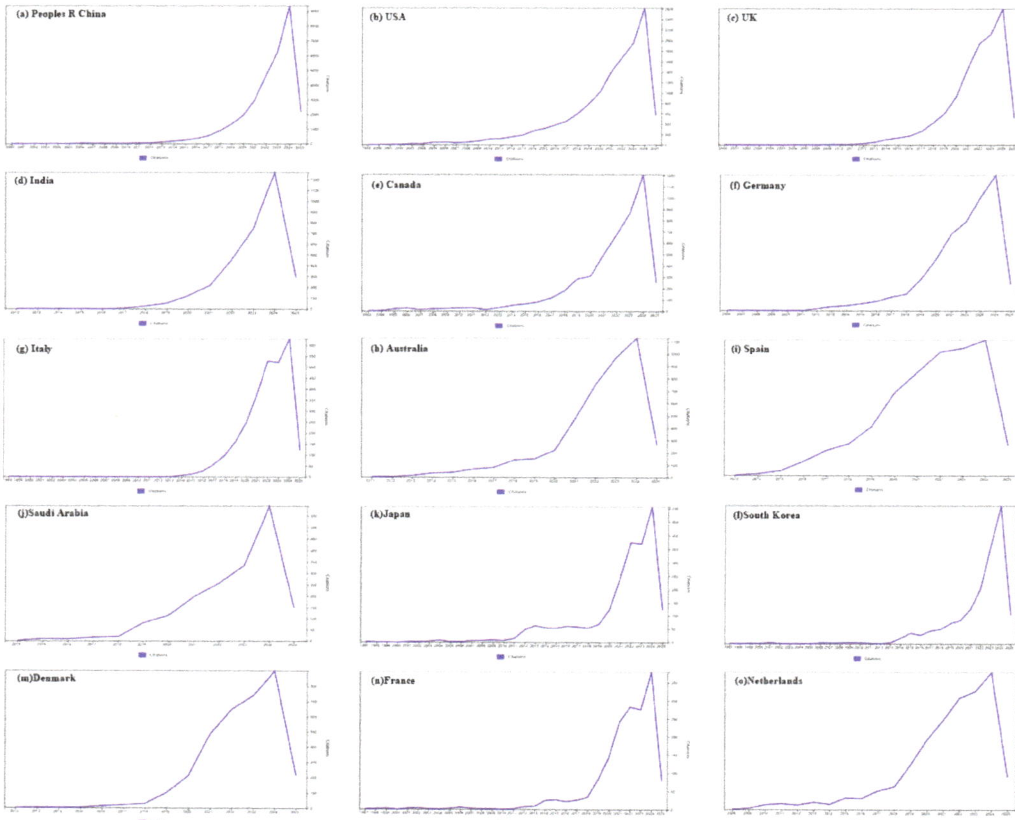

Figure 9. The citation curve of literature from various countries.

According to the data provided in Table 3, China accounts for 44.7% of the global literature volume (h-index = 84), but the average citation frequency per paper (29.08) is lower than that of the USA (41.45), the United Kingdom (47.49), Canada (46.92), and Germany (46.89). This indicates that although China has a substantial volume of research output in this field, the citation frequency of these publications is relatively low, reflecting insufficient recognition and influence in the international academic community. The citation curve of the literature from various countries is illustrated in Figure 9. Among the top 50 highly cited studies in China, technical literature constitutes the majority (about 70%), while there are relatively few studies on basic theoretical research, policy research, typical case analysis, and international perspectives, which may also lead to its low citation frequency. In addition, although China has garnered significant achievements in terms of publication volume, its centrality score is only 0.03, far lower than that of countries such as the USA (centrality = 0.24), the UK (centrality = 0.25), Canada (centrality = 0.34), Spain (centrality = 0.22), France (centrality = 0.64), etc. These data reflect that China's influence and international cooperation in the field of green transportation in regards to electric vehicle cities need to be improved, and efforts are still needed to improve research quality and enhance international cooperation and influence. This is not only a motivation for Chinese researchers, but also an inspiration for future development directions.

4. Analysis of Author Cooperation and Institution Cooperation

4.1. Author Cooperation View Analysis

The author co-occurrence analysis can identify the core authors in a certain field and the cooperation intensity between authors. The number of publications is presented in the form of the size of nodes, the thickness of connecting lines shows the cooperation intensity between authors, and the order of publication time is presented in the form of color shades. The author cooperation views of the CNKI literature and the WOS literature are shown in Figure 10.

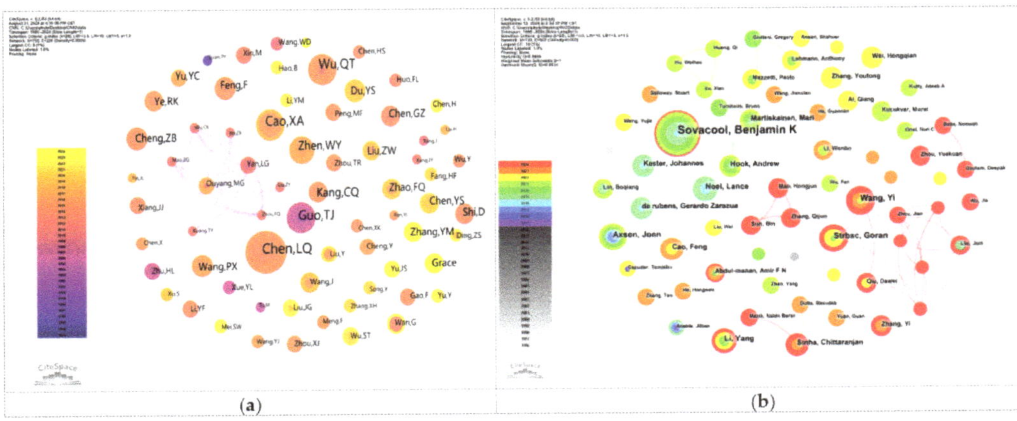

(**a**) (**b**)

Figure 10. (**a**) Author cooperation view in the CNKI literature; (**b**) author cooperation view in the WOS literature.

In the author cooperation view of the CNKI literature, the number of authors is 702 (N = 702), the number of connecting lines is 220 (E = 220), and the density of the cooperation network is only 0.0009 (density = 0.0009), indicating that there is little cooperation between authors. According to Price's law, the formula for calculating the number of core authors is $M \approx 0.749 * \sqrt{N_{max}}$, where M refers to the number of studies, and N_{max} refers to the number of publications by the author with the most literature published in the corresponding years. If the number of studies published by the author is greater than M, he is the core author. In this paper, the author with the most publications is "Chen, L.Q.", with a total of 21 papers published, and M is about 3.4. Thus, the minimum number of publications of the core author in this study is 4. The number of publications of the core authors of the CNKI literature is shown in Table 4.

Table 4. Number of publications of the core authors of the CNKI literature (top 10).

No.	Count	Centrality	First Published Year	Authors
1	21	0	2010	Chen, L.Q.
2	9	0	2004	Wu, Q.T.
3	9	0	1996	Guo, T.J.
4	9	0	2011	Cao, X.A.
5	8	0	2012	Zhen, W.Y.
6	6	0	2007	Feng, F.
7	5	0	2010	Cheng, Z.B.
8	5	0	2011	Wang, P.X.
9	5	0	2015	Du, Y.S.
10	4	0	2012	Liu, Z.W.

Through the author cooperation view of the CNKI literature, it can be found that the publication time of "Chen, L.Q.", the author with the largest number of publications, was focused on the years 2010–2012. He mainly reviewed and analyzed the development and policy support of China's new energy vehicle industry. He pointed out that the development of the new energy vehicle industry can optimize the consumption structure of transportation energy, reduce urban air pollution, and provide huge economic and social benefits [71–73].

Yan, L.G., Ouyang, M.G., Liu, Z.Y., Zhou, F.Q., Kuang, T.Y., Mao, Z.Q., Wu, C.K., and He, Z.X. formed an author group. Their publication time was mainly concentrated on the years 2004–2014, mainly reviewing and analyzing the industrialization process of energy conservation and new energy vehicles in China [74].

Scholars Zhang, Y.M., Yan, Z., and Yu, J.S. have formed a small research team. Their publication time was mainly focused on the years 2022–2023, and they primarily studied the key technologies and applications of a green microgrid for the goals of achieving peak carbon dioxide emissions by 2030 and carbon neutrality by 2060 [75]. The number of publications of the core authors in the WOS literature is shown in Table 5.

Table 5. Number of publications by the core authors in the WOS Literature (top 10).

No.	Count	Centrality	First Published Year	Authors
1	23	0	2018	Sovacool, Benjamin K.
2	9	0	2021	Wang, Yi
3	9	0	2013	Axsen, Jonn
4	9	0	2020	Strbac, Goran
5	7	0	2020	Li, Yang
6	7	0	2018	Noel, Lance
7	6	0	2018	Kester, Johannes
8	6	0	2019	Hook, Andrew
9	6	0	2023	Sinha, Chittaranjan
10	6	0	2019	Martiskainen, Mari

Through the author cooperation view, it can be observed that the author with the largest number of publications is "Sovacool, Benjamin K.". From 2018 to 2024, he has published more than 20 relevant papers. His main research fields include the analysis of factors influencing people's preferences for electric vehicles [76–79], as well as low-carbon transition [80–82].

Zhang, N., Zhang, X., Liu, X., Wang, Y., Strbac, G., Zhou, J., Liu, J. and Qiu, D.W. formed an author group, with an important position in the authors' cooperation network. They mainly studied vehicle-to-grid (V2G), electric vehicle route planning and scheduling, etc. [83,84]. In order to study the interaction mechanism in the system between V2G and energy storage charging piles, a collaborative optimization model considering the complementarity of energy storage and charging in the vehicle was proposed [85].

There are also some small-scale cooperation teams, such as Sun, B., Zhang, Q.J., and Mao, H.J., who formed a small research team. They developed a novel statistical–dynamic energy consumption prediction framework, called the vehicle energy conservation equation (VECE) [86].

4.2. Institution Cooperation View Analysis

The quality and quantity of academic papers are important indicators for evaluating and comparing research institutions, directly affecting their academic influence. The research institution cooperation views of the CNKI and WOS literature are shown in Figure 11.

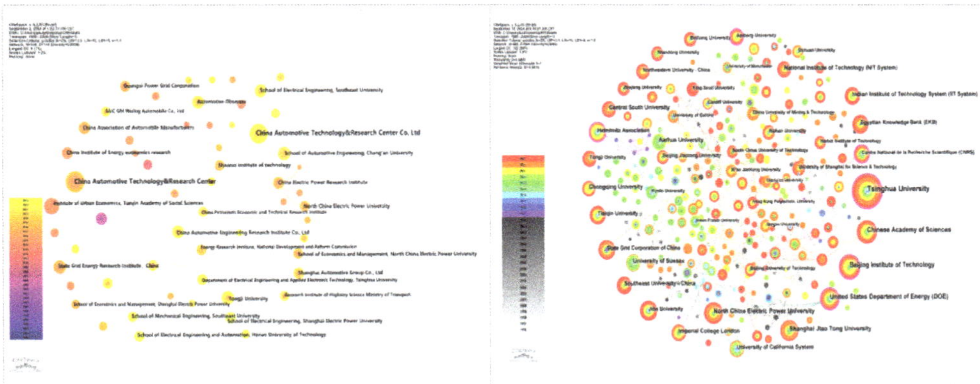

Figure 11. (**a**) Institution cooperation view of CNKI literature; (**b**) institution cooperation view of WOS literature.

It can be seen from the research institution cooperation view of the literature in the CNKI that the main research institutions include the China Automotive Technology and Research Center, the Institute of Urban Economics of Tianjin Academy of Social Sciences, the School of Electrical Engineering of Southeast University, Tongji University, etc. The publication numbers for these core research institutions are shown in Table 6.

Table 6. Number of publications for the core research institutions of the CNKI literature (top 10).

No.	Count	Centrality	First Published Year	Institution
1	19	0	1999	China Automotive Technology and Research Center, Tianjin, China
2	15	0	2019	China Automotive Technology and Research Center Co., Ltd., (CATARC), Tianjin, China
3	10	0	2010	Institute of Urban Economics of Tianjin Academy of Social Sciences, Tianjin, China
4	8	0	2015	Automotive Observer, Beijing, China
5	8	0	2012	School of Electrical Engineering of Southeast University, Nanjing, China
6	8	0	2009	China Electric Power Research Institute (CEPRI), Beijing, China
7	8	0	2011	Tongji University, Shanghai, China
8	8	0	2011	State Grid Energy Research Institute, Beijing, China
9	8	0	2010	China Association of Automobile Manufacturers, Beijing, China
10	8	0	2014	China Automotive Engineering Research Institute Co., Ltd., Chongqing, China

According to the research institution cooperation view of the literature in the CNKI, it can be found that the top research institution is the China Automotive Technology and Research Center. The publication time is focused on the years 1999–2019. This institution mainly studied the fiscal and taxation support policies for energy conservation and new energy vehicles, the research and development progress and key deployment of China's new energy vehicle and battery technology, and the interpretation of the recycling of the traction batteries used in electric vehicle-dismantling specifications [87–89].

The second-ranked research institution is the CATARC, with a total of 15 papers published from 2019 to 2024. It has a cooperative relationship with North China Electric Power University, CEPRI, Nanjing Power Supply Company, and other research institutions, forming a cooperative network of research institutions.

The third-ranked research institution is the Institute of Urban Economics of Tianjin Academy of Social Sciences, which focuses on the development and policy of the new energy vehicle industry [71]. The fourth-ranked research institution is the Automotive Observer, which mainly reviews and analyzes the countermeasures and suggestions of industry experts regarding the energy conservation and emission reduction of automobile enterprises, studying the path to achieving carbon peak and carbon neutrality goals for the automobile industry [90–92]. The goals of carbon peaking and carbon neutrality have brought new challenges to the automobile industry and provided new opportunities for the development of the new energy vehicle industry.

Through the research institution cooperation view of the literature in the WOS, it can be seen that Tsinghua University, the Chinese Academy of Sciences, the Beijing Institute of Technology, the United States Department of Energy (DOE), Shanghai Jiao Tong University, North China Electric Power University, the University of California System, Imperial College London, Southeast University—China, and the University of Sussex are the top 10 research institutions with the largest number of publications. Among them, there are six Chinese research institutions, two American research institutions, and two British research institutions, as shown in Table 7.

Table 7. Number of publications for the core research institutions in the WOS literature (top 10).

No.	Count	Centrality	First Published Year	Institution
1	84	0.19	2010	Tsinghua University
2	48	0.09	2004	Chinese Academy of Sciences
3	47	0.13	2013	Beijing Institute of Technology
4	43	0.13	2003	United States Department of Energy (DOE)
5	38	0.06	2008	Shanghai Jiao Tong University
6	36	0.02	2016	North China Electric Power University
7	31	0.07	1999	University of California System
8	30	0.03	2011	Imperial College London
9	28	0.02	2013	Southeast University—China
10	28	0.07	2005	University of Sussex

The top research institution is the Tsinghua University of China, which published 84 papers from 2010 to 2024. Especially during 2021–2024, the number of publications increased rapidly, with 20 in 2023. There are 45 domestic and foreign research institutions cooperating with Tsinghua University, and in recent years, their main research direction has been focused on the battery technology of electric vehicles [93–95].

The second-ranked research institution is the Chinese Academy of Sciences, which cooperates with 37 institutions, and it published 48 papers in totally. The publication time was mainly focused on the years 2021–2024, with a total of 33 papers. The main research focus was battery recycling for electric vehicles [96–98].

The third-ranked research institution is the Beijing Institute of Technology, which published 47 papers in 2013–2024. The publication time was mainly 2021–2024. It cooperates with 23 research institutions, and the main research focuses in the past few years have been energy management and the sharing mode of electric vehicles [99–101].

5. Keywords Analysis

5.1. *Keyword Co-Occurrence Analysis*

In the keyword co-occurrence view, the node size represents the counts of keyword co-occurrence. Keywords with high centrality and counts are the common concern of researchers for a period of time, i.e., the research hot spots. The keyword co-occurrence knowledge mappings for the CNKI and WOS literature are shown in Figure 12.

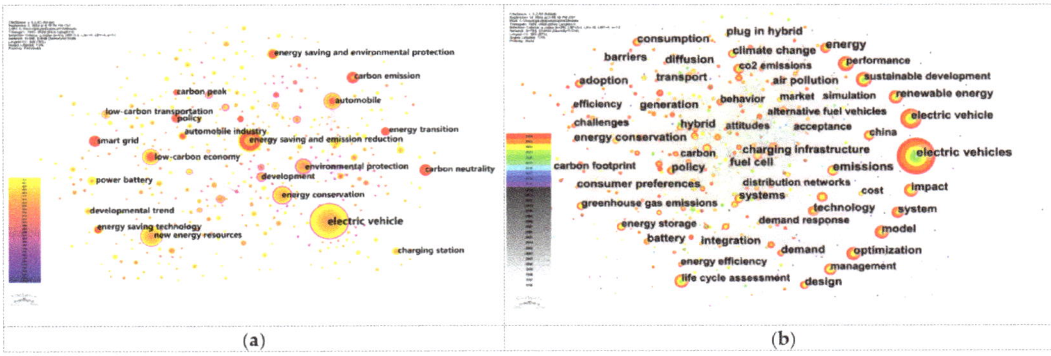

Figure 12. (**a**) Keyword co-occurrence knowledge mapping for the CNKI literature; (**b**) keywords co-occurrence knowledge mapping for the WOS literature.

From the keyword co-occurrence knowledge mapping of the CNKI literature, it can be seen that the count for "electric vehicle" is the highest, and its node is the largest; since "electric vehicle" was used as the subject when retrieving the literature, this keyword is not discussed below.

According to the count of keywords, the top 10 keywords are selected for ranking. Among them, keywords with high counts, such as "energy conservation and emission reduction", "new energy", "energy conservation", "automobile", "low-carbon economy", "environmental protection", and "low carbon" rank high. Keywords such as "energy conservation and emission reduction" (centrality = 0.18), "new energy" (centrality = 0.14), and "low-carbon economy" (centrality = 0.14) exhibit high centrality (centrality $\geq$ 0.1), which reflects the research focus in this field, to a certain extent, as shown in Table 8.

Table 8. Keywords with high counts in the CNKI literature (count $\geq$ 15).

No.	Count	Centrality	First Published Year	Keyword
1	138	0.18	2009	energy conservation and emission reduction
2	98	0.14	2009	new energy
3	87	0.09	1996	energy conservation
4	75	0.12	1993	automobile
5	67	0.14	2009	low-carbon economy
6	55	0.10	1995	environmental protection
7	38	0.04	2010	low carbon
8	33	0.03	2010	energy conservation and environmental protection
9	32	0.07	1999	development
10	32	0.05	2009	smart power grids
11	30	0.02	2021	carbon neutrality
12	28	0.02	2013	energy saving technology
13	25	0.04	2013	carbon emission
14	19	0.04	2010	power battery
15	18	0.03	2009	automobile industry
16	17	0.01	2009	policy
17	16	0.01	2009	low carbon transportation
18	15	0.01	2010	developmental trend

The term "energy conservation and emission reduction" refers to the commitment to reduce the energy consumption and pollutant emissions of electric vehicles during their production and use through technological innovation and optimized management. This is closely aligned with China's "dual carbon" goals and represents a critical direction for promoting the green development of electric vehicles [102]. New energy focuses on the exploration and utilization of cleaner and more sustainable energy sources to power electric vehicles, such as advancements in lithium battery technology and the development of hydrogen fuel cells. This focus aims to explore diversified energy solutions and reduce reliance on traditional fossil fuels [103,104]. The concept of "low-carbon economy" highlights the integration of electric vehicles into the framework of macroeconomic development. By developing the electric vehicle industry, it drives the upstream and downstream industrial chains to transition towards low-carbon development, achieving coordinated economic and environmental development [105,106]. These keywords collectively point to the core research direction of the use of electric vehicles in the field of green transportation, namely how to achieve energy conservation and emission reduction, advocate energy transformation, and promote low-carbon economic development, ultimately achieving the green and sustainable development goals of urban transportation.

It can be seen from the WOS keyword co-occurrence knowledge mapping that "model", "energy", "impact", "system", "optimization", "renewable energy", "performance", "emissions", and "sustainable development" are high-frequency keywords. Among them, the keyword, "energy" (centrality = 0.23 > 0.1) reflects the research focus in this field, to a certain extent, as shown in Table 9.

Table 9. Keywords with high counts in the WOS literature (count $\geq$ 100).

No.	Count	Centrality	First Published Year	Keyword
1	208	0.07	2005	model
2	179	0.23	2003	energy
3	161	0.05	2013	impact
4	160	0.03	2013	system
5	155	0.01	2013	optimization
6	153	0.03	2013	renewable energy
7	152	0.09	2008	performance
8	140	0.07	1996	emissions
9	139	0.02	2013	sustainable development
10	135	0.02	2015	life cycle assessment
11	130	0.02	2013	management
12	128	0.03	2010	technology
13	110	0.03	2014	policy
14	109	0.05	2014	systems
15	108	0.05	2003	design
16	104	0.02	2017	strategy
17	102	0.01	2013	hybrid
18	102	0.02	2014	China
19	102	0.03	2013	demand

5.2. Keyword Timezone Analysis

Through the steps including "pathfinder", "pruning slice networks" and "pruning the merged network", the keyword time zone analysis of the literature at this stage was carried out. The keyword time zone view of the literature provides strong support for an understanding of the development progress and hot spot changes regarding electric vehicle use in urban green transportation. Through in-depth analysis of the keywords and hot spots in the time zone view, we can more clearly grasp the development direction and trend of this field. We also can identify the year when the keyword first appeared. The size of the node is the most intuitive reflection of the importance and counts of keywords. The keyword time zone analysis of the literature is shown in Figure 13.

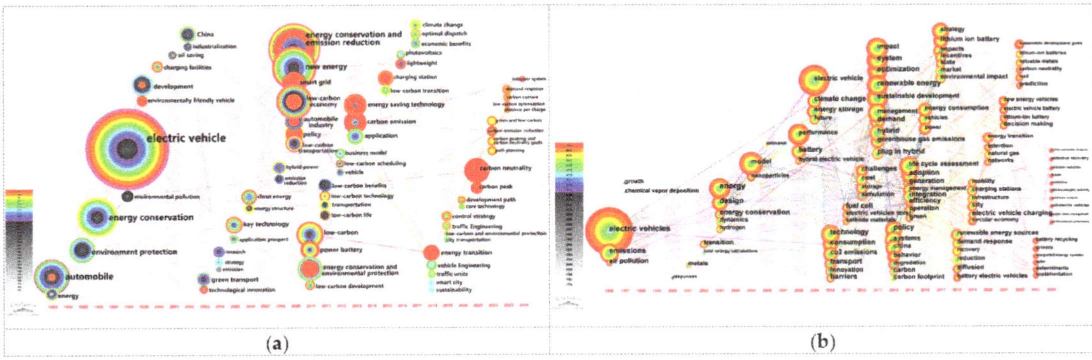

Figure 13. (**a**) Keyword time zone analysis of the CNKI literature; (**b**) keyword time zone analysis of the WOS literature.

(1) Keyword time zone analysis of the CNKI literature.

Since the number of relevant studies in the CNKI from 1980 to 1992 is small, the literature of the CNKI used to analyze the keyword time zone in this paper was published from 1993 to 2024. Through the keyword time zone analysis of the CNKI literature, it can be seen that the nodes of keywords like "electric vehicle", "energy conservation and emission reduction", "new energy", "energy conservation", "automobile", "low-carbon economy", and "environment protection" are large, showing that they appear more frequently, which reflects the research hot spots at the stage. Via CiteSpace's "Show/Hide Citation/Frequency Burst" feature, we can identify keywords such as "energy conservation and emission reduction", "smart grid", "low-carbon economy", "automobile industry", "policy", "energy conservation and environmental protection", "energy saving technology", "carbon emission", "energy transition", "carbon neutrality", "carbon peak" with a strong citation or burst rate. These keywords not only represent the latest dynamic regarding China's electric vehicle use in urban green transportation, but also reflect China's profound thinking and positive actions regarding energy, the environment, and economic development.

Since 2021, keywords such as "carbon neutrality", "carbon peak", "green and low-carbon", "path planning", "low-carbon optimization", and "indicator system" have appeared for the first time. It can be seen that Chinese scholars have been focusing on the green low-carbon path planning and the construction of a carbon emission index system for electric vehicles, with the goals of carbon peaking and carbon neutrality. For example, Zhou, C., et al. (2024) constructed a scientific and comprehensive regional carbon emission index system for electric vehicles, covering four dimensions, i.e., regional development, low-carbon transportation, vehicle operation, and power production. They measured 12 key indexes of carbon emissions from electric vehicles and selected six representative provinces in China for case verification [107]. The regional carbon emission index of electric vehicles can effectively quantify the comprehensive carbon emission level of electric vehicles and identify the advantages and disadvantages of the low-carbon development of the electric vehicle industry in different regions, as well as provide a reference for seeking the optimization path to achieve the low-carbon transition of electric vehicles at the regional level.

(2) Keyword time zone analysis of the WOS literature.

It can be seen from the keyword time zone view of the WOS literature that the keyword "electric vehicles" first appeared in 1996, and keywords like "emissions" and "air pollution" appeared for the first time this year as well, indicating that the scholars were beginning to pay attention to greenhouse gas emissions and air pollution. Since 2021, the global academic community has set off a wave for electric vehicle research in regards to urban green transportation. A series of keywords such as "lithium-ion battery", "decision making", "carbon neutrality", "battery recycling", "techno-economic analysis", and "emission reduction" appeared frequently in the WOS literature. This trend not only highlights the close attention paid by international scholars regarding environmental protection and the sustainable development of electric vehicles, but also heralds profound changes in energy structure and technological innovation for the global automobile industry.

In addition, through the keywords time zone view of the WOS literature, it can also be seen that "valuable metals" in 2021 and "lithium-ion batteries" in 2022 were the keywords showing strong citation or burst rates. As the core power source of electric vehicles, the lithium-ion battery is related to the performance, safety and cost of electric vehicles, so its importance is self-evident [108]. How to improve the energy density, prolong the cycle life, and reduce the manufacturing cost of lithium-ion batteries have become hot topics in the current research [109]. With the rapid development of electric vehicles, the demand for lithium-ion batteries has increased dramatically, which has also produced a series of problems, such as resource shortages and environmental pollution. Therefore, how to efficiently and sustainably produce, use, and recycle lithium-ion batteries is also one of the current research hot spots [110,111].

5.3. Keyword Cluster Analysis

The keywords in the literature of the CNKI and WOS are clustered by CiteSpace. The keyword cluster analysis views and peak views of the literature in the CNKI and WOS are shown in Figure 14, Figure 15 and Figure 16, respectively.

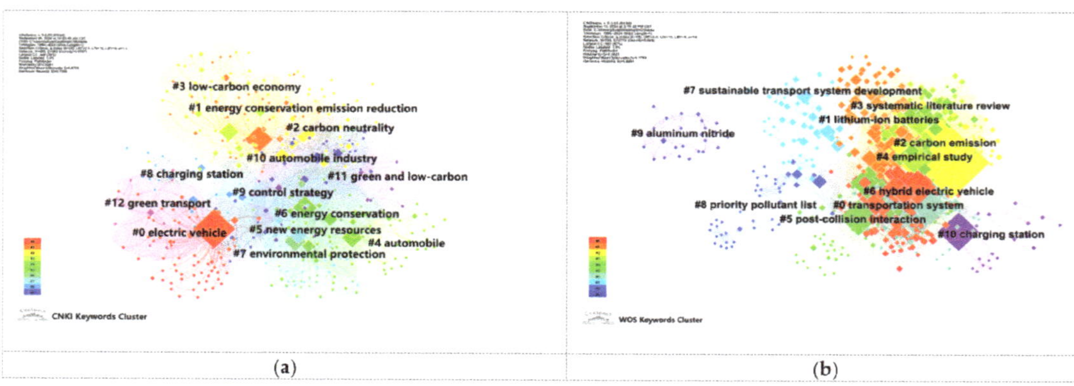

Figure 14. (**a**) CNKI keyword cluster view; (**b**) WOS keyword cluster view.

In the CNKI keyword cluster view, N = 586, E = 982, density = 0.0057, modularity Q = 0.6281 > 0.3, and mean silhouette = 0.8706 > 0.5. In the WOS keyword cluster view, N = 785, E = 2772, density = 0.009, modularity Q = 0.5829 > 0.3, and mean silhouette = 0.7769 > 0.5. It can be seen that the keyword cluster effects of the literature in both the CNKI and WOS are all superior, and the credibility is high.

According to the keyword cluster peak view, it can be seen that the time when each cluster starts, continues, and ends, as well as the heights of the peaks, indicate the activity of the clusters. Through the CNKI literature keyword cluster peak view, it can be seen that the keyword #1, "energy conservation and emission reduction" cluster started around 2005, and the peak gradually rose in 2009–2012, indicating that the studies related to the "energy conservation and emission reduction" cluster increased at that time. The peak of the keyword #2, "carbon neutrality" cluster has risen rapidly since 2021, indicating that the studies related to the cluster have increased rapidly from then on. In recent years, the Chinese government has attached great importance to climate change and has included the "twin goals"—carbon peaking and carbon neutrality—into the country's overall development strategy. In September 2020, President Xi Jinping proposed China's carbon peak and carbon neutrality goals at the 75th session of the United Nations General Assembly. In October 2021, the Chinese government issued the document titled, "Working Guidance for Carbon Dioxide Peaking and Carbon Neutrality in Full and Faithful Implementation of the New Development Philosophy". With the promotion of national policies, carbon peaking and carbon neutrality have gradually become hot topics in scientific research.

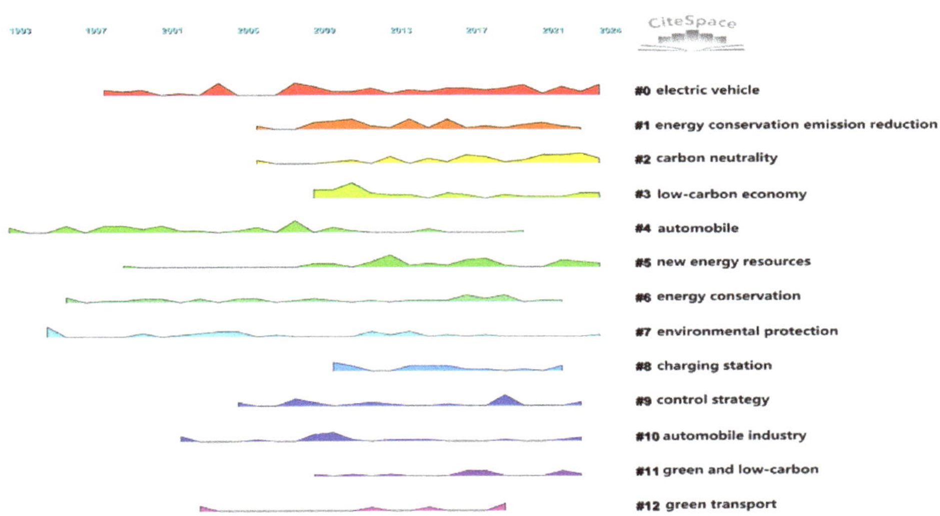

Figure 15. CNKI keyword cluster peak view.

Through observing the WOS keyword cluster peak view, it can be seen that the keyword #0, "transportation system" cluster started in 2008–2009, developed rapidly after 2014, and reached its peak in 2018–2020, indicating that research on this cluster was active in this period. The research on the keyword #1, "lithium-ion batteries" cluster started in 2010–2011. And since 2020, the number of studies relevant to this cluster has increased greatly and the peak has also risen rapidly, indicating that this cluster has become a hot spot.

Via keywords cluster analysis, it can be seen that the CNKI keyword cluster includes: "energy conservation and emission reduction", "carbon neutrality", "low-carbon economy", "new energy", "energy conservation", "environmental protection", "charging station", "green low carbon", "green transportation", and 13 other clusters. The number of WOS keyword clusters is 11, including "transportation system", "lithium-ion batteries", "carbon emission", "systematic literature review", etc.

Through the keywords cluster analysis of the literature in the CNKI and WOS, it can be divided into three main hot research directions.

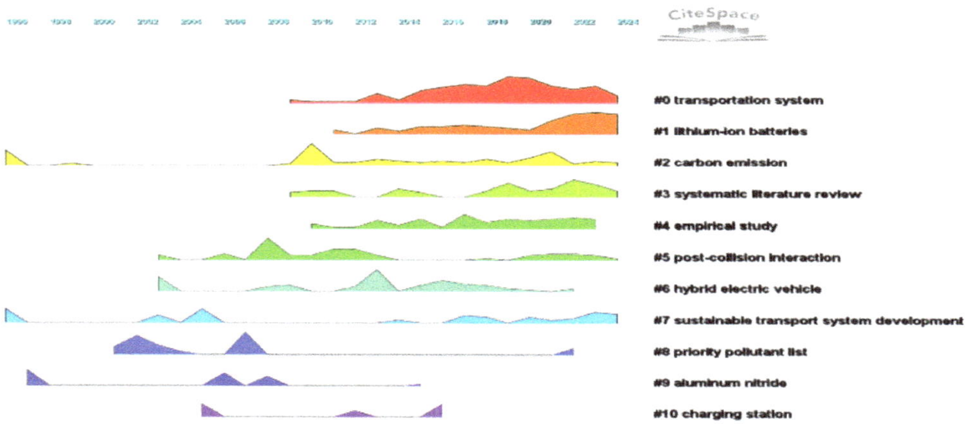

Figure 16. WOS keywords cluster peak view.

(1) Research on the infrastructure of electric vehicles.

This research field includes four CNKI keyword clusters and three WOS keyword clusters.

CNKI cluster: #0 electric vehicle, #4 automobile, #8 charging station, and #10 automobile industry. Labels related to the infrastructure research of electric vehicles also included "environmental conservation vehicle", "charging facilities", "power batteries", etc.

WOS cluster: #1 lithium-ion batteries, #6 hybrid electric vehicle, and #10 charging station.

Xue, B., et al. (2021) studied the sustainable development of small vehicle power batteries for carbon neutrality [112]. It has become an urgent task to promote the sustainable development of human society, realizing the efficient and environmentally friendly recycling and reuse of waste lithium-ion batteries (LIBs) [113]. He, H., Wei, J., and Wang, Z. (2024) analyzed the problems existing regarding the efficiency and degree of environmental protection for the charging systems of electric vehicles in China, putting forward an optimization strategy for electric vehicle charging systems [114]. Chen, H., Luo, X., and Zhang, J. (2024) predicted the demand for a charging infrastructure for electric vehicles and proposed suggestions for formulating the development plan for the charging infrastructure of electric vehicles [115].

(2) Research on electric vehicles to aid in energy conservation and emission reduction, low carbon output, and environmental protection.

This research field contains three CNKI keyword clusters and two WOS keyword clusters.

CNKI cluster: #1 energy conservation and emission reduction, #6 energy conservation, and #7 environmental protection.

WOS cluster: #2 carbon emission and #8 priority pollutant list.

Electric vehicles offer the advantages of environmental protection, energy conservation, and emission reduction, which can reduce energy consumption and emissions and play a positive role in building a green and low-carbon society. Energy conservation and environmental protection have become the main trends in automobile technology development, and studying new energy and energy conservation technology for automobiles is an effective way to solve energy shortages and reduce the environmental pollution caused by automobiles [116]. Wu, X. (2022) studied the energy conservation planning of electric vehicles based on big data technology [117]. Gong, H., et al. (2023) introduced the development status of China's new energy electric vehicles, analyzed the energy conservation and emission reduction effects of new energy electric vehicles with the help of energy consumption calculation methods and carbon emission calculation methods, and proposed the development path for new energy electric vehicles in China [118]. Yu, B., et al. (2024) proposed a carbon emission calculation method and a carbon emission flow calculation model for electric vehicles, which can track the carbon flow during the charging and discharging of electric vehicles in real time, accurately quantifying the carbon emission reductions [119].

(3) Research on electric vehicles and their contributions to the green, low carbon, and sustainable development of urban transportation.

This research field contains six CNKI keyword clusters and two WOS keyword clusters.

CNKI cluster: #2 carbon neutrality, #3 low-carbon economy, #5 new energy resources, #9 control strategy, #11 green and low-carbon, and #12 green transport.

WOS cluster: #0 transportation system and #7 sustainable transport system development.

Cities are the basic unit of human production and living, as well as the basic unit for measuring carbon emissions and environmental pollution. They are also an important part of carbon emission reduction and sustainable development [120]. In order to protect the ecological environment, all localities should integrate the concept of ecological progress and green, circular, and low-carbon development into urban transportation development planning and vigorously build a green transportation system [121]. Xuan, Y. (2024) designed a method to optimize the allocation of electric vehicle charging stations, considering low-carbon transportation, and quantitatively analyzed the impact of charging station planning schemes on transportation carbon emissions by establishing a low-carbon transportation indicator system that comprehensively took four factors, i.e., carbon emissions, transportation energy structure, transportation electric energy cleanliness, and charging station usage rate, into account [122]. Ding, L. (2023) analyzed the situation of domestic and foreign policies related to automobile carbon emissions, studied the carbon emissions of the whole life cycle of the automobile, and put forward suggestions for the low-carbon development of the automobile industry [123].

6. Analysis of the Evolution Trend in Timeline Form

The "Burst Terms" analysis feature of CiteSpace can analyze the latest research trends and the frontier of "electric vehicles". The burst terms of this study are shown in Figure 17.

Through the analysis of the burst terms and timeline, we can see that keywords such as "energy conservation and emission reduction", "energy transition", "energy conservation technology", "carbon neutrality", "carbon peak", and "carbon emissions", in the CNKI literature, and keywords like "emissions", "power system", "renewable energy", "storage", "fuel consumption", "recovery", "lithium-ion batteries", in literature of the WOS, have comprised the research hot spots and frontiers in recent years. Research using the keyword "emissions" in WOS literature extends through the whole cycle (1996–2024), with an especially a large number studies appeared in 2009–2016, and it became the research hot spot at that time. From 2021 to 2024, scholars have focused on keywords such as "recovery", "lithium-ion batteries", and "valuable metals". It can be seen that the recycling of automobile batteries and the effective reuse of valuable metals in batteries have become the research hot spots over the past three years.

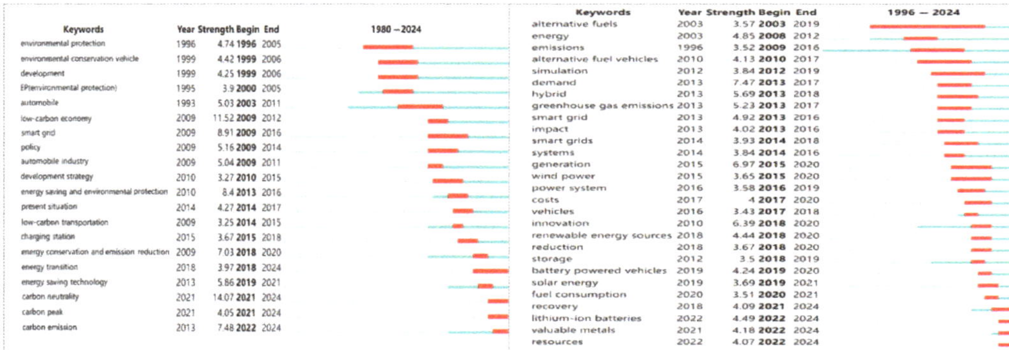

Keywords	Year	Strength	Begin	End	1980 − 2024
environmental protection	1996	4.74	**1996**	2005	
environmental conservation vehicle	1999	4.42	**1999**	2006	
development	1999	4.25	**1999**	2006	
EP(environmental protection)	1995	3.9	**2000**	2005	
automobile	1993	5.03	**2003**	2011	
low-carbon economy	2009	11.52	**2009**	2012	
smart grid	2009	8.91	**2009**	2016	
policy	2009	5.16	**2009**	2014	
automobile industry	2009	5.04	**2009**	2011	
development strategy	2010	3.27	**2010**	2015	
energy saving and environmental protection	2010	8.4	**2013**	2016	
present situation	2014	4.27	**2014**	2017	
low-carbon transportation	2009	3.25	**2014**	2015	
charging station	2015	3.67	**2015**	2018	
energy conservation and emission reduction	2009	7.03	**2018**	2020	
energy transition	2018	3.97	**2018**	2024	
energy saving technology	2013	5.86	**2019**	2021	
carbon neutrality	2021	14.07	**2021**	2024	
carbon peak	2021	4.05	**2021**	2024	
carbon emission	2013	7.48	**2022**	2024	

(**a**)

Keywords	Year	Strength	Begin	End	1996 — 2024
alternative fuels	2003	3.57	**2003**	2019	
energy	2003	4.85	**2008**	2012	
emissions	1996	3.52	**2009**	2016	
alternative fuel vehicles	2010	4.13	**2010**	2017	
simulation	2012	3.84	**2012**	2019	
demand	2013	7.47	**2013**	2017	
hybrid	2013	5.69	**2013**	2018	
greenhouse gas emissions	2013	5.23	**2013**	2017	
smart grid	2013	4.92	**2013**	2016	
impact	2013	4.02	**2013**	2016	
smart grids	2014	3.93	**2014**	2018	
systems	2014	3.84	**2014**	2016	
generation	2015	6.97	**2015**	2020	
wind power	2015	3.65	**2015**	2020	
power system	2016	3.58	**2016**	2019	
costs	2017	4	**2017**	2020	
vehicles	2016	3.43	**2017**	2018	
innovation	2010	6.39	**2018**	2020	
renewable energy sources	2018	4.44	**2018**	2020	
reduction	2018	3.67	**2018**	2020	
storage	2012	3.5	**2018**	2019	
battery powered vehicles	2019	4.24	**2019**	2020	
solar energy	2019	3.69	**2019**	2021	
fuel consumption	2020	3.51	**2020**	2021	
recovery	2018	4.09	**2021**	2024	
lithium-ion batteries	2022	4.49	**2022**	2024	
valuable metals	2021	4.18	**2022**	2024	
resources	2022	4.07	**2022**	2024	

(**b**)

Figure 17. (**a**) CNKI keywords with the strongest citation bursts; (**b**) WOS keywords with the strongest citation bursts. The red line indicates the starting and ending time when keywords have the strongest citation bursts.

CiteSpace is used to draw the timeline views of for the CNKI literature and the WOS literature, as shown in Figure 18 and Figure 19, respectively.

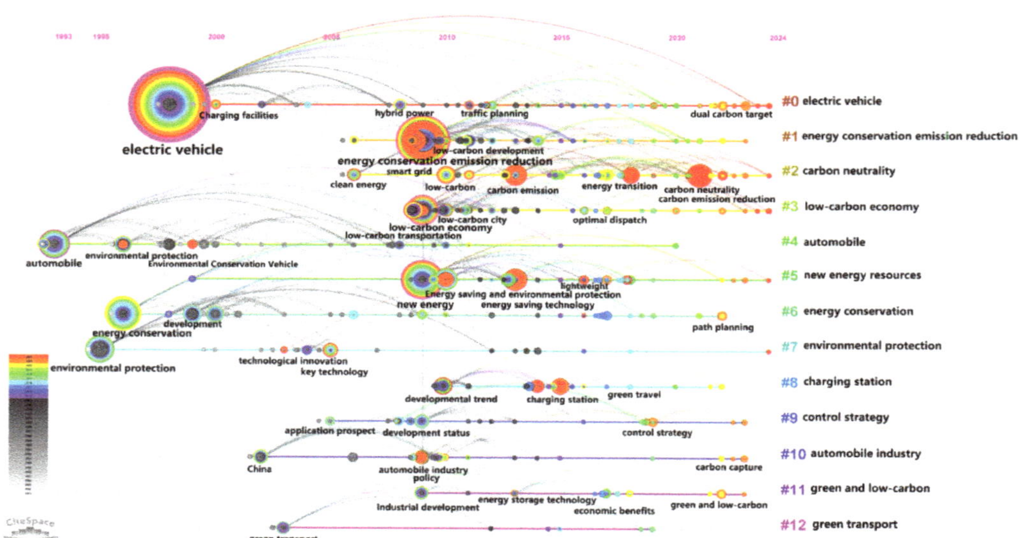

Figure 18. CNKI literature timeline view.

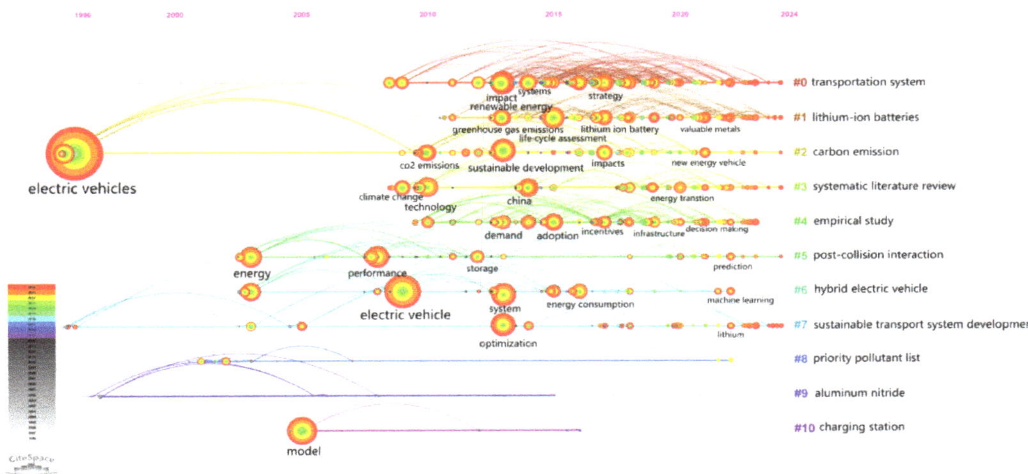

Figure 19. WOS literature timeline view.

From the timeline view, we can analyze the changes of the research hot spots for electric vehicles in urban green transportation, as well as the hot spots regarding development trends in the timeline. The location of each node represents the time that the keyword first appeared, and the more occurrences, the larger the node becomes.

7. Discussion

7.1. Publication Trends and Country Cooperation in the Literature

Through literature analysis, it can be observed that the research focus on electric vehicles in the field of green low-carbon transportation has gradually increased since 2010. This trend reflects a broad global consensus regarding the key role of electric vehicles in alleviating traffic pollution and solving the energy crisis. Especially since 2019, the annual publication volume of WOS literature in this research field has shown a significant growth trend. By 2023, the number of publications had reached 500, and in 2024, it had reached 419 in the period from January to August alone, a phenomenon that has been attributed to the growing global focus on sustainable transportation, the active promotion of government policies in various countries, and the rapid advancements in electric vehicle technology. In the analysis of publication volume at the national level, China is far ahead of other countries, with 1100 WOS publications. This remarkable achievement is closely related to the high attention paid by the Chinese government to the electric vehicle industry, the strong policy support, and the rapid growth of market demand. Despite China's leading position in terms of the number of studies focusing on electric vehicles, the centrality score is far lower than that of France, Canada, the UK, the USA, and other countries, which means that China's relevant studies need to be further strengthened in terms of international academic influence and international cooperation.

7.2. Author Cooperation and Institutional Cooperation

In terms of author cooperation, it can be seen that in the CNKI, Chen Liuqin, the author with the highest number of publications, published 21 papers, showing his importance in the field. His studies mainly focus on the development and policy support of the new energy vehicle industry, emphasizing the key role of new energy vehicles in optimizing the transportation energy consumption structure and reducing urban air pollution. In the WOS, the author group led by Sovacool, Benjamin K., whose studies involve the preference for electric vehicles and low-carbon transition, reflects the global attention on electric vehicle research, from multiple perspectives. Through the cooperation network of these core authors, it can be seen that international research cooperation is gradually increasing.

As for institutional cooperation, institutions such as the China Automotive Technology and Research Center occupy the leading positions in the ranking of publication volume in the CNKI, indicating their importance in the field. They have provided important theoretical support and policy recommendations for the development of new energy vehicles in China. In the WOS, Tsinghua University performs particularly well, ranking first in terms of the number of publications, showing its leading position in regards to electric vehicle research. Its cooperation with 45 research institutions, at home and abroad, further demonstrate its extensive influence in the international academic community. Tsinghua University's research on electric vehicle battery technology is one of the major current technical bottlenecks regarding new energy vehicles, showing its foresight and practicality. In addition, institutions such as the Chinese Academy of Sciences and the Beijing Institute of Technology have also played an important role in the field of electric vehicles, focusing on battery recycling and energy management, respectively. These studies not only provide an important theoretical basis for the sustainable development of new energy vehicles, but also provide a scientific reference for the formulation of relevant policies.

In general, through the analysis of the author cooperation view and the institutional co-operation view, it can be seen that the cooperation among Chinese scholars and institutions needs to be further strengthened. It is suggested that relevant research institutions and scholars should actively seek interdisciplinary and inter-agency cooperation and promote international exchanges and collaboration, which will help to further enhance the global influence of electric vehicle research.

7.3. The Evolution of Keyword Research via Timeline

By analyzing the literature, the research and development of China's electric vehicles in the field of urban green transportation and low-carbon transportation can be clearly divided into three stages. This phased division not only reflects the evolution of technology and policy, but also reveals the changes in China's strategic direction and academic focus on electric vehicles.

(1) 1993–2008: initial exploration stage.

At this stage, the electric vehicle industry had just started in China, and hot keywords such as "environmental protection", "environmentally friendly vehicles", "development", "environmental protection", "automobile", and "alternative fuels" appeared frequently. These keywords reflect the initial attention to and exploration of electric vehicles by scholars at that time. The government's supportive policies provided a good environment for the development of electric vehicles.

Research Focus: The research in this period mainly focuses on the basic concept, the technical route of electric vehicles, and a preliminary evaluation of market demand. At this stage, scholars began to pay attention to how to promote the development of green cars through policy support and regulation formulation.

Policy Support: The active involvement of the government provided the impetus for the budding of the electric vehicle industry. Although the technology and market were still in the initial stage, this period laid the foundation for subsequent development.

(2) 2009–2020: rapid development period.

This stage comprises the rapid development period of China's electric vehicle research and industrialization. Hot keywords such as "low-carbon economy", "smart grid", "policy", "automobile industry", "energy conservation and environmental protection", "low-carbon transportation", and "charging pile" appeared frequently. The "Ten Cities, Thousands of Vehicles" initiative, launched in 2009, marked the official start of the industrialization of new energy vehicles [124].

Market Environment: With the development of the economy and the improvement in living standards, the public's attention on energy issues and environmental pollution increased. With the advantages of low-carbon emissions, environmental protection, and high efficiency, electric vehicles became an important low-carbon transportation carrier [125]. The research at this stage focused not only on technological progress, but also on changes in policy, market, and social needs.

Industrialization Process: The electric vehicle industry achieved rapid development with the support of national policies and scientific research strength, and the relevant industrial chains were gradually improving, laying the foundation for subsequent technological innovation and market expansion.

(3) 2021–2024: high-speed development and "twin goals".

At this stage, hot keywords such as "carbon neutrality", "carbon peak", "carbon emissions", "recovery", and "lithium-ion batteries" show the close integration of China's electric vehicle research with current policies. The "twin goals" proposed in 2020 point out the direction for the development of the electric vehicle industry.

Strategic Objectives: The goals of carbon peaking and carbon neutrality make electric vehicles an important path to achieve green and low-carbon transportation [126]. Relevant research has begun to pay attention to the carbon emissions of the product life cycle, emphasizing the importance of electric vehicles in the future green development of transportation [127].

Technological innovation: At this stage, China's electric vehicle industry has made remarkable progress in regards to technological innovation. The application of new charging technology, efficient driving systems, and automatic driving technology is being rapidly promoted, enhancing the international competitiveness of Chinese brand electric vehicles.

Market Expansion: Electric vehicles not only occupy an important position in the domestic market, but are also gradually moving towards the global market, showing China's technical strength and market potential in the field.

Through the stage-by-stage analysis of China's electric vehicle research and development in the field of "urban green transportation and low-carbon transportation", it can be seen that the development of the electric vehicle industry is closely related to national policies, market demand, and technological innovation. From the initial exploration stage, to the rapid development period, and then to the current high-speed development stage, China's electric vehicle industry has formed a relatively complete ecosystem. In the future, with the deepening of the "twin goals", the electric vehicle industry will continue to usher in new opportunities and challenges in regards to technological innovation and market expansion. The academic community should continue to pay attention to the research in this field and provide a scientific basis for policy formulation and industrial development.

7.4. Research Hot Spots and Frontier Trends

Through the analysis of burst terms and timeline evolution, the research hot spots and frontier trends for electric vehicles and related fields in the WOS and CNKI literature can be clearly recognized.

In the CNKI literature, the frequency and burst rate of keywords such as "energy conservation and emission reduction", "energy transformation", "carbon neutrality", "carbon peak", and "carbon emissions" reveal the importance of this field in recent years. These keywords not only reflect China's strategic goals in addressing climate change and promoting sustainable development, but also show the in-depth study of relevant technologies and policies in academia.

Energy Conservation and Emission Reduction: The frequent occurrence of this keyword indicates that researchers are actively exploring ways to reduce energy consumption and greenhouse gas emissions through technological innovation and policy measures. This point is highly consistent with China's "twin goals", reflecting the country's emphasis on sustainable development.

Energy Transformation: With the rapid development of renewable energy, energy transformation has become the global focus. Researchers have begun to explore how to effectively promote the transformation of traditional energy to clean energy in order to achieve the dual benefits of economic development and environmental protection.

Carbon Neutrality and Carbon Peak: The emergence of these keywords marks China's active participation in global climate governance. Scholars focus on how to achieve the peak of carbon emissions and the ultimate carbon neutrality. Their research directions include policy design, the technology path, and economic incentives.

In the WOS literature, keywords such as "emissions", "power system", "renewable energy", "storage", "fuel consumption", "recovery" and "lithium-ion batteries" reveal research hot spots in the international academic community in regards to electric vehicles and related fields.

Emissions: This keyword runs through the entire research cycle from 1996 to 2024. Especially from 2009 to 2016, the number of relevant studies surged, showing that scholars paid close attention to greenhouse gas emissions during this period. This trend reflects the increasing global attention on climate change and the in-depth discussion of the potential of electric vehicles in regards to emission reduction.

Renewable Energy and Power System: The emergence of these keywords reflects the close relationship between electric vehicles and renewable energy systems. Researchers explore how to combine electric vehicles with smart grids and renewable energy to achieve more efficient energy use and a lower environmental impact.

Recovery and Lithium-Ion Batteries: From 2021 to 2024, scholarly research including keywords such as "recovery", "lithium-ion batteries", and "valuable metals" has increased significantly. This shows that with the popularity of electric vehicles, the recycling of batteries and the effective reuse of their internal valuable metals have become the research hot spots. The life cycle management of batteries, the innovation of recycling technology, and the sustainable utilization of resources are currently the focus of the academic world and the industrial community.

In summary, by analyzing the keywords and the evolution of timelines in the WOS and CNKI literature, it can be observed that the research hot spots in regards to electric vehicles and related fields are changing continually. Chinese scholarly research under the "twin goals" focuses on energy conservation, emission reduction, and carbon emission management, while the international academic community pays more attention to the technological frontiers, such as battery recycling and resource utilization. These studies not only provide theoretical support for the development of electric vehicles, but also offer a practical basis for policy formulation and technological innovation, promoting the sustainable development of the global automobile industry.

7.5. The Impact Analysis of Electric Vehicles on the Urban Transportation Network in China

This article utilizes CiteSpace and VOSviewer software to analyze the challenges faced by electric vehicles in supporting the construction of a green and low-carbon transportation network in Chinese cities.

(1) Policy Dimension: The Contradiction between Policy Incentives and Implementation Gaps.

Current Status and Impact: Policy is the core driving force behind the development of the electric vehicle industry [128]. In recent years, the Chinese government has introduced a series of policies, such as "The New Energy Vehicle Industry Development Plan (2021–2035)" [129] and "The Implementation Opinions on Further Enhancing the Service Guarantee Capacity of Electric Vehicle Charging Infrastructure" [130], aimed at accelerating the popularization of electric vehicles and the transformation of urban transportation networks through policy guidance and support. Guan, X., et al. (2025) conducted an in-depth analysis of the carbon emission reduction effects of new energy vehicles in China on 27 provinces (excluding Taiwan, Xinjiang, Tibet, Ningxia, Qinghai, Hong Kong, and Macau) using time-series data between 2016 and 2022 [131]. The research findings indicate that the carbon emission reduction effects of new energy vehicles showed an overall upward trend, with a significant acceleration since 2021. This shift may be closely related to "The New Energy Vehicle Industry Development Plan (2021–2035)" and associated policies released by the Chinese government in November 2020, which provided substantial support for the development of new energy vehicles and carbon emission reduction efforts.

Existing Dilemmas: Despite substantial policy support, some issues regarding the actual implementation process remain. For example, there exist a lack of coordination and consistency among policies, discrepancies between local and national policies, and low levels of policy alignment among different local governments, resulting in the incomplete realization of policy effects [126]. Additionally, the long-term nature and stability of the current policies need to be further strengthened to avoid market fluctuations and investment risks for enterprises [132]. At the same time, the implementation of policies lacks effective supervision and evaluation mechanisms [133], which may lead to resource waste and inefficiency.

(2) Technological Dimension: The Disconnect Between Technological Advancement and Practical Application.

Current Status and Impact: Technological advancement serves as the cornerstone of the development of the electric vehicle industry [134]. In recent years, significant breakthroughs have been made in electric vehicle technology, particularly concerning battery technology, charging technology, and intelligent management systems [135]. For instance, improvements in battery energy density and charging efficiency have led to substantial enhancements in the driving range and user convenience of electric vehicles [136]. Furthermore, the optimization of intelligent traffic management systems and route planning algorithms has provided technological support for the efficient operation of urban transportation networks [137,138].

Existing Dilemmas: Despite significant technological advancements, there are still some bottlenecks. For example, further breakthroughs in battery technology are limited by material science and manufacturing processes [139,140], while the lack of unified technical standards for the charging infrastructure has resulted in charging compatibility issues [141]. Additionally, the intelligent and connected technologies of electric vehicles still require refinement to meet the complex demands of urban transportation networks [142]. Specifically, the current mainstream lithium-ion batteries face limitations in regards to materials science for improving energy density. For example, positive and negative electrode materials should exhibit good reversibility, a high ion diffusion rate and electronic conductivity, a long cycle life, good ecological compatibility, and low cost [143]. In terms of manufacturing processes, consistency and stability issues during the production process affect battery quality and performance [144,145]. In terms of charging infrastructure, the lack of unified technical standards has led to differences in charging interfaces and protocols for different brands and models of electric vehicles, seriously affecting charging compatibility and causing great inconvenience to users [146,147].

(3) Infrastructure Dimension: The Contradiction of Challenges in the Construction and Popularization of Charging Networks.

Current Status and Impact: The improvement of infrastructure is a crucial support for the promotion of electric vehicles [148]. In recent years, China has accelerated the construction of charging stations, battery-swapping stations, and other infrastructure, forming an initial charging network that covers urban and intercity transportation [149]. For example, the construction of charging infrastructure and urban smart traffic management systems has not only improved traffic efficiency but also promoted the green transformation of urban transportation networks [150]. The continuous development of charging infrastructure provides a solid guarantee for the popularization of electric vehicles, thereby enhancing their market acceptance and user experience [151].

Existing Dilemmas: Although progress has been made in infrastructure construction, there are still some issues. Vinicius, B., et al. (2025) argue that in areas with low usage of electric vehicles, investors are reluctant to build charging stations due to insufficient charging demand. In contrast, in areas where electric vehicles are widely used, the speed for constructing charging infrastructure cannot keep up with the growth rate of vehicles, resulting in insufficient charging facilities in some areas, especially in remote areas and urban–rural areas [152]. For example, the distribution of charging stations is uneven, with insufficient charging facilities in certain areas, especially in remote regions and urban–rural fringe zones, where electric vehicle users often face inconveniences in charging [153,154]. Additionally, the maintenance and management of charging infrastructure also need to be further strengthened to improve the utilization rate and reliability of the facilities [155].

(4) Social Dimension: Challenges of Enhanced Awareness of Green Travel and the Lag in Usage Habits.

Current Status and Impact: The promotion of electric vehicles has had a profound effect on social lifestyles and travel habits [156]. With the widespread adoption of electric vehicles, public awareness and acceptance of green travel have gradually increased, leading more people to choose electric vehicles as their daily means of transportation [157,158]. Furthermore, the promotion of electric vehicles has also facilitated the development of shared mobility and smart travel, further optimizing urban transportation networks [159,160].

Existing Dilemmas: Despite the growing social acceptance, several obstacles persist. For instance, some consumers harbor concerns about the range capabilities and charging convenience of electric vehicles, which negatively impacts their purchasing decisions [161]. Additionally, the usage habits and cultural norms surrounding electric vehicles remain underdeveloped, creating an urgent need for enhanced public education and promotional efforts to facilitate societal acceptance and behavioral adaptation [162,163].

(5) Economic Dimension: Dual Challenges of Market Competitiveness and Industry Chain Improvement.

Current Status and Impact: The promotion of electric vehicles has exerted a profound impact on urban economic structures and industrial chains [16]. The continuously increasing sophistication of the electric vehicle industry chain has driven rapid development across related industries, including battery manufacturing, charging infrastructure production, and intelligent transportation management systems [164]. Additionally, the widespread adoption of electric vehicles has also catalyzed the green economy and sustainable development strategies of cities, injecting new momentum into the ecological transformation and technological upgrading of urban mobility systems [165].

Existing Dilemmas: Although electric vehicles significantly contribute to economic growth, they still face several challenges. For instance, the production and usage costs of electric vehicles remain relatively high, which somewhat limits their market competitiveness [166]. From the perspective of ownership costs, the purchase cost of an electric vehicle is generally higher than that of a traditionally fueled vehicle, mainly due to the high cost of batteries, which accounts for a large proportion of the total vehicle cost [167,168]. In terms of the economic efficiency of charging stations, the current charging station construction costs are relatively high [169], including site leasing, equipment procurement, installation, and commissioning costs. Additionally, the utilization rates of some charging stations are relatively low, particularly in remote areas and emerging regions, resulting in long investment payback periods [170]. Moreover, the overall improvement of the electric vehicle industry chain still requires further enhancement to boost overall efficiency and competitiveness, allowing it to better meet market demands [171].

(6) Environmental Dimension: The Dual Challenge of Insufficient Life Cycle Assessment and the Lack of Ecological Benefit Monitoring.

Status and Impact: The promotion of electric vehicles has generated a positive impact on environmental sustainability [172]. The widespread adoption of electric vehicles has reduced the carbon emissions from traditional fuel-powered vehicles, contributing to the achievement of carbon peak and carbon neutrality goals [173]. Furthermore, the promotion of electric vehicles has also encouraged the use of clean energy, further optimizing the energy structure in urban areas [174].

Existing Dilemmas: Despite the significant environmental benefits, several issues persist. For example, the lifecycle emissions assessment of electric vehicles has not been fully implemented, leading to a lack of systematic analysis of their environmental impacts at the production, usage, and disposal stages [175]. Although the positive effects of electric vehicles on urban ecological environments are recognized [176], there is an absence of systematic ecological benefit evaluation and research on their impact on residents' health and quality of life, which constrain public awareness and undermine the effectiveness of promotion policies [177–179].

8. The Path of Electric Vehicles to Build an Urban Green Transportation Network

8.1. Exploring China's Solutions for Building a Green Transportation Network

Based on scientometric methodologies, this paper systematically deconstructs the knowledge foundations, evolutionary trajectories, and multidimensional bottlenecks in the electric vehicle-driven development of green and low-carbon urban transportation networks in China and finally innovatively constructs a six-dimensional analytical framework of "policy-driven–technological innovation–infrastructure support–social response–economic reconstruction–ecological collaboration". After in-depth research and comprehensive analysis of the relevant literature, at home and abroad, the authors propose the

Chinese plan for how to build an urban green transportation network, specifically regarding the application of electric vehicles. The program will be discussed from several dimensions, such as policy support, technological innovation, infrastructure construction, social acceptance, economic benefits, and environmental protection, as shown in Figure 20.

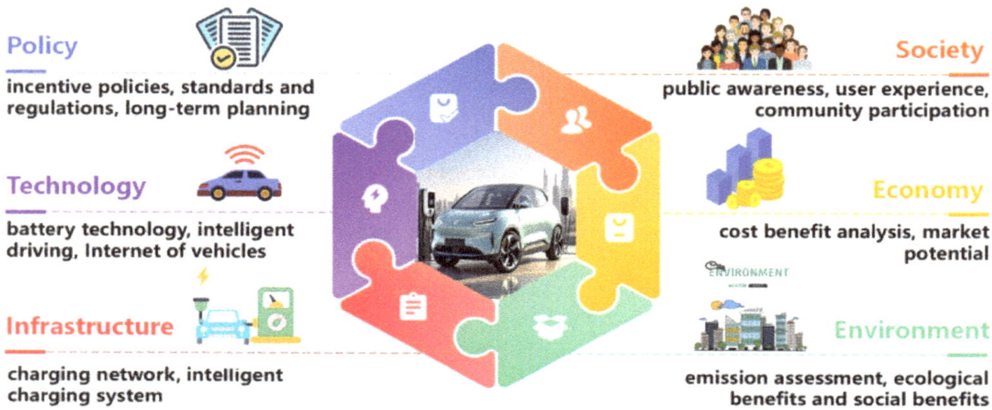

Figure 20. China's plan for electric vehicles to help build an urban green transportation network (six-dimensional analytical framework).

8.1.1. Policy Support

Incentive Policy: Wang, S., et al. (2024) [180] selected panel data from 20 provinces and cities from 2010 to 2020 to examine the implementation effects of incentive policies on the promotion and application of new energy vehicles in China from multiple perspectives. The results indicate that incentive policies can effectively promote the adoption of new energy vehicles, but the impact of local subsidy policies varies across different regions. It is suggested that governments, at all levels, should issue a series of preferential policies, including providing car purchase subsidies, exemption from purchase tax, reduction in parking fees for electric vehicle buyers, etc. to reduce the cost of purchasing and using electric vehicles and stimulate market demand. Financial institutions are encouraged to provide low-interest loans, i.e., green funds to guide social capital to expand participation in the industrial development of electric vehicles.

Standards and Regulations: Standards and regulations related to electric vehicles should be established to ensure product quality and safety and facilitate the healthy development of the market, e.g., establishing technical standards and safety regulations to ensure the compliance of electric vehicles in the market.

Long-term Planning: A medium and long-term green transportation development plan should be formulated, clarifying the status and role of electric vehicles in the urban green transportation network.

8.1.2. Technological Innovation

Battery Technology Breakthroughs: In order to solve the bottleneck faced by current battery technology, government departments at all levels should increase research and development (R&D) investment, focusing on breakthroughs in key core technologies for electric vehicles and power batteries. Advancements in these technologies will help improve the energy density, charging speed, and lifespan of batteries, thereby reducing the overall cost of electric vehicles and lowering the threshold for users. Specifically, government agencies and research teams should ramp up R&D efforts in regards to new battery materials, actively exploring novel battery systems such as solid-state batteries, lithium–sulfur batteries, and lithium–silicon batteries. For instance, Shchurov, N.I., et al. (2021) point out that silicon-based anodes exhibit significant application potential due to their high theoretical capacity (approximately ten times that of graphite) [181]. However, to fully leverage the advantages of silicon materials, several critical issues need to be addressed, including enhancing the electrical conductivity of silicon, managing the volume expansion during charging and discharging, and overcoming the difficulties involved in the large-scale production of silicon nanowire materials. Once these issues are resolved, silicon nanowires hold broad application prospects in the field of power batteries.

Intelligent Driving and Internet of Vehicles: It is recommended that government at all levels and relevant enterprises increase their support to promote the rapid development of autonomous driving technology and Internet of vehicles technology to further enhance the safety and convenience of electric vehicles, which can improve the users' travel experience. For example, the improvement of autonomous driving technology will enable vehicles to respond more intelligently to various complex road conditions and reduce human errors in operation, thereby enhancing overall driving safety. And the development of vehicle networking technology will enable vehicles to communicate efficiently with other vehicles and transportation infrastructure. This information sharing can effectively reduce traffic congestion and improve road use efficiency.

8.1.3. Infrastructure Construction

Charging Network Layout: It is suggested that government at all levels should further improve the construction and planning of charging infrastructure and facilitate the efficient interaction between electric vehicles and smart grids. For example, charging infrastructure should be built in parking lots, bus stations, and expressway service areas, especially in key areas such as residential areas, commercial areas, and office areas so as to ensure that electric vehicle users can charge their vehicles conveniently, anytime and anywhere.

Intelligent Charging System: Using the Internet of things technology and big data analysis, the intelligent charging management system should be constructed to realize the optimal allocation and scheduling of charging resources, improve the charging efficiency, and reduce the user wait time. Moreover, we should encourage the use of renewable energy sources such as solar energy and wind energy in charging facilities to enhance the green attributes of electric vehicles. Referring to the research ideas of Wang, K., et al. (2025) [182], we consider coordinating with the integrated power distribution system in the construction of intelligent charging systems. This includes optimizing the layout and operation strategy of charging facilities to improve the resilience and stability of the entire energy system. For example, in the face of emergencies or power supply fluctuations, the coordinated management of intelligent charging systems ensures the normal charging of electric vehicles, while minimizing the impact on the overall energy system.

8.1.4. Social Acceptance

Public Awareness: Government at all levels should strengthen their education and publicity activities to enhance citizens' awareness and participation in green travel, forming form a green transportation and green travel culture with the participation of the whole population.

User Experience: There should be a focus on the optimization of the user experience, improving charging convenience, endurance, and vehicle performance to improve user satisfaction. The convenience of the charging facilities directly affects the public's acceptance of electric vehicles, and a good user experience can further enhance users' recognition of and trust in electric vehicles. Enterprises should be encouraged to participate in the promotion and application of electric vehicles, and enterprises' innovation and practice regarding green travel should be supported, striving to form a joint force comprising government, enterprises, and the public.

8.1.5. Economic Benefits

Cost–Benefit Analysis: When evaluating the cost of using electric vehicles, it is recommended to comprehensively compare electric vehicles with traditional fuel vehicles, which includes many aspects such as purchase costs, maintenance costs, and fuel costs. Through detailed calculation and comparison, we can understand the advantages and disadvantages of electric vehicles in terms of economy so as to provide a strong basis for consumers and enterprises to make decisions.

Market Potential: Electric vehicles exhibit huge potential in different markets, but each market is different. It is suggested that the government and enterprises should deeply analyze the characteristics of each market, identify the target user groups, and understand needs and preferences of each one. On this basis, the corresponding market strategy is formulated to maximize the market share and brand influence of electric vehicles.

8.1.6. Environmental Protection

Emission Assessment: In order to fully understand the impact of electric vehicles on the environment during their life cycle, it is recommended that the government and related companies regularly conduct detailed emission assessments. This assessment process covers all stages of electric vehicles, from production, to use, to scrap, aiming to deeply analyze their comprehensive impact on the environment. Through this assessment, the public can better understand the impact of electric vehicles on the environment at all stages of their life cycle, thus providing a scientific reference for improving and optimizing the production and use of electric vehicles.

Ecological Benefits: Zhao, X., et al. (2024) [183] used the spatial econometric model to investigate the impact of promoting new energy vehicles on carbon emissions. The research findings indicate that a 1% increase in new energy vehicle sales in a city would reduce local carbon emissions by 0.096%. The positive impact of electric vehicles on the urban ecological environment is obvious, especially in terms of improving air quality, reducing noise pollution, and mitigating the urban heat island effect. In order to better understand and quantify these effects, it is recommended that the government and related enterprises regularly conduct a systematic analysis and evaluation of the urban ecological benefits. First of all, regular monitoring of air quality changes, especially in areas with a high popularization rate for electric vehicles, can help us understand the actual effect of electric vehicles on reducing harmful gas emissions. This not only helps to develop more effective environmental policies but also provides a scientific basis for the public to enhance their trust and support for electric vehicles. Secondly, the assessment of noise pollution is equally important. By comparing the noise levels of electric vehicles and traditional fuel vehicles in different areas of the city, relevant departments can recognize the contribution of electric vehicles to reducing urban noise so as to promote more urban planning and traffic management measures to further improve the living quality of residents. In addition, research on the urban heat island effect should also be included in the assessment. By analyzing the impact of the use of electric vehicles on urban temperature changes, the government can better formulate strategies to cope with climate change and promote sustainable urban development.

Social Benefits: It is recommended that the government and relevant enterprises regularly analyze and evaluate the positive impact of electric vehicle promotion on residents' health, quality of life, and socioeconomic development. First, the impact of electric vehicle promotion on residents' health can be quantified by monitoring air quality, residents' health data, and the incidence of related diseases. Second, electric vehicles produce less noise and can improve urban residents' living quality. Thus, it is recommended to obtain an understanding of the impact of electric vehicles on people's daily lives through resident surveys and satisfaction assessments. Third, the development of the electric vehicle industry can create employment opportunities and promote the innovation and progress of related technologies. Therefore, governments and enterprises can regularly analyze the economic contribution of the electric vehicle industry chain, including manufacturing, sales, charging facilities construction, and maintenance and evaluate its role in promoting the local economy.

8.2. Analysis of Implementation Difficulty and Potential Impact in Cities of Different Sizes

Large cities: Taking Beijing as an example, as a megacity, it benefits from strong financial resources in terms of policy support. However, the complexity of urban planning and the difficulty of interdepartmental coordination poses significant challenges. Nevertheless, in large cities, the ownership of electric vehicles is generally high, and the demand is very robust. For example, the number of new energy vehicles in Beijing has exceeded 1 million [184]. Once the charging infrastructure is completed, the high utilization rate of these facilities will have a significant impact on reducing carbon emissions and alleviating traffic congestion.

Medium-sized cities: Taking Shijiazhuang as an example, policy support is relatively flexible, but financial resources are limited. This results in subsidies and R&D investments that cannot compared to those of large cities. In order to promote development, Shijiazhuang City plans to optimize and improve the layout of charging piles before 2025 and to initiate the Bulinke New Energy Charging Pile and Energy Storage Project in the Economic and Technological Development Zone in order to consolidate the pilot achievements of the comprehensive electrification of vehicles in the public sector [185].

Small cities: Taking Tang County in Hebei Province as an example, policy formulation is simple, but fiscal revenue is low, restricting the sustainability and strength of policy support. Meanwhile, due to the small population and industrial scale, there is insufficient support for the electric vehicle industry, making it difficult to promote. To address this issue, Tang County plans to accelerate the construction of charging infrastructure in 2025 and implement consumption policies to comprehensively promote the trade-in of old cars for new ones [186].

8.3. Differentiated Implementation Pathways for Green Transportation Networks

Large, medium-sized, and small cities each possess unique advantages and face distinct challenges regarding the development of the electric vehicle industry. Cities should formulate corresponding policies and strategies based on their specific characteristics to promote the healthy and sustainable growth of the electric vehicle sector.

(1) Implementation path for large cities.

Policy Support: The government should enhance interdepartmental coordination, establish long-term and stable policies, and integrate resources from various stakeholders.

Technological Innovation: Governments and R&D institutions should increase investment in the research and development of core technologies related to electric vehicles, particularly in regards to battery technology and autonomous driving. Additionally, enterprises and research organizations should be encouraged to strengthen international collaboration to attract top global talents and advanced technologies, thereby promoting the high-end development of the industry.

Infrastructure Construction: The government should accelerate the construction of high-speed and efficient charging networks and promote intelligent charging and battery swapping technologies to enhance the user experience.

(2) Implementation path for medium-sized cities.

Policy Support: Local governments should actively seek funding from higher authorities and formulate targeted policies for electric vehicles that align with local industrial characteristics. For example, Shijiazhuang City can leverage its strengths in the steel industry to formulate supporting policies.

Technological Innovation: Local governments should actively collaborate with research institutions in large cities to facilitate technology transfer, introduce advanced technologies, and cultivate local talent.

Infrastructure Construction: Local governments should rationally lay out the charging networks based on urban development plans. Priority should be given to constructing charging facilities in core areas, public service zones, and transportation hubs of cities to meet the growing demand for electric vehicle charging.

(3) Implementation Path for Small Cities.

Policy Support: The government should actively seek support and funding from the national and provincial levels and guide local enterprises to participate in supporting services related to the electric vehicle industry.

Technological Innovation: The government should encourage local enterprises to strengthen technical exchanges and cooperation with developed regions and actively send professional personnel to study advanced technologies in developed regions.

Infrastructure Construction: The government should rationally plan the layout of charging facilities based on actual needs. It is recommended to adopt a gradual development strategy, focusing on distributed and small-scale charging facilities to reduce construction costs and improve resource utilization efficiency.

9. Conclusions, Research Limitations, and Research Prospects

9.1. Conclusions

This paper uses analysis tools, specifically, CiteSpace, to review and perform a visual analysis of 2650 studies in the China National Knowledge Infrastructure (CNKI) database and 2460 studies in the Web of Science Core Collection (WOS) database. The results show the following: (1) In the CNKI, the number of publications regarding electric vehicles in urban green transportation increased rapidly from 2008 to 2011, and the annual number of publications over the past three years fluctuated around 150. In the WOS, the number of annual publications in this research field has shown a rapid growth trend. Since 2019, it has increased sharply, reaching 500 in 2023. (2) In terms of the literature in the CNKI, the number of cooperative studies between authors and the number of research teams are small, and the degree of cooperation between authors is low, with a dispersed research state. The literature in the WOS also displays this problem. (3) In recent years, the research hot spots in this field in the CNKI have mainly focused on "energy conservation and environmental protection", "energy conservation and emission reduction", "carbon neutrality", "carbon peak", "carbon emissions", and other fields. In the WOS, the research hot spots include "recovery", "lithium-ion batteries", "valuable metals", etc. (4) With the rapid development of science and technology and the general improvement in environmental awareness, people are paying more attention to low-carbon and green transportation. (5) This paper explores the impact of electric vehicles on urban traffic networks in China and systematically analyzes their innovative role. We have innovatively constructed a six-dimensional analytical framework of "policy-driven–technological innovation–infrastructure support–social response–economic reconstruction–ecological collaboration" and proposed strategies and pathways for electric vehicles to facilitate the construction of green transportation networks in cities.

The development of green transportation is an important part of building a low-carbon city. As a representative of green transportation, electric vehicles have great development prospects and advantages. The government, enterprises, and all social circles should work together to promote the transformation of urban transportation into a low-carbon and environmentally friendly entity. Therefore, the government should continue to increase the support for the electric vehicle industry and accelerate the popularization of electric vehicles in cities through policy guidance and technological innovation, jointly building a greener, healthier, and more harmonious low-carbon city.

9.2. Research Limitations

Although this study employs scientometric methods to systematically analyze the application of electric vehicles in urban green transportation, it still displays certain limitations. The data, sourced primarily from the CNKI and Web of Science Core Collection databases, may lead to incompleteness. Industry reports, internal corporate research, and other materials which might contain crucial information on market dynamics, technological innovation practices, and other aspects of electric vehicles, were not included in the analysis. The absence of such data could affect the comprehensive analysis of certain research directions. For instance, when analyzing the development of the electric vehicle industry chain, the lack of industry report data might hinder an accurate understanding of the actual progress in some niche areas. Therefore, the findings of this study may not fully encompass all aspects of the field. Future research should consider expanding data collection channels to obtain more comprehensive and accurate conclusions.

9.3. Prospects for Future Research

(1) Research on vehicle intelligence.

Under the major trend of green transportation, the development of intelligent electric vehicles will become a new trend. With the rapid development of mobile communication, Internet of things, and other technologies, the vehicle information system is advancing to be more intelligent and Internet-based. In the future, the intelligent vehicle network technology will be further improved, and it is believed that the deep interaction between the vehicle system and the cloud will create a better intelligent green transportation experience [187].

(2) Research on lightweight materials for vehicles.

In order to cope with increasingly stringent emission regulations, the lightweight design of automobiles has become one of the future trends. The use of advanced materials such as carbon fiber and aluminum alloy can significantly reduce vehicle weight and improve energy efficiency so as to increase the mileage of electric vehicles [188].

(3) Research on autonomous driving technology.

Autonomous driving technology is the hot spot in the research of electric vehicle technology. Through the integration of advanced sensors, artificial intelligence algorithms, and big data analysis technology, the autonomous navigation and decision making of vehicles will be realized. In the future, with the continuous optimization of algorithms and breakthroughs in sensor technology, autonomous vehicles will be able to more accurately identify and perceive the surrounding environment, thereby achieving a higher level of autonomous driving [189].

(4) Research on power battery technology innovation.

Researchers are working to develop higher energy density, longer lifespan, and safer battery technologies to meet the growing demand for electric vehicle endurance [190]. In addition, the charging technology is also continually advancing, aiming to shorten the charging time and improve the charging efficiency to improve the user experience.

(5) Research on urban transportation planning.

Urban transportation planning is very important for promoting the development of green transportation. Future research should focus on how to rationally plan green transportation networks through intelligent transportation systems to improve urban transportation efficiency and reduce traffic congestion and carbon emissions [191]. At the same time, it is also necessary to consider how intelligent transportation systems can be combined with urban planning, land use, and environmental protection policies to achieve a more comprehensive green transportation network.

These research hot spots jointly promote the continuous progress of electric vehicle technology and provide solid technical support for future urban green transportation.

Author Contributions: Conceptualization, D.L. and A.D.L.; methodology, D.L. and A.D.L.; data curation, Y.G.; investigation, D.L. and Y.G.; writing—original draft preparation, D.L., A.D.L. and Y.G.; writing—review and editing, D.L., A.D.L. and Y.G.; supervision and submission, D.L. and Y.G. All authors have read and agreed to the published version of the manuscript.

Funding: This paper is supported by the 2025 Annual Social Science Development Research Project of Hebei Province (20250602004).

Data Availability Statement: The data presented in this study are available on request from the corresponding author.

Conflicts of Interest: The authors declare no conflicts of interest.

References

1. Li, K.; Tian, X.; Zheng, S.; Jiang, S. Research on new energy vehicle infrastructure construction in heze under the background of "double carbon". *Automob. Appl. Technol.* **2023**, *48*, 14–18. [CrossRef]
2. Ma, L. The digital and intelligent transformation of the "dual carbon" industry: Mechanisms, challenges and responses: From the perspective of three elements of digitization and intelligence. *Ecol. Econ.* **2024**, *40*, 5–54.
3. Qu, R.; Ye, J.; Zhang, X. Hot spots and trends in green and low-carbon building retrofit research in the context of urban renewal. *Green Build.* **2024**, *6*, 69–77.
4. Liu, S.; Li, W.; Liu, D. Green Travel is the Necessary Path for the Green and Low-Carbon Transformation of Urban Transportation. China Communications News, 11 August 2023. Available online: https://www.zgjtb.com/2023-08/11/content_369962.html (accessed on 11 August 2023).
5. Bradshaw, C. The Valuing of Trips. In *The Walk n Roll City, Second Car-Free Cities Conference, Toronto, January 1992*; Revised Sep 1994; Prepared for Ottwalk and the Transportation Working Committee of the Ottawa–Carleton Round-Table on the Environment; Transportation & Infrastructure Committee: Toronto, ON, Canada, 1994.
6. Yang, T.; Peng, J.; Yu, M.; Ying, F. Green Transportation Planning in New Town of China: Take Naning South New Town as Example. *Urban Transp. China* **2021**, *19*, 58–64. [CrossRef]
7. Lu, H. Approaches Towards Realization of Urban GreenTransportation. *Urban Transp. China* **2009**, *7*, 23–27. [CrossRef]
8. Ma, Z.; Gao, Y.; Hu, D.; Wang, J.; Ma, F.; Xiong, Y. Green transportation level measurements and spatial-temporal evolution characteristics of urban agglomeration transportation systems. *J. Tsinghua Univ. (Sci. Technol.)* **2022**, *62*, 1236–1250. [CrossRef]

9. Wang, Y.; Tian, Z.; Xue, B.; Li, H.; Ma, T. Review on Optimization Methods of Green TransportationSystem in Urban Built Environment. *Shanghai Urban Plan. Rev.* **2023**, *6*, 11–17.
10. Yu, H.; Yang, Q. Evaluation of the Development Level of Green Transportation in National Central Cities. *Sustainability* **2024**, *16*, 7270. [CrossRef]
11. Gao, X.; Ding, C.; Hou, X.; Duan, D. Green transportation technology innovation in China city system: Dynamies and determinants. *Geogr. Sci.* **2025**, *45*, 119–129. [CrossRef]
12. Lu, H. Current Situation analysis and countermeasures of green and low carbon transport transformation in China. *China Sustain. Trib.* **2024**, *Z2*, 53–57.
13. Wang, J.; Hao, Y.; Zhao, J.; Liu, R.; Wang, S.; Tian, S. Connotation and Development Path of Green Transportation in the Context of Urban Low-Carbon Transformation. *Urban Transp. China* **2024**, *22*, 52–57. [CrossRef]
14. Zhao, Y.; Wang, X. Low-carbon emission reduction potential and impact analysis of electric vehicles. *Heilongjiang Environ. J.* **2022**, *35*, 120–121.
15. Zhou, Q. Research on the development trend of intelligent transportation system and its optimization strategy. *People's Public Transp.* **2024**, *14*, 20–22. [CrossRef]
16. Wang, Y.; Xu, W.; Hu, Q. Integration and coordination in the development of China's electric vehicle industry: Mechanisms and policy implications. *China Econ.* **2024**, *19*, 2–30. [CrossRef]
17. The Communist Party of China Central Committee and the State Council, Working Guidance for Carbon Dioxide Peaking and Carbon Neutrality in Full and Faithful Implementation of the New Development Philosophy. Available online: https://www.gov.cn/zhengce/2021-10/24/content_5644613.htm (accessed on 24 October 2021).
18. The Central People's Government of the People's Republic of China. Action Plan Carbon Dioxide Peaking Before 2030. Available online: https://www.gov.cn/gongbao/content/2021/content_5649731.htm (accessed on 24 October 2021).
19. Sun, Y.; Zhuang, H. Research on the Rule of Law Guarantee for Green Transportation under the Carbon Peaking and Carbon Neutrality Goals. *J. Kunming Univ.* **2023**, *45*, 69–78. [CrossRef]
20. Tong, Z.M.; Guan, S.; Zhang, Q.G.; Cao, X.E. Data-driven energy efficient speed planning for battery electric industrial vehicles: Forklift as a case study. *J. Clean. Prod.* **2024**, *443*, 140923. [CrossRef]
21. Lu, P.Y.; Hamori, S.; Sun, L.; Tian, S.R. Does the electric vehicle industry help achieve sustainable development goals?-evidence from China. *Front. Environ. Sci.* **2024**, *11*, 1276382. [CrossRef]
22. Kramer, J.; Petzoldt, T. A matter of behavioral cost: Contextual factors and behavioral interventions interactively influence pro-environmental charging decisions. *J. Environ. Psychol.* **2022**, *84*, 101878. [CrossRef]
23. Sebastián, M.G.; Guede, J.R.; Juárez-Varón, D. Analysis of factors influencing attitude and intention to use electric vehicles for a sustainable future. *J. Technol. Transf.* **2024**, *49*, 1347–1368. [CrossRef]
24. Wicki, M.; Brückmann, G.; Bernauer, T. What do we really know about the acceptance of battery electric vehicles?—Turns out, not much. *Transp. Rev.* **2023**, *43*, 62–87. [CrossRef]
25. Vega-Perkins, J.; Newell, J.P.; Keoleian, G. Mapping electric vehicle impacts: Greenhouse gas emissions, fuel costs, and energy justice in the United States. *Environ. Res. Lett.* **2023**, *18*, 014027. [CrossRef]
26. Ji, Y.B.B.; Dong, J.C.; Fei, X.; Wang, G.; Fei, X. Research on carbon emission measurement of Shanghai expressway under the vision of peaking carbon emissions. *Transp. Lett.-Int. J. Transp. Res.* **2023**, *15*, 765–779. [CrossRef]
27. Pucci, P. Spatial dimensions of electric mobility-Scenarios for efficient and fair diffusion of electric vehicles in the Milan Urban Region. *Cities* **2021**, *110*, 103069. [CrossRef]
28. Singh, D.; Paul, U.K.; Pandey, N. Does electric vehicle adoption (EVA) contribute to clean energy? Bibliometric insights and future research agenda. *Clean. Responsible Consum.* **2023**, *8*, 100099. [CrossRef]
29. Shang, H.R.; Sun, Y.T.; Meng, F.X. Life cycle assessment of atmospheric environmental impact on the large-scale promotion of electric vehicles in China. *Resour. Environ. Sustain.* **2024**, *15*, 100148. [CrossRef]
30. Klos, M.J.; Sierpinski, G. Strategy for the Siting of Electric Vehicle Charging Stations for Parcel Delivery Service Providers. *Energies* **2023**, *16*, 2553. [CrossRef]
31. Cai, Y.C.; Zhang, J.; Wang, C.L. An Analytical Framework for Assessing Equity of Access to Public Electric Vehicle Charging Stations: The Case of Shanghai. *Sustainability* **2024**, *16*, 6196. [CrossRef]
32. Tuffour, J.P.; Ewing, R. Can battery electric vehicles meet sustainable energy demands? Systematically reviewing emissions, grid impacts, and coupling to renewable energy. *Energy Res. Soc. Sci.* **2024**, *114*, 103625. [CrossRef]
33. Zhao, X.; Han, J. How Is Transportation Sector Low-Carbon (TSLC) Research Developing After the Paris Agreement (PA)? A Decade Review. *Sustainability* **2025**, *17*, 2261. [CrossRef]
34. Chen, C. CiteSpace II: Detecting and visualizing emerging trends and transient patterns in scientific literature. *J. Am. Soc. Inf. Sci. Technol.* **2006**, *57*, 359–377. [CrossRef]

35. Chen, C. Science mapping: A systematic review of the literature. *J. Data Inf. Sci.* **2017**, *2*, 1–40. [CrossRef]
36. Van Eck, N.; Waltman, L. Software survey: VOSviewer, a computer program for bibliometric mapping. *Scientometrics* **2010**, *84*, 523–538. [CrossRef]
37. Arruda, H.; Silva, E.R.; Lessa, M.; Proença, D., Jr.; Bartholo, R. VOSviewer and bibliometrix. *J. Med. Libr. Assoc. JMLA* **2022**, *110*, 392. [CrossRef] [PubMed]
38. What Is Data Visualization? Definition, Examples, and Learning Resources. Available online: https://www.tableau.com/learn/articles/data-visualization (accessed on 20 March 2025).
39. What Is Tableau. Available online: https://www.tableau.com/why-tableau/what-is-tableau (accessed on 20 March 2025).
40. He, Y. How the Complexity of Policy Mix Affects Policy Effectiveness—An Analysis Based on the Promotion of New Energy Vehicles. *J. Public Manag.* **2024**, *21*, 41–55. [CrossRef]
41. Wang, L.; Zheng, Q. Innovation Capability and Network Evolution of China's New Energy Vehicle Industry from a Policy Chain Perspective. *Trop. Geogr.* **2025**, *45*, 361–373.
42. Ajzen, I. The theory of planned behavior. *Organ. Behav. Hum. Decis. Process.* **1991**, *50*, 179–211. [CrossRef]
43. Biresselioglu, M.; Demirbag Kaplan, M.; Yilmaz, B. Electric mobility in Europe: A comprehensive review of motivators and barriers in decision making processes. *Transp. Res. Part A Policy Pract.* **2018**, *109*, 1–13. [CrossRef]
44. Egbue, O.; Suzanna, L. Barriers to widespread adoption of electric vehicles: An analysis of consumer attitudes and perceptions. *Energy Policy* **2012**, *48*, 717–729. [CrossRef]
45. Fornell, C.; Larcker, D. Evaluating Structural Equation Models with Unobservable Variables and Measurement Error. *J. Mark. Res.* **1981**, *18*, 39. [CrossRef]
46. Graham-Rowe, E.; Gardner, B.; Abraham, C.; Skippon, S.; Dittmar, H.; Hutchins, R.; Stannard, J. Mainstream consumers driving plug-in battery-electric and plug-in hybrid electric cars: A qualitative analysis of responses and evaluations. *Transp. Res. Part A Policy Pract.* **2012**, *46*, 140–153. [CrossRef]
47. Helveston, J.; Liu, Y.; Feit, E.; Fuchs, E.; Klampfl, E.; Michalek, J. Will subsidies drive electric vehicle adoption? Measuring consumer preferences in the U.S. and China. *Transp. Res. Part A Policy Pract.* **2015**, *73*, 96–112. [CrossRef]
48. Krupa, J.S.; Rizzo, D.M.; Eppstein, M.J.; Brad Lanute, D.; Gaalema, D.E.; Lakkaraju, K.; Warrender, C.E. Analysis of a consumer survey on plug-in hybrid electric vehicles. *Transp. Res. Part A Policy Pract.* **2014**, *64*, 14–31. [CrossRef]
49. Rezvani, Z.; Jansson, J.; Bodin, J. Advances in consumer electric vehicle adoption research: A review and research agenda. *Transp. Res. Part D Transp. Environ.* **2015**, *34*, 122–136. [CrossRef]
50. Sierzchula, W.; Bakker, S.; Maat, K.; van Wee, B. The influence of financial incentives and other socio-economic factors on electric vehicle adoption. *Energy Policy* **2014**, *68*, 183–194. [CrossRef]
51. Skippon, S.; Garwood, M. Responses to battery electric vehicles: UK consumer attitudes and attributions of symbolic meaning following direct experience to reduce psychological distance. *Transp. Res. Part D Transp. Environ.* **2011**, *16*, 525–531. [CrossRef]
52. Wang, S.; Wang, J.; Li, J.; Wang, J.; Liang, L. Policy implications for promoting the adoption of electric vehicles: Do consumer's knowledge, perceived risk and financial incentive policy matter? *Transp. Res. Part A Policy Pract.* **2018**, *117*, 58–69. [CrossRef]
53. Kumar, R.R.; Alok, K. Adoption of electric vehicle: A literature review and prospects for sustainability. *J. Clean. Prod.* **2020**, *253*, 119911. [CrossRef]
54. Chan, C.C. The State of the Art of Electric, Hybrid, and Fuel Cell Vehicles. *Proc. IEEE* **2007**, *95*, 704–718. [CrossRef]
55. Clement-Nyns, K.; Haesen, E.; Driesen, J. The Impact of Charging Plug-In Hybrid Electric Vehicles on a Residential Distribution Grid. *IEEE Trans. Power Syst.* **2010**, *25*, 371–380. [CrossRef]
56. Geels, F.W. A socio-technical analysis of low-carbon transitions: Introducing the multi-level perspective into transport studies. *J. Transp. Geogr.* **2012**, *24*, 471–482. [CrossRef]
57. Guo, S.; Zhao, H. Optimal site selection of electric vehicle charging station by using fuzzy TOPSIS based on sustainability perspective. *Appl. Energy* **2015**, *158*, 390–402. [CrossRef]
58. Lund, H.; Kempton, W. Integration of renewable energy into the transport and electricity sectors through V2G. *Energy Policy* **2008**, *36*, 3578–3587. [CrossRef]
59. Nykvist, B.; Nilsson, M. Rapidly falling costs of battery packs for electric vehicles. *Nat. Clim. Change* **2015**, *5*, 329–332. [CrossRef]
60. Offer, G.J.; Howey, D.; Contestabile, M.; Clague, R.; Brandon, N.P. Comparative analysis of battery electric, hydrogen fuel cell and hybrid vehicles in a future sustainable road transport system. *Energy Policy* **2010**, *38*, 24–29. [CrossRef]
61. Sovacool, B.K.; Hirsh, R.F. Beyond batteries: An examination of the benefits and barriers to plug-in hybrid electric vehicles (PHEVs) and a vehicle-to-grid (V2G) transition. *Energy Policy* **2009**, *37*, 1095–1103. [CrossRef]
62. Yong, J.Y.; Ramachandaramurthy, V.K.; Tan, K.M.; Mithulananthan, N. A review on the state-of-the-art technologies of electric vehicle, its impacts and prospects. *Renew. Sustain. Energy Rev.* **2015**, *49*, 365–385. [CrossRef]

63. Ciez, R.E.; Whitacre, J.F. Examining different recycling processes for lithium-ion batteries. *Nat. Sustain.* **2019**, *2*, 148–156. [CrossRef]
64. Dunn, J.B.; Gaines, L.; Kelly, J.C.; James, C.; Gallagher, K.G. The significance of Li-ion batteries in electric vehicle life-cycle energy and emissions and recycling's role in its reduction. *Energy Environ. Sci.* **2015**, *8*, 158–168. [CrossRef]
65. Harper, G.; Sommerville, R.; Kendrick, E.; Driscoll, L.; Slater, P.; Stolkin, R.; Anderson, P. Recycling lithium-ion batteries from electric vehicles. *Nature* **2019**, *575*, 75–86. [CrossRef]
66. Hawkins, T.; Gausen, O.; Strømman, A. Environmental impacts of hybrid and electric vehicles—A review. *Int. J. Life Cycle Assess.* **2012**, *17*, 997–1014. [CrossRef]
67. Hawkins, T.; Singh, B.; Majeau-Bettez, G.; Strømman, A. Comparative Environmental Life Cycle Assessment of Conventional and Electric Vehicles. *J. Ind. Ecol.* **2012**, *17*, 53–64. [CrossRef]
68. Nordelöf, A.; Messagie, M.; Tillman, A.; Ljunggren Söderman, M.; Van Mierlo, J. Environmental impacts of hybrid, plug-in hybrid, and battery electric vehicles—What can we learn from life cycle assessment? *Int. J. Life Cycle Assess.* **2014**, *19*, 1866–1890. [CrossRef]
69. Notter, D.A.; Gauch, M.; Widmer, R.; Wäger, P.; Stamp, A.; Zah, R.; Althaus, H. Contribution of Li-Ion Batteries to the Environmental Impact of Electric Vehicles. *Environ. Sci. Technol.* **2010**, *44*, 6550–6556. [CrossRef]
70. Zackrisson, M.; Avellán, L.; Orlenius, J. Life cycle assessment of lithium-ion batteries for plug-in hybrid electric vehicles—Critical issues. *J. Clean. Prod.* **2010**, *18*, 1519–1529. [CrossRef]
71. Chen, L. Policy Support of the Development of New Energy Vehicles Industry. *J. Gansu Adm. Inst.* **2010**, *3*, 11–19.
72. Chen, L. The problems faced by the development of the new energy vehicle industry. *Auto Ind. Res.* **2011**, *6*, 2–9.
73. Chen, L. Inventory of China's new energy vehicle policies. *High-Technol. Commer.* **2012**, *7*, 2–48.
74. Wang, H.; Shi, H.; Chen, P.; Ou, Y. Analysis on the progress of energy saving and new energy vehicle industrialization in China based on a database. *J. Automot. Saf. Energy* **2014**, *5*, 294–297.
75. Zhang, Y.; Yao, J.; Wang, J.; Chen, L.; Xi, P.; Zhu, M.; Wang, Y.; Fu, W.; Wang, K.; Ge, K.; et al. Key Technologies and Application of Green Building Microgrids for Dual-Carbon Goals. *Constr. Sci. Technol.* **2023**, *11*, 114–117. [CrossRef]
76. Sovacool, B.K.; Kester, J.; Noel, L.; de Rubens, G.Z. The demographics of decarbonizing transport: The influence of gender, education, occupation, age, and household size on electric mobility preferences in the Nordic region. *Glob. Environ. Change-Hum. Policy Dimens.* **2018**, *52*, 86–100. [CrossRef]
77. Chen, C.; de Rubens, G.Z.; Noel, L.; Kester, J.; Sovacool, B.K. Assessing the socio-demographic, technical, economic and behavioral factors of Nordic electric vehicle adoption and the influence of vehicle-to-grid preferences. *Renew. Sustain. Energy Rev.* **2020**, *121*, 109692. [CrossRef]
78. Sovacool, B.K.; Kester, J.; Noel, L.; de Rubens, G.Z. Income, political affiliation, urbanism and geography in stated preferences for electric vehicles (EVs) and vehicle-to-grid (V2G) technologies in Northern Europe. *J. Transp. Geogr.* **2019**, *78*, 214–229. [CrossRef]
79. Lee, D.; Mcdermott, M.H.; Sovacool, B.K.; Isaac, R. Toward just and equitable mobility: Socioeconomic and perceptual barriers for electric vehicles and charging infrastructure in the United States. *Energy Clim. Change* **2024**, *5*, 100146. [CrossRef]
80. Sovacool, B.K. Who are the victims of low-carbon transitions? Towards a political ecology of climate change mitigation. *Energy Res. Soc. Sci.* **2021**, *73*, 101916. [CrossRef]
81. Sovacool, B.K.; Martiskainen, M.; Hook, A.; Baker, L. Decarbonization and its discontents: A critical energy justice perspective on four low-carbon transitions. *Clim. Change* **2019**, *155*, 581–619. [CrossRef]
82. Sovacool, B.K.; Turnheim, B.; Hook, A.; Martiskainen, M. Dispossessed by decarbonisation: Reducing vulnerability, injustice, and inequality in the lived experience of low-carbon pathways. *World Dev.* **2020**, *137*, 105116. [CrossRef]
83. Qiu, D.; Wang, Y.; Ding, Z.; Strbac, G. Graph Reinforcement Learning for Carbon-Aware Electric Vehicles in Power-Transport Networks. *IEEE Trans. Smart Grid* **2024**, *15*, 3919–3935. [CrossRef]
84. Qiu, D.; Wang, Y.; Zhang, N.; Strbac, G.; Kang, C. Decarbonising the GB Power System via Numerous Electric Vehicle Coordination. *IEEE Trans. Power Syst.* **2024**, *39*, 5880–5894. [CrossRef]
85. Tan, Q.; Shan, Z.; Guo, M.; Ding, Y.; Liu, Y. Optimizing supply-demand balance with the vehicle to grid system: A case study. *J. Energy Storage* **2024**, *97*, 112900. [CrossRef]
86. Sun, B.; Zhang, Q.; Mao, H.; Li, Z. Validation of a statistical-dynamic framework for predicting energy consumption: A study on vehicle energy conservation equation. *Energy Convers. Manag.* **2024**, *307*, 118330. [CrossRef]
87. Huo, L.; Qu, J.; Liu, B. International experience and policy recommendations on financial and tax support policies for energy conservation and new energy vehicles. *Automob. Parts* **2013**, *23*, 22–25.
88. Ma, N. Progress and Key Deployment of Research and Development of New Energy Vehicles and Battery Technology in China. *Technol. Innov. Appl.* **2017**, *13*, 55.

89. Xie, Y.; Yu, H.; Zhang, T.; Ou, Y. Interpretation of GB/T 33598-2017: Recycling of Traction Battery Used in Electric Vehicle—Dismantling Specification. *Battery Bimon.* **2018**, *48*, 53–55. [CrossRef]
90. Zheng, J. Liu Yu: Car companies should become pioneers in energy conservation and emission reduction. *Automot. Obs.* **2021**, *6*, 30–31.
91. Li, L. Shen Ya Nan: Consumer Willingness to Purchase is Key. *Automot. Obs.* **2021**, *6*, 36–37.
92. Chen, X. The "dual carbon" Path of the Automotive Industry. *Automot. Obs.* **2021**, *10*, 55–57.
93. Miao, Y.; Liu, L.; Zhang, Y.; Tan, Q.; Li, J. An overview of global power lithium-ion batteries and associated critical metal recycling. *J. Hazard. Mater.* **2022**, *425*, 127900. [CrossRef]
94. Chen, Q.; Lai, X.; Gu, H.; Tang, X.; Gao, F.; Han, X.; Zheng, Y. Investigating carbon footprint and carbon reduction potential using a cradle-to-cradle LCA approach on lithium-ion batteries for electric vehicles in China. *J. Clean. Prod.* **2022**, *369*, 133342. [CrossRef]
95. Feng, X.; Zhang, F.; Feng, J.; Jin, C.; Wang, H.; Xu, C.; Ouyang, M. Propagation dynamics of the thermal runaway front in large-scale lithium-ion batteries: Theoretical and experiment validation. *Int. J. Heat Mass Transf.* **2024**, *225*, 125393. [CrossRef]
96. Wang, J.; Jia, K.; Ma, J.; Liang, Z.; Zhuang, Z.; Zhao, Y.; Li, B.; Zhou, G.; Cheng, H. Sustainable upcycling of spent LiCoO2 to an ultra-stable battery cathode at high voltage. *Nat. Sustain.* **2023**, *6*, 797–805. [CrossRef]
97. Ren, Z.; Li, H.; Yan, W.; Lv, W.; Zhang, G.; Lv, L.; Sun, L.; Sun, Z.; Gao, W. Comprehensive evaluation on production and recycling of lithium-ion batteries: A critical review. *Renew. Sustain. Energy Rev.* **2023**, *185*, 113585. [CrossRef]
98. Nie, S.; Cai, G.; Huang, Y.; He, J. Deciphering stakeholder strategies in electric vehicle battery recycling: Insights from a tripartite evolutionary game and system dynamics. *J. Clean. Prod.* **2024**, *452*, 142174. [CrossRef]
99. Huang, R.; He, H.; Zhao, X.; Gao, M. Longevity-aware energy management for fuel cell hybrid electric bus based on a novel proximal policy optimization deep reinforcement learning framework. *J. Power Sources* **2023**, *561*, 232717. [CrossRef]
100. Huang, R.; He, H. Naturalistic data-driven and emission reduction-conscious energy management for hybrid electric vehicle based on improved soft actor-critic algorithm. *J. Power Sources* **2023**, *559*, 232648. [CrossRef]
101. Zou, P.; Zhang, B.; Yi, Y.; Wang, Z. How does travel satisfaction affect preference for shared electric vehicles? An empirical study using large-scale monitoring data and online text mining. *Transp. Policy* **2024**, *146*, 59–71. [CrossRef]
102. Aierken, A. Research on the Energy Saving and Emission ReductionEffect and Development Path of New Energy ElectricVehicles. *Auto Time* **2022**, *14*, 125–126.
103. Guo, Q.; Jin, L.; Feng, Y.; Li, S. Research on the Application of New Energy and Energy-saving Technologies for Automobiles. *Auto Time* **2024**, *16*, 116–118.
104. Wang, Y.; Su, L.; Gu, J.; Hong, L. Multi-scenario low-carbon optimization scheduling study for urban distribution networks considering electric vehicle charging demand. *Water Resour. Hydropower Eng.* **2024**, *55*, 32–44. [CrossRef]
105. Li, Y.; Gao, Z.; Wang, K.; Zhou, L.; Wu, X. Uncertainty unit commitment optimization method based on clean development mechanism. *Acta Energiae Solaris Sin.* **2023**, *44*, 368–375. [CrossRef]
106. Li, X. Research on the Development Issues and Countermeasures of New Energy Vehicles in the Context of a Low-Carbon Economy. *Car Test Rep.* **2023**, *18*, 62–64.
107. Zhou, C.; Wang, Y.; Liu, W.L.; Liu, W.S.; Zhang, Y. Construction and application of regional carbon emission index systm for electric vehicles. *Coal Econ. Res.* **2024**, *44*, 105–109. [CrossRef]
108. Xu, Z.; Zhang, C.; Sun, B.; Liu, S. The electric-thermal coupling simulation and state estimation of lithium-ion battery. *J. Energy Storage* **2023**, *58*, 106431. [CrossRef]
109. Yang, X.; Xie, Z.; Lu, X.; Wei, M.; Tan, X.; Ling, H.; Li, Y. Research on the utilization of ultra-long carbon nanotubes in lithium-ion batteries based on an environment-friendly society. *Environ. Sci. Pollut. Res.* **2023**, *30*, 56003–56015. [CrossRef] [PubMed]
110. Li, J. Economic analysis of retired batteries of electric vehicles applied to grid energy storage. *Int. J. Low-Carbon Technol.* **2023**, *18*, 896–901. [CrossRef]
111. Wang, K.; Hu, T.; Shi, P.; Min, Y.; Wu, J.; Xu, Q. Efficient Recovery of Value Metals from Spent Lithium-Ion Batteries by Combining Deep Eutectic Solvents and Coextraction. *ACS Sustain. Chem. Eng.* **2022**, *10*, 1149–1159. [CrossRef]
112. Xue, B.; Chang, Z.; Wu, H.; Hu, Y.; Huang, K.; Yu, Y. Research on Sustainable Development of Small Vehicle Power Battery for Carbon Neutrality. *Adv. New Renew. Energy* **2021**, *9*, 443–448. [CrossRef]
113. Liu, Z.; Liu, G.; Cheng, L.; Gu, J.; Yuan, H.; Chen, Y.; Wu, Y. Development of sustainable and efficient recycling technology for spent Li-ion batteries: Traditional and transformation go hand in hand. *Green Energy Environ.* **2024**, *9*, 802–830. [CrossRef]
114. He, H.; Wei, J.; Wang, Z. Research on Efficiency and Environmental Protection of Electric Vehicle Charging System. *Car Test Rep.* **2024**, *9*, 149–151.
115. Chen, H.; Luo, X.; Zhang, J. Study on Demand Prediction and Development Planning of Eleetric Vehicle Charging Infrastructure. *Electr. Eng.* **2024**, *12*, 33–35.

116. Fan, L.; Liu, W.; Lu, L. Based on the analysis of Automotive New Energy and Energy Saving Technology Application. *Auto Time* **2022**, *24*, 127–129.
117. Wu, X. Research on Energy Saving of EV Based on Big Data. *Automot. Dig.* **2022**, *12*, 1–6. [CrossRef]
118. Gong, H.; Zhong, H.; Sun, M.; Xie, X. Energy Conservation and Emission Reduction Effect and Development of New-energy Electric Vehicles. *Energy Energy Conserv.* **2023**, *12*, 80–84. [CrossRef]
119. Yu, B.; Lei, X.; Shao, Z.; Jian, L. V2G Carbon Accounting and Revenue Allocation: Balancing EV Contributions in Distribution Systems. *Electronics* **2024**, *13*, 1063. [CrossRef]
120. Sun, X. Low Carbon City on Perspective of Carbon-emissions Reduction. *Appl. Energy Technol.* **2019**, *7*, 17–20.
121. Li, S. Prospect of Application of Pure Electric Bus in Small and Medium-sized Cities Based on Green Transportation Mode. *Logist. Sci-Tech* **2019**, *42*, 91–93. [CrossRef]
122. Xuan, Y.; Fan, L.; Sun, Z.; Jiang, J.; Chen, D.; Deng, K.; Wang, M. An optimal allocation method for electric vehicle charging stations considering low carbon transportation. *Zhejiang Electr. Power* **2024**, *43*, 69–79. [CrossRef]
123. Ding, L. The Current Situation and Development Suggestions for Carbon Emissions Management in the Domestic and Overseas Automotive Industry under the Background of "Peak Carbon Dioxide Emissions and Carbon Neutrality". *Shanghai Auto* **2023**, *7*, 4–9. [CrossRef]
124. Cheng, G.Y.; Gao, Z.Q. Policy Proposals on Accelerating the Independent Development of China's Electric Vehicle Industry. *Forum Sci. Technol. China* **2010**, *6*, 64–66. [CrossRef]
125. Pan, H. Research on the Promotion and Application Strategy of Urban New Energy Vehicles Based on Low Carbon Transportation Development: A Case Study of Shenzhen. *Auto Ind. Res.* **2015**, *06*, 14–19.
126. Guo, J.F.; Zhang, X.M.; Chao, Q.; Gu, F. Electric vehicles contribute to China's energy security and carbon peaking and carbon neutrality. *Bull. Chin. Acad. Sci.* **2024**, *39*, 397–407. [CrossRef]
127. Mo, L.; Liu, B.; Shi, H.; Ma, N. Research on Carbon Emission Assessment of Automotive Products Based on Life Cycle Assessment and Suggestions for Low Carbon Development. *Automot. Dig.* **2022**, *11*, 57–62. [CrossRef]
128. Li, W.; Zou, Y.; Zhu, C. Electric Vehicle Revolution: New Track for Industrial Competition among Major Countries. *Int. Econ. Rev.* **2023**, *4*, 93–117.
129. General Office of the State Council of the People's Republic of China. New Energy Vehicle Industry Development Plan (2021–2035). Available online: https://www.gov.cn/zhengce/content/2020-11/02/content_5556716.htm (accessed on 2 November 2020).
130. National Development and Reform Commission, PRC. Implementation Opinions on Further Enhancing the Service Guarantee Capability of Electric Vehicle Charging Infrastructure. Available online: https://www.gov.cn/zhengce/zhengceku/2022-01/21/content_5669780.htm (accessed on 21 January 2022).
131. Guan, X.; Tian, Z.; Si, H.; Ren, Y. Research on the Carbon Emission Reduction Effect and Influencing Factors of New Energy Vehicles in China for Carbon Neutrality. *Soft Sci.* **2025**, 1–18. Available online: https://link.cnki.net/urlid/51.1268.g3.20250310.1757.01081 (accessed on 22 March 2025).
132. Chen, C.; Lin, Z.; Huang, C.; Zhao, J.; Ou, S. Review of Decision-Making Analysis for Transportation Energy Transition. *J. South China Univ. Technol. (Nat. Sci. Ed.)* **2025**, *53*, 32–48.
133. Chen, W.; Huang, Z. WTO Compliance of US's Clean Vehicle Tax Credit: Suggestions on China's Initiatives to the Reform of the Rules on New Energy Subsidies. *Bus. Econ. Law Rev.* **2024**, *6*, 30–46.
134. Wang, J. Challenges and Strategies for High-Quality Development of China's Electric Vehicle Industry Under Technological Transformation. *Enterp. Econ.* **2024**, *43*, 15–24. [CrossRef]
135. Hong, J.; Liang, F.; Yang, J.; Li, K. New energy vehicle industry and technology development status. *Sci. Technol. Rev.* **2023**, *41*, 49–59.
136. Zeng, J.; Zhang, J.; Zhang, X. Evolution and trend of the global power battery industry from economic perspective. *Battery Bimon.* **2024**, *54*, 855–860. [CrossRef]
137. Shen, Y.; Yuan, X.; Zhao, S.; Meng, B.; Wang, Y. Energy-saving Optimization Control for Connected Automated Electric Vehicles: State of the Art and Perspective. *Acta Autom. Sin.* **2023**, *49*, 2437–2456. [CrossRef]
138. Zhu, B.; Jia, S.; Zhao, J.; Han, J. Review of Research on Decision-making and Planning forAutomated Vehicles. *China, J. Highw. Transp.* **2024**, *37*, 215–240. [CrossRef]
139. Shi, P.; Shan, Z.; Zhu, H.; Hai, B.; Wang, L.; Lu, F. A Review of Integrated Design Technologies for New Energy Vehicle Power Battery Systems. *China Mech. Eng.* **2025**, 1–19. Available online: http://kns.cnki.net/kcms/detail/42.1294.TH.20250117.1359.019.html (accessed on 23 March 2025).
140. Editorial Office of China Journal of Highway and Transport. Review on China's Automotive Engineering Research Progress: 2023. *China J. Highw. Transp.* **2023**, *36*, 1–192. [CrossRef]

141. Guo, Y.; Qin, W.; Wu, L.; Chen, Y.; Cheng, G. How is the "rapid progress" possible?-An analysis based on the technological catch-up in China's new energyvehicle industry. *Stud. Sci. Sci.* **2025**, 1–20. [CrossRef]
142. Liu, Z.; Wei, Y. Research on Key Core Technologies for "Lane Changing and Overtaking" in China's Automotive Industry Evidence from the Intelligent Connected Vehicle Industry. *Sci. Res. Manag.* **2025**, 1–21. Available online: http://kns.cnki.net/kcms/detail/11.1567.G3.20250108.1339.004.html (accessed on 23 March 2025).
143. Cao, D. Technology status and development trend of anode materials for lithium-ion batteries. *Pet. Refin. Eng.* **2024**, *54*, 1–7. [CrossRef]
144. Zhai, X.; Sun, X.; Jiang, T.; Bei, X.; Yang, H. A Review on Production Strategy for All-Solid-State Batteries. *Automot. Dig.* **2022**, *2*, 31–35. [CrossRef]
145. Chen, S.; Weng, S.; Zeng, G.; Wang, M. Summary of Lithium-ion Battery Production Process Specifications and Safety Performance Testing Methods. *Guangzhou Chem. Ind.* **2024**, *52*, 4–6.
146. Zhang, B.; Xie, N.; Wang, Y.; Chen, H.; Cao, D. Research on Battery Swapping Standard System for Electric Commercial Vehicle. *China Auto* **2025**, *35*, 86–93.
147. Xu, G. Standardizing the dimensions and communication interfaces of power batteries to promote the development of the new energy vehicle industry. *Auto Maint. Repair* **2024**, *21*, 70–75. [CrossRef]
148. Hu, M.; Li, D.; Wei, L. Research on pricing and charging infrastructure joint investmentstrategies for new energy electric vehicles. *Syst. Eng.-Theory Pract.* **2025**, 1–24. Available online: http://kns.cnki.net/kcms/detail/11.2267.N.20250214.1745.016.html (accessed on 23 March 2025).
149. Xing, W.; Wu, W. The Development Logic, International Competition and Prospects of China's New Energy Vehicle Industry. *J. Xinjiang Norm. Univ. (Ed. Philos. Soc. Sci.)* **2025**, *46*, 123–139. [CrossRef]
150. Ling, S.; Guo, J.; Li, Y.; Ma, S. How should Charging Infrastructure Subsidies Assess the Situation? Analysis Based on Multi-party Dynamic Game. *Chin. J. Manag. Sci.* **2024**, *32*, 290–300. [CrossRef]
151. Li, D.; Lin, W.; Gao, W.; Shang, C. Optimal electric vehicle charging infrastructure subsidy with consumer range anxiety. *Syst. Eng.-Theory Pract.* **2025**, 1–24. Available online: http://kns.cnki.net/kcms/detail/11.2267.N.20240606.1429.015.html (accessed on 23 March 2025).
152. Vinicius, B.; Leonardo, B.; Bruno, H.; Tiago, S.; Jorge, V.; Benedito, D. Life cycle assessment comparison of electric and internal combustion vehicles: A review on the main challenges and opportunities. *Renew. Sustain. Energy Rev.* **2025**, *208*, 114988. [CrossRef]
153. Zhao, X.; Li, X. Empirical evidence of the impact of industrial policies on the charging infrastructure industry and their mechanisms. *China Popul. Resour. Environ.* **2024**, *34*, 47–57.
154. Zhang, H.; Liu, Y.; Tan, X.; Qi, Y.; Yang, L.; Jia, M. Research on the impact of leap-forward development of new energy vehicles onurban traffic congestion. *Chin. J. Manag. Sci.* **2025**, 1–19. [CrossRef]
155. Shen, M.; Jiang, Z.; Zhou, Q.; Yin, J. EV Charging Station as Emerging Urban Infrastructure: New Servicefeatures and planning insights. *City Plan. Rev.* **2024**, *48*, 16–27.
156. Hu, J.; Yan, H.; Zhang, N.; Huang, B.; Chen, R.; Chen, N.; Wang, D.; Lu, J. Cognitive System and Key Issues of Green and Low-Carbon ProductionMode and Life Style. *Electr. Power* **2023**, *56*, 246–254.
157. Wang, N.; Tian, H.; Guo, J. Literature Review of Dynamic Relocation Problem for Electric Shared Vehicles. *J. Tongji Univ. (Nat. Sci.)* **2023**, *51*, 1469–1478.
158. Xu, X.; Ma, Q.; Wang, W. An Analysis of the Direct Network Effect on New Energy Vehicle Promotion-Based on Multiagent Modeling and Simulation. *J. Hunan Univ. (Soc. Sci.)* **2024**, *38*, 41–50. [CrossRef]
159. Ma, J.; Zhang, Y.; Duan, Z.; Tang, L. Research review on behavior strategies of electricvehicles considering charging demands. *J. Traffic Transp. Eng.* **2024**, *24*, 66–79. [CrossRef]
160. Li, M.; Sun, J.; Fu, Y.; Zhao, B. Sharing electric vehicle scheduling and service pricing optimizationbased on battery swapping. *J. Chongqing Univ. Technol. (Nat. Sci.)* **2024**, *38*, 169–180.
161. Lv, R.; Li, B.; Li, M.; Mo, R.; Fang, S.; Guo, M.; Lu, B.; Fu, C. Heterogeneous Impaet of Public Charging Infrastructure on Electric Vehicle Purchases. *J. Technol. Econ.* **2023**, *42*, 143–154.
162. Zhao, X.; Mao, Y.; Li, X. Energy prices and new energy vehicle consumption: A perspective based on consumer expectations. *Soft Sci.* **2025**, 1–14. Available online: http://kns.cnki.net/kcms/detail/51.1268.G3.20250113.1448.010.html (accessed on 23 March 2025).
163. Sun, L.; Feng, Z.; Han, M.; Deng, Q. Cost-optimized Elastic Charging Strategy for Electric Vehicles. *J. Chin. Comput. Syst.* **2024**, *45* 132–138. [CrossRef]
164. Li, Y.; Ouyang, M.; Zhao, Z. Charging Mechanism, System Configuration and Promotion Pathway of Vehicle to Grid for Electric Vehicles. *Proc. CSEE* **2024**, *44*, 6920–6940. [CrossRef]

165. Wu, T.; Huang, K.; Liu, Z.; Jiang, W. Review on the integrated capacity of transportation and power networks. *J. Automot. Saf. Energy* **2024**, *15*, 634–649.
166. Zhang, L.; Xia, Y.; Zu, L. The optimal number of the charging station and government subsidies considering the electric vehicle's adoption target. *Chin. J. Manag. Sci.* **2025**, 1–35. [CrossRef]
167. Zhu, C.; Liu, D.; Teng, X.; Zhang, G.; Yu, D.; Liu, S.; Hu, N. Comparative Analysis and Forecast Research on Comprehensive Economy of New Energy Vehicles. *Automot. Eng.* **2023**, *45*, 333–340. [CrossRef]
168. Zhang, Y.; Tang, X. Development Status, Problems and Countermeasures of China's New Energy Vehicle Industry under the "Dual Carbon" Goal. *China Resour. Compr. Util.* **2024**, *42*, 148–153. [CrossRef]
169. Lin, J.; Zhang, D.; Liu, Y.; Wang, Z. Robust Optimization of Hierarchical Site Selection Location of Centralized Charging Station Based on Uncertain Demand. *J. Chongqing Jiaotong Univ. (Nat. Sci.)* **2024**, *43*, 86–95. [CrossRef]
170. Yuan, H.; Xu, X.; Yan, Z.; Fang, C.; Liu, J. Review of centralized EV charging and battery swappingfacility planning and optimal scheduling. *Power Syst. Prot. Control* **2024**, *52*, 157–174. [CrossRef]
171. Ding, C.; Zhang, M.; Sun, L. China-Europe New Energy Vehicle Industry Disputes: Status Quo, Causes and Prospects. *Chin. J. Eur. Stud.* **2024**, *42*, 36–62.
172. Liu, J.; Li, Z.; Li, C.; Li, Y. Interaction between electric vehicles and power-transportation coupled networks: Current status, challenges and development trends. *J. Electr. Power Sci. Technol.* **2024**, *39*, 12–24. [CrossRef]
173. Yu, X.; Ji, Z.; Ji, L.; Niu, D.; Xu, X. Analysis on Carbon Emission Reduction Potential of Electric Vehicles in China under Goal of Carbon Neutrality and Carbon Peaking. *Smart Power* **2024**, *52*, 25–31.
174. Zhu, M.; Ji, J.; Jin, S.; Ji, Y.; Bie, Y. A state-of-the-art review on the integrated development technology of electric vehicles and clean energy. *J. Automot. Saf. Energy* **2024**, *15*, 1–19.
175. Chen, Y.; Liu, Q.; Xing, Y.; Zhang, C.; Zhang, S. Life cycle assessment and uncertainty analysis of intelligent electric vehicles: A case study of BYD Han EV. *Acta Sci. Circumstantiae* **2024**, 1–14. Available online: https://link.cnki.net/urlid/11.1843.x.20241218.0925.002 (accessed on 23 March 2025).
176. Zhu, J.; Xie, C.; Zhang, D.; Lan, J.; Zhou, J. Review of Electric-Carbon Coupling Market Studies: Status, Challenges, and Sustainability Perspectives. *Electr. Power Constr.* **2025**, *46*, 158–173.
177. Song, X.; Deng, C.; Shen, P.; Qian, Y.; Xie, M. Environmental lmpact and Carbon Footprint Analysis of Pure Electric Vehicles Based on Life Cycle Assessment. *Res. Environ. Sci.* **2023**, *36*, 2179–2188. [CrossRef]
178. Gao, D.; Wang, Y.; Zheng, X.; Yang, Q. Electric vehicle charging status monitoring and safety warning method based on deep learning. *Electr. Mach. Control* **2023**, *27*, 122–132. [CrossRef]
179. Yang, L.; Yu, B.; Feng, Y. Life cycle assessment of electric vehicle carbon emissions: A case study of passenger vehicles in China. *China Popul. Resour. Environ.* **2023**, *33*, 113–124.
180. Wang, S.; Kuai, L.; Zhao, L. lmpact of Incentive Policies on the Promotion of New Energy Vehicles in China. *J. Technol. Econ.* **2024**, *43*, 64–74. [CrossRef]
181. Shchurov, N.I.; Dedov, S.I.; Malozyomov, B.V.; Shtang, A.A.; Martyushev, N.V.; Klyuev, R.V.; Andriashin, S.N. Degradation of Lithium-Ion Batteries in an Electric Transport Complex. *Energies* **2021**, *14*, 8072. [CrossRef]
182. Wang, K.; Xue, Y.; Shahidehpour, M.; Chang, X.; Li, Z.; Zhou, Y. Resilience-Oriented Two-Stage Restoration Considering Coordinated Maintenance and Reconfiguration in Integrated Power Distribution and Heating Systems. *IEEE Trans. Sustain. Energy* **2025**, *16*, 124–137. [CrossRef]
183. Zhao, X.; Li, X.; Zhao, Q. Does the new energy vehicles adoption reduce carbon emissions? A perspectiveof spatial spillover effect. *J. Arid Land Resour. Environ.* **2024**, *38*, 1–8. [CrossRef]
184. 2025 Government Work Reports of Beijing. Available online: http://bj.people.com.cn/n2/2025/0122/c14540-41116991.html (accessed on 23 March 2025).
185. 2025 Government Work Reports of Shijiazhuang. Available online: https://www.sjz.gov.cn/columns/a42a569b-80e5-43b6-bd7a-f0d09c01f221/202501/24/e56405fb-183b-409b-a5d6-0de287eb45ad.html (accessed on 23 March 2025).
186. 2025 Government Work Reports of Tang County. Available online: https://www.tangxian.gov.cn/content-903-181449.html (accessed on 23 March 2025).
187. Yu, J.; Guo, P.; Tu, L.; Xue, Y.; Zhang, H. Development Status and Suggestions for the Coordinated Development of Intelligent Connected Vehicles and Smart City Infrastructure. *Chin. Overseas Archit.* **2024**, *7*, 70–74. [CrossRef]
188. Wang, J. Research on the Design and Manufacture of Hghtweight New Energy Special Vehicles. *Agric. Mach. Using Maint.* **2023**, *08*, 33–35. [CrossRef]
189. Guo, J. The Current Status and Future Development Trends of Autonomous Driving Technology. *Spec. Purp. Veh.* **2024**, *8*, 46–48. [CrossRef]
190. Li, H.; Chen, L. Development of Key Material System for Solid-State Batteries. *Strateg. Study CAE* **2024**, *26*, 19–33. [CrossRef]

191. Zhang, Q. Sustainable Urban Road Design and Green Transportation Development. *Theor. Res. Urban Constr.* **2024**, *22*, 22–24. [CrossRef]

energies

MDPI

Article

A Novel Multilevel Inverter Topology Generating a 19-Level Output Regulated by the PD-PWM Method

Sofia Lemssaddak [1,2], Abdelhafid Ait Elmahjoub [1], Mohamed Tabaa [2,*], Adnane El-Alami [1] and Mourad Zegrari [1]

1 Digital Engineering for Leading Technology and Automation Laboratory (DELTA Lab), ENSAM Casablanca, Hassan II University of Casablanca, Casablanca 20360, Morocco; sofia.lemssaddak@gmail.com (S.L.)

2 Multidisciplinary Laboratory of Research and Innovation (LPRI), Moroccan School of Engineering Sciences (EMSI), Casablanca 20250, Morocco

* Correspondence: m.tabaa@emsi.ma

Abstract: Traditional multilevel inverter topologies, such FC, NPC, and CHB, have a few significant disadvantages. They need a great number of parts, which raises the complexity, expense, and switching losses. Furthermore, their intricate control schemes make voltage balancing and synchronization challenging. Lastly, under some circumstances, they experience severe harmonic distortion, necessitating the inclusion of expensive filters to enhance signal quality. This paper proposes a novel multilevel converter topology that uses the phase-disposition PWM (PD-PWM) technique to control a 19-level output. This new configuration maintains performance comparable to the CHB-MLI reference while using fewer switches, simplifying control, and reducing costs. Our approach is based on extensive simulations conducted in the MATLAB Simulink environment, with results compared to the CHB-MLI. A low-pass filter is added to improve the output voltage quality, reducing the THD% to 1.33%. This strategy offers several advantages, including simpler control, lower costs, increased reliability, and higher-quality output. The system was replicated using MATLAB Simulink and validated through hardware-in-the-loop (HIL) testing. The HIL method ensures real-world testing without causing damage to the hardware. The integrated system includes sensors and necessary hardware for a comprehensive energy management solution.

Keywords: PD-PWM: phase-disposition pulse width modulation; CHB-MLI: cascaded H-bridge multilevel inverter; THD: total harmonic distortion; hardware in the loop

1. Introduction

One of the main pillars in the movement toward climate neutrality is electric cars, or EVs [1]. Since every EV charger needs an inverter to power the car, the system is essential. Numerous topologies of inverters are available, and multilayer inverters (MLIs) are the subject of more and more research due to developments in academia and industry [2,3]. Due to their capacity to generate high-quality voltage waveforms with lower total harmonic distortion (THD), multilevel converter topologies have become more and more common in recent years [4–9]. Numerous advantages result from this trait, such as increased energy efficiency, less strain on electrical components, and fewer running expenses [10–14]. Additionally, by aiding in the decarbonization process, most notably by integrating hydrogen into transportation, multilevel converters might indirectly help alternative energy transport systems. These converters, for instance, can be employed in hydrogen rail systems to increase energy conversion, optimize power supply, and boost on-board electrical system efficiency [12]. However, to manage increasing voltage levels, traditional multilevel converter topologies like NPC, FC, and CHB are becoming more complicated and need many components. This study suggests a novel multilevel converter architecture to overcome these issues. It can provide harmonic performance on par with the CHB design while utilizing a notably smaller number of switches. Even with fewer switches, the new design uses the same control method (PD-PWM) to produce a voltage harmonics rate comparable to the CHB topology.

The structure of this article is as follows: A general survey of fundamental multilevel converter topologies is presented in Section 2, and the generalized configuration of the suggested topology is presented in Section 3. This section also covers the control of the modulation method (PWM), as well as the states and operating mode of a 19-level single-phase sinusoidal inverter. Section 4 presents hardware-in-the-loop (HIL) testing, a method used in the development and testing of complex process systems. A low-pass filter is proposed in Section 5 to improve the circuit and minimize harmonic content and compare the number of switches and THD of the proposed topology with traditional topologies, and also with the results found experimentally. The final section presents future prospects and conclusions.

2. State of the Art: Multilevel Converter Topologies

To fully understand multilevel converters, we must first understand their topologies. These topologies come in a variety of forms, all of which are intended to provide an excellent voltage waveform at the output. The authors of papers [15–17] have thoroughly examined the various MLI topologies, including active neutral-point-clamped (ANPC), cascaded multilevel inverter (H-bridge), flying capacitor topology (FC), and diode-clamped multilevel inverter (NPC), without overlooking the new topologies with component reduction. The development of converter topologies since 1970 is depicted in Figure 1. The novel multilevel converter topology introduced by [16] demonstrates significant improvements in energy efficiency and reduced impact on the supply network, which aligns with the goals of our PD-PWM approach. By optimizing the switching strategy and minimizing component count, their work supports the advancements in control techniques that we explore in our study [18] provide an in-depth analysis of the Packed E-Cell (PEC) converter topology, including its operational principles and experimental validation. Their work underscores the potential of advanced topologies like PEC to enhance performance and efficiency, complementing our exploration of the PD-PWM technique for multilevel converters. By incorporating such recent advancements, our study aims to build on these innovations to further optimize control strategies and improve converter performance. Ref. [19] discusses advanced methods for power flow control and energy utilization in AC electrified railways, which highlights the ongoing efforts to optimize energy efficiency in complex systems. Similarly, our paper on the PD-PWM technique for multilevel converters explores innovative strategies to enhance performance and efficiency, demonstrating how advanced control techniques can significantly improve system operation in various applications for railway traction systems. Ref. [20] suggests a multi-port power conditioning (MP-PC) device that uses an architecture based on a full-bridge modular converter (full-bridge MMC). In this application, multilevel converters—like MMC—are essential for ensuring accurate voltage regulation and power balancing while facilitating the smooth integration of renewable energies into the rail network. To enhance power quality in multi-zone railway systems, Ref. [21] focuses on multilevel converters, such as the modular five-branch full-bridge converter (FB5B-MMC).

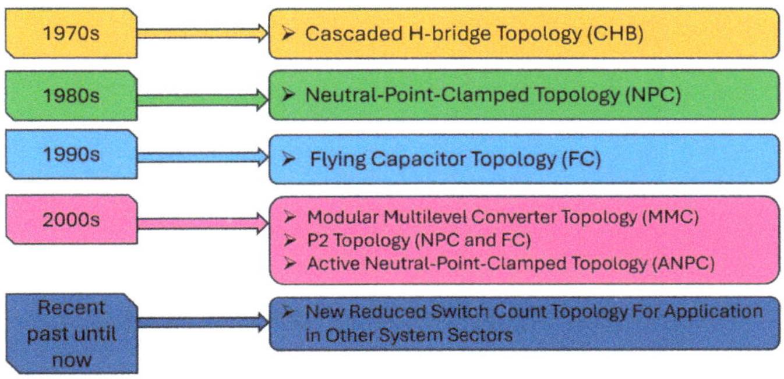

Figure 1. The development of different MLI topologies.

2.1. Diode-Clamped Multilevel Inverter (NPC)

Power conversion systems frequently employ active neutral point (ANP) multilevel converters as their converter topology [18]. The same goal of this kind of converter was initially applied in 1980 to lower the rate at which the converter introduces TDH harmonics into electrical applications. The two topologies—three and five levels—that will be covered in more detail below are depicted in Figure 2.

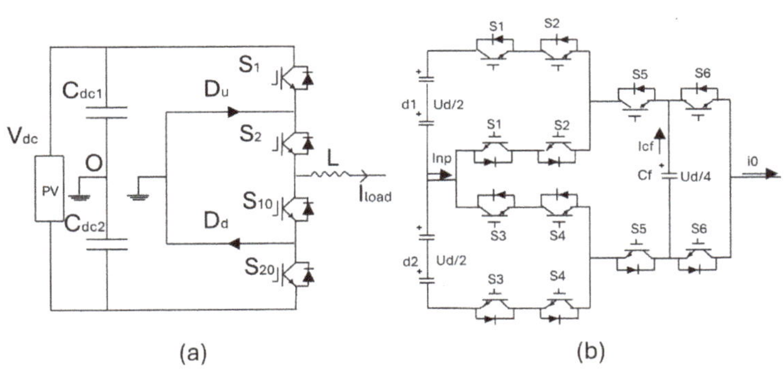

Figure 2. NPC converter: three levels [18] (**a**); five levels [22] (**b**).

For a three-level NPC converter (see Figure 3), we have three possible operating stages giving three different levels, as shown in Table 1.

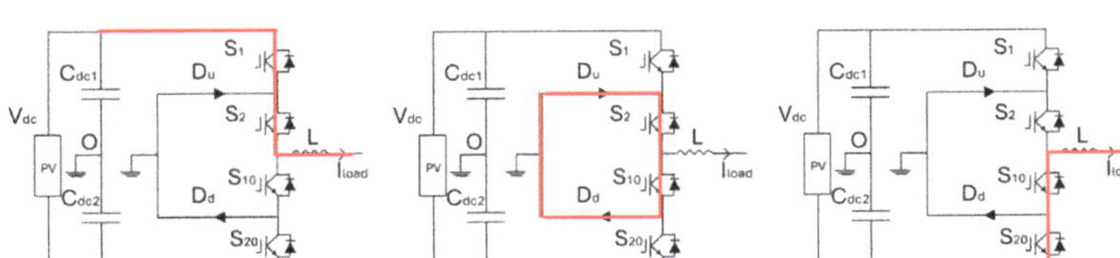

Figure 3. Three different levels of inverter.

Table 1. Possible status of the three-level NPC inverter.

T_1	T_2	T_3	T_4	V_{out}
1	1	0	0	$V_{DC/2}$
0	0	1	1	$V_{-DC/2}$
0	1	1	0	0

We have five possible operational phases, or five separate levels, for the five-level NPC converter. Switch operation details are displayed in Table 2. By eliminating the need for passive filters in this context, thanks to a simple control mechanism, the NPC architecture offers several advantages. These include producing a high-quality waveform from three levels onwards, providing better spectral performance compared to a conventional three-phase inverter [22]. A DC/AC inverter with an LC filter is cascaded with a boost converter to synchronize the grid frequency with the inverter [23]. However, when the number of levels exceeds three, voltage balancing across capacitors becomes increasingly complex, as it is influenced by the load's power factor and modulation index. Consequently, the number of diodes increases unnecessarily with additional levels, making power flow control for each converter more challenging [24].

Table 2. Possible status of the five-level NPC inverter.

S1	S2	S3	S4	S5	S6	S7	S8	V_{out}
1	1	1	1	0	0	0	0	$V_{DC/2}$
0	1	1	1	1	0	0	0	$V_{DC/4}$
0	0	1	1	1	1	0	0	0
0	0	0	1	1	1	1	0	$-V_{DC/4}$
0	0	0	0	1	1	1	1	$-V_{DC/2}$

2.2. Capacitor-Clamped Multilevel Inverter (FC)

As a direct replacement for the NPC architecture, the FC (flying-capacitor multilevel inverter) topology was introduced in 1992 [25]. One benefit of this design is that it can address the issue of loopback diodes in multilevel NPC inverters [26]. The term "flying-capacitor inverter" refers to this topology's usage of capacitors rather than diodes. The drawbacks of diodes, including conduction losses and reliability issues, are removed by using capacitors. The various voltage levels needed for inverter functioning are created in large part by flying capacitors [26]. The architecture of the three-level FC inverters is shown in Figure 4.

A method for active voltage balancing that makes use of valley current detection and constant effective duty cycle management is presented in the research work conducted by [26]. By balancing the voltage between the capacitors, this technique maximizes the dependability and efficiency of the inverter. A thorough analysis of switched-capacitor inverters, including flying-capacitor inverters, is given in the contribution in [23]. Compared to conventional NPC inverters, flying-capacitor multilevel inverters have a few advantages. Voltage balancing techniques and a thorough description of these inverters are included in [27–29], which facilitates a better understanding of their functioning and directs future research in this area.

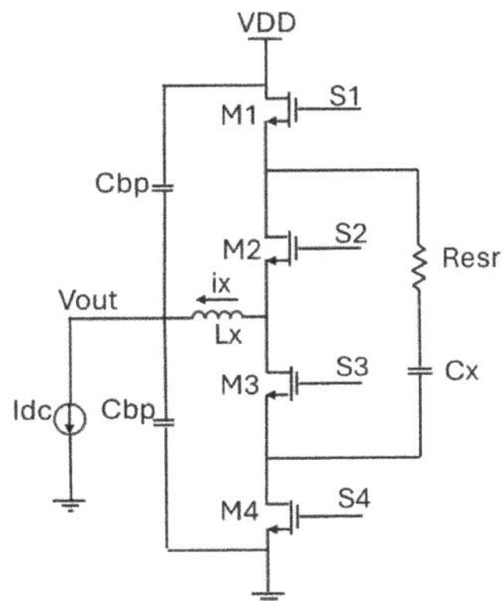

Figure 4. Three-level FC converters [26].

2.3. *Cascaded Multilevel Inverter (H-Bridge)*

Multilevel inverters frequently employ H-bridge converter topologies. Unidirectional multilevel chargers find applications in electric vehicles [30]. They provide excellent voltage wave quality and great performance. To increase their effectiveness and efficiency, researchers have conducted a few experiments in this regard. For instance, the authors conducted an intriguing study on power balancing in modular multilevel converters and H-bridge cascade converters with an emphasis on imbalanced operations. With a focus on solar system applications, Ref. [31] suggests a novel multilevel inverter topology based on H-bridge cascade converters and the extended quasi-Z-boost current source. There are benefits to this topology in terms of cost, dependability, and efficiency. Paper [6] presents a comparative analysis of multilevel converters, including those based on H-bridge converters. This paper investigates multilevel converter applications, control systems, and both traditional and sophisticated topologies. Ref. [32] provides yet another in-depth analysis of multilevel converters with parallel connectivity. The many topologies of parallel multilevel converters are examined in this review, with an emphasis on the benefits and drawbacks of each. It also looks at these converters' applications and control strategies. A thorough understanding of H-bridge and multilevel converters is provided by Refs. [33–36], which include comprehensive details on topologies, control systems, and related applications.

2.4. *Active Neutral-Point-Clamped (ANPC)*

The application of active neutral-point-clamped converters, or ANPCs, has increased interest in power conversion. Reduced switching losses and enhanced output wave quality are two benefits of ANPC converters [37,38]. Five-level ANPC converter design and simulation have received a lot of attention in the literature, which shows how promising they are for practical use [39]. Extensive study has also been conducted on the control of active multilevel converters. For instance, a five-level, six-switch ANPC converter has been effectively controlled using sliding mode control. This control strategy produces robustness to disturbances and accurate reference tracking performance, as shown in Figure 5.

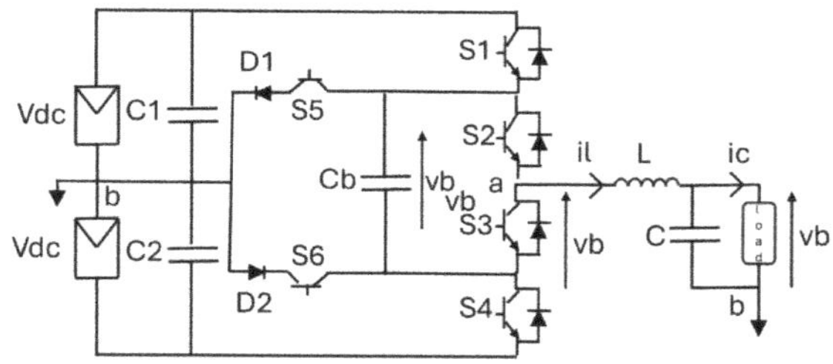

Figure 5. The 6S-5L-ANPC inverter proposed in [39].

Furthermore, Ref. [40] provides an overview of the most recent developments in the multilevel converter sector. The development of active multilevel converter technology was aided by the knowledge and experience those experts and researchers shared. ANPC converters hold great potential as energy conversion solutions. Power electronics researchers are becoming more and more interested in these technologies, as seen by studies into their design, simulation, control, and performance.

2.5. New Topologies with Reduced Commutations

Ref. [41] presents a comparative analysis of reduced-switching multilevel inverter topologies. The authors of this study looked at several inverter topologies and assessed how well they performed in terms of the quantity of switches needed.

Comparably, Figure 6 is a study of multilevel inverter topologies with fewer switches and is provided in Ref. [35]. This paper provides an outline of these novel topologies' benefits as well as the advancements made in the field. The authors of a different publication [42] suggested a new cascaded hybrid inverter with fewer switches by utilizing switched capacitors for home loads and renewable energy sources. This study offers a creative way to cut down on the number of switches required in multilevel inverters. Furthermore, a case study was conducted to build and analyze a new multilevel inverter topology that utilized the PDPWM approach and had fewer switches [43,44]. By lowering the number of switches needed while keeping output quality good, this research helps multilevel inverters operate better. A novel converter topology with nine levels and nine switches and optimal modulation control was introduced by the authors of a different study [45]. The significance of creating novel topologies and cutting-edge modulation approaches to raise inverter quality and efficiency was underlined.

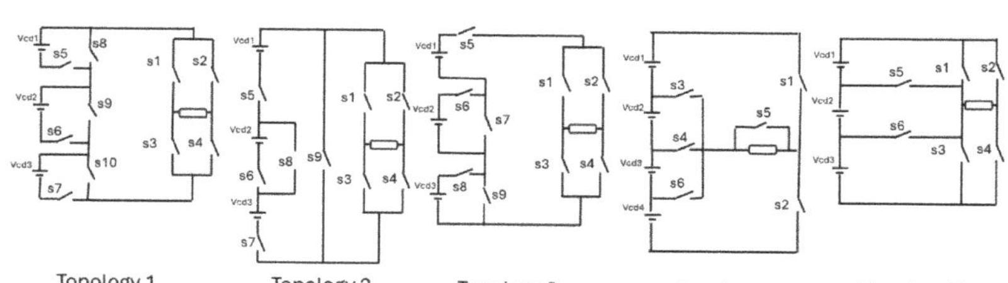

Figure 6. The five topologies proposed by [35].

3. Proposed Topology

3.1. Circuit Description

Multilevel converters with fewer components provide several benefits, such as faster system integration, reduced costs, increased dependency, and simpler control. The suggested example shows that it is possible to achieve high output levels while using the fewest possible components. While having fewer switches often simplifies the design and execution of the control algorithm, reducing the number of switches in a multilevel converter while enhancing performance will give various advantages [46]. One of these advantages is control simplicity. This will help to simplify and maybe strengthen control by facilitating modulation, voltage level management, and switch coordination [47]. Additionally, cutting back on the number of switches helps save the cost of buying, installing, and maintaining individual components [48]. This can be especially helpful in applications where costs are a concern, not to mention the reliability guaranteed, as having fewer components lowers the chance that any one of them would fail, which can raise the system's total reliability [49].

In critical applications where system availability is crucial, this can be especially significant. It may be simpler to integrate a multilevel converter into a larger system by using an architecture with fewer switches for electronic design and component layout [50,51]. The suggested structure, which generates a 19-level output voltage waveform using 9 switches, 5 diodes, and 3 DC power sources, is depicted in Figure 7. Several topologies in the literature, including the CHB-MLI topology, which calls for 36 switches and 9 DC power sources, are not like this method. Unequal DC voltage sources are included in our suggested topology. The "level generation part" is the process of switching S1, S2, S3, S4, and S5 to create the various levels in the output waveform. The "polarity generation part" is the process of flipping S6, S6′, S7, and S7′ to reverse the polarity of the output waveform. Five diodes (D1, D2, D3, D4, and D5) are also incorporated into the circuit to guarantee appropriate current flow and prevent short circuits.

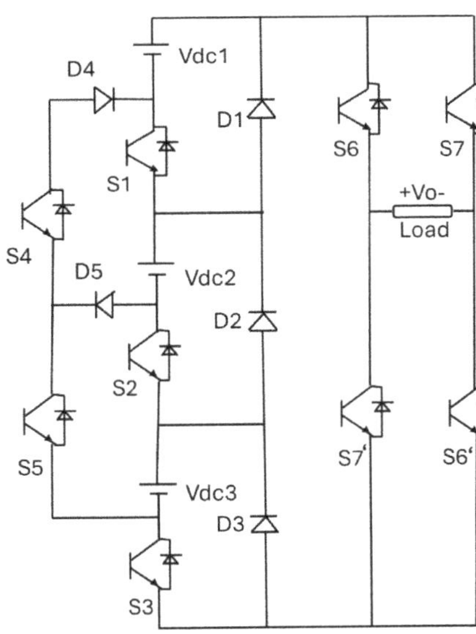

Figure 7. Proposed new topology.

3.2. Operating Modes and States

The various operating modes of the proposed topology are illustrated in Figure 7, which provides an overview of how the circuit operates under different conditions. Each half-cycle—positive and negative—is divided into multiple distinct modes, as detailed in Figure 8(A1–I1) for the positive half-cycle and (A2–I2) for the negative half-cycle. For each mode, the red line in the diagrams indicates the current flow direction, always providing a clear representation of system operation. A complete cycle comprises nineteen distinct operating modes: nine modes during the positive half-cycle, nine modes during the negative half-cycle, and a single mode corresponding to the zero period where no output voltage is generated. Each of these modes is responsible for generating a specific level of the output voltage, thereby enabling the precise control of the multilevel inverter's output waveform. The switching states for all modes are detailed in Table 3, where the binary values "1" and "0" indicate the ON and OFF states of the switches, respectively. Modes 1 to 9 correspond to the positive half-cycle, Mode 10 represents the zero period, and Modes 11 to 19 define the negative half-cycle. A notable feature of the proposed topology is the symmetry in the switching sequence for switches S1, S2, S3, S4, and S5 across both half-cycles, which simplifies the control strategy and ensures consistent performance. Moreover, the complementary operation of switches S6 and S7 is a key aspect of the topology. During the positive half-cycle, switch S6 is active, while S7 remains off. Conversely, during the negative half-cycle, S7 is active, and S6 remains off. This alternating operation not only enhances the efficiency of the circuit but also ensures proper current direction and accurate level generation. The suggested topology's output voltage of nineteen levels will have many benefits, especially for applications that demand harmonic suppression and good output quality. An ideal sinusoid-like output voltage waveform is more closely approximated by a high number of levels. In applications where power quality is crucial, like power supply systems, this lowers harmonics and enhances voltage quality. It is important to keep in mind that output harmonics can be reduced by using multilevel converters. Lower harmonic distortion levels can be attained by increasing the number of levels, which is crucial for fulfilling power quality requirements. Because multilevel converters can provide output waveforms with cleaner characteristics, they might be more appropriate for use with renewable energy sources like wind turbines and solar panels. The various switching states of the suggested topology are displayed in Table 3, which integrates redundant vectors and should enhance system performance in a few ways, including a decrease in harmonic distortions through improved output waveform optimization, which helps to reduce harmonics and enhance output voltage quality. Through loss distribution and thermal management, switching losses can be distributed more effectively, localized heating can be decreased, and component life can be extended, not to mention that enhanced control flexibility allows the system to apply sophisticated modulation strategies that can be modified to fit specific operating conditions.

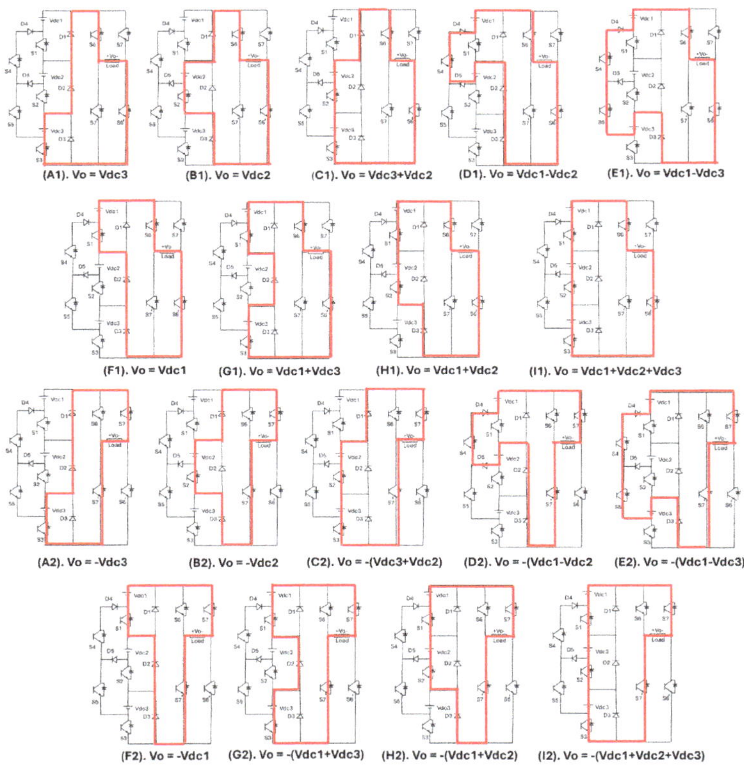

Figure 8. Device conduction states for generating levels during the positive half-period (A1 to I1) and the negative half-period (A2 to I2).

Table 3. The states of the commutations of the proposed topology.

Mode	V_{out}	S1	S2	S3	S4	S5	S6	S7
1	$9E = V_{dc1} + V_{dc2} + V_{dc3}$	1	1	1	0	0	1	0
2	$8E = V_{dc1} + V_{dc2}$	1	1	0	0	0	1	0
3	$7E = V_{dc1} + V_{dc3}$	1	0	1	0	0	1	0
4	$6E = V_{dc1}$	0	0	0	1	1	1	0
5	$5E = V_{dc1} - V_{dc3}$	0	0	0	1	1	1	0
6	$4E = V_{dc1} - V_{dc2}$	0	0	0	1	0	1	0
7	$3E = V_{dc2} + V_{dc3}$	0	1	1	0	0	1	0
8	$2E = V_{dc2}$	0	1	0	0	0	0	1
9	$1E = V_{dc3}$	0	0	1	0	0	1	0
10	0	0	0	0	0	0	0	0
11	$-1E = -V_{dc3}$	0	0	1	0	0	0	1
12	$-2E = -V_{dc2}$	0	1	0	0	0	0	1
13	$-3E = -(V_{dc2} + V_{dc3})$	0	1	1	0	0	0	1
14	$-4E = -(V_{dc1} - V_{dc2})$	0	0	0	1	0	0	1
15	$-5E = -(V_{dc1} - V_{dc3})$	0	0	0	1	1	0	1
16	$-6E = -V_{dc1}$	0	0	0	0	0	0	1
17	$-7E = -(V_{dc1} + V_{dc3})$	1	0	1	0	0	0	1
18	$-8E = -(V_{dc1} + V_{dc2})$	1	1	0	0	0	0	1
19	$-9E = -(V_{dc1} + V_{dc2} + V_{dc3})$	1	1	1	0	0	0	1

3.3. The PD-PWM Control Strategy

Derivative converter modulation technique selection is based on application requirements, including control complexity, output quality, reaction time, and efficiency. Simpler techniques like PWM or SPWM [52] are appropriate for general-purpose applications whereas more sophisticated techniques like SVPWM or PSM are better for high-efficiency systems. Designing an ideal system requires an understanding of the trade-offs.

The modulation approach selected has a direct impact on how effective the suggested remedy is. The phase-disposition modulation (PD-PWM) technique was chosen following a careful analysis [53]. Power switches in multilevel inverters are frequently inspected using this method. The phase and frequency of the triangle carriers are maintained constant in the PD-PWM that is used [42]. Phase-disposition pulse width modulation (PD-PWM) is a widely used modulation technique in multilevel inverters due to its simplicity and effectiveness in generating high-quality output voltages. In PD-PWM, all carrier signals are aligned in phase and distributed uniformly across the amplitude range of the reference signal. For a 19-level inverter, 18 carrier signals are used, each corresponding to a voltage step in the inverter. These carriers are compared with the reference sinusoidal waveform to produce switching signals that control the power switches in the inverter. The primary role of PD-PWM in a 19-level multilevel inverter is to achieve a high-resolution output voltage with minimal distortion. By aligning the carriers in the same phase, PD-PWM effectively reduces low order harmonics, resulting in a significant improvement in the power quality of the output.

This is particularly important in high-power applications, where total harmonic distortion (THD) must be minimized to meet stringent standards. The high number of levels naturally lowers THD, and PD-PWM further enhances this by ensuring smooth transitions between voltage levels. Additionally, PD-PWM simplifies the implementation process, especially in digital control platforms like DSPs or FPGAs, reducing computational complexity and making it ideal for efficient and reliable operation. The PD-PWM technique was first implemented and tested in MATLAB/Simulink R2024b to validate its performance in controlling a 19-level multilevel inverter like in Figure 9. The model comprised a sinusoidal reference signal and 18 uniformly distributed carrier signals aligned in phase. These carriers were compared to the reference waveform using logical comparison blocks to generate switching signals for the power electronic switches in the inverter. The inverter topology employed in the simulation was a cascaded H-bridge configuration, with each H-bridge connected to a DC source. The stepped output voltage waveform produced by the inverter closely approximated a sinusoidal waveform, significantly reducing harmonic content. Performance analysis was conducted using Fast Fourier Transform (FFT) tools to evaluate the total harmonic distortion (THD).

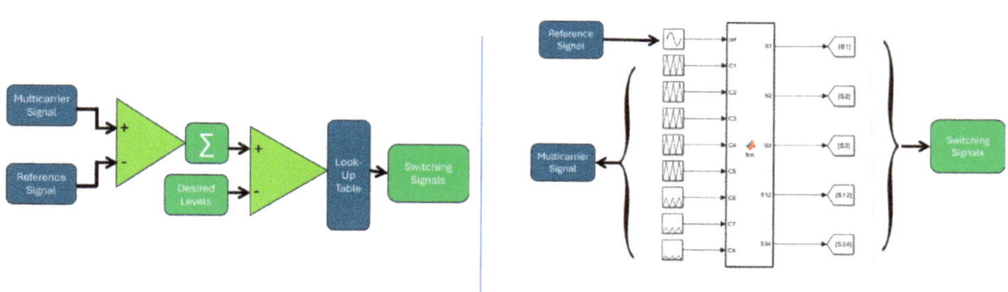

(a) (b)

Figure 9. PD-PWM control: block diagram [54] (**a**); PWM in MATLAB with function (**b**).

The PD-PWM technique was selected for this work due to its simplicity of implementation in controllers and its excellent compatibility with multilevel inverter topologies. Compared to more advanced methods such as Space Vector PWM (SVPWM), PD-PWM offers reduced computational complexity, fast real-time performance, and effective harmonic suppression, making it particularly suitable for hardware-constrained applications.

Figure 10 shows our reference signal compared with the 18 carriers we used. The PD-PWM method demonstrated excellent harmonic mitigation, with results showing a low THD and improved output voltage quality. Various scenarios, including changes in load, modulation index, and carrier frequency, were simulated to assess system stability and efficiency. The results confirmed the robustness of the PD-PWM technique in ensuring smooth transitions between voltage levels and maintaining stable operation across different operating conditions. This section provides a comprehensive overview of the proposed circuit, its various operating modes, and the control strategy based on PD-PWM.

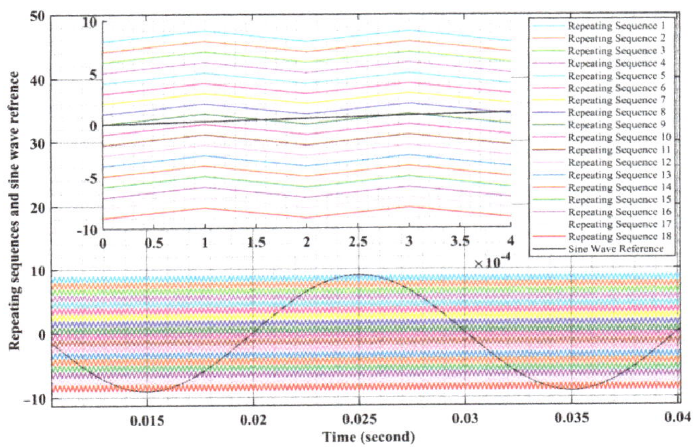

Figure 10. PD-PWM.

3.4. Total Harmonic Distortion (THD) and MTTF Calculation

Total harmonic distortion (THD) is an important metric for assessing the caliber of the inverter output voltage. It is described by the following expression and quantifies the existence of harmonic components in relation to the fundamental component. In [55], the following equation defines it:

$$\mathrm{THD} = \frac{\sqrt{\sum_{n=2}^{\infty} V_n^2}}{\mathrm{V1}} * 100\% \tag{1}$$

where V_n is the RMS value of the output voltage's nth harmonic component and V_1 is the RMS value of the fundamental component, which is usually 50 Hz.

Increasing the number of voltage levels in multilevel inverters, especially those that use phase-disposition PWM (PD-PWM), greatly enhances the approximation of a sinusoidal waveform, which lowers harmonic content. A condensed estimate of THD under optimal circumstances is as follows:

$$\mathrm{THD} = \frac{1}{\sqrt{(2m-1)}} \tag{2}$$

where *m* is the output waveform's number of voltage levels. This demonstrates that THD falls as *m* rises, which is consistent with our design objective.

This work's goal is to provide a novel multilevel inverter topology that achieves a very low THD while reducing the number of components. Reduced motor and transformer losses, less electromagnetic interference, and increased system dependability and efficiency are just a few advantages of lower THD. These features are especially beneficial for applications involving renewable energy and industry.

THD reduction offers several benefits, including less electromagnetic interference, which improves inverter stability; less strain on electronic components, which extends their lifespan; and reduced motor and transformer losses, which boost efficiency. Because of these improvements, our architecture is particularly suitable for industrial and renewable energy applications where minimizing harmonic distortion is necessary to maximize system efficiency. This study shows that, in comparison to traditional options, our design significantly reduces THD due to improved switching control and an optimized component count.

This analytical expectation is supported by the experimental and simulated data shown in Section 5. When compared to traditional structures, our suggested topology shows a significant decrease in THD, confirming the efficiency of the control technique and the optimized architecture.

Our topology has a lower failure rate since it requires fewer components. And longer life and improved robustness result from having fewer MOSFETs and capacitors. In [56], a converter's failure rate is often calculated by adding the failure rate of its essential parts:

$$\lambda_{total} = \sum_{i=1}^{n} \lambda_i \tag{3}$$

where the failure rate of the i-th component (MOSFET, diode, capacitor, inductor, etc.) is represented by λ_i. The topology's total number of components is denoted by n.

The following formula can be used to determine the Mean Time to Failure (MTTF) after the overall failure rate (λ_{total}) has been determined:

$$MTTF = \frac{1}{\lambda_{total}} \tag{4}$$

where $MTTF$ is expressed in hours.

These component failure rates are typically represented as "Failures In Time," or FIT, which is the number of failures per 10^9 h of operation. These figures are derived from standard approximations for electronic power parts:

- MOSFETs: λ(MOSFET) = 50 FIT;
- Diodes: λ(diode) = 20 FIT;
- Capacitors: λ(Capacitors) = 5 FIT;
- Inductors: λ(inductor) = 2 FIT.

To compare the failure rate of our topology with that of traditional topologies, we find the following:

- MTTF for CHB topology (36 MOSFETs, 0 diodes): 555,556 h;
- MTTF for NPC topology (36 MOSFETs, 306 diodes): 126,263 h;
- MTTF for FC topology (36 MOSFETs, 153 capacitors): 389,864 h;
- MTTF for our proposed topology (9 MOSFETs, 3 diodes): 1,960,784 h.

Our suggested topology provides a lifetime that is 3.5 times longer than that of the CHB topology, almost 15 times longer than that of the NPC topology, and 5 times longer than that of the FC topology.

This demonstrates how our topology decreases the number of crucial parts (capacitors, diodes, and MOSFETs), significantly lowering the failure rate and boosting system dependability.

4. Hardware-in-the-Loop

Hardware-in-the-loop (HIL) simulation or testing is a method utilized in the development and testing of intricate process systems [57]. HIL simulation offers an efficient platform by integrating the complexity of the controlled plant into the testing environment. This complexity is introduced into the test and development phases by incorporating a mathematical representation of all relevant dynamic systems, known as the "plant simulation". The HIL process has been in existence for less than 15 years. The increasing adoption of the HIL process across industries is primarily propelled by two key factors: time to market and system complexity. HIL simulation constitutes a form of real-time simulation where input and output signals exhibit time-dependent values identical to those of the actual process. HIL necessitates collaboration between the host and target systems. Typically conducted in laboratory settings, HIL is employed on the ground to conveniently and safely evaluate prototype controllers under various operational loads and conditions. The system is controlled by the DSP F28379D board [58].

The real-time implementation of the PD-PWM control strategy was carried out using the TMS320F28379D digital signal processor (DSP) and validated via a hardware-in-the-loop (HIL) platform. The PD-PWM logic developed in Simulink was converted to C code using Embedded Coder and loaded onto the DSP. The HIL system emulated the physical inverter and load, providing a safe and controlled environment to test the PD-PWM algorithm under realistic conditions.

The DSP and the PWM modules were utilized to generate high-resolution switching signals, which were fed into the HIL platform. The real-time feedback from the HIL system, including voltage levels, harmonic distortion, and dynamic responses, enabled fine-tuning of the PD-PWM algorithm. The results showed that the technique effectively minimized THD, maintained voltage balance across the inverter stages, and ensured stable performance under varying operating conditions. Figure 11 demonstrates the schematic of the system proposed using HIL.

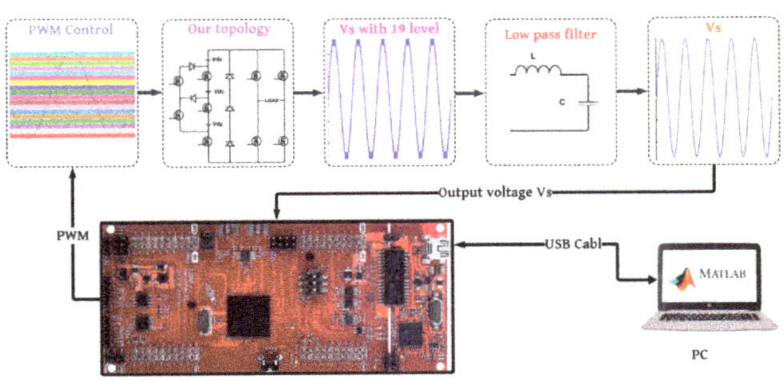

Figure 11. The proposed system using HIL.

The hardware-in-the-loop (HIL) approach, utilizing the DSP F28379D, is instrumental in controlling a multilevel converter using the PD-PWM technique. The DSP F28379D, with its dual-core architecture and high-resolution PWM capabilities, acts as the real-time controller, enabling the precise modulation of pulse width signals. To implement HIL with MATLAB-Simulink R2024b, follow these steps:

1. Model Development in Simulink: Begin by creating a detailed Simulink model of the multilevel converter and the PD-PWM control algorithm. Ensure that the model

includes all relevant components, such as the converter topology, switching signals, and control logic.

2. Code Generation: Use MATLAB's Embedded Coder to generate C code from your Simulink model. This code will be deployed onto the DSP F28379D. Configure the code generation settings to match the target hardware specifications, including the processor 306 type and peripheral interfaces.
3. Configure DSP with Code Composer Studio (CCS): Open the generated code in Code Composer Studio (CCS), the integrated development environment for the DSP F28379D. Set up the project in CCS, configuring the DSP's CPU cores, clock settings, and peripherals (such as the ADC for sensor inputs and the PWM module for controlling the converter).
4. Simulink Real-Time Interface: Establish communication between MATLAB-Simulink and the DSP F28379D by setting up the Simulink Real-Time (SLRT) or external mode. This allows you to run the Simulink model in real-time while monitoring and tuning parameters on the fly.
5. HIL Simulation: Connect the DSP F28379D to the PC running MATLAB-Simulink via a JTAG debugger (Texas Instruments, Dallas, TX, USA) or USB interface. Start the HIL simulation, where the DSP controls the multilevel converter model running in Simulink. The DSP processes real-time data from the model, and the PWM signals are generated and fed back into the simulation, allowing for closed-loop testing.
6. Testing and Validation: Monitor the system's performance in real-time through Simulink. Adjust control parameters as needed to optimize performance. Validate the control strategy by comparing the HIL results with the expected outcomes from the purely simulated model.
7. Iterative Refinement: Make sure the control algorithm satisfies performance objectives, such as lowering total harmonic distortion (THD) and speeding up response time, by using input from HIL testing. The system will be prepared for deployment in a real-world setting after validation.

This process ensures that the control strategy for the multilevel converter is rigorously tested and optimized in a safe and controlled environment before being applied to actual hardware, minimizing risks and improving system reliability.

5. Results and Discussion

5.1. Simulation Results

The simulation of the proposed topology was carried out in the MATLAB/Simulink environment on an Intel CORE i7 10TH GEN to study and compare its performance. Table 4 shows the parameters chosen for the simulations, as well as the modulation carried out using the PD-PWM strategy, with a fundamental frequency of 50 Hz used for the reference, to use 18 carriers of 5 kHz. Figure 12 shows the reference and control signal using the PD-PWM strategy for switches S1, S2, S3, S4, and S5. Hybrid bridges S6 and S7 control four switches at the same time, two switches for each bridge.

Table 4. Simulation parameters.

Fundamental Frequency	Load	DC	Carrier Frequency
50 Hz	10 Ω	$V_{dc1} = 180$ V $V_{dc2} = 60$ V $V_{dc3} = 30$ V	5 kHz

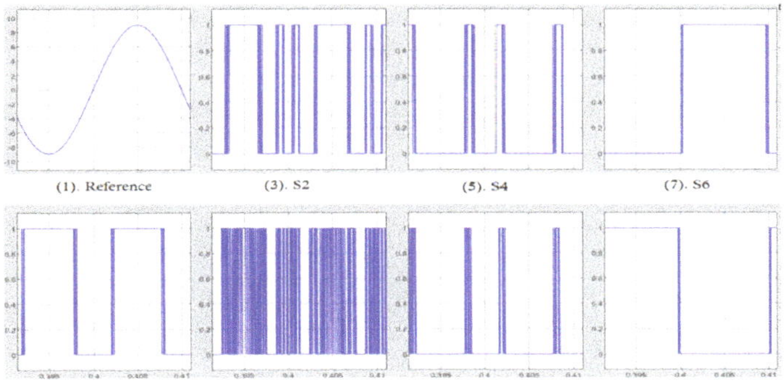

Figure 12. Gate sequence across the switches for 19-level inverter (2, 3, 4, 5, 6, 7, and 8). Switching control of S1, S2, S3, S4, S5, S6, and S7.

In conducting a comprehensive examination of the proposed new topology, our initial step involved a thorough comparison with 19-level CHB-MLIs under identical conditions to those of our topology. To energize the CHB-MLIs, we employed nine DC sources, each boasting a voltage rating of 30 V. Accompanying this setup were 36 switches, as depicted in Figure 13, portraying the schematic layout of the CHB converter necessary to achieve 19 levels. The depiction of this converter schematic not only reveals the intricacies of the circuit but also underscores the considerable number of components involved, including the 36 switches and 9 DC voltage sources.

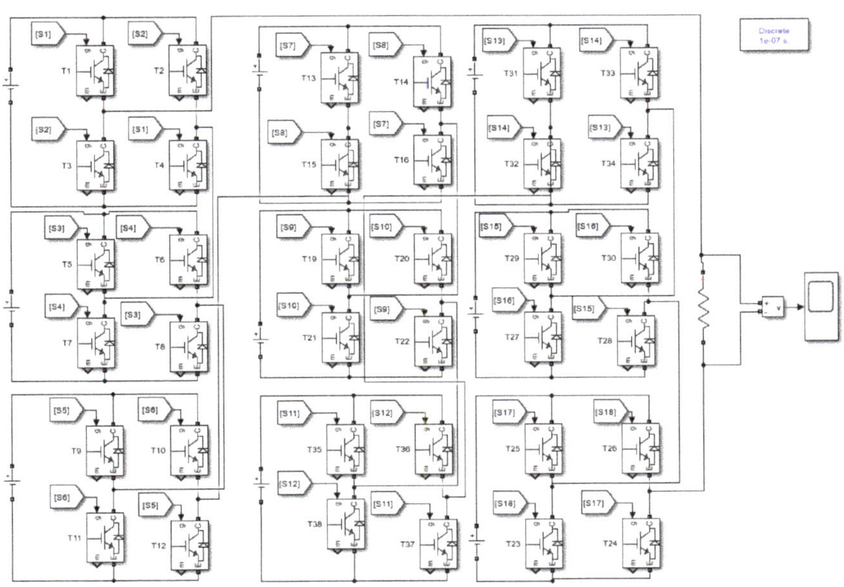

Figure 13. The simulation of CHB converter-19L on MATLAB Simulink.

The complexity of the circuitry, coupled with the substantial count of components, poses significant challenges in both operation and implementation, particularly concerning cost and maintenance considerations. Managing such a circuit demands meticulous attention to detail and incurs notable expenses, not only in terms of initial setup but also in ongoing maintenance requirements. Thus, while the 19-level CHB-MLI configuration may offer certain advantages, its practical realization remains daunting due to its inherent complexity and the associated logistical challenges.

Figure 14 shows the simulation of the new topology, in which only nine switches and three DC power sources (30 V, 60 V, and 180 V) were used to implement a 19-level converter.

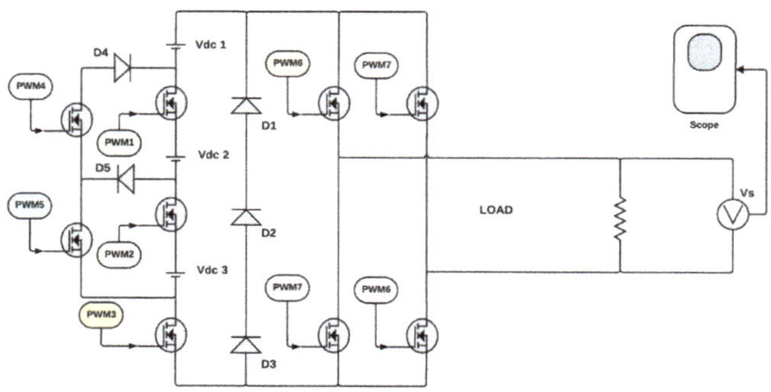

Figure 14. The simulation of our new topology on MATLAB Simulink 19L.

Figure 15 provides a visual representation of the simulated models for both the conventional CHB-MLI and the newly proposed topology. These models depict the output voltage waveforms, offering insights into their respective performance characteristics. Additionally, accompanying the visual representations, FFT (Fast Fourier Transform) analysis was conducted, revealing the frequency spectrum of the output signals. This comprehensive analysis aids in assessing the efficacy and suitability of the proposed topology compared to the established CHB-MLI configuration. This significant reduction in the number of components makes the solution feasible and testable, and we note that the output voltage reaches its maximum 269.2 V and a THD% of 6.23%.

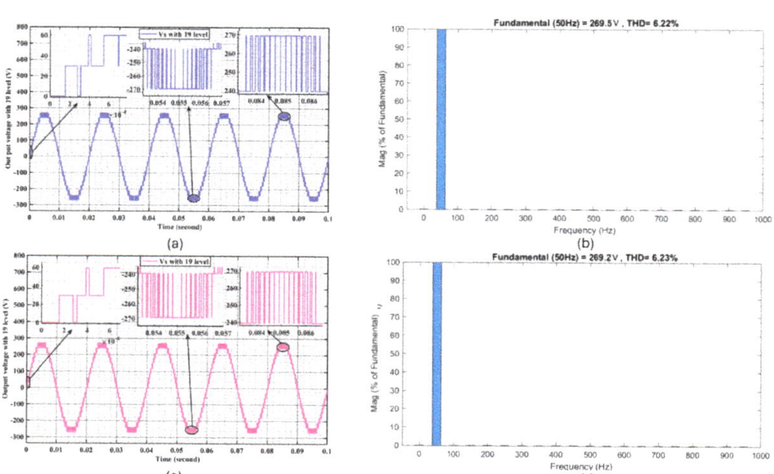

Figure 15. Comparison between the results of the 19-level inverter implemented with CHB and the 19-level inverter using our topology: output signal of CHB (**a**); THD% of CHB (**b**); output signal of the new topology (**c**); THD% of the new topology (**d**).

5.2. Improving Performance

In enhancing the output signal quality as referenced in [41,42], our strategy involves the incorporation of an LC output filter. When a single-phase converter is coupled with an LC output filter, as shown in Figure 16, it can be characterized by a set of differential equations. The fundamental principle underlying an LC low-pass filter mirrors that of an RC low-pass filter, albeit utilizing an inductor (L) and a capacitor (C).

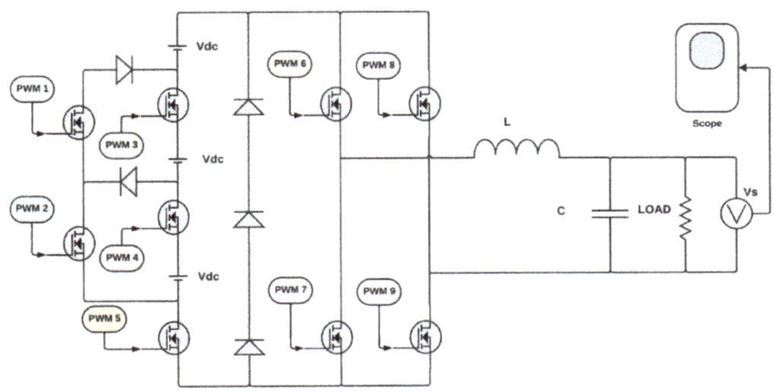

Figure 16. Our new technology with integrated LC filter.

Often referred to as a second-order low-pass filter, this configuration is commonly utilized for attenuating high-frequency components from a signal, thereby refining the overall output waveform fidelity and reducing unwanted noise interference.

$$Z_{RC} = \frac{Z_R * Z_C}{Z_R + Z_C} = \frac{R}{1 + j\omega RC} \tag{5}$$

$$Z_L = j\omega L \tag{6}$$

The signal can flow through the filter with little attenuation at low signal frequencies because the Zc capacitor's impedance is large in comparison to the ZL coil's impedance. However, the Zc capacitor's impedance decreases with increasing frequency, allowing the filter to block high-frequency signals.

$$\frac{V_s}{V_e} = \frac{Z_{RC}}{Z_{RC} + Z_L} = \frac{1}{1 + j\frac{\omega}{\omega_0}} \equiv \frac{\mathrm{A}}{1 + j\frac{\omega}{\omega_0}} \tag{7}$$

with

$$\omega_0 = \frac{1}{\sqrt{LC}} \tag{8}$$

In our proposal, we assume an ideal generator scenario where the generator is considered perfect. Specifically, we assume a resistance value of R = 50 Ω. In this context, the transfer function, a crucial aspect of system characterization, is frequently depicted in various representations for analytical purposes:

$$\frac{di^2}{dt^2} + \frac{Rdi}{Ldt} + \frac{1}{LC}i = 0 \tag{9}$$

Solution:

$$i(t) = [A\ cos(\omega st) + B\ sin(\omega st)]\ exp(-\mu t) \tag{10}$$

We suppose µ high oscillation in 1 ms:

Then,

$$L = \frac{RQ}{\omega_0} = 2 * 10^{-2}H \tag{11}$$

and

$$C = \frac{1}{RQ\omega_0} = 2 * 10^{-8} \tag{12}$$

where

- V_s is the filter output voltage.
- V_e is the filter input voltage.
- L is the inductance.
- C is the capacitance.
- R is the resistance.
- ω_0 is the natural angular frequency.
- ω is the angular frequency of the signal.
- Q is the quality factor.
- μ is the damping coefficient.
- i(t) is the current.

A continuous control law cannot be designed using the converter's discrete output signal, which has nineteen possible values. To resolve this issue and facilitate management of the system being examined, the average model of the proposed topology can be expressed in Figure 17.

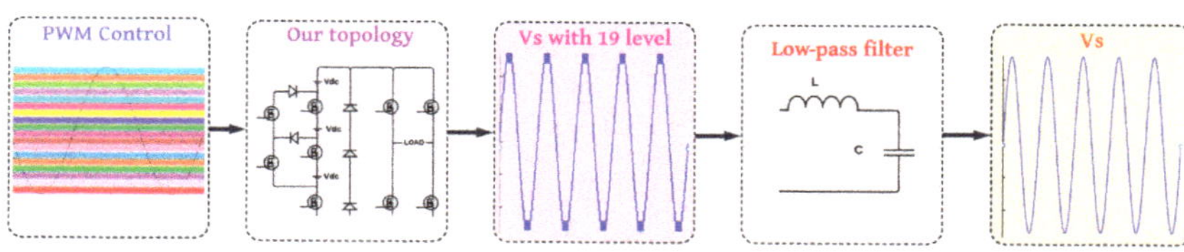

Figure 17. LC filter integration.

The considerable alteration observed in the percentage of total harmonic distortion (%THD) after the integration of an LC filter is clearly shown in Figure 18. This strategy proves both practical and rational for reducing harmonic content and improving the fidelity of the output signal.

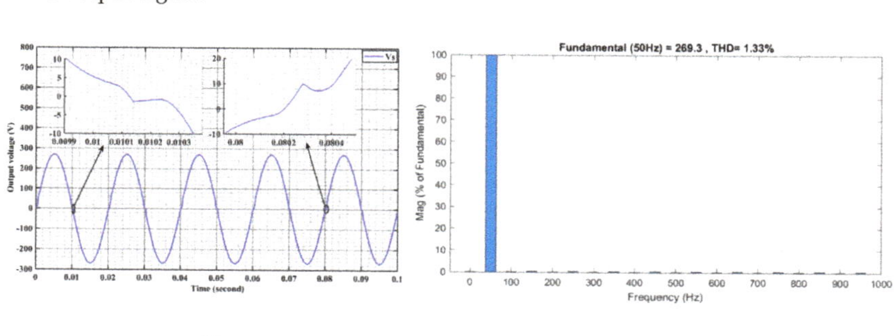

Figure 18. Output signal after low-pass filter integration (**a**); THD% (**b**).

5.3. Experimental Validation Using HIL with DSP F28379D

To validate the proposed 19-level inverter topology under realistic operating conditions, a hardware-in-the-loop (HIL) platform was employed using the Texas Instruments DSP TMS320F28379D (Texas Instruments, Dallas, TX, USA). This experimental setup enabled the testing of the PD-PWM control strategy and real-time performance analysis of the system while ensuring safety and flexibility in tuning the control parameters.

Figure 19 presents the experimental output waveform measured during the HIL. The blue curve represents the signal generated by the 19-level inverter and includes high-frequency components and slight irregularities, effectively simulating the real-world behavior of the converter. The red dashed line shows the ideal sinusoidal reference, used as a benchmark. As can be observed, the experimental signal closely follows the reference, reaching a peak amplitude of approximately 311 V, which confirms that the inverter produces a nearly ideal single-phase AC voltage.

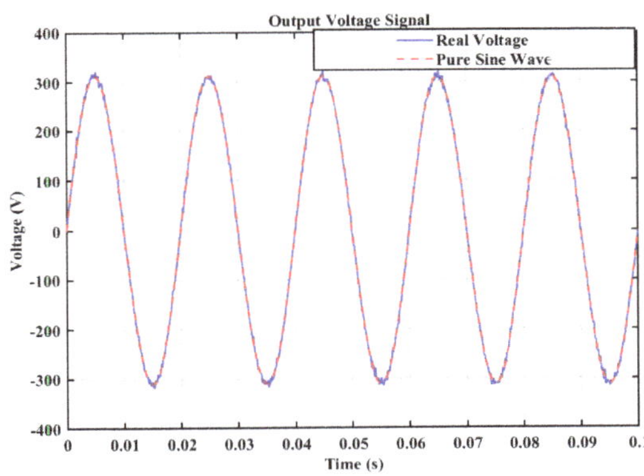

Figure 19. Experimental output waveform measured during the HIL.

Minor fluctuations around the waveform crest and trough are attributable to real-time switching events and measurement noise, which are typical in practical inverter operation. This behavior further confirms the effectiveness of the PD-PWM strategy, which was implemented on the DSP using Embedded Coder from MATLAB/Simulink and loaded via Code Composer Studio (CCS).

The HIL validation demonstrates

- The accurate tracking of the reference waveform with minimal deviation;
- The reliable generation of 19 discrete levels with proper voltage shaping;
- Realistic experimental behavior replicating practical inverter operation.

These results confirm that the proposed topology is not only functionally viable in simulation but also performs robustly when tested in a real-time embedded environment, achieving low total harmonic distortion (THD) and meeting voltage output quality standards.

To further evaluate the waveform quality, an FFT-based harmonic analysis was conducted on the output voltage captured during the HIL test. The spectrum revealed a strong fundamental component at 50 Hz, with an amplitude close to 269.3 V. Additional harmonics—primarily, the 3rd, 5th, and 7th—were present, reflecting typical switching-related distortions. The total harmonic distortion (THD) was calculated to be approximately 2.20% in Figure 20, confirming that the multilevel structure combined with the PD-PWM control technique successfully generates a high-fidelity voltage waveform suitable for power quality-sensitive applications. The comparison of simulation results (before and after the addition of an LC filter) with experimental data derived from the HIL platform with a DSP controller is summarized in Table 5. With only a minor increase in peak voltage, the data clearly demonstrates that the installation of an LC filter lowers the THD% from 6.2% to 1.3% in simulation. The THD% in the experimental setup is 2.2%, which is still quite near to the simulated value and validates the usefulness of the suggested topology and filtering approach to non-optimal parts and parasitic effects in the hardware.

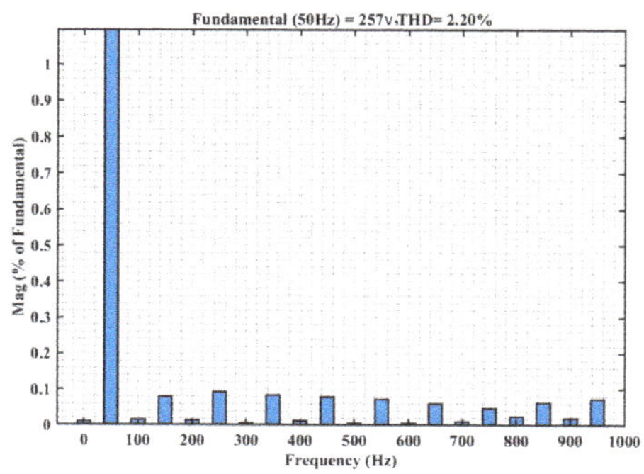

Figure 20. THD of experimental result.

Table 5. Simulated and experimental results (with and without LC filter) are compared.

Parameter to Be Compared	Before Adding an LC Filter	After Adding an LC Filter	Experimental Using HIL with DSP
THD%	6.2%	1.3%	2.2%
Peak (V)	269.2 V	269.3 V	257 V

Table 6 and Figure 21 compare and evaluate various multilevel converter topologies. This analysis clearly shows that our proposed topology is the best. With only a few switches required, it provides 19 levels. Clearly, when the number of levels is considerable, other topologies require many more components than the CHB, NPC, and FC topologies. That is why they are difficult, if not impossible, to implement, and for the NPC and FC topologies, we could not even run the simulation to compare them, given the large number of diodes, capacitors, and DC sources. Refs. [51,52,59,60] used almost the same number of switches, even though they had more components—our topology has the lowest THD%.

Our topology is therefore distinguished by the following features: very low THD (total harmonic distortion), even with many levels; optimal use of components, such as diodes and capacitors, limiting hardware requirements; and fewer switches, simplifying design and reducing costs. With its high efficiency and low harmonic distortion, the suggested multilevel inverter topology is especially ideal for new applications in plug-in hybrid electric vehicles (PHEVs) and electric vehicles (EVs), where system compactness and power quality are crucial needs [61].

Table 6. Comparison between different MLI topologies.

Topologies	N_{LEVELS}	N_{DC}	N_S	N_D	N_C	THD%
CHB-MLI	19	9	36	0	0	6.20%
NPC [4]	19	18	36	306	0	-
FC [5]	19	18	36	0	153	-
[51]	19	2	9	6	2	7.40%
[52]	19	2	8	4	0	7.19%
[62]	15	3	7	3	0	3.02%
[59]	19	3	13	0	0	3.89%
[60]	19	2	13	0	2	3.61%
[63]	17	2	12	0	4	4.83%
[64]	17	2	8	0	2	8.49%
[65]	15	4	12	0	0	5.50%
[66]	15	4	12	0	0	6.09%
[67]	7	1	8	1	2	5.68%
[68]	9	4	10	0	0	19.6%
[69]	11	1	12	0	2	-
[70]	21	4	12	0	0	12.64%
[71]	15	3	10	0	0	5.45%
[72]	21	3	10	0	0	3.49%
Proposed	**19**	**3**	**9**	**5**	**0**	**1.33%**

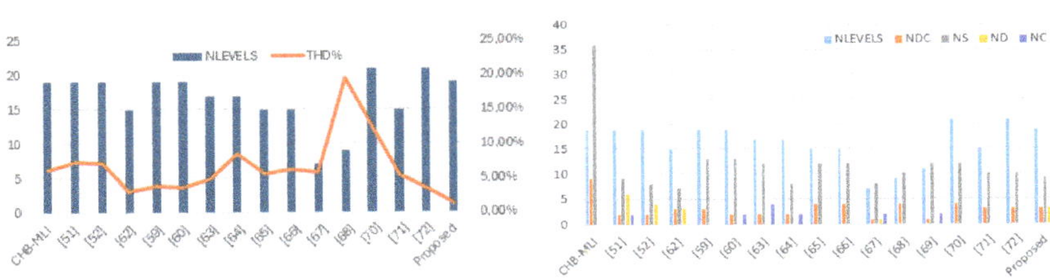

Figure 21. A graphical comparison of topology types [51,52,59,60,62–72].

6. Conclusion and Perspectives

This study presents a revolutionary multilevel converter architecture that can produce an output waveform with 19 levels. The suggested topology, which is managed by the PD-PWM technique, significantly reduces the total system complexity and the number of power switches needed, which lowers implementation costs and simplifies control. This novel converter's whole design and operation were thoroughly documented, and MATLAB Simulink simulations were used for rigorous validation. Additionally, the feasibility of implementing the suggested architecture in hardware was confirmed by experimental validation using HIL with DSP F28379D, which is a crucial step towards practical deployment. To further improve output voltage accuracy and system stability, future work will concentrate on creating a comprehensive closed-loop control system that can dynamically optimize converter settings in response to operating conditions. Overall, the study's findings show that the suggested topology, which combines a lower component count, superior output, and outstanding adaptability for real-world uses, is a promising option for next-generation multilevel inverters.

Performance is further improved by adding a low-pass LC filter, which lowers the output voltage's THD to 1.33%.

Author Contributions: Conceptualization, S.L.; Methodology, M.T.; Software, S.L.; Investigation, M.T.; Resources, M.T. and A.E.-A.; Writing—original draft, S.L.; Visualization, A.E.-A.; Supervision, A.A.E., M.T. and M.Z. All authors have read and agreed to the published version of the manuscript.

Funding: This research received no external funding.

Data Availability Statement: The original contributions presented in the study are included in the article, further inquiries can be directed to the corresponding author.

Conflicts of Interest: The authors declare no conflict of interest.

Abbreviations

PD-PWM	Phase-disposition pulse width modulation
CHB-MLI	Cascaded H-bridge multilevel inverter
THD	Total harmonic distortion
NPC	Diode-clamped multilevel inverter
CHB	Cascaded multilevel inverter
FC	Flying capacitor topology
ANPC	Active neutral-point-clamped
MLI	Multilevel inverter
PV	Photovoltaic

References

1. Etxegarai, G.; Camblong, H.; Ezeiza, A.; Lie, T.T. Design of Three Electric Vehicle Charging Tariff Systems to Improve Photovoltaic Self-Consumption. *Energies* **2024**, *17*, 1806. [CrossRef]
2. Eswar, K.N.D.V.S.; Doss, M.A.N.; Vishnuram, P.; Selim, A.; Bajaj, M.; Kotb, H.; Kamel, S. Comprehensive Study on Reduced DC Source Count: Multilevel Inverters and Its Design Topologies. *Energies* **2023**, *16*, 18. [CrossRef]
3. Chen, J.; Hu, H.; Wang, M.; Ge, Y.; Wang, K.; Huang, Y.; Yang, K.; He, Z.; Xu, Z.; Li, Y.R. Power Flow Control-Based Regenerative Braking Energy Utilization in AC Electrified Railways: Review and Future Trends. *IEEE Trans. Intell. Transp. Syst.* **2024**, *25*, 6345–6365. [CrossRef]
4. MVijeh, M.; Rezanejad, M.; Samadaei, E.; Bertilsson, K. A General Review of Multilevel Inverters Based on Main Submodules: Structural Point of View. *IEEE Trans. Power Electron.* **2019**, *34*, 9479–9502. [CrossRef]
5. Ahessab, H.; Hakam, Y.; Gaga, A.; El Haddadi, B. Design and Simulation of an Intelligent Grid-Connected MPPT Inverter with Battery Storage Using ANN Algorithm. In *Advances in Electrical Systems and Innovative Renewable Energy Techniques;* Springer: Cham, Switzerland, 2024.
6. Balal, A.; Dinkhah, S.; Shahabi, F.; Herrera, M.; Chuang, Y.L. A Review on Multilevel Inverter Topologies. *Emerg. Sci. J.* **2022**, *6*, 185–200. [CrossRef]
7. Puthiyapurayil, M.R.M.K.; Nasirudeen, M.N.; Saywan, Y.A.; Ahmad, W.; Malik, H. A Review of Open-Circuit Switch Fault Diagnostic Methods for Neutral Point Clamped Inverter. *Electronics* **2022**, *11*, 3169. [CrossRef]
8. Rana, R.A.; Patel, S.A.; Muthusamy, A.; Lee, C.w.; Kim, H.-J. Review of Multilevel Voltage Source Inverter Topologies and Analysis of Harmonics Distortions in FC-MLI. *Electronics* **2019**, *8*, 1329. [CrossRef]
9. Maheswari, K.T.; Bharanikumar, R.; Arjun, V.; Amrish, R.; Bhuvanesh, M. A comprehensive review on cascaded H-bridge multilevel inverter for medium voltage high power applications. *Mater. Today Proc.* **2021**, *45 Pt 2*, 2666–2670. [CrossRef]
10. Prasad, K.N.V.; Kumar, G.R.; Kiran, T.V.; Narayana, G.S. Comparison of different topologies of cascaded H-Bridge multilevel inverter. In Proceedings of the 2013 International Conference on Computer Communication and Informatics, Coimbatore, India, 4–6 January 2013; pp. 1–6. [CrossRef]
11. Hathiyaldeniye, T.; Karawita, C.; Bagen, B.; Pahalawaththa, N.; Annakkage, U.D. Contrôleurs optimaux pour améliorer la récupération transitoire des onduleurs suiveurs de réseau connectés à des réseaux électriques faibles. *IEEE Open Access J. Power Energy* **2022**, *9*, 161–172. [CrossRef]
12. Espinosa, E.; Veillon, M.; Melin, P.; Baier, C.; Munoz, J.; Espinoza, J.; Hernandez, J. A Staircase Modulation for Aometric Inverter Operating with Equals Fundamental Voltage and Mini-mum THD. In Proceedings of the IECON 2022—48e Conférence Annuelle de l'IEEE Industrial Electronics Society, Bruxelles, Belgique, 17–20 October 2022; pp. 1–7. [CrossRef]

13. Kavitha, R.; Premalatha, K.; Kumar, M.M.; Kumutha, S.; Manimaran, T.N. Line Voltage THD Minimization in Multilevel Inverter using Particle Swarm Optimization. In Proceedings of the 2023 Fifth International Conference on Electrical, Computer and Communication Technologies (ICECCT), Erode, India, 22–24 February 2023; pp. 1–5. [CrossRef]
14. Dhanamjayulu, C.; Meikandasivam, S. Implementation and Comparison of Symmetric and Asymmetric Multilevel Inverters for Dynamic Loads. *IEEE Access* **2018**, *6*, 738–746. [CrossRef]
15. Bettawar, P.A.; Punam, S.R. An inclusive review on various multilevel converter topologies for a grid connected photo-voltaic system. *Int. Res. J. Eng. Technol.* **2019**, *6*, 307–311.
16. Choudhury, S.; Bajaj, M.; Dash, T.; Kamel, S.; Jurado, F. Multilevel inverter: A survey on classical and advanced topologies, control schemes, applications to power system and future prospects. *Energies* **2021**, *14*, 5773. [CrossRef]
17. Poorfakhraei, A.; Narimani, M.; Emadi, A. A review of multilevel inverter topologies in electric vehicles: Current status and future trends. *IEEE Open J. Power Electron.* **2021**, *2*, 155–170. [CrossRef]
18. Barbosa, P.; Steimer, P.; Meysenc, L.; Winkelnkemper, M.; Steinke, J.; Celanovic, N. Active Neutral-Point-Clamped Multilevel Converters. In Proceedings of the 36e Conférence IEEE 2005 des Spécialistes de L'électronique de Puissance, Dresde, Allemagne, 16 June 2005; pp. 2296–2301. [CrossRef]
19. Sebaaly, F.; Kanaan, H.Y.; Moubayed, N. Three-level neutral-point-clamped inverters in transformerless PV systems—State of the art. In Proceedings of the MELECON 2014—2014 17th IEEE Mediterranean Electrotechnical Conference, Beirut, Lebanon, 13–16 April 2014; pp. 1–7. [CrossRef]
20. Ounejjar, Y.; Al-Haddad, K. A novel high energetic efficiency multilevel topology with reduced impact on supply network. In Proceedings of the 2008 34th Annual Conference of IEEE Industrial Electronics, Orlando, FL, USA, 10–13 November 2008; pp. 489–494.
21. Sharifzadeh, M.; Al-Haddad, K. Packed E-Cell (PEC) converter topology operation and experimental validation. *IEEE Access* **2019**, *7*, 93049–93061. [CrossRef]
22. Hema, P.; Sri, N.S.; Sakthivel, A.; Kaliamoorthy, M. Design and Simulation of Five Levels Active Neutralpoint Clamped Inverter. *J. Energy Environ. Sustain.* **2017**, *4*, 25–30. [CrossRef]
23. Wang, K.; Zheng, Z.; Li, Y.; Liu, K.; Shang, J. Neutral-Point Potential Balancing of a Five-Level Active Neutral-Point-Clamped Inverter. *IEEE Trans. Ind. Electron.* **2013**, *60*, 1907–1918. [CrossRef]
24. Jayakumar, V.; Chokkalingam, B.; Munda, J.L. A comprehensive review on space vector modulation techniques for neutral point clamped multi-level inverters. *IEEE Access* **2021**, *9*, 112104–112144. [CrossRef]
25. Ghias, A.M.Y.M.; Pou, J.; Ciobotaru, M.; Agelidis, V.G. Voltage-balancing method using phase-shifted PWM for the flying capacitor multilevel converter. *IEEE Trans. Power Electron.* **2014**, *29*, 4521–4531. [CrossRef]
26. Kesarwani, K.; Stauth, J.T. Resonant and multi-mode operation of flying capacitor multi-level DC-DC converters. In Proceedings of the 2015 IEEE 16th Workshop on Control and Modeling for Power Electronics (COMPEL), Vancouver, BC, Canada, 12–15 July 2015; pp. 1–8. [CrossRef]
27. Stillwell, A.; Candan, E.; Pilawa-Podgurski, R.C.N. Active voltage balancing in flying capacitor multi-level converters with valley current detection and constant effective duty cycle control. *IEEE Trans. Power Electron.* **2019**, *34*, 11429–11441. [CrossRef]
28. Barzegarkhoo, R.; Forouzesh, M.; Lee, S.S.; Blaabjerg, F.; Siwakoti, Y.P. Switched-Capacitor Multilevel Inverters: A Comprehensive Review. *IEEE Trans. Power Electron.* **2022**, *37*, 11209–11243. [CrossRef]
29. Marzo, I.; Sanchez-Ruiz, A.; Barrena, J.A.; Abad, G.; Muguruza, I. Power Balancing in Cascaded H-Bridge and Modular Multilevel Converters Under Unbalanced Operation: A Review. *IEEE Access* **2021**, *9*, 110525–110543. [CrossRef]
30. Hakam, Y.; Gaga, A.; Elhadadi, B. Exploring the state of electric vehicles: An evidence-based examination of current and future electric vehicle technologies and smart charging stations. *Energy Rep.* **2024**, *11*, 4102–4114. [CrossRef]
31. Shirmardp, B.; Choupanp, R.; Zakerianp, A.; Nazarpourp, D.; Daryoosh, N. A New Multilevel Inverter Topology Based on Cascade H-Bridge and Extended Boost Quasi Z-Source for Photovoltaic System Applications. In Proceedings of the 7th Conference on Renewable, Clean and Efficient Energies, Online, 21 May 2015; Available online: https://hal.science/hal-02545355 (accessed on 8 February 2025).
32. Estévez-Bén, A.A.; Alvarez-Diazcomas, A.; Rodríguez-Reséndiz, J. Transformerless multilevel voltage-source inverter topology comparative study for PV systems. *Energies* **2020**, *13*, 3261. [CrossRef]
33. Fang, J.; Blaabjerg, F.; Liu, S.; Goetz, S.M. A Review of Multilevel Converters with Parallel Connectivity. *IEEE Trans. Power Electron.* **2021**, *36*, 12468–12489. [CrossRef]
34. Van, C.M.; Xuan, T.N.; Hoang, P.V.; Trong, M.T.; Cong, S.P.; Van, L.N. A Generalized Space Vector Modulation for Cascaded H-bridge Multi-level Inverter. In Proceedings of the 2019 International Conference on System Science and Engineering (ICSSE), Dong Hoi, Vietnam, 20–21 July 2019; pp. 18–24. [CrossRef]

35. Khan, A.; Ahmad, M.; Bhatti, M.A.; Ijaz, M.A.; Ullah, S. A Comparative Study of Multilevel Inverter Typologies with Reduced Switches. In Proceedings of the 2019 International Conference on Engineering and Emerging Technologies (ICEET), Lahore, Pakistan, 21–22 February 2019; pp. 1–5. [CrossRef]
36. Majdoul, R.; Touati, A.; Aitelmahjoub, A.; Zegrari, M.; Taouni, A.; Ouchatti, A. A Nine-Switch Nine-Level Voltage Inverter New Topology with Optimal Modulation Technique. In Proceedings of the 2020 International Conference on Electrical and Information Technologies (ICEIT), Rabat, Morocco, 4–7 March 2020; pp. 1–6. [CrossRef]
37. Zeng, J.; Lin, W.; Liu, J. Switched-Capacitor-Based Active-Neutral-Point-Clamped Seven-Level Inverter with Natural Balance and Boost Ability. *IEEE Access* **2019**, *7*, 126889–126896. [CrossRef]
38. Zheng, G.; Chen, Y.; Kang, Y. A Modular Multilevel Converter (MMC) Based Solid-State Transformer (SST) Topology with Simplified Energy Conversion Process and Magnetic Integration. *IEEE Trans. Ind. Electron.* **2021**, *68*, 7725–7735. [CrossRef]
39. El-Alami, A.; Kazi, M.A.; Baraka, I.; Majdoul, R.; AitElmahjoub, A. Sliding Mode Control of Six-Switch Five-Level Active Neutral Point Clamped (6S-5L-ANPC). In *Smart Applications and Data Analysis*; Hamlich, M., Bellatreche, L., Siadat, A., Ventura, S., Eds.; SADASC 2022. Communications in Computer and Information Science; Springer: Cham, Switzerland, 2022; Volume 1677. [CrossRef]
40. Wang, Y.; Aksoz, A.; Geury, T.; Ozturk, S.B.; Kivanc, O.C.; Hegazy, O. A review of modular multilevel converters for stationary applications. *Appl. Sci.* **2020**, *10*, 7719. [CrossRef]
41. Priya, M.; Ponnambalam, P.; Muralikumar, K. Modular-multilevel converter topologies and applications—A review. *IET Power Electron.* **2019**, *12*, 170–183. [CrossRef]
42. Omer, P.; Kumar, J.; Surjan, B.S. A Review on Reduced Switch Count Multilevel Inverter Topologies. *IEEE Access* **2020**, *8*, 22281–22302. [CrossRef]
43. Rezaei, M.A.; Nayeripour, M.; Hu, J.; Band, S.S.; Mosavi, A.; Khooban, M.-H. A New Hybrid Cascaded Switched-Capacitor Reduced Switch Multilevel Inverter for Renewable Sources and Domestic Loads. *IEEE Access* **2022**, *10*, 14157–14183. [CrossRef]
44. Chappa, A.; Gupta, S.; Sahu, L.K.; Gautam, S.P.; Gupta, K.K. Symmetrical and Asymmetrical Reduced Device Multilevel Inverter Topology. *IEEE J. Emerg. Sel. Top. Power Electron.* **2021**, *9*, 885–896. [CrossRef]
45. Vikash, P.; Vijayaraja, L.; Kumar, S.G. 21 Level Asymmetric Inverter without Inversion Circuit with Reduced Switch Count. In *PREPARE@ u® | IEI Conferences*; 2021; Available online: https://preprint.prepare.org.in/index.php/iei/article/view/189 (accessed on 8 February 2025). [CrossRef]
46. Abdulhamed, Z.E.; Esuri, A.H.; Abodhir, N.A. New topology of asymmetrical nine-level cascaded hybrid bridge multilevel inverter. In Proceedings of the 2021 IEEE 1st International Maghreb Meeting of the Conference on Sciences and Techniques of Automatic Control and Computer Engineering MI-STA, Tripoli, Libya, 25–27 May 2021; IEEE: New York, NY, USA, 2021; pp. 430–434.
47. Reyes-Severiano, Y.; Aguayo-Alquicira, J.; León-Aldaco, S.E.D.; Carrillo-Santos, L.M. Comparative analysis of PD-PWM technique in the set: Multilevel Inverter-Induction motor. *Ing. Investig. Tecnol.* **2020**, *21*. [CrossRef]
48. Chakir, F.; EL Magri, A.; Lajouad, R.; Kissaoui, M.; Chakir, M.; Bouattane, O. Design and analysis of a new multi-level inverter topology with a reduced number of switches and controlled by PDPWM technique. *Int. J. Electr. Comput. Eng. Syst.* **2023**, *14*, 593–600. [CrossRef]
49. Xue, C.; Zhou, D.; Li, Y. Finite-Control-Set Model Predictive Control for Three-Level NPC Inverter-Fed PMSM Drives with LC Filter. *IEEE Trans. Ind. Electron.* **2021**, *68*, 11980–11991. [CrossRef]
50. Qureshi, M.R.; Mahar, M.A.; Larik, A.S. Harmonic Analysis and Design of LC Filter for a Seven-level Asymmetric Cascaded Half Bridge Multilevel Inverter. *Int. J. Electr. Eng. Emerg. Technol.* **2020**, *3*, 52–58.
51. Sagvand, F.; Siahbalaee, J.; Koochaki, A. An Asymmetrical 19-Level Inverter with a Reduced Number of Switches and Capacitors. *Electronics* **2023**, *12*, 338. [CrossRef]
52. Mohammed, M.F.; Qasim, M.A. Single Phase T-Type Multilevel Inverters for Renewable Energy Systems, Topology, Modulation, and Control Techniques: A Review. *Energies* **2022**, *15*, 8720. [CrossRef]
53. Kumar, S.S.; Sivachitra, M.; Meenakumari, R.; Gowsiha, G.; Kokila, S.; Santhiya, R. Study and Implementation of 19-Level Multilevel Inverter and its Onfigurations. *Ann. Rom. Soc. Cell Biol.* **2021**, *25*, 1809–1815.
54. Thakre, K.; Mohanty, K.B.; Kommukuri, V.S.; Chatterjee, A.; Nigam, P.; Gupta, S.K. Modified cascaded multilevel inverter for renewable energy systems with less number of unidirectional switches. *Energy Rep.* **2022**, *8*, 5296–5304. [CrossRef]
55. Bektaş, E.; Aldabbagh, M.M.; Ahmed, S.R.; Hussain, A.S.T.; Taha, T.A.; Ahmed, O.K.; Ezzat, S.B.; Hashim, A.M. Enhancing Harmonic Reduction in Multilevel Inverters using the Weevil Damage Optimization Algorithm. *J. Robot. Control (JRC)* **2024**, *5*, 717–722.
56. Hang, L.; Subramaniam, U.; Bayrak, G.; Moayedi, H.; Ghaderi, D.; Minaz, M.R. Influence of a proposed switching method on reliability and total harmonic distortion of the quasi Z-source inverters. *IEEE Access* **2020**, *8*, 33088–33100. [CrossRef]

57. Faria, P.; Vale, Z. Realistic Load Modeling for Efficient Consumption Management Using Real-Time Simulation and Power Hardware-in-the-Loop. *Energies* **2023**, *16*, 338. [CrossRef]
58. Hakam, Y.; Gaga, A.; Tabaa, M.; Elhadadi, B. Intelligent Integration of Vehicle-to-Grid (V2G) and Vehicle-for-Grid (V4G) Systems: Leveraging Artificial Neural Networks (ANNs) for Smart Grid. *Energies* **2024**, *17*, 3095. [CrossRef]
59. Dhanamjayulu, C.; Kaliannan, P.; Padmanaban, S.; Maroti, P.K.; Holm-Nielsen, J.B. A new three-phase multi-level asymmetrical inverter with optimum hardware components. *IEEE Access* **2020**, *8*, 212515–212528. [CrossRef]
60. Iqbal, A.; Siddique, M.D.; Reddy, B.P.; Maroti, P.K.; Alammari, R. A new family of step-up hybrid switched capacitor integrated multilevel inverter topologies with dual input voltage sources. *IEEE Access* **2020**, *9*, 4398–4410. [CrossRef]
61. Zhang, Y.; Xu, S.; Song, Y.; Qi, W.; Guo, Q.; Li, X.; Kong, L.; Chen, J. Real-Time Global Optimal Energy Management Strategy for Connected PHEVs Based on Traffic Flow Information. *IEEE Trans. Intell. Transp. Syst.* **2024**, *25*, 20032–20042. [CrossRef]
62. Antar, R.K.; Hussein, T.A.; Abdullah, A.M. Design and implementation of reduced number of switches for new multilevel inverter topology without zero-level state. *Int. J. Power Electron. Drive Syst.* **2022**, *13*, 401. [CrossRef]
63. Bin Arif, M.S.; Mustafa, U.; Ayob, S.B.M.; Rodriguez, J.; Nadeem, A.; Abdelrahem, M. Asymmetrical 17-level inverter topology with reduced total standing voltage and device count. *IEEE Access* **2021**, *9*, 69710–69723. [CrossRef]
64. Dhanamjayulu, C.; Prasad, D.; Padmanaban, S.; Maroti, P.K.; Holm-Nielsen, J.B.; Blaabjerg, F. Design and implementation of seventeen level inverter with reduced components. *IEEE Access* **2021**, *9*, 16746–16760. [CrossRef]
65. Fahad, M.; Siddique, M.D.; Iqbal, A.; Sarwar, A.; Mekhilef, S. Implementation and analysis of a 15-level inverter topology with reduced switch count. *IEEE Access* **2021**, *9*, 40623–40634. [CrossRef]
66. Anand, V.; Singh, V. Compact symmetrical and asymmetrical multilevel inverter with reduced switches. *Int. Trans. Electr. Energy Syst.* **2020**, *30*, e12458. [CrossRef]
67. Gopal, Y.; Kumar, Y.N.V.; Kumari, A.; Prakash, O.; Chowdhury, S.; Almehizia, A.A. Reduced Device Count for Self-Balancing Switched-Capacitor Multilevel Inverter Integration with Renewable Energy Source. *Sustainability* **2023**, *15*, 8000. [CrossRef]
68. Memon, R.; Mahar, M.A.; Larik, A.S.; Shah, S.A.A. Design and Performance Analysis of New Multilevel Inverter for PV System. *Sustainability* **2023**, *15*, 10629. [CrossRef]
69. Ali, M.; Tariq, M.; Lin, C.-H.; Chakrobortty, R.K.; Alamri, B.; Alahmadi, A.; Ryan, M.J. Operation of a UXE-Type 11-Level Inverter with Voltage-Balance Modulation Using NLC and ACO-Based SHE. *Sustainability* **2021**, *13*, 9035. [CrossRef]
70. Arulappan, A.T.A.; Selvaraj, M.; Aladian, A. SHE PWM based 21 level inverters with hardware analysis. *Int. J. Power Electron. Drive Syst. (IJPEDS)* **2024**, *15*, 993–1000. [CrossRef]
71. Srivastava, A.; Chauhan, A.; Tripathi, A. Performance Comparison of Asymmetrical Multilevel Inverter with Different Switching Techniques. *Energies* **2025**, *18*, 715. [CrossRef]
72. Khasim, S.R.; Dhanamjayulu, C.; Padmanaban, S.; Holm-Nielsen, J.B.; Mitolo, M. A novel asymmetrical 21-level inverter for solar PV energy system with reduced switch count. *IEEE Access* **2021**, *9*, 11761–11775. [CrossRef]

MDPI AG
Grosspeteranlage 5
4052 Basel
Switzerland
Tel.: +41 61 683 77 34

Energies Editorial Office
E-mail: energies@mdpi.com
www.mdpi.com/journal/energies

www.ingramcontent.com/pod-product-compliance
Lightning Source LLC
Chambersburg PA
CBHW061607220926
51850CB00005B/446